Neuroscience of Attention

Attention is critical to our daily lives, from simple acts of listening to a conversation or reading, to the more demanding situations of trying to concentrate in a noisy environment or driving on a busy roadway. This book offers a concise introduction to the science of attention, featuring real-world examples and fascinating studies of clinical disorders and brain injuries. It introduces cognitive neuroscience methods and covers the different types and core processes of attention. The links between attention, perception, and action are explained, along with exciting new insights into the brain mechanisms of attention revealed by cutting-edge research. Learning tools – including an extensive glossary, chapter reviews, and suggestions for further reading – highlight key points and provide a scaffolding for use in courses. This book is ideally suited for graduate or advanced undergraduate students as well as for anyone interested in the role attention plays in our lives.

Joseph B. Hopfinger is a Professor of Psychology and Neuroscience at the University of North Carolina at Chapel Hill and co-Editor-in-Chief of the journal *Cognitive Neuroscience*. He has also received the *Brain Research* Young Investigator Award.

Cambridge Fundamentals of Neuroscience in Psychology

Developed in response to a growing need to make neuroscience accessible to students and other non-specialist readers, the *Cambridge Fundamentals of Neuroscience in Psychology* series provides brief introductions to key areas of neuroscience research across major domains of psychology. Written by experts in cognitive, social, affective, developmental, clinical, and applied neuroscience, these books will serve as ideal primers for students and other readers seeking an entry point to the challenging world of neuroscience.

Books in the Series

The Neuroscience of Expertise, by Merim Bilalić

The Neuroscience of Adolescence, by Adriana Galván

The Neuroscience of Suicidal Behavior, by Kees van Heeringen

The Neuroscience of Creativity, by Anna Abraham

The Neuroscience of Addiction, by Francesca Mapua Filbey

The Neuroscience of Sleep and Dreams, 2e, by Patrick McNamara

The Neuroscience of Intelligence, 2e, by Richard J. Haier

The Cognitive Neuroscience of Bilingualism, by John W. Schwieter and Julia Festman

Fundamentals of Developmental Cognitive Neuroscience, by Heather Bortfeld and Silvia A. Bunge

Neuroscience of Attention, by Joseph B. Hopfinger

Cognitive Neuroscience of Memory, 2e, by Scott D. Slotnick

Cognitive and Social Neuroscience of Aging, 2e, by Angela Gutchess

Introduction to Human Neuroimaging, 2e, by Hans Op de Beeck and Chie Nakatani

Neuroscience of Attention

Joseph B. Hopfinger
University of North Carolina at Chapel Hill

Shaftesbury Road, Cambridge CB2 8EA, United Kingdom

One Liberty Plaza, 20th Floor, New York, NY 10006, USA

477 Williamstown Road, Port Melbourne, VIC 3207, Australia

314–321, 3rd Floor, Plot 3, Splendor Forum, Jasola District Centre,
New Delhi – 110025, India

103 Penang Road, #05–06/07, Visioncrest Commercial, Singapore 238467

Cambridge University Press is part of Cambridge University Press & Assessment,
a department of the University of Cambridge.

We share the University's mission to contribute to society through the pursuit of
education, learning and research at the highest international levels of excellence.

www.cambridge.org
Information on this title: www.cambridge.org/highereducation/isbn/9781316513293

DOI: 10.1017/9781009072434

When citing this work, please include a reference to the DOI 10.1017/9781009072434

First published 2025

A catalogue record for this publication is available from the British Library

A Cataloging-in-Publication data record for this book is available from the Library of Congress

ISBN 978-1-316-51329-3 Hardback
ISBN 978-1-009-07338-7 Paperback

Additional resources for this publication at www.cambridge.org/Hopfinger

To

William, Michael, and Jenni

Thank you for all your support and for helping my attention always return to what matters most.

Contents

Figures

Preface

Attention has been discussed for millennia, but only in the last few decades have the methods of cognitive neuroscience been able to probe into the human brain to reveal the neural mechanisms underlying this critical mental process. Current topics in the popular press, such as stories of distracted driving, the effects of video games on students' concentration, and the frustratingly long lines at airport security checkpoints, highlight a few real-world examples of the importance of attention. New industries have arisen over the past few years promising to enhance our attention through a variety of "brain training" techniques. Between the implementation of distracted driving laws, the treatment options for attention disorders, and the push to enhance our brains, understanding the brain mechanisms of attention is more important than ever.

This book is intended for audiences including graduate students, advanced undergraduate students, and laypeople or scientists interested in the history and current research into the neural mechanisms of attention. It starts with the historical background of early theories of attention, presents seminal cognitive neuroscience studies that have revealed core processes of attention, and discusses future directions of research into the brain mechanisms of attention that have real-world implications. This book is aimed to fill a gap between the broad but brief coverage of attention in most undergraduate cognitive psychology textbooks and the in-depth but narrow focus of research articles and books that address only a select aspect of attention.

How to Use This Book

The first few chapters of this book provide a foundation for understanding current research on attention by explaining the history of attention research and the strengths and limitations of the many cognitive neuroscience methods that are used in the attempts to uncover the neural mechanisms of attention processes. The subsequent chapters are organized around major themes in attention research, such as voluntary versus involuntary influences on attention, the distinct process of controlling attention versus the effects of that control, and the role of neural rhythms in the allocation of attention across time. The different deficits and disorders of attention are compared, and research into the plasticity and training of attention systems is discussed, along with the presentation of new models of perception and attention. Each chapter can be read as a standalone account of an area of attention research, but the chapters also contain cross-references to related material in other chapters, thus serving to extend and reinforce each other in an integrated way.

Since many different techniques and types of equipment are used to probe into the brain basis of attention, a full chapter is dedicated to explaining the various methods of cognitive neuroscience. As will become clear, there is no single perfect method, as each has strengths as well as critical limitations. Thus, an understanding of neural mechanisms can only come by integrating findings from across these varied methods. A critical goal of this book is to help readers understand how each method works, so that experimental findings can be evaluated in terms of what each result can or cannot tell us about the processes of attention.

Each chapter starts out with a list of learning objectives and ends with a summary of key points and review questions. Boxes within each chapter highlight controversies or recent trends in the research being done on the topic of that chapter, and a short list of suggested "Further Readings" is provided for readers wanting to dive deeper into seminal papers or in-depth reviews of the topic. A Glossary is also included to provide quick access to definitions and explanations of important terms. The chapters are written to be accessible to those without expert knowledge, but they also present some of the most critical research findings on each topic and provide extensive citations. Even in a book dedicated just to attention there is not room to describe all the exciting research being done, but the citations and suggested readings provide recommendations for where interested readers can obtain further knowledge of specific areas of research. Thus, this book could be used as a standalone textbook in upper-level undergraduate courses, as part of advanced graduate seminars focused on attention or cognitive neuroscience, or simply as a means for anyone interested in this topic to gain a deep understanding of how the brain enables the multifaceted processes of attention.

1   What Is Attention?

Learning Objectives

- Identify key figures in the history of attention research
- Compare and contrast basic functions ascribed to attention
- Describe classic experimental paradigms used to study attention
- Understand how attention research developed over time and its current directions

1.1   "Everyone Knows What Attention Is . . ."

In probably the most famous quote in the history of attention research, the preeminent psychologist William James (1890), shown in Figure 1.1, began his description of attention with the phrase "Everyone knows what attention is." This presumed familiarity has led to the ubiquity of the concept of attention in our vernacular but has also complicated research into the topic. The problem is that although everyone "knows" what attention is, there are a variety of mental processes that are subsumed under this term. A focus of this book will be to differentiate the neural mechanisms that make up the many different components of attention, as well as to explain why these varied processes are lumped together under the powerful umbrella term of "attention."

James' extended quote touches upon several of the most relatable aspects of attention and conveys some of the subjective "feel" of attention:

Everyone knows what attention is. It is the taking possession by the mind, in clear and vivid form, of one out of what seem several simultaneously possible objects or trains of thought. Focalization, concentration of consciousness are of its essence. It implies withdrawal from some things in order to deal effectively with others, and is a condition which has a real opposite in the confused, dazed, scatterbrained state which in French is called *distraction*, and *Zerstreutheit* in German. (William James (1890), *Principles of Psychology*)

The following sections of this chapter will introduce and provide background on the core processes highlighted in James' quote, including the unitary focus and selectivity of attention ("withdrawal from some things in order to deal effectively with others"), the phenomenological feeling of peak alertness ("a real opposite in the confused, dazed, scatterbrained state"), and the association with conscious awareness ("taking possession by the mind"). In the century since James' quote, many other processes associated with attention have become the focus of intense

Figure 1.1 William James. *Source:* Image from MS Am 1092 (1185), Houghton Library, Harvard University (public domain).

research efforts. As discussed later, these include the ability to sustain performance on a task for prolonged periods of time, the binding of multiple features together in object perception, and the increasingly common misperception that we can accomplish more by multitasking. Before moving on to discuss the role of attention in many aspects of our everyday lives (Chapter 2), this chapter will provide historical background into the research that set the stage for current cognitive neuroscience studies that are revealing the brain mechanisms of attention.

1.2   Prehistory of Attention Research: Philosophy and Psychology Precursors

Attention, as a phenomenon, was of interest to philosophers long before psychology became a scientific field of experimental research. As Williams James' quote suggests, it seems that everyone knows what attention is, even without empirical research into the topic. Attention has, however, been particularly hard to pin down when it comes to defining exactly what it is and understanding the brain mechanisms that support it. There are writings, from as early as the seventeenth century, showing that philosophers were grappling with just what attention is (Figure 1.2). In one of the earliest works to note the importance of attention, Nicolas Malebranche discussed how attention is critical because without such a mechanism, our perceptual apparatus would overwhelm our minds with a flood of information. Malebranche wrote: "It is therefore necessary to look for means to keep our perceptions from being confused and imperfect. And, because, as everyone knows, there is nothing that makes them clearer and more distinct than attentiveness, we must try to find the means to become more attentive than we are" (Malebranche, 1674, as translated in Nadler, 1992). Malebranche was also among the first to suggest that we don't have direct access to the external world itself, but rather just to our

Figure 1.2 Picture of writing with a quill pen. There has been interest in the topic of attention for centuries, with written records from philosophers dating back to the seventeenth century. *Source:* Getty Images; Creative #: 466268089; credit: aluxum.

mental representations of the world. His writings reveal that the importance of attention for focusing the mind was appreciated centuries ago, long before scientific experiments began to investigate the brain mechanisms underlying these critical processes.

Another aspect of attention that was of interest to early philosophers was the link between sensation and consciousness. In the early 1700s, Gottfried Wilhelm Leibniz developed the concept of *apperception*, which held that there was a stage of processing at which current sensations were linked to previous experiences. According to this view, attention was a critical link that allowed sensory experiences to move into conscious awareness. Furthermore, this ability to link sensation to memory was thought to allow us to form a concept of "self" that experiences these things. Leibniz thus thought of attention and the linking of sensation to *consciousness* as central to who we are. The eighteenth-century philosopher Johann Friedrich Herbart explored the idea of *involuntary* influences on attention and consciousness. He suggested that unconscious processes linking dormant concepts in the mind to the information currently being attended could allow those additional concepts to break through to consciousness. Furthermore, in his theories on educational practices, Herbart emphasized the effectiveness of tailoring instruction to account for these processes for deepening the understanding of what was being taught (reviewed in Kenklies, 2012).

In the late nineteenth century, William James wrote his famous book that covered a variety of psychological processes and that included the quote presented at the start of this chapter. In addition to the important concepts highlighted in that quote, James also differentiated between multiple types of attention in his book. He distinguished between attention to stimuli currently impinging upon the sensory organs ("sensorial attention") versus attention to representations in the mind that were not physically present ("intellectual attention"). This latter type of attention has become an area of renewed interest, especially in recent studies of mind wandering. Indeed, James suggested that the ability to control this wandering had important implications, noting that "the faculty of voluntarily bringing back a wandering attention, over and over again, is the very root of judgment, character, and will" (James, 1890). In addition to this potential index of ones' character, James was more specific about the effects of attention, suggesting that it "makes us: a) perceive, b) conceive, c) distinguish, d) remember better than otherwise we could." (We will discuss each of these effects of attention in more detail in Chapter 5.) Later in the text, James added a fifth effect of attention – the shortening of reaction time – that related to the earlier work of Franciscus Donders and opened the possibility that attention could be investigated empirically. Following these early advances, however, the topic of attention was largely neglected in the following 50 years (see Box 1.1).

Franciscus Donders, a nineteenth-century doctor and professor of physiology, had an important and lasting impact on the study of cognitive processes. Much of his fame in the

Box 1.1 Why attention disappeared in the early twentieth century

The quote from William James presented at the beginning of this chapter is one of the most famous quotes in all of psychology. With eloquence and depth, it describes the multifaceted functions of attention in a way that continues to resonate with people over a hundred years later. This could have been the spark that ignited interest in the cognitive abilities of the mind and spurred a flood of research into understanding the mechanisms of attention. Instead, almost no progress was made for decades. In much of the 60 years following the publication of James' *Principles of Psychology*, attention was ignored as a topic of scientific research. Why? In a word, **"behaviorism."** For the first half of the twentieth century, behaviorism dominated the field of psychology. Although not created to directly oppose William James' focus on describing the functions of the mind ("functionalism"), behaviorism as a school of thought didn't have room for the mental concepts James described. Behaviorism developed in part as a reaction to "structuralism," the school of thought that largely dominated psychological research in the late 1800s. Developed initially by Wilhem Wundt for use in the first experimental psychology lab, and expanded and championed by Edward Titchener, structuralism relied upon the method of *introspection*. As a method of research, introspection in the 1800s was quite different from how it is considered today. When we refer to introspection now, we typically mean that we're simply doing some self-reflection or thinking about our own thoughts. In psychology labs in the 1800s, however, introspection was defined in a much more restrictive sense. Introspective reports in those studies were acceptable only if they met certain strict criteria for describing the mental experience of a perception (e.g., its quality, intensity, duration, or clearness) without relying on simple verbal labels of the items' physical attributes (i.e., what Titchener referred to as the "stimulus error"). Furthermore, different labs had somewhat different criteria for what was acceptable as an introspective report, which made replication across labs difficult. Although structuralism intended to probe many of the mental functions that we continue to find important to this day

Box 1.1 **(cont.)**

(e.g., perception, thinking, emotions), the subjective nature of what constituted a "correct" introspective account frustrated other researchers interested in psychology. John Watson, B. F. Skinner, and other psychologists of the time felt that the reliance on subjective reports in structuralism was a fatal flaw. Instead, they suggested that one must focus exclusively on overt, objective, easily identifiable, and unambiguous behavior if one wanted to make progress in understanding psychology. Thus, behaviorism was born in the early twentieth century and enjoyed success, as experimental results were easily replicated across labs. This ease of replication, along with its quantitative results and its rapid progress in advancing the understanding of learning processes, helped behaviorism to dominate psychology research. For the next 50 years, research into the human mind was largely restricted to what behaviorists believed were tractable issues, which did not include ill-defined mental processes such as attention. As developed by Watson and Skinner, behaviorism was largely "anti-mentalistic." According to this viewpoint, mental phenomena like attention and consciousness were not worthy of scientific study, because such concepts did not have distinct and easily observable behaviors that could be unambiguously measured in a quantitative manner. Fearing that investigation into such concepts would only lead back to the inconclusive and idiosyncratic results that structuralism produced, the behaviorists strongly argued that psychology need only concern itself with overt behavior. Thus, for much of the decades from 1900 to 1950, concepts such as attention, language, and consciousness were largely ignored in psychology research. Despite William James' cogent description of the complex, multifaceted mental experience of attention, it took a revolution (the "cognitive revolution" – see Section 1.3) to allow attention to become a focus of scientific study.

area of psychology comes from his work on vision and the eye. In the mid-1800s he described processes of vision (e.g., refraction, accommodation, convergence) and eye abnormalities (e.g., astigmatism, presbyopia) in ways that are still useful today. His most critical contribution to the study of attention, however, wasn't related to his groundbreaking work on the eye, nor was it achieved through theories or experiments on attention. What Donders did that had a huge impact on attention research, and all of cognitive psychology, was to introduce the concept of **mental chronometry** – studying the timing of mental events. Donders had the critical insight that, although mental events could not be directly observed, the effects of mental events on overt behavior could be observed, and in highly precise and replicable ways if measured correctly. Implicit in his approach was the critical assumption that mental events take *time*, in stark contrast to a prevalent idea at that time that thoughts proceed at infinite speed. Donders' methods were based on the ideas that the amount of time required for a specific mental process should be consistent whenever engaged and that this time could be precisely quantified. In 1868, Donders published a study in which he used reaction times to show a consistent difference in the time it took subjects to respond to simple-response tasks compared to choice-response tasks (Donder, 1868/1969). In one of these, subjects were to make a rapid manual response to the appearance of a spot of light. In the simple-response task, subjects made the same response (e.g., button press with one hand) regardless of the color of the light; in the choice-response task, subjects responded differently to a red light versus a white light (e.g., right hand response to red light, left hand to a white light). Donders proposed that he could thus isolate a particular mental process by comparing two conditions, in which the process of interest was present in one task but removed from the other. Critically, all other aspects of the stimuli and task were the same, thus only the added process of interest should be different. This design was later termed the **subtractive method**, because by subtracting subjects' response times between the two conditions (e.g., choice task minus simple task), the resulting difference would be the time needed to perform that specific mental operation (e.g., the process of deciding which hand to respond with). Through precise measurement of response times, he was thus able to quantify how long that choice process took (154 ms in that particular experiment). It should be noted that the aspect of Donders' approach referred to as "pure insertion," in which an additional mental event can be inserted into a task without affecting in any way the other processes involved, has been shown in subsequent work to be an oversimplifying assumption, because there are often interactions with other mental processes when any process is added. More advanced analytical methods, however, can account for these interactions, so that meaningful measures of the timing of mental events can be obtained. Donders' development of mental chronometry was a critical step in allowing scientists to study the mental processes that philosophers had theorized about for centuries but that psychologists had avoided because there hadn't been a reliable and quantifiable way to measure such processes. With mental chronometry and the strategy of isolating mental events through comparison of well-controlled conditions, the pieces were in place to begin the scientific study of attention. But before cognitive psychology could flourish as a field, the dominating school of thought in psychology in the late nineteenth and early twentieth centuries, behaviorism, had to be challenged.

1.3   The Cognitive Revolution (1950s)

Psychology research during the early decades of the twentieth century was dominated by **"behaviorism"** and its focus on investigating psychology and learning through highly controlled studies in nonhuman animals. As opposed to the sometimes-vague concepts and variable methods associated with Titchener's structuralism, dominant in the late nineteenth century, behaviorism provided clear aims and reliable and replicable experimental results. Behaviorism provided important insights into the learning process, from Ivan Pavlov's classical conditioning experiments in the late 1800s (e.g., a dog salivating at the sound of a bell, after it has learned that the bell is associated with delivery of food) to B. F. Skinner's (1938) work on operant (or instrumental) conditioning (i.e., using reinforcements or punishments to strengthen, reduce, or shape a response). Behaviorist approaches to the study of learning dominated the field of psychology in the first half of the twentieth century, and strong adherents of this approach such as John Watson and B. F. Skinner proposed an anti-mentalistic view of psychology. According to this view, it was unnecessary to consider unobservable mental events, because the behaviorist study of stimulus–response contingencies and schedules of reinforcement was all that was needed to understand human psychology and behavior.

By the mid-1950s, however, doubts were beginning to arise about whether behaviorism was a sufficient means to understand human psychology. These doubts were fueled by a number of different events at that time. The "cognitive revolution" in psychology refers to the seismic shift in the focus of psychology research from the anti-mentalistic behaviorist tradition that dominated psychology research through most of the early twentieth century toward an interest in the internal mental states associated with thinking and the cognitive processes of attention, memory, language, and decision-making.

One highly significant event was World War II. As part of efforts to enhance the effectiveness and safety of troops, learning principles from behaviorist studies were applied to soldiers' training. However, the actions of soldiers and pilots during the war revealed that aspects of the training didn't always translate well to the battlefield. Well-trained soldiers would sometimes make mistakes in the field – or in the cockpit – that weren't easily explained by behaviorist theories. This resulted in a renewed interest in the other mental events, such as attention, which had long been theorized to play a crucial role in cognition but which had been largely ignored by behaviorists through the first few decades of the 1900s.

During the mid-1900s, researchers were also finding results that were at odds with core principles of behaviorism. For instance, in the cheekily titled article "Misbehavior of Organisms" (a play on Skinner's famous book from 1938 titled *The Behavior of Organisms*) the authors (Breland & Brelend, 1961) start off by saying, "There seems to be a continuing realization by psychologists that perhaps the white rat cannot reveal everything there is to know about behavior." They proceeded to report on multiple cases in which animals, of multiple species, were not behaving in the ways they had been trained, even though operant conditioning principles had been strictly followed. They describe numerous cases in which the training was initially effective in one specific context, but that over time, or in other circumstances, the animals began "misbehaving" and not performing the trained actions. The authors realized that these cases of misbehavior usually involved the animals reverting to natural instincts, and according to behaviorist principles, instincts, or any behavior that is not learned through conditioning, should not be able to trump the conditioned behavior. In one example, chickens were trained to "bat" a baseball that was set up on a tiny baseball field, complete with toy players and an outfield fence. Next to the field was an open cage, and the chicken was trained to go to one side of the cage and pull a string that was attached to a small bat, which would swing and hit a ball that was rolled out at one end of the field. If the batted ball rolled through to the other end and hit the fence, the chicken would receive a reward, which they picked up from a feed hopper at the other end of the cage. The chickens quickly learned to perform this behavior, pulling the string to swing the bat and then moving straight to the feed hopper to collect their reward. But when the environment was changed slightly and the outer cage was removed (the string to pull and the feed hopper remained), the chickens no longer performed as trained. Instead, the sight of the moving ball would excite them, and they would immediately chase it around, pecking at it. In trial after trial, they missed out on any chances for a reward, consistently chasing after and pecking at the ball. The change in environment had seemingly brought out something in the chickens that had never been conditioned by the experimenters but had a strong effect on the behavior, despite it never resulting in a food reward. Such findings revealed that even strict and highly controlled conditioning procedures were insufficient to fully understand behavior, especially outside of the laboratory. Together with the reports from World War II of soldiers not being able to perform operations in battle situations that they were highly trained to do, this suggested that something critical was missing from the behaviorist approach. Many argued that this missing piece was the investigation of the internal and "hidden" cognitive processes of the mind.

Another influential event in the 1940s and 1950s was the invention of the computer, along with the rapid development and application of this technology (Figure 1.3). Computers could

Figure 1.3 Computer, circa 1954. The first transistorized computer in the USA, named TRADIC (for TRAnsistor DIgital Computer or TRansistorized Airborne DIgital Computer). It was built by Jean Howard Felker (at left). *Source:* Image retrieved from https://commons.wikimedia.org/wiki/File:TRADIC_computer.jpg (public domain).

be used to perform complex tasks, conducting thousands of calculations with speed and accuracy. With further developments, higher-order information processing tasks (e.g., data processing, manufacturing control, vote tabulations) that used to be done by human workers were taken over by computers. With the realization that computers were performing some tasks as well as humans, it was natural to consider if the human mind might work like a computer. As opposed to the behaviorist viewpoint that an understanding of the human mind required only a complete understanding of stimulus inputs, schedules of feedback, and overt responses, computers needed specific instructions. These programming instructions, or code, had to be highly detailed, specific, and complete. The information processing that a computer performed was entirely dependent upon the code programmed into it. This led psychologists to question the behaviorists' exclusion of innate processes in the human brain and to wonder if it might be possible to discover the "code" that could explain how humans think. They realized that a new focus of research – into the internal mental events of cognitive processes – was needed.

Advancements in computer technology also led to the development of artificial intelligence (AI) research. In 1950, Alan Turing wrote an influential paper discussing the potential use of an "imitation game" to assess the possibility of intelligence in a machine (Turing, 1950). Later called the "Turing Test," the procedure involved asking questions of a machine and seeing if it answered in the same way as a human. Turing began his article with the question "Can machines think?" but then posed the more tractable question of whether a machine could respond in a way indistinguishable from a human. Importantly, the machine was not being tested on whether it would produce correct answers, which could be programmed into it, but rather if it perfectly mimicked human responses. For our purposes, it is interesting that Turing chose to proceed without defining "thinking" or "intelligence." Indeed, his article preceded the establishment of the research field of cognitive psychology, which would go on to investigate those very issues. Therefore, Turing cleverly avoided arguments over the way those complex terms could have been defined and instead stressed that if the machine could answer questions in a way indistinguishable from a human, then it would have attained that aspect of human thinking. Of course, the artificial machine had to be created by hand, with individual pieces of hardware for different functions and many lines of programming code to specify processing steps. The implication was that if we could program an intelligence, then human thinking could be separable into discrete pieces as well, raising the need for a science to investigate exactly what those pieces are. In 1955, Allan Newell, Herbert Simon, and John Clifford Shaw created an AI program that was able to prove mathematical theorems just as well as talented mathematicians (reviewed in Feigenbaum & Feldman, 1963). This provided strong new support for the idea that even complex human thought could be reduced to the manipulation of bits (many, many bits) of information according to a set of formal rules. With this as a sort of proof of concept that mental events could be represented in this way, the doors were opened to a new way of thinking about mental processes in the human brain, and experimental research in cognitive psychology began in earnest.

Another event in the 1950s that highlighted the limitations of behaviorism was a scathing review of B. F. Skinner's book *Verbal Behavior* (1957). Skinner had set out to explain how behaviorist theories of learning could explain human behavior, even something as complex as language (i.e. "verbal behavior"). Almost 2 years after the publication of the book, Noam Chomsky, a prominent linguist, wrote a lengthy review countering Skinner's arguments (Chomsky, 1959).

One of Chomsky's core criticisms was that Skinner's account of how children acquired language was not feasible; it suffered from a "poverty of stimulus" – meaning that children weren't exposed to nearly enough linguistic information (stimuli) to explain their immense knowledge of language systems and syntax. The explanation, according to Chomsky, was that we must have an innate system in our brains that organizes the relatively sparse linguistic input we're exposed to and allows us to understand the complexities of language, despite the relatively impoverished input. Chomsky called this innate system "universal grammar." As described earlier, the strong influence of instincts in animals undermined behaviorist theories that held that behavior could be explained by conditioning principles alone. Here, Chomsky essentially made the point that something as important as language required a "language instinct" (e.g., Pinker, 1994), further exposing the flaw in the behaviorist approach. Chomsky's review hit a resonant chord with scientists who were interested in the processes of human thinking and were finding the theories and methods of behaviorism to be insufficient. These scientists thus began to focus on internal mental processes and to develop the fields of cognitive science, cognitive psychology, and cognitive neuroscience.

Before completely leaving the topic of behaviorism, it must be acknowledged that the learning principles discovered through these methods were clearly important findings that remain highly relevant to this day. Furthermore, many experimental procedures and designs developed by behaviorists are in use today, such as in important studies of addiction and the brain basis of some psychological disorders. But the strict exclusion of any research into internal mental events by early behaviorists has thankfully been put to rest. It should also be noted that although B. F. Skinner largely dismissed the importance of trying to study internal mental events, even he noted a few important aspects of attention. In his influential book *Science and Human Behavior* (Skinner, 1953), he commented on the role of attention in mediating between a stimulus and a response: "But attention is more than looking at something or looking at a class of things in succession ... Attention is a controlling relation – the relation between a response and a discriminative stimulus" (p. 123). Since behaviorists were very much concerned with stimulus–response contingencies, it is quite the admission that something as vague as attention was given such a critical role. Skinner additionally noted that attention may be important because of the way in which it controls processing of a stimulus: "The control exerted by a discriminative stimulus is traditionally dealt with under the heading of attention. This concept reverses the direction of action by suggesting, not that a stimulus controls the behavior of an observer, but that the observer attends to the stimulus and thereby controls it" (p. 123). Finally, Skinner also commented that there are *involuntary* mechanisms that affect where our attention is focused: "Nevertheless, we sometimes recognize that the object 'catches or holds the attention' of the observer" (p. 122). The interplay of voluntary and involuntary mechanisms on attention will be covered in detail in Chapter 6 of the present book.

1.4   The Cocktail Party

As with most topics in the field of cognitive psychology, research into attention began in earnest in the 1950s. One of the first areas of research within the domain of attention centered on understanding "the cocktail party phenomenon." What Colin Cherry (1953), Neville Moray

Figure 1.4 Cocktail party. *Source:* Photograph by Lance Cpl. Jacob D. Osborne, released by the United States Marine Corps with the ID 100914-M-4756O-557 (public domain).

(1959), and Anne Treisman (1960) found interesting and compelling about the gathering of people at cocktail parties (Figure 1.4) was that it highlighted a core feature of attention: In the cacophony of many different and overlapping voices and sounds, we're able to focus on the one person we're trying to converse with, ignoring all the other sounds vying for our attention. Although the initial forays into attention research were motivated by this real-world phenomenon, the scientists understood that strict experimental controls were necessary to be able to draw conclusions about the processes involved. Thus began the ongoing process of designing experiments that could isolate a critical cognitive process while still relating to the real-world situations that motivated the question in the first place. As will be covered in the next chapter, there are many real-world situations that highlight the critical roles that attention plays in our daily lives. Starting with the initial experiments on the "cocktail party phenomenon," there began a tension that continues to this day between the desire to carefully control extraneous variables in an experiment and the desire to maintain high ecological validity. The concern is that as the laboratory experiments control for multitudes of factors to isolate a cognitive process, that process may no longer be engaged in the same manner as it is during the real-world situations that motivated the study. Cognitive experiments can sometimes appear to be far removed from the situations we observe in our daily lives. However, even the most eloquent descriptions of how the cognitive processes work in the real world, such as Williams James' descriptions of attention, cannot provide a complete understanding of the *mechanisms* of those processes. As we'll see throughout this book, cognitive experiments have significantly advanced our understanding of the pieces of cognition, and bit by bit we move forward in solving the puzzle of *how* attention works in the brain.

Colin Cherry (1953) was the first to refer to the "cocktail party effect" when reporting results from his dichotic listening experiments. Developed by Donald Broadbent (1952), dichotic listening paradigms had the subject listen to two simultaneous streams of speech while needing to attend and respond to just one of them. As noted earlier, World War II was a contributing factor to the cognitive revolution, and Broadbent's original experiments were motivated by his work in the Royal Air Force. Specifically, Broadbent noted difficulties in communication that arose when people, such as air control and gunnery personnel, were presented with multiple streams of speech at once. The dichotic listening paradigm allowed for a well-controlled version of this to be run in the laboratory. Cherry (1953) utilized a version of this paradigm along with a "shadowing task" in which the subject had to repeat aloud – or "shadow" – in real time an attended stream of speech while also being presented with a second stream of speech to ignore. To control for volume and head position, subjects in Cherry's shadowing experiments would wear headphones in which one ear would receive one stream of speech, while the other ear received the second stream. Although it may seem that shadowing an ongoing stream of speech in real time would be quite difficult, subjects quickly learn do it well, and this paradigm then provides a measure of the speed at which subjects are able to shadow accurately and exactly when they make mistakes. Even with a simultaneous stream of speech being presented to the other ear, Cherry's subjects were able to shadow the attended ear quite well. Furthermore, subjects reported not being aware of what was being said in the unattended ear. This provided one of the first well-controlled experiments showing that when a person is presented with two equally strong inputs, it is their internal focus of attention that determines what information rises to conscious awareness. Cherry's experiment showed that attention could work at the level of selecting one sensory "channel" – in this case, one ear over another. This experiment also tested the extent to which unattended information was processed in the brain by manipulating the type of input to the ignored ear. Cherry found that even when the ignored stream changed in terms of the language being spoken (e.g., from English to German) midway through the experiment, subjects were unaware of the change. Most subjects were also completely unaware when the ignored speech was changed to be reversed speech. Cherry did find that two types of changes were noticed in the unattended ear, however: (1) when the unattended speech changed to a (nonspeech) pure tone; and (2) when the unattended speech changed from a male to a female speaker. Overall, Cherry found that only major physical changes in the unattended stimuli rose to the level of awareness, and that changes in the meaning of the speech did not.

Cherry's experiment began the investigations into a core aspect of attention: the stage of processing at which attention modulates sensory inputs. Subsequently, this has been referred to as the "Levels of Processing" debate, and proponents of **early selection** suggested that attention affects sensory processing at a very early stage. Broadbent's (1958) "filter model" was the first of these theories. According to Broadbent's theory, attention acted as a strict filter, allowing only the information from the attended channel to proceed to higher levels of processing. The mechanism of attention in this model was to filter out – or stop processing – all but the attended channel. Based on results from the dichotic listening paradigms, it appeared that attention worked quite early in the stages of processing, selecting information from channels based on very basic physical attributes (e.g., left vs. right ear; high- vs. low-pitched speech). Since subjects in Cherry's and Broadbent's experiments couldn't remember or report on the meaning of the words in an unattended channel, early selection theories held that the attention filter was located before the stage of processing that included analysis of meaning (i.e., "semantic" analyses). Broadbent's theory is also of historical significance because it was among the first to explicitly describe human cognition using an **information processing model**, which attempted to break down complex cognitive processes into a sequence of simpler functions occurring in series.

It is interesting to note that at the time when Broadbent proposed his filter theory of attention, the automatic telephone exchange system had only recently been implemented and installed throughout his home country of England. These automatized telephone exchanges meant that, for the first time, people could place direct calls to one another; previously, people working as telephone operators were required to connect calls. In that earlier system, telephone operators were a bottleneck through which calls were selected to be completed, with each operator being able to assist only one caller at a time. If the operators were busy, a caller needed to wait before their call could be put through. The importance of this new automated technology for being able to avoid a bottleneck in processing is mirrored by Broadbent's theory of attention, in which attention acts much like the telephone operator in determining what information will be selected to proceed. Broadbent's theory also stressed the important idea of capacity limitations in the human mind. The reason that attention is necessary, according to this view, is because the mind cannot process all possible inputs at once. Broadbent's theory suggested that the bottleneck occurs at the transition from a stage of processing with very large capacity to one with very limited capacity. This idea of capacity limits, and the concept of mental resources, would prove to be highly influential throughout many areas of cognitive psychology research. It is interesting to note that, although we now reference Broadbent as establishing one of the first cognitive theories of attention, he often described his theory as one of selective *perception*. Indeed, in a later article, he laments the vagueness of the term "attention," saying "it has also been used as a theoretical concept, a mysterious asset or energy which is sometimes attached to human functions and sometimes not. This use of attention . . . is not very helpful" (Broadbent, 1982, p. 253).

Whereas Broadbent's early-selection theory accounted well for Cherry's (1953) results, Moray (1959) performed a clever variation of the shadowing task and found results that challenged that model. Moray performed a series of experiments that were similar to Cherry's original experiments, but in some conditions he inserted the subject's own name into the unattended stream of speech. According to Broadbent's filter model, the words in the unattended channel are not processed to the level of meaning; the unattended stimuli should be filtered out completely before reaching that stage of processing. Since all stimuli in the unattended channel should have been filtered out, the meaning of those unattended words should make no difference. However, Moray's subjects noticed their name being spoken in the unattended channel. Moray (1959) suggested that "subjectively 'important' messages, such as a person's own name, can penetrate the block" of the attention filter. This result was a problem for early-selection theories of attention. Some researchers then suggested that all stimuli must actually be processed to the level of meaning, with other factors determining what rises to conscious awareness and whether the subject would be able to recall the information later. Theories along these lines, in which all sensory processing proceeds automatically through multiple levels, with attention acting at higher stages of processing (e.g., after the level of semantic identification), are often referred to as "**late-selection theories**." Deutsch and Deutsch (1963) proposed such a model, and they suggested that the level of attentional selection may be a function of both the state of *alertness* in the subject and the "importance" of the stimuli.

Although Moray's 1959 study was instrumental in showing that at least some unattended information was being processed to a high level, it is interesting to note that not all subjects noticed their name being spoken. Only a third of the subjects in his experiment noticed their name in the unattended channel. In addition, the number of critical data points was low, because that experiment had a small sample size (12 subjects) and each subject could only be asked once if they had noticed their name. Moray also didn't report on the accuracy of shadowing, so the third of subjects who noticed their name may have been able to do so because they just happened to be sampling the unattended channel when their name occurred. Indeed, this latter point was a critical argument used by early-selection theorists to critique Moray's results. The argument is that some subjects may have switched their attention to the unattended ear (out of boredom, or curiosity, or simply because they lost focus) *before* their name had been presented. Thus, they may have been conscious of their name because they were already sampling that channel of input when their name happened to occur. More recently, Wood and Cowan (1995) replicated Moray's study with a larger sample size, and they also conducted a careful analysis of shadowing accuracy. This study replicated the result that about a third of participants noticed their name in the unattended channel. Most critically, Wood and Cowan found that in those subjects who noticed their name, shadowing accuracy had remained high before the name was presented, providing evidence that those subjects were still attending to the correct channel when their name was presented in the unattended channel. Furthermore, the subjects who noticed their name showed poor shadowing accuracy for a word or two *after* their name. This result provided evidence that their name, even when unattended, captured their attention away from the to-be-attended stream for a short period of time before they could reorient back to the correct stream. Finally, although we may have the sense, in our everyday lives, that we always notice when someone nearby says our name, laboratory experiments such as that by Wood and Cowan suggest that we most often don't notice this. And, of course, we're not aware of all the times we don't notice our name being spoken. We can only be consciously aware of the times this does capture our attention – which is rarer than we might think. The fact that our name occurring in an unattended channel can be processed to its level of meaning is a problem for strict early filter models because only inputs to the attended channel should be processed to such a high level. However, the fact that we often don't notice our name is a problem for strict late-selection theories because this means that most of the time our name is not getting through to consciousness, so there would have to be additional selection

mechanisms at play besides the late filter. Therefore, a different conceptualization of attention was needed.

Anne Treisman (1960) proposed a different model of attention based on her results from another clever variation of the dichotic listening/shadowing task. In her task, the two streams of speech were sometimes switched between the ears in the middle of sentences, unbeknownst to the subjects. Treisman found that subjects would briefly follow the story and shadow the wrong ear for a word or two before realizing that they were shadowing the wrong ear, at which point they would switch back to shadowing the speech from the correct ear. This was a critical finding, because Moray's (1959) results could have been due to the subjects' own names just being exceptionally unique stimuli that could get through the attentional filter because they are so strongly associated as important over years of personal experience. But Treisman's results showed that words that by themselves were not special in any way (e.g., "table," "jumping," "increased") could break through the attentional filter and rise into consciousness if their meaning fit precisely with the sentence currently being processed. This provided new evidence that even these ordinary words were being processed to a high semantic level. Treisman proposed a new theory, in which attention worked as an "attenuating filter." A critical advance of Treisman's attenuating filter theory was that attention was explained as *modulating* processing, not acting as a strict gate that either let information through or completely halted it. According to Treisman's model, attention could enhance or suppress inputs but did not completely filter any out. Thereby, unattended inputs would be relatively weakened but could still progress to higher stages. Depending on the mental state of the listener and the relatedness of concepts currently active in memory, even weak inputs could rise to the level of awareness. Although this model was important for advancing theories of attention beyond strict filter theories, it leaves unanswered the critical question of exactly *when* and *where* within the stages of processing attention can act to attenuate or enhance processing. This is exactly where cognitive neuroscience studies have proved integral, and we will return to this "levels of processing" debate in Chapter 5, where the effects of attention on neural processing are discussed.

1.5   The Eyes and Attention

Predating the cognitive revolution and the development of attention research as a scientific field, Hermann von Helmholtz was conducting important investigations into the structure and components of the eye that would lead to new theories of perception and, by extension, attention. Helmholtz's research in the mid-1800s still has direct impact today, as instruments that he invented to study the eye, such as the ophthalmoscope, are common tools used by optometrists during routine eye exams (Figure 1.5). His research into sensation has also had lasting impact, such as his accurate description of our color vision as being supported by three types of cones in the retina. His theories on depth and motion perception were among the first to discuss how top-down processing plays a critical role in visual perception, an idea that has recently been revitalized in predictive coding models (which we will cover in more detail in Chapter 9). Helmholtz also stressed the distinction between

Figure 1.5 Ophthalmoscope. Left: Ophthalmoscope from 1860. *License:* CC BY 4.0; *Credit:* Ophthalmoscope designed by Richard Liebreich circa 1860. Wellcome Collection. Right: Modern-day ophthalmoscope in use during eye exam. *Source:* Image by SPC Jeffrey Hernandez, public domain, via Wikimedia Commons.

sensation and perception, believing that the former was a more tractable scientific issue because perception was affected by various idiosyncratic factors.

While Helmholtz is most famous for his enduring work on the eye, he is also responsible for one of the first published reports that *attention* is not completely linked to the *eyes*. In a demonstration, in which he served as his own subject, Helmholtz (1896) found that perception depended on where his internal attention was focused, not simply on where his eyes were fixated (Helmholtz, 1896; translated by Nakayama & Mackeben, 1989). Looking straight ahead into a dark box in which a sheet with printed letters was at one end, Helmholtz lit a spark to briefly illuminate the sheet. He found that the letters he perceived varied as a function of where he had decided to focus his attention *before* the spark of light. In describing his experiments, Helmholtz noted that "by a voluntary kind of intention, even without eye movements, and without changes of accommodation, one can concentrate attention on the sensation from a particular part of our peripheral nervous system and at the same time exclude attention from all other parts" (Helmholtz, 1894; translation by Warren & Warren, 1968). Helmholtz's demonstration that attention could be moved independently from the eyes is now referred to as **covert attention**, in contrast to **overt attention**, in which attention moves along with the head or eyes. As will be covered in later chapters, the link between attention and the eyes is strong, but Helmholtz demonstrated that attention is not simply where the eyes are looking. Furthermore, Helmholtz described the process of covert attention as being an *effortful* process driven by a volitional choice ("by a voluntary kind of intention"). As we will cover in later chapters, this foreshadows the distinction between voluntary and involuntary processes of attention. Finally, his description of "excluding attention from all other parts" foreshadows the idea that attention is a limited resource that determines which small part(s) of our sensory inputs are being fully processed at any one time.

Whereas Helmholtz's work was a critical step in showing that attention could be dissociated from the eyes, recent work is using other information from the eyes to investigate attention and cognitive processes. For example, researchers have begun to use microsaccades and pupil diameter as additional measures of attention. Microsaccades refer to the tiny movements of the eyes during periods of fixation. Even when the eyes appear to be still during fixations, highly precise tracking of the eyes reveals these tiny movements. Recent studies have found that microsaccades during a period of fixation can index the direction of *covert* shifts of attention (e.g., Engbert & Kliegl, 2003; Hafed & Clark, 2002). In a recent study, Barnhart and colleagues found that the onset times of microsaccades were correlated with task difficulty and that the direction of microsaccades indexed covert shifts of attention (Barnhart, Martinez-Conde, Macknik, Costela, & Goldinger, 2019). In addition to microsaccades, other research has found that pupil diameter varies with attentional focus and load. Privitera, Carney, Klein, and Aguilar (2014) found that microsaccades provided a good index of the direction of covert attention, whereas pupil dilation was related to visual detection and decision processes. Vilotijević and Mathôt (2023) have recently proposed that pupil dilation is related to the need to detect information in the periphery, finding increases in pupil size when subjects need to attend covertly to information outside the fixated region.

These recent reports correlating attention with subtle effects of the eyes have reinvigorated interest in **motor theories of attention**. Hafed and Clark (2002), in their initial study linking microsaccades to covert attention, suggested that microsaccades may be triggered by "subliminal" activation of the oculomotor system. This idea harks all the way back to the late nineteenth century, when philosophers were first contemplating the relation of attention to overt movements of the eyes and head. Alexander Bain (1888) suggested that attention could be explained as motor plans that were not quite carried out. More recently, Rizzolatti, Riggio, Dascola, and Umiltá (1987) proposed their premotor theory of attention, which holds that "the program for orienting attention either overtly or covertly is the same, but in the latter case the eyes are blocked at a certain peripheral stage" (Rizzolatti et al., 1987, p. 37). Motor theories of attention will be discussed in more detail in the chapters on the control of attention (Chapter 7) and predictive coding models of attention (Chapter 9). These theories have some compelling supporting evidence from studies of spatial attention but are not as clearly applicable in the cases when attention is focused on nonspatial features (e.g., color, sound, smell, or internal thoughts) that aren't associated with overt movements toward locations in space.

Finally, recent research into a different mechanism of the eye, pupil dilation, relates to yet another topic that Helmholtz commented on over 150 years ago: the ability to maintain attentional focus. Kahneman (1973) described research using pupil dilation to monitor a subject's level of arousal, which figured prominently in his theory of attention. More recently, Smallwood and colleagues confirmed that pupil diameter was associated with the onset of visual images during periods of focused attention but *not* during periods of **mind-wandering** (Smallwood, Brown, Tipper, Giesbrecht, Franklin, Mrazek, Carlson, & Schooler, 2011). These results were replicated in another study that controlled for luminance changes, finding again that pupil dilation tracked with stimulus onset during focused attention but *not* during mind wandering (Kang, Huffer, & Wheatley, 2014). This critical distinction between focused attention and mind-wandering is reminiscent of Helmholtz noting:

> The natural unforced state of our attention is to wander around to ever new things, so that when the interest of an object is exhausted, when we cannot perceive anything new, then attention against our will goes to something else. . . . If we want attention to stick to an object, we have to keep finding something new in it, especially if other strong sensations seek to decouple it. (Helmholtz, 1867, p. 770; translation in Hohwy, 2013)

As described by Helmholtz, sustaining attention can be very difficult – at times, it takes a great deal of effort. At other times, however, it remains transfixed with ease. We may happily gaze on a beautiful painting for an extended time. In the latter case, however, we're typically appreciating different details or aspects of the work, or taking in its meanings and depths of complexity. In that way, although our attention at one level may be consistently on the painting, at another level we're paying attention to different parts of the painting or linking it in our minds to other knowledge or experiences we've had. As Helmholtz noted, we are in that case "finding something new in it." When we need to pay attention to something that doesn't naturally excite our sense of wonder, however, the effort required to sustain attention is abundantly clear to us. There has recently been renewed interest in understanding what is going on in our brains when our minds are wandering. In Chapter 2, we'll return to this topic when we cover real-life situations, such as our ability to sustain attention on dry textbook material for a class, or when driving long miles on empty highways, or when naval officers monitor the repetitive sweeps across their sonar screens.

1.6   The "Spotlight" of Attention

A common metaphor for attention has been a "spotlight." Similar to how a spotlight illuminates one region of a theater stage and makes it easier to perceive the actors and actions at that location, attention can be thought of as enhancing our perception of objects at a certain location (Figure 1.6). The beam of a flashlight on a nighttime stroll ensures that we can clearly see the path ahead, while the surrounding areas remain shrouded in darkness. These examples illustrate a couple of important aspects of attention. One attribute is that attention can be allocated according to spatial location. Similar to a spotlight or flashlight, the focus of attention can be allocated to a specific region of space, enhancing processing at that spot regardless of what's there. Another attribute that the spotlight metaphor exemplifies is that attention is "*selective*," meaning that it can enhance only a relatively small portion of the world around us. This selective nature of attention further implicates the notion of "resources" in the brain.

These different aspects of attention can be studied with a simple "cuing" task, sometimes called the "Posner cuing paradigm," in reference to Michael Posner, who developed the task in the 1980s (Posner, 1980). The task was designed to be a simple paradigm that would allow attention to be studied across ages, from young children to older adults, as well as in nonhuman animals. In terms of cognitive neuroscience, the applicability to nonhuman animals was important because Posner developed this task with an interest in finding the neural basis of

Figure 1.6 Spotlight of attention analogy and the "zoom-lens" model. The middle three images show an extreme version of the spotlight analogy, in which only a small part of the visual world is being attended and the rest is relatively suppressed. The image at far right illustrates the zoom-lens model, in which the power to boost what's in the focus of attention is diminished as the focus is spread more broadly over a wider area, compared to a stronger boost when the attended region is smaller and all resources are more tightly focused, as shown in the middle three images.

attention but before the human neuroimaging methods of positron emission tomography (PET) and functional magnetic resonance imaging (fMRI) were available. The task was simple enough to be performed by nonhuman primates, making it possible to investigate neural mechanisms through single-unit recordings. Those studies, along with the many human electrophysiology and neuroimaging studies conducted using this task, will be covered in more detail in Chapter 5. Here, we discuss the basic task and its implications for understanding the mechanisms and processes of attention.

In this task, the visual display is quite simple, usually consisting of just a fixation spot in the center of the screen, flanked by two outlined squares to the left and right of fixation (Figure 1.7). The subject's task is to respond, as quickly as possible, to a target stimulus that will occur within the box on either one side or the other. Depending on the task, the subject may simply be detecting the stimulus or may be discriminating some detail of it. In the simple detection task, the subject is asked to press a button as quickly as possible when they perceive the stimulus. In the discrimination version, the subject is asked to respond as quickly and accurately as possible once they have determined the critical feature of the target stimulus. Typically, there are two options that the subject must choose between: for example, discriminating its location (left or right), size (large or small), or color (red or green). What is critical to this task is that before the target appears the subject is "cued" in some way to shift their attention to one of the two locations. This can be done with a peripheral cue at the location to be attended (e.g., the box at one location may brighten briefly, or flash off and on) or a central cue that instructs subjects where to attend on the current trial (e.g., an arrow at fixation pointing to one location). The responses to targets can then be compared for targets that occurred at the location that was cued (i.e., "valid" target trials, in terms of being validly cued) versus targets that occurred at the other location (i.e., "invalid" or "uncued" location targets).

Experiments using this paradigm showed that valid targets are responded to faster and more accurately than invalid targets. In addition to these two types of trials, Posner also implemented "neutral" cue trials, in which the cue did not instruct subjects where to attend. Examples of "neutral" cues could be a double-headed arrow at fixation (pointing to both possible target

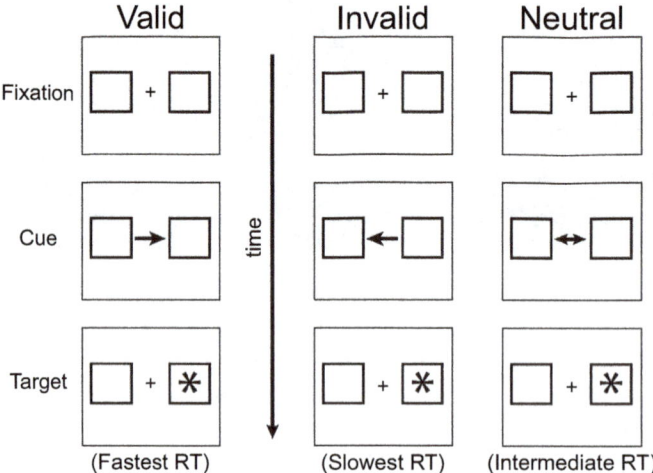

Figure 1.7 Posner cuing task. Trial sequence for the three main types of trials in a cuing task. The target (asterisk) can occur in either the left or right box. On "valid" trials, a cue (arrow) correctly directed attention to the location where the target occurred. On "invalid" trials, the cue incorrectly directed attention to the wrong side. On "neutral" trials, the cue gave no information about the location of the target. In a typical cuing paradigm, the cue is predictive, in that it correctly predicts the target location on ~75% of trials.

locations) or a simple blink of the fixation cross (indicating the trial had started but not which location to attend). Critically, these neutral trials include a cue event, to ensure that arousal and alertness are equated when comparing neutral cues to directional cues. In that way, the difference in response speed and accuracy between directional cues and neutral cues would be due to the effect of orienting spatial attention. Having these three types of trial (valid, invalid, neutral) is critical for providing evidence that attention involves both costs and benefits. As William James described in his famous quote, attention "implies withdrawal from some things in order to deal effectively with others" (James, 1890). Posner's original experiment, and many replications since then, have shown that attention produces not only benefits (i.e., faster responses to validly cued targets compared to neutral trials) but also costs (i.e., slower responses on invalidly cued targets compared to neutral trials; reviewed in Posner, 2016).

Results from these paradigms have shown that the allocation of spatial attention directed by a cue stimulus significantly affects the time it takes for a subject to respond to the subsequent target. But what is attention actually doing? Is it *speeding up* the progression of processing through the multiple stages? Or enhancing the *quality* of processing in some stage (or stages) so that the stimulus is clearer and decisions can be made quicker? And which specific qualities or stages of processing between the initial sensory input and final motor response are being modulated? The question of what attention does has been the subject of philosophical musings for centuries. In the seventeenth century, John Locke defined consciousness as the specific part of perception that we're mindful of, and this foreshadowed the debates about levels of attentional selection (Locke, 1689). As we'll cover in subsequent chapters, the relation of attention to consciousness remains an area of active debate in philosophy, with a major topic being whether attention is necessary and sufficient for consciousness. In psychology, research into "inattentional blindness" (Mack & Rock, 1998) and "change blindness" (Rensink, O'Regan, & Clark, 1997) is continuing to advance our knowledge of the role of attention in conscious awareness. While much interest in attention has been focused on how it affects access to consciousness, the cuing paradigm highlights that attention is more than just a gate to awareness. The subjects are conscious of the target stimulus on all trials, but their manual responses are significantly affected by where attention was at the onset of the stimulus. Therefore, attention is doing something more than just allowing access to awareness. This relates to the subjective feeling that attended stimuli are somehow clearer than unattended stimuli. Research studies have investigated this issue and found that attention enhances perception in a number of ways, including visual acuity, contrast sensitivity, and perceptual quality (Anderson & Druker, 2013; Carrasco, 2018; Kwak, Hanning, & Carrasco, 2023).

A final critical piece relating to the "spotlight" metaphor regards the size of the spotlight. Research has suggested that the spotlight of attention isn't a fixed size but varies in extent, similar to how some flashlights allow you to adjust the size of the beam. This led Eriksen and St. James (1986) to propose the "**zoom-lens model**" of attention. In addition to accounting for different size windows of attention, this model proposes that the degree to which attention boosts processing is a function of the area over which attention is spread (Figure 1.6, rightmost panel). When focused on a smaller region, the effects of attention are greater than if it is widely distributed over a larger region. Similar to how the luminance power of an adjustable flashlight is constant but can be spread over a large area or tightly focused on a smaller region, the power of attention to boost processing is similarly dependent on the size of the area being attended. The initial work that led to the zoom-lens model came from a new task developed by Eriksen and colleagues (Eriksen & Eriksen, 1985; Eriksen & St. James, 1986). In each trial of this task, the display was composed of a circular array of letters, which could include target and nontarget letters. One set of target letters required one response (e.g., right-hand button press), while another set required a different response (e.g., left-hand button press). On each trial, the locations to be responded to were "cued" to the subject by a line underneath the locations to be monitored shortly before the array of letters was presented. (Note that the cue in this paradigm is different from the cue in the Posner paradigm: In the Posner cuing paradigm the cue is predictive of where a target is likely to occur, but the subject must respond to the target regardless of its location; in the Eriksen paradigm, the cue is instructive and subjects only respond to stimuli at the location of the cues.) This allowed the researchers to control the size of the area that was relevant on each trial (e.g., increasing the size from one to two to three contiguous locations). When more than one location was cued, the locations were always contiguous. One of the cued locations would always have one of the target stimuli, and the other cued locations would have neutral distractors (letters that were not associated with any motor response). The results showed monotonically increasing response times as more locations were cued, suggesting decreasing intensity of attention at a given location as the attentional focus was being distributed across a wider region. Of critical importance to this paradigm, "incompatible distractors" (stimuli associated with the opposite manual response to the target letter's response) could appear at the noncued locations. Although these locations should be ignored, the presence of incompatible distractors in the display slowed responses compared to when only neutral letters were in the display. This additional slowing was interpreted as showing that the stimuli at the irrelevant locations were being processed to the level of generating motor plans, which would compete with the production of the correct motor response. Critically, Eriksen and colleagues found that as the size of the attentional window (defined as the number of cued locations) *increased*, the ability of attention to boost the attended information and suppress the irrelevant information was *decreased*. Specifically, the wider the attentional window, the greater the response slowing due to incompatible distractors (controlled for physical distance). This and subsequent research provided new evidence that the spotlight of attention can be expanded or shrunk but that the power of attention effects within the spotlight is inversely related to its size. As described in Box 1.2, another interesting line of research has explored whether the spotlight can be split into multiple beams.

Box 1.2 Splitting the Spotlight?

The spotlight model and the zoom-lens model have in common the assumption of a unitary focus of attention, which fits with our everyday experiences of focusing on just one thing at a time. However, whereas the work of Posner, Eriksen, and their colleagues supported a single unitary focus of attention, other researchers proposed that the beam of attention can be split into separate parts. Castiello and Umiltà (1992) designed a series of experiments in which attention was cued to two locations – one in the right visual hemifield, the other in the left – and the size of the cued region was varied independently to be a large- or small-sized box. The authors argued that if attention was a unitary beam that had to be spread over both hemifields, then the size of the attended boxes shouldn't affect the subjects' response times, because attention has to be spread widely to cover both hemifields regardless of the size of the box in each hemifield. The results, however, showed that reaction times were dependent on the box sizes, suggesting that attention was being focused separately in the two hemifields. This study left open the question of whether attention could be split into many beams, and the authors noted that the ability to split attention into two beams in their design could be related to the two hemispheres of the brain. Since their displays had just one attended box in each hemifield, a possible explanation for their findings was that each hemisphere of the brain could have its own beam of attention. Even if we typically end up with a single focus of consciousness, the idea was that at some level each hemisphere of the brain might be able to have a beam of its own. Indeed, this idea was tested later in "split-brain" patients who had their corpus callosum cut to relieve intractable epilepsy and therefore had disconnected left and right brain hemispheres. Luck, Hillyard, Mangun, and Gazzaniga (1989) found that these patients performed visual search tasks at a rate faster than healthy controls when stimuli were distributed across both visual fields but not when all stimuli were in a single hemifield. Furthermore, the patients' search rate was twice as fast when the stimuli were distributed across both hemifields compared to being in just one. Thus, it appears that there may be at least two independent beams of attention, one for each brain hemisphere. However, there is still debate about whether there can be separate beams of attention in the healthy, intact brain. McCormick, Klein, and Johnston (1998) used the basic design of Castiello and Umiltà's (1992) experiments but added "probe" stimuli that had to be reported in the space between the two attended boxes. The

Box 1.2 (cont.)

results showed that the space in between the two attended boxes also enjoyed the benefits of attention, supporting the view that there is just one focus of attention that can be spread out as proposed in the zoom-lens model. Whereas the original spotlight metaphor stressed the enhancement of processing within the spotlight, more recent work has focused on the inhibition of information outside of the spotlight. Researchers have found that, under some circumstances, intervening locations between attended locations can be ignored (e.g., Kramer & Hahn, 1995), and that the ability to *suppress* intervening locations may be flexible enough to allow for numerous beams of attention (Awh & Pashler, 2000). As Castiello and Umiltà (1992) discussed in their original article, however, there is also the difficulty of assessing whether attention is truly split between multiple locations at a given instant or if a single beam of attention is rapidly switching back and forth between attended locations.

While spotlight and zoom-lens models of attention can account for how we attend to spatial locations ("spatial attention"), those models aren't designed to account for other types of attention. Specifically, research has shown that we're able to attend to nonspatial features (e.g., color, shape, direction of motion). At least for visual features, these will usually be located at a specific region of space, so some accounts of "feature-based attention" include mechanisms of spatial attention as part of their models. But other types of feature-based attention are not well explained by a simple spotlight metaphor. For example, nonvisual features may not be linked to space. For example, we may be attuned to a certain sound in the environment (e.g., a song on the radio or the tweeting of a bird), and we can attend to it over other sounds in our environment without needing to localize it in space. Or we may catch a whiff of a scent that we notice and that brings memories into our consciousness without us needing to localize where the scent originated. Indeed, spotlight models don't account well for the mechanisms of selection for when we attend to our internal thoughts. Finally, research into "object-based attention" has shown that attention spreads across objects in a way that is at odds with simple spotlight or zoom-lens models. These different types of attention (i.e., spatial, feature, object) will be covered in more detail in the chapters on control of attention (Chapter 7) and the effects of attention (Chapter 5).

1.7   The "Glue" of Attention

Whereas the spotlight and related models sought to account for how attention is focused and distributed in space, other research has focused on what attention *does*. As mentioned in the previous section, there is evidence showing that attention can speed up responses to targets and enhance perceptual clarity of a given stimulus. However, others have argued that attention has to do more than that. In the real world, we are constantly exposed to many stimuli and a host of varied features as opposed to simple cuing paradigms that present a single target stimulus. In

Figure 1.8 Attention as glue? Attention is sometimes referred to as the "glue that binds," since it is thought to help bind features together onto objects. *Source:* Getty Images; credit: Vladimir Godnik. Creative #: 723503761.

addition to her important "attenuating filter theory" described earlier, Anne Treisman proposed a highly influential theory that suggested another vital role for attention. According to Treisman and Gelade's (1980) **feature integration theory (FIT)**, attention is the glue that attaches features to an object in our perceptions (Figure 1.8). In this theory, features from across the visual scene are initially processed independently in brain areas concerned with multiple attributes of perception (e.g., color, shape, size, motion). Binding all of an object's features together, however, requires attention. According to FIT, when attention is focused on a spatial location, then all of the features at that specific location are bound together and the perception of the complete object can rise to awareness. Brain regions processing the different features respond, at a "preattentive" stage of processing, to all of the stimuli in the scene. Those features, however, would essentially be free-floating until attention was focused on a location, at which point all of the features at that location could be bound to the object. Treisman's theory thus addressed the "binding problem": the question of how we know what colors, sounds, smells, and so on, are part of which objects. According to FIT, spatial attention is the critical glue that solves this binding problem. A particularly compelling piece of evidence supporting FIT concerns "illusory conjunctions." According to FIT, the brain should be able to preattentively process features in the visual scene before attention binds them to objects. Treisman and Schmidt (1982) found that when a display of colored objects is presented to subjects too quickly for them to orient their attention to the spatial locations of the objects, subjects report seeing combinations of features that hadn't been combined. For example, in a display with a blue triangle and a yellow circle, the subject might report seeing a yellow triangle. The subjects only reported features (shapes, sizes, colors) that were present somewhere within the display, showing that they were processing the stimuli at some level. Without having time to focus attention on the locations, however, the features weren't bound to the correct objects, leading to the "illusory conjunction" of features. Recent work investigating the conditions under which illusory conjunctions are found (e.g., Henderson & McClelland, 2020) and the related deficits in Balint's syndrome patients (Friedman-Hill, Robertson, & Treisman, 1995) will be discussed in the subsequent chapters.

The idea of feature processing being done in a "preattentive" stage of processing has also been influential in theories and research investigating the role of attention in visual search. Research into visual search typically contrasts "feature search" and "conjunction search." Feature search describes the situation in which the target search can be completed by simply locating a unique feature. For example, searching for a red umbrella among many black umbrellas would constitute a feature search because you simply need to shift attention to the one item that is different on a salient feature dimension from all the rest of the items. On the other hand, an example of a conjunction search would be looking for a red glove in a pile of red mittens and black gloves; in this case, neither a specific color nor shape feature defines the target, but rather the conjunction of the color red and the shape of the glove defines the target. Feature search tasks can be done quickly, regardless of the number of distractors, because none of the distractors share the key feature that defines the target. Conjunction searches, however, become progressively longer the more distractors there are. FIT argues that the slowing of such searches represents evidence that attention has to be allocated to each item successively in order to bind the different features together and let the subject correctly perceive each item with all of its features. Through a somewhat circular logic, these sorts of search tasks have been used to define what constitutes a feature: specifically, visual attributes that "pop out" and produce "flat search slopes" (i.e., the reaction time plot shows a constant, flat line as the number of distractors increases). Regardless of what attributes can pop out, however, the enduring strength of FIT is that there are conjunctions of features that take more time to find, suggesting that attention plays a role in binding those features together.

It should be noted that FIT has been criticized on a number of points. Regarding the assumption that basic features are processed in parallel at a "preattentive" stage, it has been argued that attention may still affect processing at these early stages as well (Becker, Atalla, & Folk, 2020). As opposed to being preattentive, it may be that different mechanisms of attention could be at play at different levels. A related criticism is that FIT associates attention only with this binding process, without including mechanisms of attentional selection at other stages of processing (e.g., enhancing perceptual processing of attended stimuli). In addition, the assumption of parallel processing of simple features across the visual scene may be more limited than originally assumed (reviewed in Liesefeld & Müller, 2020), as resource limitations have been reported at earlier levels as well. Finally, the original FIT didn't include the influence of context or of gestalt grouping principles on the binding process. Although the original FIT was specifically modeled to explain visual attention, the idea that a function of attention (maybe *the* primary function) is to bind features means that this mechanism should extend to other sensory modalities as well. Yet, as described by Spence and Frings (2020), there are significant challenges to adapting FIT to cover auditory and tactile attention, as well as the binding of features across senses in cross-modal attention. These challenges include extending the model beyond *spatial* attention, identifying the core *features* in other modalities, and defining what an *object* is in the olfactory and auditory domains. Nevertheless, FIT has proven to be a highly influential theory of one function of attention, and there continues to be a wealth of important research motivated by this model of attention (reviewed in Wolfe, 2020).

1.8   Attention as "Cognitive Control"

In 1935, John Ridley Stroop published an article – little regarded at the time – that would come to be one of the most highly replicated experiments of all time. A favorite demonstration used in psychology classes even today, the "Stroop effect" illustrates the mental effort required to overcome automatic processes. In the classic version of the paradigm, color words (e.g., "blue," "red," "green," "brown") are printed in different-colored ink, and the subject has to name the color of the ink as quickly as possible, ignoring the printed word. The difficulty arises when the ink is a different color than the color word – for example, when the word "red" is printed in blue ink and the subject should respond "blue." In his original paper, Stroop found that subjects were significantly slower in that condition compared to simply naming the ink color of an object (e.g., a small blue square). In subsequent studies, the different trial types have been labeled "incongruent" (when the ink color was different than the color word), "congruent" (when the color word and the ink were the same color), and "neutral trials" (when the words are not colors). Many studies have now replicated the critical result that subjects (and students in classroom demonstrations!) make errors and are slowest on the incongruent trials. These findings are taken as evidence that reading printed words is an automatic process that is difficult to stop. When the written word should be ignored so that we can concentrate on just the ink color, we can feel the effort involved in stopping ourselves from reading the text and making ourselves focus on the unusual task of reporting the color of the ink. Whereas subjects are slower in the incongruent trials compared to the neutral and congruent trials, they are fastest to respond on congruent trials. This latter speeding of times is also thought to be due to the automaticity of the reading process – when the ink and the word match, the correct response has already been generated by seeing the word, and the ink simply confirms that the subject can go ahead with that automatically generated response plan.

The Stroop effect highlights a different aspect of attention than what has been discussed in the preceding sections: that of "cognitive control" or "executive control." These terms refer to mental processes that allow for cognitive flexibility. Attention is needed in situations, such as in the Stroop task, when an automatic response (e.g., reading of the word) must be suppressed and we are forced to effortfully engage a different process. The final execution of a single response, amid multiple possible ways to respond, is another instance of attention serving as a bottleneck on processing. As opposed to the earlier descriptions of attention as filtering incoming sensory information so as not to overwhelm our consciousness and focus, the bottleneck here is the limit of a single response. As in those earlier descriptions, when the mind has to make a choice, the mechanisms of that process are ascribed to attention. Furthermore, the decision of what response to focus on can also affect the allocation of attention in the ways described earlier (e.g., focusing on different features of the stimuli).

Although it has since become a highly influential paradigm, Stroop's original (1935) study was rarely cited and was largely ignored for the first quarter century after its publication. This was in large part due to when it was published – at a time when behaviorism dominated psychology research. Behaviorists in general weren't interested in the mental processes involved in attention and cognitive control. Following the cognitive revolution and the development of the field of cognitive psychology in the 1960s, however, Stroop's paper was rediscovered, and its ability to simply and efficiently measure cognitive control was finally appreciated. Although Stroop used words in his original study to capitalize on the overlearned and automatic process of reading, many variants of his task have been developed in the succeeding decades utilizing other types of stimuli. Indeed, the fundamental process at the core of the Stroop task is not reading but the control that needs to be exerted over an automatic process. Therefore, researchers have developed different tasks in which an automatic process has to be overcome. For example, in a "numerical Stroop task" incongruent trials could involve saying the number of digits present when the printed number is different (e.g., needing to respond "five" to the display "3 3 3 3 3," or "one" to a display consisting of "5"). In a "spatial Stroop task" subjects need to respond "up" or "down" to indicate the location of an arrow presented above or below the fixation cross, and they are slowed in the condition in which the arrow is pointed in the direction incongruent with its spatial location. There are variations of the task used in social psychology research, such as the "emotional Stroop" task, in which words (e.g., "happy," "sad") are printed on faces showing emotions that can be congruent or incongruent with the words (Agustí, Satorres, Pitarque, & Meléndez, 2017). In all of these variations, the key element is that an effortful control process – attention – must be used to ignore the irrelevant (and salient) information and to focus on the attribute that must be reported.

It is also noteworthy that the processes assessed by Stroop tasks have been found to be relevant in many clinical disorders. For example, reviews have concluded that the interference control measured in Stroop tasks is consistently compromised in attention deficit hyperactivity disorder (ADHD; Lansbergen, Kenemans, & van Engeland, 2007), and reduced cognitive inhibition is also reliably observed in some forms of schizophrenia (Westerhausen, Kompus, & Hugdahl, 2011). Results from emotional Stroop tasks have shown some ability to dissociate generalized anxiety disorder compared to social phobia, suggesting a possible aid to diagnosis (Becker, Rinck, Margraf, & Roth, 2001). In addition, a recent meta-analysis suggested that the Stroop task was more effective than other tests of executive functions in discriminating between Alzheimer's disease and healthy aging (Guarino, Favieri, Boncompagni, Agostini, Cantone, & Casagrande, 2019).

Related to the idea that attention can act at different stages to resolve processing bottlenecks and conflicting plans, Daniel Kahneman's theory of attention and effort stressed the idea that resources and capacity limits are critical for understanding attention (Kahneman, 1973). According to Kahneman's model, various tasks compete for a pool of resources, with the amount of resources partially dependent on arousal and effort. Depending on the nature of the tasks and the difficulty of each, there may not be enough resources available to complete each one, and attention is then necessary to allocate processing resources. Although recent research has questioned the premise of attention being equated with effort (Bruya & Tang, 2018), the idea that task difficulty and resources together determine the stage of processing at which attention works has been supported through more recent theories of "attentional load" (e.g., Lavie & Tsal, 1994), which will be covered in more detail in Chapter 5.

Another highly influential paradigm to measure cognitive control was developed by Barbara and Charles Eriksen in the 1970s (Eriksen & Eriksen, 1974). This task, now typically referred to as the "Eriksen flanker task," was similar to the Stroop task in that it manipulated congruent and incongruent stimuli, but it could also be used to investigate questions about spatial attention. In the most common version of the Eriksen flanker task, a row of five arrowheads appears, and the subject must discriminate the direction of the central arrowhead and press a button corresponding to that direction. Critically, in this task, there are two different possible response options corresponding to the two different directions of the arrows (e.g., leftward pointing and rightward pointing). On each trial, the "flanking" arrows (the two arrows on each side of the central arrow) either all point in the same direction as the central arrow or they all point in the opposite direction. Hence, there are "congruent trials" in which the flankers are in the same direction as the central target arrow (e.g., > > > > >) and "incongruent trials" in which the flankers point in the opposite direction from the target arrow (e.g., < < > < <). On congruent trials all stimuli in the display evoke the same motor response, but on incongruent trials the flankers evoke a different motor plan than the central target. Eriksen and Eriksen's original study, and numerous replications, found that subjects were significantly slower on the incongruent trials, suggesting that attention was unable to completely filter out the nearby (flanking) distractors and that generating an incompatible motor plan caused conflict at the level of response execution. As in the Stroop task, attention is needed to help resolve the conflict and allow the correct motor plan to be executed. The Eriksen flanker task can also include a neutral condition in which the central arrow is flanked by meaningless stimuli, such as a series of dashes that have not been associated with any motor response (e.g., − − > − −). Using this neutral condition has revealed costs and benefits, in that incongruent flankers slow responses and congruent flankers speed up responses compared to the neutral condition. In addition to testing how attention is involved when there are conflicting motor plans, the Eriksen flanker task can also be used to investigate the extent and distribution of spatial attention, since the spacing and size of the stimuli can be easily manipulated.

Note that in their original study Eriksen and Eriksen (1974) used letters as stimuli, and subjects were trained to respond to one set of letters with one button press, a different set of letters with a different button press, and to make no response to a third set of letters. This allowed the researchers to have flexibility in the assignment of letter–response matchings, and it let them examine conflict or congruence across multiple different stimuli. These were important controls for making the case that the conflict was arising at a level beyond simple object identification, because multiple different physical stimuli could be associated with the same or different motor responses. Other variants of the flanker paradigm include using numbers or color patches, and some studies use a vertical alignment of stimuli. Most current applications of the Eriksen flanker task, however, use the horizontal arrow stimuli described earlier, since this implementation requires little training and can be completed fairly quickly. This simple design has also allowed it to be used widely across ages, from developmental studies (reviewed in Ridderinkhof, Wylie, van den Wildenberg, Bashore, & van der Molen, 2021) to aging research (Duque, Petitjean, & Swinnen, 2016), and it has been used to assess executive control processes

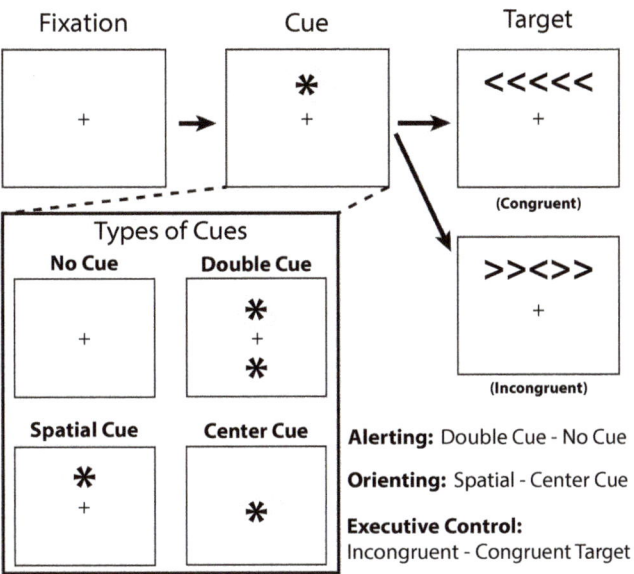

Figure 1.9 Attention network task. The trial sequence is shown in the top row, along with the two different types of target display (congruent and incongruent); note that the target to be responded to is always the central arrow. The four different types of cues are shown in the middle insert. The types of trials used to measure the three aspects of attention (alerting, orienting, and executive control) with the attention network task are defined in the bottom right. Target displays can occur above or below fixation.

in schizophrenia (Hong, Yang, Wang, Li, Ding, & Sheng, 2021), ADHD (Mullane, Corkum, Klein, & McLaughlin, 2009), and depression (Klawohn, Santopetro, Meyer, & Hajcak, 2020). Given the generally robust effects, it has also been used to examine how cognitive control is influenced by substance addictions (e.g., Nikolaou, Field, Critchley, & Duka, 2013).

Finally, a more recently developed task is called the attention network task (ANT). The ANT was designed to assess multiple functions of attention in a single paradigm (Fan, McCandliss, Sommer, Raz, & Posner, 2002). The ANT essentially combines the Posner cuing paradigm with the Eriksen flanker task, plus it adds a measure of alerting. Specifically, the ANT is meant to test three main functions of attention: alerting, orienting, and executive control. As shown in Figure 1.9, the task begins with the brief presentation of a "cue" on most trials, which is followed a short time later by a target display. The target display is a version of the classic Eriksen flanker paradigm, with a central target surrounded by distractors. The comparison of trials with congruent flankers versus incongruent flankers provides the measure of executive control. Note that some versions also include trials with neutral flankers (e.g., – – < – –) to allow for analysis of the separate costs and benefits of the flanking information. In the original version of the ANT, the "spatial cue" was 100% predictive of where the target would occur (e.g., the target was validly cued on 100% of trials), but recent versions have added a portion of "invalid" trials, in which the target display occurs at the opposite location. These newer versions allow the experimenter to measure the *reorienting* of attention after an invalid cue, in addition to the standard measure of orienting in the ANT (spatial vs. center cue trials). Finally, on some trials, no cue stimulus is presented (i.e., "no-cue" trials). The comparison of these no-cue trials to trials containing the presence of a cue but not indicating a location (e.g., double cue trials, center cue trials) provides a measure of alerting. The ANT has proven to be an efficient and highly effective way to assess these different functions of attention. This task has been used across many populations, and a new database of studies promises to enhance researchers' ability to interpret patterns of results across ages, groups, and trainings (Arora, Lawrence, & Klein, 2020). Many cognitive neuroscience studies have used the ANT to reveal the neural bases and mechanisms of alerting, orienting, and executive control, and these will be discussed in turn throughout the following chapters.

CHAPTER SUMMARY

- Attention as a concept has been of interest to philosophers for centuries, but empirical research into the mechanisms of attention only began in earnest after the "cognitive revolution" in the 1950s.
- Dichotic listening studies motivated by the "cocktail party phenomenon" led to some of the first theories of attention, including the early filter, late filter, and attenuating filter models.
- Classic theories of the distribution of attention have included "spotlight," "zoom-lens," and "feature-based" models.
- Feature integration theory holds that a primary function of attention is to bind features together onto objects.
- The attention network task can be used to measure three main mechanisms of attention: alerting, orienting, and executive control.

REVIEW QUESTIONS

Describe the events that led to the cognitive revolution and that motivated the rise of empirical research into the mechanisms of attention.

Explain the different functions and varieties of attention and describe how classic experimental paradigms are used to test those different mechanisms.

Discuss the relation of attention to the eyes in terms of theories and types of attention.

Compare and contrast executive control with other processes of attention.

FURTHER READINGS

James, W. (1890). *The Principles of Psychology*. Harvard University Press.
- This book contains the quote used at the start of this chapter, along with in-depth descriptions of attention and other cognitive processes, in the *functionalist* style that James championed.

Posner, M. I. (2016). Orienting of attention: Then and now. *Quarterly Journal of Experimental Psychology*, *69*(10), 1864–1875.
- In this article, Posner describes his development of the attention cuing paradigm and reviews the first 25 years of its use in cognitive psychology and cognitive neuroscience.

Treisman, A., & Gelade, G. (1980). A feature-integration theory of attention. *Cognitive Psychology, 12*, 97–136.
  - This article describes the highly influential theory that first defined a core function of attention as being the binding of features to objects. It also proposes that simple features are analyzed in a preattentive stage of processing.

Wood, N., & Cowan, N. (1995). The cocktail party phenomenon revisited: How frequent are attention shifts to one's name in an irrelevant auditory channel? *Journal of Experimental Psychology. Learning, Memory, and Cognition, 21*(1), 255–260.
  - This article reviews the classic "cocktail party" experiments and reports a new study that provides a more precise measure of our own name capturing our attention.

2 Attention in Everyday Life

Learning Objectives

- Identify real-world situations in which attention plays a critical role
- Describe how understanding the mechanisms of attention can enhance our safety
- Explain the role of attention (and distraction) in studying
- Understand how the varieties of misdirection practiced by magicians relate to the different component processes of attention
- Appreciate different functions of attention as experienced in our daily lives

2.1 Distracted Driving

One of the most common examples of the role of attention in our everyday lives is distracted driving. Whether we are the driver of the vehicle or simply a passenger, every time we go somewhere in a motor vehicle our safety depends on the ability of the person behind the wheel to pay attention. In the fast-moving environment of modern roads, even a temporary distraction can lead to fatal consequences. Students learning to drive spend hours in the classroom learning the *rules* of the road, and much of the initial "behind-the-wheel" instruction is aimed toward understanding *how to control* the vehicle. While these are important and necessary steps, novice drivers are not usually given training in how to ensure optimal *attention*. A danger in this regard is that people are usually overconfident in their ability to "multitask." This overconfidence leads to resistance to laws that try to force people into behaviors that research studies have shown enhance their safety.

While the variety of things that can potentially distract a driver are numerous, one item in particular has been shown to consume attention to a dangerous degree: the ubiquitous cell phone (Figure 2.1). Today's phones are used for so many things that we rarely go extended periods without picking them up to do something, but driving is one activity that is clearly safer when we don't use our phones. Even in the early days of cell phones, when the major activity they were used for was simply conversation, researchers began to worry about the effects of engaging in such conversations on road safety. Reports from traffic accidents revealed that talking on cell phones was associated with a large number of accidents (Redelmeier & Tibshirani, 1997). In addition, naturalistic observation studies showed that there was a correlation between talking on a cell

Figure 2.1 Distracted driving. A driver being distracted by something on their phone.
Source: Getty Images; credit: Arterra/Contributor; Editorial #: 923335016.

phone and not stopping correctly at red lights and stop signs (Strayer, Drews, & Crouch, 2006). Of course, correlation doesn't imply causation, so those reports of associations could not directly assess whether talking on a cell phone caused poor driving. The alternative explanation was that people who are natural risk-takers may be the same people who take the risk of talking on a phone while driving. Therefore, psychologists studying attention have designed experiments to try to better assess whether the cell phone conversation itself was *causing* the detriment in driving performance. In one of the first well-controlled laboratory experiments investigating this issue (Strayer & Johnston, 2001), participants were randomly assigned to either a group that had to participate in a cell phone conversation or a group that did not. In order not to endanger any lives, the experiment was done in a laboratory, which reduces real-world validity (aka, **ecological validity**), and in this case the "driving" was performed using a joystick to keep a dot steady on a screen, and the "braking" involved pushing a button when a specified target appeared. The results of the study showed a significant effect of the phone conversation, as the subjects assigned to this condition showed significantly worse performance on both the "driving" and "braking" parts of the task. These early results have since been replicated multiple times (reviewed in Caird, Willness, Steel, & Scialfa, 2008), including in realistic diving simulators in which the participant sits within a shell of a car, with a normal steering wheel and pedals, and views the roadway via computer monitors where the windshield and rearview mirrors would be. Although these experiments cannot ever be safely conducted on real roadways amid other people driving cars, the results of experiments with highly realistic driving simulators provide robust evidence that our ability to drive safely is significantly impaired whenever we engage in a cell phone conversation.

The studies mentioned earlier represent an important step in showing that paying attention to cell phone conversations likely causes poor driving. These studies also took care to address potential confounds. Specifically, an objection to initial claims of the dangers of driving with cell phones was that it was simply the *holding* of the phone that was dangerous because the driver couldn't keep both hands on the steering wheel. The early studies thus compared conversations that were done using handheld phones versus hands-free systems and found that carrying on conversations in either way significantly impaired driving performance. Other objections included that the observed impairment from cell phone conversations could be simply due to the need to listen intently. According to this logic, it would make no more sense to ban cell phone use while driving than to ban listening to the radio or a book on tape. Research has shown, however, that carrying on an active conversation impairs driving ability far more than simply listening to the radio or a book on tape (Strayer & Johnston, 2001), presumably because of the additional social pressure to not lose track of a conversation with another person. Relatedly, conversations with passengers in the same car are not as detrimental to the driver's ability to focus on the road, presumably due to the shared experience of the driving environment; in other words, the passengers can see when traffic or road conditions worsen and pause their conversations appropriately. More recently, concerns have been raised that the complexity of new dashboard displays may be a source of distraction (Box 2.1).

Much of the research on distracted driving was specifically aimed at testing whether cell phone usage was dangerous, and the results helped lead to the enactment of laws and recommendations to keep people safe on roadways. However, this line of research also highlights some of the core processes of attention. For example, these studies address *how* attention can be *allocated*. The concept of "multitasking" assumes that we can pay attention to multiple things at once. While research has suggested that in some circumstances we can divide our attention to some extent across different *stimuli* (see Chapter 5), we do not seem to be very good at being able to divide our attention between *tasks*. Rather, in most situations in which "multitasking" is supposedly being done, we are actually switching (sometimes quite rapidly) between tasks, not performing multiple tasks simultaneously. Thus, "multitasking" seems to be a misnomer, as it would be more accurate to describe it as "rapid switching." Whereas the term "multitasking" seems to imply that attention can be on two or more tasks at once, "rapid switching" would make it clear that attention is only on one task at a time. This difference in terminology could have a real effect on whether and when people decide to do two tasks together. If someone believes they really can "multitask," they may not see a danger in using their cell phone while they're driving, whereas if they realize that they're actually switching attention back and forth, it may become clearer to them that it's a bad idea to take attention away from driving.

The ability to switch between tasks highlights another important mechanisms of attention: the ability to *disengage* from one thing and move on to something else. As it relates to driving, the ability to disengage attention has been linked to the "moth effect" – being drawn toward bright lights at night and not disengaging from them (Kitamura & Matsunaga, 1994). Anecdotal reports have suggested that some crashes may be caused by an impaired ability to disengage attention after impaired drivers become focused on flashing stimuli (Green, 2006). Research on driving has provided evidence that bright, dynamic stimuli are especially attractive to a driver's attention at night (Gruner & Ansorge, 2017). Beyond driving, this relates to the general attentional process of disengagement; the neural mechanisms of this process will be discussed in more detail in Chapter 7.

Box 2.1 **Distracted by your own car?**

In classic horror films, a particularly hair-raising moment is when the protagonist realizes that the threatening calls are coming from *within their own house*! While much of the research into distracted driving has focused on how we are dangerously distracted by cell phones, human-factors psychologists have become increasingly interested in how our own cars may be making it harder for us to drive safely. With advances in technology, the past decade has seen a significant increase in the amount of information that can be presented to the driver, mostly via visual displays. These visual displays often include information that is not relevant to driving, such as screens showing the radio station/artist/song we're listening to, or a menu of options for selecting audio input or speaker output. There's often a digital menu for controlling the air-conditioning /heating options across a variety of zones and another for giving us details about the status of different vehicle systems. While all of these may vie for our attention, it's relatively straightforward to assign all non-driving-related information a lower priority, so that we can deal with them at a safer time. However, the advancements in display technology and the assumption that more information is always better have led to a concerning level of complexity in the displays that we actively use when driving. The main display in most vehicles is no longer limited to a large analog speedometer, an RPM gauge, and maybe a couple of lights to indicate that your high beams are on or that your fuel is low. These displays now contain a great deal more information – often being updated in real time – such as aspects of the internal environment of the car, outside road conditions, or even real-time weather updates. The display may beep and warn you not to change lanes because another car or pedestrian is too close, or it may scold you if you've started to change lanes without signaling. It may keep you updated on how many people in the car have their seatbelts on and whether or not your antiskid system is active, maybe even suggesting times when you should turn it on or off. Human-factors engineers are needed here because, with the proliferation of information being displayed, there's a real danger that at least some of the symbols and alerts will require the driver to pay more extensive attention to them in order to decode and understand the meaning of what's being presented. Indeed, trying to pack more information into a limited display space can result in symbols being used that are less than intuitive. As noted earlier, research has shown that conducting phone conversations while driving is dangerous not because of the physical holding of the phone or the addition of auditory information into the environment but because such conversations require attention, taking mental resources away from the critical task of driving safely. Without research into how to make these new dashboard displays simple and clear so as not to require a prolonged dwelling of attention, the distraction that endangers us could be coming from *inside our own car*!

Finally, research into cell phone use while driving also highlights that attentional *resources* are *shared* between quite different types of task. The visual–motor task of driving may seem quite distinct from the processing involved in carrying on a conversation or listening to a story. However, research into driving and cell phone use provides evidence that there is a common pool of attention resources, and neither task can be completed well if these resources are sufficiently drained. In addition to the findings that cell phone conversations impair driving performance, Radeborg, Briem, and Hedman (1999) found that driving can also disrupt an ongoing conversation. This trade-off between either driving well or encoding the conversation well illustrates that there is a shared resource between these mental operations. It also reminds us that attention is critically important for fully comprehending language and conversations. When our attention has briefly strayed away from the person speaking to us, we might become aware that they had continued speaking, but the details of what they just said can be entirely lost to us. As discussed in the next section, this role of attention in understanding language extends beyond spoken language and includes reading comprehension as well.

2.2   Reading (and Rereading!)

At one time or another, everyone has had the experience of reading along at a good clip and suddenly breaking out of a daydream to realize that we have no idea what we just read. Amazingly, at the moment we realize our lapse of attention, we may be "reading" a sentence at some considerable distance from the last sentence we remember reading! Seemingly on autopilot, our eyes have done their job of looking from one word to the next in a typical reading pattern, and we've often even progressed down the page – but none of that content made it through to our consciousness because our attention was entirely focused on internal thoughts. As this scenario illustrates, a great deal of neural processing occurs outside of our conscious awareness, and attention appears to play a gatekeeper role in helping to select the information that will rise to the level of consciousness. Another aspect of attention this scenario reveals is its phasic nature. Try as we might to simply sustain a constant focus of attention on the material we're reading, our minds are apt to wander. The variation in our ability to sustain attention arises from many factors. These can include our level of interest in the material being read, as well as factors unrelated to the material, including the time of day and our level of arousal. Too low of a level of arousal may lead to general sleepiness and daydreaming, whereas too high a level of arousal may make it hard to sit still as our mind races to new and more stimulating thoughts. As will be discussed in future chapters, the brain basis of **sustained attention** and the mechanisms by which only a subset of information processing is selected for access to conscious awareness and higher levels of processing are active areas of research.

Reading also provides evidence that attention is not strictly where the eyes are. Studies have shown that after some period of time fixing on one word in a sentence (with that time being determined by linguistic properties including the frequency of that word's use in the language), we begin to pay attention to upcoming words in the sentence *before* we move our eyes there. This "parafoveal preview" refers to getting a sneak peak of the word or words coming up next in the sentence. Researchers have shown that this parafoveal processing is distributed asymmetrically around the word currently being fixated, with attention extending outward in the direction that reading occurs. In languages that are read from left to right, such as English, the parafoveal processing extends in a rightward direction, for as many as 15 characters to the right of fixation, as well as three to four characters to the left (Rayner, 1998). In languages that are read right to left (e.g., Arabic, Hebrew), this parafoveal preview extends in the leftward direction (Fischer & Weber, 1993). Studies of reading have shown that the linguistic and semantic properties of the words in the periphery affect where the eyes will land next (O'Regan, 1979). The movements of attention and the eyes during reading are determined not only by the physical properties of the upcoming word(s), but also by the meaning of the words. For example, common words such as "the" are more likely to be skipped when reading than are words of the same length that are more informative (e.g., "ate" or "met"). Although the eyes move rapidly during reading, the movements of attention take into account higher-level properties of the stimuli outside the area being fixated. This reveals that even when the eyes are moving very quickly, such as during reading, attention is not simply tied to where the eyes are fixated.

Regarding how we sometimes execute the eye movements of "reading" while being unaware of what we're reading, it is even more complicated than the eyes simply moving along normally, seemingly on autopilot. As Reichle, Reineberg, and Schooler (2010) found, there are subtle differences in how the eyes move during normal reading versus "mindless reading." When attention isn't being paid to the text (i.e., during mindless reading), the lexical and linguistic properties of the words have less of an effect on eye movements and fixation patterns. The eyes are still moving across the page in a manner quite *close* to the usual manner of reading, but the linguistic properties of the words aren't modulating the eye movements as much as they usually do, suggesting that attention is needed to fully process the stimuli at the level of word meanings. This highlights that the relation between attention and language is complex and bidirectional, and that it's important to understand exactly when and where attention modulates neural processing. Attention is necessary for understanding an ongoing conversation and for reading lines of text; therefore, attention clearly affects the processing of language. In the other direction, language can also have a significant influence on attention; specifically, the allocation of attention. How long we dwell on a word and whether we'll skip a less informative upcoming word depend on the lexical and semantic properties of the word. Language affects how we move our eyes when reading and also what concepts are becoming primed to capture our attention in the near future. Language has a strong effect on what topics and schemas are active in our consciousness, and related concepts appear to have preferential access to our attention and awareness. As will be covered in more detail in Chapter 6, there are many interacting factors that determine what captures our attention at any given moment in time.

2.3   Studying and Remembering

As any student can attest, the efficiency of studying varies greatly, and it would be wonderful to get more out of the hours spent trying to learn new material (Figure 2.2). One of the major factors determining our later recall is how much attention we allocate when first learning the material. Numerous studies have shown that memory is significantly worse when attention is divided, versus focused, at the time of initial learning (Craik, Govoni, Naveh-Benjamin, & Anderson, 1996; Foerde, Knowlton, & Poldrack, 2006). However, although the research is

Figure 2.2 Students studying at a library. *Source:* Getty Images; credit:
Dejan Marjanovic; Creative #: 1698863376.

clear, people greatly overestimate their ability to multitask effectively, and this likely comes into play in the classroom. With our ever-present cell phones nearby, it's easy to be distracted in the classroom by the buzz of an incoming text message. If we give in to the urge to check the message, we may get drawn into a quick conversation, and it may be many minutes before we re-engage with what the teacher is presenting. Even if we simply check the message for a moment just to see what's happening without any intention to respond, the removal of attention from the class material still means lost time when trying to re-engage with the lecture. Similarly, if we're taking notes on a laptop, the many competing options for what we can do on that device impact our ability to efficiently learn what we're in the classroom to learn. We may be tempted to "just quickly" check something on the web, or to put a note in our calendar about something unrelated that just popped into our mind. But even these brief diversions disengage us from the class, and we may miss critical information and have to work harder to get ourselves back up to speed. The interaction of attention and memory processes will be explored in more detail in Chapter 6, but in terms of the present topic, there is abundant research showing that dividing attention during study has detrimental effects on our ability to remember that material later (Naveh-Benjamin, Guez, & Sorek, 2007).

Avoiding distraction can be difficult enough, even in the immediate moment when we're trying to pay attention to the teacher in front of us during a lecture, but studying outside of the classroom involves other factors as well. When we want to settle in for a good study session, there are the added choices of *where* to study – alone or surrounded by other people – and whether we want to study with music or in silence (Figure 2.2). Students often express a strong opinion on the latter choice, with some feeling that listening to music would obviously make their studying more difficult, while others are just as sure that they study best while listening to music. The reasons for these opposing opinions highlight some of the key processes of attention that will be covered in this book. On the one hand, the research showing that we're generally quite poor at multitasking would support those students who are adamant that music is detrimental for studying and that quiet is best. According to this viewpoint, any additional noise in the environment just means we have to work harder to filter it out or be distracted from the goal of studying because our attention will be partially allocated to the music we're hearing. If that's true, however, why do so many students strongly believe that music helps them? At least part of the answer to this relates to the idea that attention is an *active process that requires resources*. We need sufficient mental resources in order to do what attention does: to boost the processing of the information we want to attend to (in this case encoding the class material better) and to suppress the processing of irrelevant information we need to ignore (e.g., other people, random noises). In regard to the latter, it would seem that the simplest way to avoid irrelevant information would be to not expose oneself to it (in other words, simply to not turn on music). However, the concept of limited mental resources provides a different perspective on this situation. If one is not mentally alert, staring at pages of notes may not be effective for encoding that information into memory. In that mental state, we may be hard pressed to stay focused on the material we're trying to study, instead being distracted by any small noises in the environment or even just our own passing thoughts because we lack the mental resources to engage our attention processes fully and to suppress irrelevant information. Listening to music, for some people, may help by giving the attention system a boost, increasing the resources available by raising overall alertness. The mental "pick-me-up" gained from listening to music may provide the extra resources needed to better focus attention on the material being studied and suppress potential distractions.

Another potential benefit of listening to music while studying may be that the music drowns out other sounds in the environment that could otherwise distract us. Certainly, turning up music loud enough can mask other noises in the environment, but why doesn't the music itself then become a distraction? Recent research into the effects of memory and expectation on the allocation of attention may provide an important clue. Memorially unique stimuli have a stronger influence on the allocation of attention (Chanon & Hopfinger, 2008; Parks & Hopfinger, 2008), and predictive coding models (covered in more detail in Chapter 9) suggest that predictability and expectation play significant roles in attentional capture. According to the logic of predictive coding theories, if one chooses songs that are very well-known, there may be less likelihood of distraction because the brain has a highly accurate model of what auditory stimuli are upcoming, greatly reducing the likelihood that attention will be captured by those sounds. In this way, listening to your own choice of music may be an effective way to control the environment – the sounds of the music drown out other sounds in the environment that could have been distracting and yet those specific songs are predictable enough to not trigger involuntary capture of attention away from the task being performed. Of course, the style of music may also be important. As noted in the earlier section on distracted driving, comprehending speech can be very attention-demanding. The link between language and attention would strongly suggest that listening to instrumental music should be more beneficial when studying because lyrics of any sort could potentially interfere with studying, at least when the material being studied is written or spoken.

The idea that music may be able to boost task performance was a driving force behind the *Music While You Work* radio program aired by the British Broadcasting Company (BBC) from 1940 through 1967. The program, begun during World War II, was aimed most specifically at increasing the productivity of factory workers. Parameters for the program specified that the music should be upbeat, of a consistent volume and melody, and consist of familiar pieces. Jazz, for example, was not typically included because it often deviates from a standard melody. Overall, the parameters used to choose the music illustrate intuitions regarding a number of the aspects of attention currently being studied in laboratories. The preference for using familiar pieces and avoiding music that deviated from a standard melody or volume relates to current research into how memory and predictions affect the capture of attention (covered in more detail in Chapters 6, 8, and 9). The restriction only to use upbeat music relates to alertness and that attention relies on a pool of resources that seems to be increased with arousal levels. Another interesting restriction was that music that included clapping was avoided because of the potential danger of factory workers clapping along, thus taking their hands off their work!

Returning to the topic of students and studying, another choice that students make is *where* to study. Why some students prefer to study in crowded libraries or coffee shops while others prefer isolated, quiet areas may relate to the different mechanisms of attention and selection. As opposed to the situation in which a single nearby conversation can be very difficult to ignore, the cacophony of voices in a crowded environment can actually be less distracting. Although there is more noise overall in a crowded space, it's not as likely that any one meaningful thread of conversation will automatically capture our attention; no single conversation "pops out" because there are so many overlapping conversations. As discussed in Chapter 1, the "cocktail party phenomenon" shows that attention is needed to fully follow one conversation in the midst of many voices. And as discussed earlier, attention is necessary for fully processing and comprehending speech. Thus, when there are multiple streams of speech overlapping and a student is highly focused on the material they're studying, any single conversation is unlikely to get processed to the level of meaning. Furthermore, according to predictive processing models, the schema for a crowded coffee shop includes the presence of many different voices and conversations; therefore, nearby conversations don't violate any expectations of what should be happening and don't capture attention in the way they would in a different context (e.g., at a nearby table in an otherwise quiet library setting).

Overall, these strong and opposing opinions about how best to study highlight another important concept that is only beginning to be investigated: individual differences in attention. It is becoming clear that in order to fully understand cognitive functions such as attention we will need to account for individual differences in the component mechanisms of those processes and how these may change over the lifespan and with training (which will be covered in more detail in Chapter 7). Outside of the classroom, a real-world situation illustrating the importance of attention to memory is when we must remember the name of a person we've just met (Box 2.2).

Box 2.2 "I'm sorry, but what was your name again?"

A common frustration when meeting new people is remembering their names. We want to show that we're interested in what they're saying and respond appropriately, but we may forget the person's name, sometimes almost immediately after they're done saying it! While this is often chalked up to "not being good at remembering names," it's not really a problem of recall at all, but a problem with attention. These occurrences show – sometimes with embarrassing consequences – that if our attention falters at all during the initial processing of a stimulus, we're unlikely to have encoded critical details into memory. A great deal of research has investigated the critical role of attention in encoding new information into memory. Why some of us continue to be so bad at paying attention to names reveals another aspect of attention: its singular focus. We simply cannot pay attention to more than one thing at a time very well. Often when meeting a new person we don't want to ignore the rest of what they're saying, and we may want to pay particular attention to what they look like so that we will recognize them in the future. But that attention being paid to the topic of conversation or to details about the person's appearance means we haven't left ourselves with enough attentional resources to fully encode their name into our memory, dooming us – once again – to need to beg forgiveness for our faulty "memory" and ask them to remind us of their name . . . again.

2.4   Airport Security Screening

One of the most disliked aspects of modern air travel is the security screening process – from waiting through the long, slow-moving lines to the discombobulation that results from having ourselves and all our belongings inspected (Figure 2.3). The reason for the slow lines and the need to remove items from our bags relates directly to the limits of attention. Ensuring that nothing dangerous passes through is a very demanding task, and procedures have been enacted to help ensure that screening agents are able to pay attention to each person and to each bag in turn. Similar to how attention serves as a bottleneck, reducing all of the inputs from all of our senses to the very small portion we're actively aware of, airport security checkpoints are designed to limit how much information the screeners must process at each moment. Checking all the items on all the people and in all the bags can't be done simultaneously, and even when done one piece at a time errors occur if it's done too quickly. Therefore, the bottleneck we experience at airport security checkpoints is designed to ensure that the screeners process only a limited amount of information at a time. Although this greatly slows down the checking process, it is necessary given what we know about the limits of attention.

Experiments are helping us appreciate the demanding nature of visual search tasks and how it depends on cognitive and attentional resources. The job of the airport security screeners – and the difficulty in detecting the items of interest – relates to the different types of search processes: feature search versus conjunction search. Feature search refers to the situation in which we are attempting to find a stimulus that has a unique feature (e.g., a specific color, motion,

Figure 2.3 Long lines at an airport security screening checkpoint. *Source:* Getty Images; credit: Andy Cross/Contributor; Editorial #: 528468594.

orientation, or size) that is clearly different than all the other stimuli in the scene. During this type of search, the object we're searching for (i.e., our target) seems to pop out effortlessly. We don't need to engage in an effortful and voluntary search process; rather, the object with that unique feature automatically rises into our consciousness and we quickly orient to it. In contrast, a conjunction search refers to the situation – more common in everyday life – in which we are looking for an object that shares features with many different stimuli in the environment. For example, in experiments testing conjunction search, the subject may be asked to find a red-colored "T" within an environment that has other red-colored letters as well as T's of different colors. This type of search is much slower, as the subject needs to allocate attention individually to each item to determine whether it contains the exact conjunction of features (e.g., red color and T shape) that defines the target. Returning to the airport example, the security screener's job is difficult in large part because it is *not* a feature search – there is not a simple, single feature defining the target that can automatically pop out. Instead, they perform the much more demanding process of allocating attention from one item to the next to ensure they correctly identify all the items, since the prohibited items they are searching for share multiple features with many of the safe items that make up the vast majority of items they're viewing.

The screener's task is also made much more difficult because objects look quite different on the screening display. The X-ray scans are displayed on a video screen, reducing the actual 3D configuration of the items in our bags to a 2D representation, with objects overlapping in the same 2D space (Figure 2.4). In addition, security scanners often project the images via a color-coding scheme based on the absorption of X-rays and density of the material, so objects

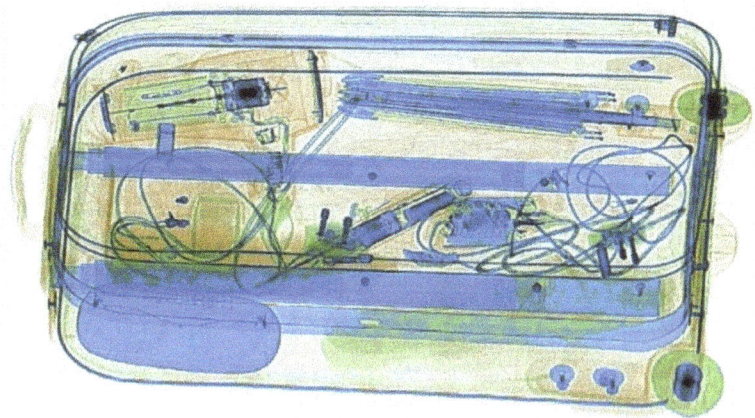

Figure 2.4 Airport security X-ray screening image of a carry-on bag. Objects are color-coded by X-ray absorption and density, as described in the text. Adapted from Muhl-Richardson et al. (2021), *Cognitive Research* (Springer Nature). Creative Commons CC BY 4.0.

do not appear in the colors in which we normally see them. For example, low-density materials (e.g., cloth, organic substances) will typically appear orange, intermediate-density materials in green to light blue, and the densest materials (e.g., metals) are coded dark blue or black. In addition, an object may look quite different based on what items are overlapping with it, thus requiring attention to dwell longer in order to identify each object before moving on to the next set of overlapping pieces. Identifying each of the objects in the bags is thus slowed down by the need to mentally separate the overlapping and unusually colored pieces in the 2D display and to configure those pieces into recognizable objects. This need to mentally create the objects from impoverished perceptual information increases the difficulty, further limiting the speed at which attention – and therefore the screening – can proceed.

Finally, the necessity to do this search process over and over and over again means that significant mental effort is required to sustain such a high level of attention for those prolonged periods of time. This process has been investigated in studies of **sustained attention** – the consistent maintenance of a highly attentive state over long periods. The security screener's task of searching for items that are very rarely (or never) present creates a situation in which it is difficult to sustain a high level of attention. Indeed, studies have found that longer shift durations are associated with worse target detection performance (Meuter & Lacherez, 2016).

Many of the difficulties in airport screening are relevant to *health* screenings as well (Wetter, 2013). For example, in trying to detect cancer, a radiologist will review magnetic resonance imaging (MRI) or computed tomography (CT) scans for what may be very small deviations from the normal variance in a scan. Sometimes a more highly salient feature (e.g., larger, brighter, or unusually shaped) may capture attention, even if it's not a harmful abnormality, potentially leading to missing a more subtle but diagnostically important abnormality (Adamo, Cox, Kravitz, & Mitroff, 2019). While radiologists are trained to try to ensure that such misses don't happen, training methods may be further enhanced by incorporating results from studies of attention and visual search. In addition, researchers are exploring ways in which artificial intelligence (AI) can be used to help in this regard. This highlights that human attention systems are limited, and that in some critical situations it may be necessary to integrate advancements in computer vision and machine learning. Overall, these real-life situations underscore the critical need to understand the attention systems of the human mind so that we can understand the limits of these systems and the ways we can enhance and supplement them to ensure that critical jobs can be completed as effectively and efficiently as possible.

2.5   Radar and Sonar Monitors

While we can't help but be very aware of airport security screening, there's another vital job being performed every time we fly that we don't often think about: air traffic control. A major part of air traffic control work is to monitor radar screens and keep track of airplanes and flight paths simultaneously (Figure 2.5). Some of the difficulties of this job, however, are fundamentally different than those of the security screeners. Those screeners are essentially looking for a needle in an ongoing sequence of haystacks, needing to process and then ignore most of the nontarget information they're viewing. Air traffic controllers, on the other hand, have a more limited amount of information to process overall, but all of that information is important and can't be ignored. Thus, whereas the security screener's job is to efficiently filter out irrelevant items, the air traffic controller must attend to multiple critical items and try to keep them all in mind. This relates to studies – discussed in Chapters 5 and 6 – on how we

Figure 2.5 Air traffic controller and displays. *Source:* Getty Images; credit: Horacio Villalobos/Contributor; Editorial #: 1466101810.

allocate our attention, sometimes spreading it over multiple items, and the interacting mechanisms of working memory and attention.

While at times the air traffic controller's job relies on this ability to stay aware of and track multiple flights simultaneously, at other times the danger is losing one's focus on the task. In contrast to the nerve-racking demands of tracking many flights at once, the challenge during quieter times may be to stay vigilant. **Vigilance** refers to the ability to stay attentive and mentally alert even after extended periods without anything of interest occurring (i.e., when the task has become "boring"). Some of the earliest research in vigilance was motivated by radar operators working in the Royal Air Force during World War II. Norman Mackworth developed techniques to test subjects' abilities to detect infrequent stimuli over prolonged periods (e.g., 2 hours) of monitoring displays, finding a notable decrease in performance after 30 minutes on task (Mackworth, 1948). The attentional demands illustrated in air traffic control can be compared to the critical work of sonar technicians on naval ships and submarines. By intently monitoring changes in sonar signals, these technicians are able to track the movements of other underwater vessels. As is the case for air traffic controllers, the job of the sonar technician may vacillate between times of high stress and periods of boredom. Being able to maintain a steady level of vigilance and alertness can be key to performing these jobs well. In that respect, it is interesting to note that the mandatory retirement age for air traffic controllers in the USA is 56 years, considerably younger than the standard retirement age of 65 years for most other jobs. This reflects that air traffic control is a highly demanding job, with no room for error, and that attention abilities may decline as we get older. In regards to attention and aging, large-sample-size studies have provided evidence that some attention abilities peak when people are in their early 40s and then decline through old age (Fortenbaugh, DeGutis, & Germine, 2015). Much current research is focused on understanding which attention abilities decline with age and whether attention training can counteract some of those declines.

2.6   Video Games and "Brain Training"

Considering the difficulties inherent in attentionally demanding jobs such as those described in the preceding sections, as well as the decline in attention functions with natural aging, there is great interest in the possibility of enhancing attention through cognitive training. Implicit in these efforts is the belief that the requisite attentional skills are malleable and may be able to undergo plasticity-related changes with training. This potential malleability is also a key factor in ongoing debates regarding kids and video games (Figure 2.6). On one side, critics of video games argue that the rise in rates of attention deficit hyperactivity disorder (ADHD) and the shortening of attention spans may be due to kids spending more and more time playing video games. On the other side, proponents suggest that video games can enhance some cognitive abilities. Although diametrically opposed as to whether their effects are good or bad, both sides agree that habitually playing video games can trigger – or at least modulate – changes in the brain mechanisms of attention.

Some of the initial work on this topic examined whether "expert" video game players (i.e., people who have played video games many hours per week for years) exhibit differences in their

Figure 2.6 Kids playing video games. *Source:* Getty Images; credit: Edwin Tan #: 1299948398.

attention and cognitive skills compared to individuals who never play video games. These studies found that "expert" video game players had faster reaction times and scored better on tests of visual perception and attention (e.g., Dye, Green, & Bavelier, 2009), and these findings have been replicated many times (reviewed in Sampalo, Lázaro, & Luna, 2023). Of course, this correlation doesn't show that playing video games *causes* enhanced attention and perception. It could instead be that people who naturally have exceptional visual and attention abilities are drawn toward video games and play them frequently because they're good at them. In these longitudinal studies, we don't know whether the expert video game players were already better than most people at attention and visual perception before they started playing video games. Therefore, well-controlled laboratory studies, in which individuals who don't normally play video games are randomly assigned to either a game-playing group or a control condition, are critical to assessing the possible causal link between video games and enhanced attention abilities.

In well-controlled laboratory studies, the subjects randomly assigned to play video games are typically found to outperform subjects in the control condition, who do not play video games, on subsequent attention and perception tests. Meta-analyses have shown that there is a consistent benefit for the game-playing group in randomized controlled studies, although it is smaller than the effects seen when comparing habitual expert video game players to novices (Bediou, Adams, Mayer, Tipton, Green, & Bavelier, 2018). Potential reasons for the size of the effect being smaller in the randomized controlled experiments include the relatively short training in the laboratory studies, the specific games that are assigned (which may not match a person's interest), the relatively small sample sizes, and publication bias (favoring studies that find significant effects over studies that fail to find effects). The finding of enhanced cognition after video game "training" is not universal, however, and some studies have failed to find significant improvements in cognitive skills even with 20+ hours of video game playing, suggesting that some of the differences between "experts" and nongamers may be due to natural differences in skills or to the much more extensive amount of time (i.e., years) that the experts have been playing video games (e.g., Boot, Kramer, Simons, Fabiani, & Gratton, 2008). Even with these caveats, however, the significant effects of video game playing in most random assignment studies suggest a causal link from playing video games to better cognitive performance on some tasks.

Important questions still remain, though, regarding the effects of video games on cognition. Among these are: How far does the training transfer? In other words, training on a video game naturally improves performance in that game, but does that mean that core cognitive processes are being improved in a way that can translate to other situations? The enhanced performance across a number of different attention and cognitive tests suggests that the improvements do transfer, at least to some degree (e.g., Bediou et al., 2018). However, since many tests of visual attention are performed on computer screens and subjects are asked to respond as quickly as possible to small details that they are viewing, those tests of cognition are somewhat similar to many video games. Therefore, it is important to understand the consequences of video game playing on cognitive and social functioning outside of the lab, in the real world. Relatedly, although many studies have shown improvements on attention tasks after video game training, could there be negative consequences for any cognitive or psychological abilities that aren't being tested? Trying to test for all the potential effects of video game playing on all possible psychological processes and real-world outcomes isn't feasible; however, cognitive neuroscience methods allow us to track the effects of video game playing on underlying neural connectivity patterns and evoked neural activity. As will be discussed in more detail in Chapter 7, studies using these methods can directly assess the effects of video games on the brain and therefore can provide new insights into which psychological processes are most likely to be affected. This research is also being used to help design games that specifically target the cognitive processes and brain networks that may be most affected at different ages, from young children to aging adults. Indeed, a recent review suggests that video games may be effective at both helping to diagnose and treating some forms of ADHD (Peñuelas-Calvo, Jiang-Lin, Girela-Serrano, Delgado-Gomez, Navarro-Jimenez, Baca-Garcia, & Porras-Segovia, 2022). That meta-analytic review found evidence that video games could help differentiate subtypes of ADHD during diagnosis, and that some therapies that included video game playing enhanced cognitive skills and decreased ADHD symptoms.

Although there may be potential benefits of video game playing for some individuals, it should also be noted that there is a danger of becoming addicted to online video games. The 2022 revision of the International Classification of Diseases (ICD-11) includes "Gaming disorder" (World Health Organization, 2022), and the 2022 version of the Diagnostic and Statistical Manual of Mental Disorders (DSM-5TR; American Psychiatric Association, 2022) includes a condition called "Internet Gaming Disorder" (IGD). There remains debate about whether internet gaming should be classified as an addiction (Zastrow, 2017), but some of the criteria for diagnosing IGD are quite similar to those of other forms of addiction. For example, these criteria include experiencing withdrawal symptoms, needing to spend increasing amounts of time playing to satisfy the urge, loss of interest in other activities, and deceiving family members about frequency of use. Some early research suggested that the prevalence of IGD was relatively low in the population (1% or lower; Przybylski, Weinstein, & Murayama, 2017), but more recent research suggests that the rate could be much higher, potentially even higher than other behavioral addictions (e.g., gambling). Gao, Wang, and Dong (2022) presented a meta-analysis that found an IGD prevalence rate of almost 10% for adolescents and young adults. Furthermore, it should be noted that the rate of IGD appeared to rise following the COVID-19 pandemic (Alimoradi, Lotfi, Lin, Griffiths, & Pakpour, 2022), possibly related to the increased amount of working and schooling from home as well as heightened anxiety during that time. There remains debate about whether IGD should be considered a separate *disorder* or whether it would be more accurate to consider it a *symptom* of other disorders (e.g., anxiety and depression). Defining IGD as a clinical disorder could potentially help those who are suffering ill effects from their gaming, but on the other hand it could potentially stigmatize playing video games, even for individuals who are not experiencing bad consequences.

2.6.1   Training to Multitask

As mentioned in Section 2.1, we are inherently bad at multitasking, despite our overconfidence in being able to do so. Box 2.3 describes an early study that provided evidence that with enough training on a very specific task even complex skills can become automated. More recent research has suggested that video games may be effective, in a more limited way, in training subjects to "multitask." In most situations, the improvements in performance after such training are not because both tasks can now be done simultaneously, but rather because subjects learn to more quickly and efficiently switch between different tasks. As described earlier, a critical mechanism of attention is the ability to disengage attention from its current focus and reorient it elsewhere. If this particular skill can be enhanced through training, it would shorten the time needed to switch back and forth rapidly between tasks. Some researchers have suggested that video games may be especially effective at training this skill because games include unpredictable events, and players must learn to switch their focus and tasks frequently throughout the game. Research on the effects of video game playing in older adults found that training on single tasks using video games led to some short-term improvements on those tasks, but that training subjects on two tasks together using video games produced more robust improvements that were longer lasting (Anguera, Boccanfuso, Rintoul, Al-Hashimi, Faraji, Janowich, Kong, Larraburo, Rolle, Johnston, & Gazzaley, 2013). Furthermore, as will be covered in more detail in Chapter 7, electroencephalography (EEG) measures revealed that the multitasking training in older adults resulted in frontal theta activity patterns resembling the EEG measures of young adults performing those tasks (Anguera et al., 2013). These EEG results provide evidence that the multitasking training may help older adults learn to better utilize the attentional control mechanisms necessary for rapidly disengaging and switching attention across tasks.

Box 2.3 Multitasking in a courtroom?

We would probably feel shocked and a bit dismayed if we encountered a courtroom stenographer who was reading a novel while they were transcribing testimony during an important trial. The stenographer's job of creating, in real time, a written record of everything being said by the judge, lawyers, and witnesses is a task that requires a high degree of sustained attention. In a classic experiment testing a related skill, Spelke, Hirst, and Neisser (1976) investigated whether subjects could accurately take dictation while reading a separate story for comprehension. The purpose of the study was in part to understand the difference between controlled processes, thought to be dependent on attention, and automatic processes, thought to be independent of attention. Furthermore, the researchers were interested in the transition by which an effortful, controlled process becomes an automatic one. The subjects were trained to write down lists of words in real time as they were being dictated while also reading short stories. The size of the sample was extremely small (N = 2), in part due to the very large number of training sessions that were required (5 sessions every week over 17 weeks, for a total of 85 sessions). At first, as would be expected, the subjects' performance when trying to perform both of these tasks concurrently was significantly impaired relative to the pretraining measure of performing each task alone. Reading speed was much slower and handwriting quality was markedly impaired when the subjects first began the challenging multitasking experiment. Remarkably, however, after training, the subjects were able to complete both tasks to near single-task performance levels. The subjects were able to read at a normal pace and with good comprehension while accurately taking legible dictation. This research showed the degree to which even effortful tasks, such as writing down verbatim what someone else is saying, can become relatively automatic with enough practice. However, it should be noted that the dictation was for lists of words, not a conversation being processed to the level of meaning. While the study showed that the ability to directly transcribe words as they are being spoken can be automated, the ability to link those words together into a meaningful conversation still depends on attention. As shown in other research into "multitasking," attention is still a limiting factor for *comprehension*. Only a very small amount of information can be attended at any one time, and the processing of stimuli outside of attention is limited to only superficial features. Without attention, neural processing doesn't seem to advance to higher levels of language comprehension and stable memory traces aren't typically formed. Furthermore, upon closer inspection, most "multitasking" would more accurately be called "rapid switching," as subjects appear to alternate between tasks rather than doing multiple tasks simultaneously. Interestingly, Watson and Strayer (2010) found that a tiny fraction of people may be "supertaskers" – people who actually performed *better* when doing multiple tasks. However, the tasks in their experiment were fairly simple, so the performance of the "supertaskers" in that study may just relate to the optimal level of arousal needed to perform tasks, with the addition of a second task essentially boosting alertness in subjects for whom the original task may have been too easy to hold peak attention. In addition, the authors stress that the overwhelming majority of people are not "supertaskers"; almost everyone's performance suffers when multitasking. Furthermore, this same research group subsequently found that the people who most strongly *believe* that they can multitask well are often the *worst* at it (Sanbonmatsu, Strayer,

Box 2.3 (cont.)

Medeiros-Ward, & Watson, 2013). Therefore, on the road, in the courtroom, or throughout our daily lives, it's important to remember that our safety and the quality of our work are much improved when we pay attention to just one task at a time.

2.7   "Joint Attention"

As social animals, we are keenly aware of what the people around us are paying attention to. Although we can't always know the content of another person's internal attention state, we can easily observe what they seem to be paying attention to by way of noticing where they are looking. As discussed earlier, shifts of "overt attention," such as head and eye movements, are an excellent index of a person's current focus of attention in most real-world situations. This is not usually the case in laboratory experiments in which we attempt to isolate attention and dissociate it from where the subject is looking. In Chapter 7, the involuntary nature of orienting in direct response to someone else's eye movements ("social-gaze orienting") will be discussed in more detail, along with insights into the brain mechanisms of this process gained from cognitive neuroscience experiments. Here, however, we'll briefly discuss what is known about "joint attention," a state that arises frequently throughout our lives in which the focus of our attention is shared with another person or group of people. Everyday conversations are greatly facilitated by this type of joint attention; simply by following the gaze of the person we're talking with we gain insight into what's on their mind. In such situations, explicitly describing every item we're referring to isn't necessary, and would actually seem quite peculiar. This ability to pick up on another person's focus of attention seems to be an entirely automatic process, not requiring effort or a conscious decision on our part. Indeed, studies have shown that we can't help but orient our own attention when a conversational partner suddenly turns their head to look at something (Moore & Dunham, 1995). Even without the relatively large movement of someone's head turning, the more subtle movements of a person's eyes can just as strongly trigger a reflexive orienting of our own attention (Friesen & Kingstone, 1998).

The importance – and hardwired nature – of this type of shared attention has been illustrated in compelling recent research by Fitch, Lieberman, Luyster, and Arunachalam (2020). This research provided new insights into how joint attention is critically involved in young children's ability to learn language. In their study, 2-year-old children were exposed to new words by way of observing the actions and conversation of two nearby adults. Overhearing words in this way is thought to be an important means by which children learn new words, especially during the incredible period of language development that occurs around this age. In their experiment, the researchers created two different conditions for overhearing the critical new word: (1) during joint attention; or (2) without joint attention. During the joint attention condition, the two adults were attending to each other and to the object being described by the new word; in the condition without joint attention, only one of the adults was looking at the object while the other adult was looking down at a notepad in their lap throughout the entire conversation

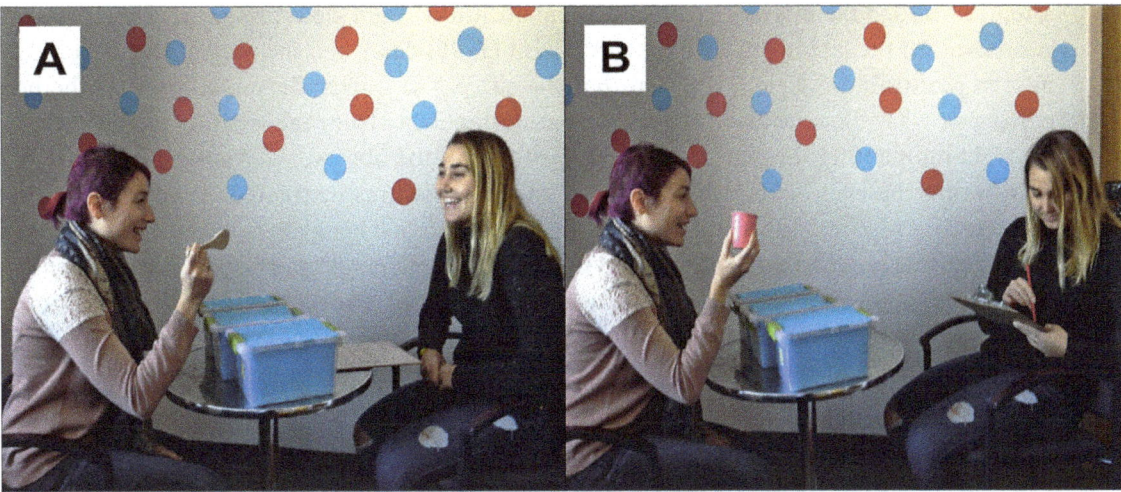

Figure 2.7 Joint attention. Illustration of the toddler's perspective of conversations between two adults in the "joint attention" condition (left frame, labeled "A") and the "without joint attention" condition (right frame, labeled "B"). Reprinted from Fitch et al. (2020), *Journal of Experimental Child Psychology*, with permission from Elsevier.

(Figure 2.7). The results revealed a robust effect of joint attention on language learning, as toddlers learned the new words when both adults in the conversation were attending to the objects. Those words were significantly less likely to have been learned, however, by toddlers in the "without joint attention" condition. When the two adults having the conversation were not jointly attending to the object, the toddlers didn't learn that new word, despite the object being clearly visible to the toddler when the word was spoken. Joint attention thus appears to be critically important in how we interpret the interactions of those around us. In the case of young children learning language, it appears that observing joint attention between conversation partners strengthens the link between an object and the sounds of the language used to describe that object. Beyond the strong effects of joint attention in orienting us to what another person is talking about, it has been suggested that attention plays a critical role in other aspects of social interactions as well (Capozzi & Ristic, 2018). For example, experiments have found that subjects judge individuals who consistently exhibit reliable gaze signals (i.e., looking at the objects being spoken about) as more trustworthy and give them more altruistic help in subsequent interactions (Rogers, Bayliss, Szepietowska, Dale, Reeder, Pizzamiglio, Czarna, Wakeley, Cowen, & Tipper, 2014).

2.8   Magic

From famous stage magicians to street-side hustlers, the ability to amaze and trick people rests largely on how well these performers can control an audience's attention. Sleight of hand and highly practiced motor skills are obviously important, but being able to move someone's attention and hold it away from a critical location often determines whether the trick is successful. In the late nineteenth century, Binet (1896) described the central role that attention held in performing magic. He noted that the success of a trick depended on the whether the performer was able to draw the observer's attention away from the location of the trick at just the right moment. He also described how the focus of the observer's *attention* was more critical than where the *eyes* were fixated, noting "attention is thus distracted ... rendering invisible a spectacle which is perfectly visible to all eyes" (p. 564). This idea, well-utilized by magicians for years, was pursued and investigated by cognitive psychologists over 100 years after Binet's writings when Mack and Rock (1998) published a series of studies on the link between attention and consciousness, coining the term "inattentional blindness." These experiments showed that even stimuli within the "spotlight" of attention and near fixation may not rise to the level of conscious awareness if our attention is tightly focused on a different object within that spotlight. Magicians manipulate many cognitive processes, including memory, judgment, and the suspension of belief. However, most tricks start with attempting to control the audience's attention. Here, we use the example of magic tricks and the insights from magicians to illustrate some of the mechanisms and types of attention.

Gustav Kuhn (2015) describes magic as an "old art," and he notes that magicians were manipulating attention long before cognitive psychology was even established as a field of scientific study. Indeed, a primary theme in an early twentieth-century mail-order course for aspiring magicians (Tarbell, 1927/1971) was the need to direct the audience's attention. Tarbell highlighted that the audience will automatically follow the gaze of the magician, a concept that predates, by decades, cognitive studies into joint attention (aka, shared attention or social-gaze orienting). Tarbell's course also explained that it can be useful to repeat a specific action multiple times so that it no longer captures or holds the audience's attention, at which point it can be used to hide a critical action in plain sight. Critically, this method of making an audience lose interest in an action is a more subtle form of manipulating attention than simply creating a large, salient distraction. A highly obvious distraction, such as snapping one's fingers or suddenly releasing a flying dove, is very effective at moving an audience's attention, but such events allow the audience to realize that the magician has attempted to distract them at exactly that moment. By gradually making an action uninteresting – by repeating it a few times without anything unusual happening – the magician can succeed in diverting attention away from that action without the audience realizing their attention is being manipulated. For example, if a magician wants to amaze the audience by having a coin disappear from their hand and be found sitting on top of a volunteer's shoulder, the movement of the magician's hand to the volunteer's shoulder (to place the coin there) will be largely ignored by the audience if the magician has recently, and repeatedly, reached over and given the volunteer a friendly pat on the shoulder during their initial conversation. During one of these friendly pats, the coin is placed on the volunteer, but only after the audience and the volunteer have gotten used to the magician patting the shoulder, such that they stop paying attention to that particular action. This demonstrates one of the more subtle and involuntary aspects of attention: how our "interest" affects to what, and for how long, we pay attention. We'll review neuroscience investigations into these aspects of attention in the chapters on involuntary attention (Chapter 6) and the control of attention (Chapter 7). In Chapter 9, we will discuss newer predictive-coding models of perception and attention that stress that much of our conscious experience is driven by top-down predictions of the world around us. The ability of magicians to hide actions and stimuli, oftentimes in plain view, relates to how our minds use top-down models of the world when allocating our attention. The repetition of an action by the magician creates the expectation that the action is "normal," and the audience thus loses interest in it, allowing the magician to use that action to hide a critical step in the execution of a trick.

The magician's manipulation of attention is multifaceted and complex, and it touches on many of the core mechanisms of attention that cognitive neuroscientists are attempting to elucidate at the neural level. Some of the terms used by magicians to describe their craft include "overt misdirection," "covert misdirection," "passive misdirection," "active misdirection," and "time misdirection" (Macknik, King, Randi, Robbins, Teller, Thompson, & Martinez-Conde, 2008). The distinctions between these forms of misdirection in magic correspond well with distinctions made in cognitive neuroscience studies of attention. For example, the distinction between "overt" and "covert" misdirection applies directly to the concept of overt versus covert shifts of attention. Helmholtz (1894) was the first to illustrate that attention could be shifted covertly (without any movement of the head or eyes), in contrast to the overt shifting of attention when turning to look at an object of interest. Magicians use different techniques of "overt misdirection" to induce observers to *look at* an object that is away from where the trick is taking place. This can obviously be a powerful technique for hiding actions that the magician wants to perform out of direct gaze. However, at other times magicians may instead manipulate the mind's eye without needing to ensure any change in eye gaze (i.e., *covert attention*). This covert misdirection can make the trick seem even more amazing because it can seem that the magic occurred right in front of our eyes. As Binet wrote in the nineteenth century, and as Mack and Rock (1998) later showed experimentally, we can be blind to stimuli even if we're looking directly at them if our attention is somewhere else. Thus, the effectiveness of magic tricks nicely illustrates the power of attention and the dissociation between attention and where we are looking.

The concepts of "active misdirection" and "passive misdirection" refer to *how* the magician is controlling attention, and this relates to an active field of research investigating what controls our attention (covered in Chapter 7). "Active misdirection" includes when a magician directly tells the audience where to pay attention – for example, by telling them to carefully watch what they're doing with an object in one hand while their other hand is performing the critical action for the trick. Active misdirection can be as "simple" as asking a volunteer to pick a card and remember it – while the volunteer is studying their card, they aren't able to pay attention to anything else the magician may be doing with the remaining cards. This form of active misdirection uses what cognitive psychologists call "endogenous" or "voluntary" attention, because the shift of attention is under *internal*, voluntary control. This can be contrasted with what magicians call "passive misdirection," or what cognitive psychologists call "exogenous" attention, in which attention is driven in an *involuntary* manner by something outside ("exo") of the person. In magic, this can be done by the sudden appearance or movement of an unexpected object (e.g., a dove that suddenly appears and flies away). The magician need not tell the audience to attend to the object but can simply rely on the audience members' natural, involuntary reflexes to cause their attention to be captured by the salient new

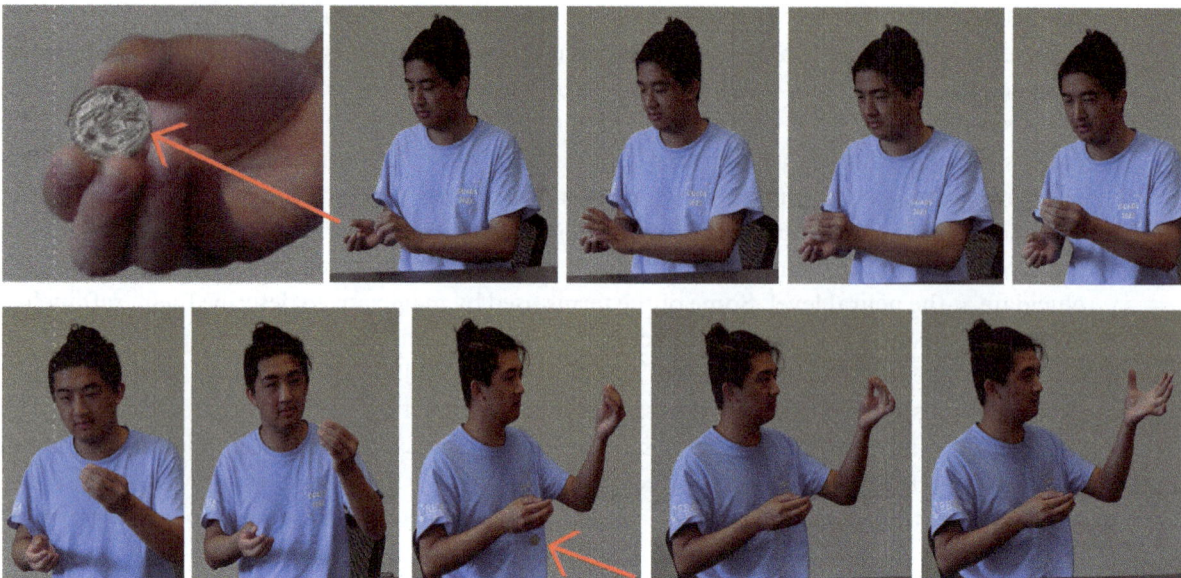

Figure 2.8 Sequence of events for the disappearing coin trick. Starting at the upper left, the magician shows the audience the coin in his right hand. He then appears to take the coin with his left hand, although the coin actually remains cupped in the original hand, subtly hidden from view. The audience believes the left hand is holding the coin and follows the movement of that hand as the magician moves it away from his body. The magician further ensures the audience is paying attention to the left hand by turning and looking directly at that hand, inducing a strong automatic shift of attention (i.e., joint attention). While the audience's attention is strongly focused on the outstretched left hand, the magician is able to drop the coin from his right hand (indicated by the arrow in the bottom row) without the audience noticing it. After revealing that the coin is not in the expected hand (shown in the last frame), the magician would proceed to open his other (right) hand to reveal that the coin is not in that hand either, leaving the audience amazed and bemused.

object, ensuring attention is kept away from the critical maneuvers the magician is performing elsewhere. As illustrated in Figure 2.8, passive misdirection can be guided in multiple ways. And as discussed in the previous section, joint attention is a powerful and involuntary reflex by which we orient our own attention to where another person suddenly looks. The magician can thus assume that when they look toward something in one of their hands, the audience is also attending to that hand, leaving the magician's other hand free to do the critical moves of the trick undetected (such as dropping a coin, as in Figure 2.8). As illustrated in Figure 2.8, although highly skilled sleight of hand is critical for keeping the coin hidden in the original hand, the manipulation of attention is equally critical to allowing the coin to be dropped in plain view without being seen!

Misdirection can be further nuanced, as indicated by the following magicians' saying: "A big move covers a small move" (Macknik et al., 2008). This axiom conveys that there is constant *competition* for attention, and that the focus of attention will likely rest, at any given moment, on the most salient item currently present. So, if a magician needs to take an action that might normally draw attention, they need to ensure that an even bigger action captures attention somewhere else at that moment. This relates to an influential theory of attention, the "biased-competition model of attention" (Desimone & Duncan, 1995), which helps us to understand how attention is allocated in the real world. Related to the concepts of competition and limited attention resources, experimental studies of magic tricks have shown that not all actions are equal when it comes to manipulating attention. Otero-Millan, Macknik, Robbins, and Martinez-Conde (2011) compared the effectiveness of two different types of motion used for misdirection. In their experiment, they utilized a disappearing coin trick, similar to that shown in Figure 2.8. In their experiment, Otero-Millan and colleagues compared two conditions that varied in the path the hand took after supposedly grabbing the coin. In one condition the hand moved in a simple, short, and straight line to its ending point, while in the other the hand took a curved (arc-shaped) route. Eye-tracking results showed that subjects were less likely to look back to the original hand, potentially discovering the coin hidden there, when the grabbing hand took a curved path away from the original hand compared to when it took a simple, straight path. There is debate about what causes this effect, but the experiment showed that relatively small differences in the type of motion can substantially affect the degree to which attention is held and how effective the misdirection might be. The idea that a more complex route may enhance the deception relates to theories of perceptual load (Lavie & Tsal, 1994). These theories posit that there are limited perceptual and attentional resources, and that these resources are allocated *in full* at each moment. Thus, if attentional resources are fully engaged by a more difficult task (e.g., following the movement of the hand along a curved path), there are no attentional resources left over to process other stimuli. When the task is easier, however (e.g., tracking a simple, straight path), there are unused attentional resources left over that are then automatically allocated to other parts of the scene. Related to the coin drop experiment, one possible explanation is that the observer cannot attend to the original hand (which is still holding the coin) when doing the more difficult task of tracking the other hand through a more complicated motion.

A final type of misdirection utilized by magicians is "temporal misdirection," in which the critical manipulation is the *timing* of the events. Specifically, by inserting delays between the critical action of a trick and the reveal of the trick, observers have a harder time linking the cause and the effect. What's important here, and what relates to theories of attention more generally, is the "disengagement" of attention. The magician, through temporal misdirection, is attempting to ensure that the observer's attention is fully disengaged from the initial event and focused elsewhere by the time the reveal occurs. Since the action that produced the trick is no longer being attended and held in mind, it's more difficult to link it to the trick that has just occurred, obscuring the trick and further enhancing the sense of wonder. This relates to the concept of attentional disengagement that was initially proposed to explain the behavior of neuropsychological patients who suffer from the syndrome known as "attentional neglect." These patients have intact vision but tend to ignore items in the visual field contralateral to their brain lesion (e.g., patients with right-hemisphere lesions ignore items in their left visual field). Critically, the patients have the ability to see items in their "bad" visual field but tend to ignore those items whenever their attention is focused on items in their "good" visual field. In other words, these patients have a hard time *disengaging* their attention from any items in their good visual field in order to move attention to items in their bad visual field. The behavior of these patients and the neural underpinnings of attentional neglect are discussed in more detail in Chapter 4.

The magician's use of temporal misdirection also relates to recent research exploring the *holding* of attention – how long attention *dwells* on an item. As will be discussed more in Chapters 6 and 8, there are involuntary mechanisms at work that affect attentional dwell time and the phasic nature of attention. For example, magicians have learned to make use of "off-beat" moments. In musical terms, the off-beat refers to what's playing in between the standard beats or during the "weak" parts of the song. In magic, this can involve utilizing the time between salient events to perform critical actions while the audience's attention is waning. For example, the punchline of a joke or the reveal of an initial magic trick is often followed by a short period of relaxed attention from the audience. Attending to the buildup of a joke or carefully tracking the performance of an initial magic trick requires sustained attention from the audience. When that joke or trick has ended, there is often a short period during which the audience's attention is relaxed. This brief period of reduced attention allows the magician time to perform critical actions for an upcoming trick while the audience is not on high alert. This also relates to laboratory experiments on the "attentional blink," in which subjects are unaware of a stimulus presented at the center of their gaze if their attention was consumed with the processing of an event that occurred just moments before. This phasic nature of our attention, and the mechanisms that determine how long attention is held, will be covered more in Chapter 8, when we discuss the temporal properties of attention.

CHAPTER SUMMARY

- Throughout our daily lives, attention is critical to our ability to safely navigate our world, carry on conversations with those around us, and encode memories.
- Research on distracted driving reveals the limited resources of attention and our inability to multitask effectively.
- How and where we choose to study may be influenced by our individual differences in the attention mechanisms of alertness, attentional capture, and suppression of potential distractions.
- Jobs such as airport security screening and air traffic control highlight processes of attention such as visual search and vigilance.
- Studies of video game playing suggest a degree of plasticity in some mechanisms of attention that could be used in targeted ways to help train children, adults, and elderly subjects who may have deficits in different aspects of attention.
- Joint attention research shows that social signals can have powerful effects on the allocation of attention, starting in infancy.
- The theories and practice of magic touch on many of the core processes of attention, including overt versus covert orienting, voluntary versus involuntary control, and the holding versus disengagement of attention.

REVIEW QUESTIONS

Describe a few of the ways in which different jobs and our everyday activities reveal that attention is a *limited resource.*

Contrast the *voluntary versus involuntary* mechanisms of attention involved in reading, studying, and airport security screening.

Using examples from across the sections of this chapter, explain the relation between attention and the eyes.

Discuss why well-controlled laboratory research is needed even though correlational studies of distracted driving and video game expertise find effects and even though magicians have been effectively manipulating attention for centuries.

FURTHER READINGS

Bediou, B., Adams, D. M., Mayer, R. E., Tipton, E., Green, C. S., & Bavelier, D. (2018). Meta-analysis of action video game impact on perceptual, attentional, and cognitive skills. *Psychological Bulletin*, *144*(1), 77–110.
  - This article reviews the results of multiple studies on the effects of playing action video games and includes both longer-term cross-sectional studies ("experts" who have chosen to play action video games for years versus novices) and shorter-term random assignment experiments.

Friesen, C. K., & Kingstone, A. (1998). The eyes have it! Reflexive orienting is triggered by nonpredictive gaze. *Psychonomic Bulletin & Review*, *5*(3), 490–495.
  - This is one of the first articles to report that social gaze cuing can be triggered in an involuntary manner.

Macknik, S. L., King, M., Randi, J., Robbins, A., Teller, Thompson, J., & Martinez-Conde, S. (2008). Attention and awareness in stage magic: Turning tricks into research. *Nature Reviews Neuroscience*, *9*(11), 871–879.
  - The authors, who include both psychologists and magicians, review magic theory and discuss how magic acts can provide insights into the mechanisms of attention.

Muhl-Richardson, A., Parker, M. G., Recio, S. A., Tortosa-Molina, M., Daffron, J. L., & Davis, G. J. (2021). Improved X-ray baggage screening sensitivity with "targetless" search training. *Cognitive Research: Principles and Implications*, *6*(1), 33.
  - The authors discuss theories of attention and visual search in relation to baggage screening and present promising results from a new training method.

Strayer, D. L., & Drews, F. A. (2007). Cell-phone-induced driver distraction. *Current Directions in Psychological Science*, *16*(3), 128–131.
  - This article reviews research showing that cell phones are correlated with poor driving in the real world and discusses results from well-controlled laboratory studies suggesting that cell phone use causes distraction.

3 Investigating the Brain

Methods of Cognitive Neuroscience

Learning Objectives

- Identify the strengths and limitations of each cognitive neuroscience method
- Describe the relation between neural activity and blood flow measures
- Compare and contrast the spatial and temporal resolutions of these methods
- Describe the methods that manipulate neural activity
- Understand why there is no one "best" method of cognitive neuroscience

3.1 A Brief History of Localizing Cognitive Functions

Long before the technological advancements that allowed neural activity to be measured within the living human brain, the cognitive changes observed after brain injury fascinated scientists and led to new areas of research. Across many areas of psychology, the first insights regarding how mental processes work have come from patients who suddenly lose some cognitive function, often in ways that are surprising and hard to fathom. For millennia, humans have realized that the brain was critical for life, but it was only relatively recently in human history that the brain was identified as the seat of cognition. In one of the earliest known references to the brain, an Egyptian medical treatise from the seventeenth century BCE (see Figure 3.1) explained that serious head injuries, in which the matter inside the skull was badly damaged, were likely to be fatal (Minagar, Ragheb, & Kelley, 2003).

This medical treatise provides evidence that even thousands of years ago people knew that the brain was critical for *life*; however, it wasn't thought that the brain was important for *thinking*. Indeed, when the Egyptian pharaohs were prepared for mummification after death, great effort was taken to preserve the body, and critical organs were carefully removed and placed in special crucibles to ensure these important parts would be available in the afterlife. The brain, however, was scooped out through the nose and tossed in the trash. These esteemed leaders were painstakingly prepared for the afterlife, with all the critical parts of their body ready for their use, except for their brain! The one part of our anatomy that, today, we would most want to preserve was simply discarded. It wasn't until around the third century BCE that cognition was

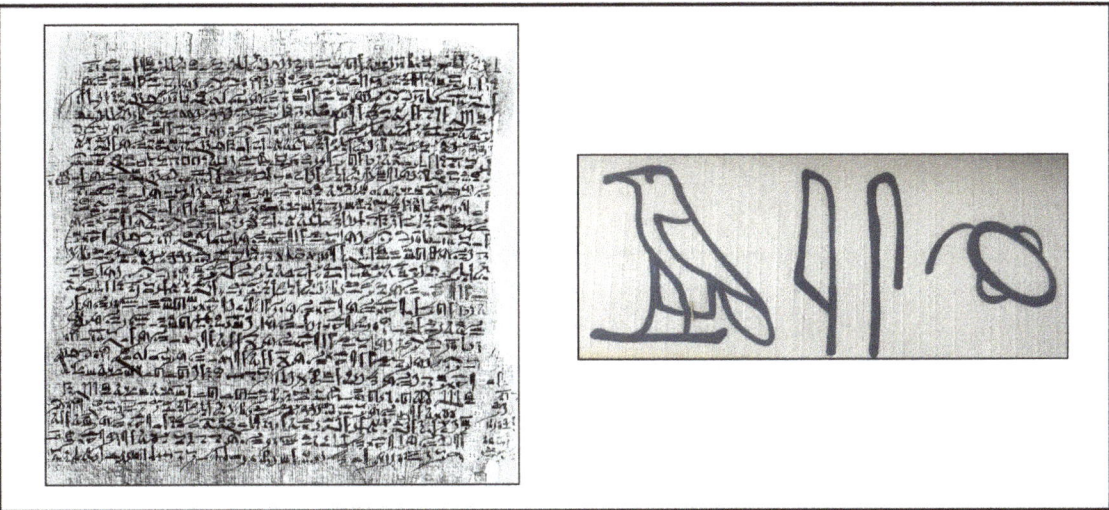

Figure 3.1 Earliest known written reference to the "brain." Left: Plate from Edwin Smith surgical papyrus, seventeenth century BCE, describing a medical treatment plan for patients. *Source:* The Edwin Smith surgical papyrus: published in facsimile and hieroglyphic transliteration with translation and commentary in two volumes by James Henry Breasted. Credit: Wellcome Collection. Attribution 4.0 International (CC BY 4.0). Right: Artist rendering of the series of hieroglyphic symbols that referred to the organic material inside the human skull (i.e., the brain). Drawing courtesy of Jennifer Hopfinger.

finally linked to something inside the skull (reviewed in Green, 2003). But even then, it wasn't the brain itself but rather the fluids inside the brain that were considered the seat of thinking. Possibly reflecting Hippocrates' theory that health was determined by the balance of sacred fluids throughout the body, it was assumed that cognition was supported by the clear fluids contained in the well-protected ventricles of the brain. This "ventricular doctrine" was popular for over a thousand years, and over time it came to include the idea that different cognitive functions could be located to different areas of the brain. By the early 1500s CE, different cognitive functions were being attributed to different ventricles. In one version of this approach (shown in Figure 3.2), information from all the senses was considered to go directly to the first ventricle, where perception occurred. Higher-order cognitive processes like reasoning and judgment were then considered to take place in the second ventricle, and finally information was believed to be stored as memories in the third ventricle.

Not until the mid-1500s did it finally become clear that the material of the brain, not the fluids within and around it, supported mental and cognitive processes. Even then, it was initially the *white matter* of the brain that was thought to be responsible for mental processes, while the gray matter was still regarded as just a protective covering – hence the use of the term *cortex*, meaning "bark" or "outer covering," to refer to that outer surface of the brain. By the 1700s, scientists had begun to realize that the cortex itself was the likely seat of mental events; however, there was debate regarding whether every mental function relied on all parts of the cortex equally, or whether different parts of the brain were responsible for different mental processes.

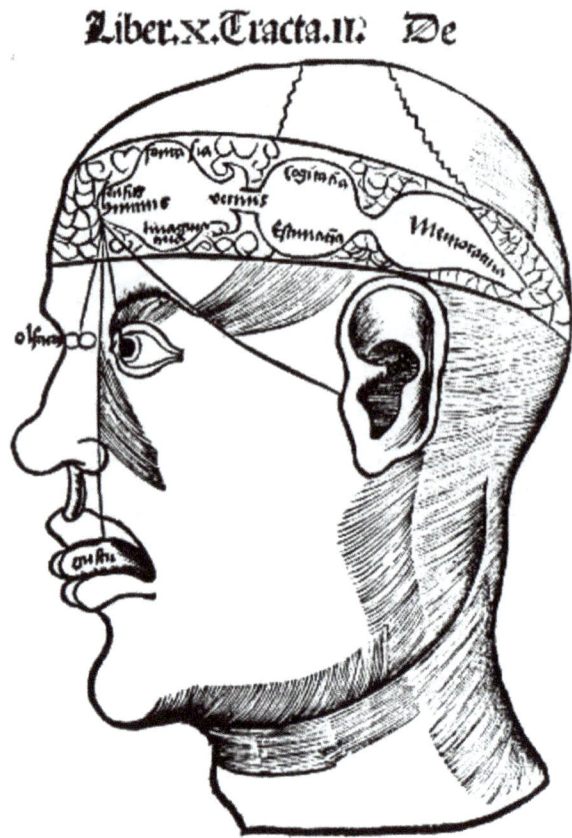

Figure 3.2 Illustration of the ventricular doctrine, circa 1512. *Source:* Brain (https://commons.wikimedia .org/wiki/File:Brain,_G_Reisch.png) by G. Reisch, 1512. Public domain.

Globalists held that the whole brain acted together during every cognitive function, whereas *localizationists* believed that different mental functions were localized to different parts of the brain. As noted earlier, even some proponents of the ventricular doctrine had proposed a degree of localizationism, in that different mental functions were linked to different ventricles. Franz Joseph Gall's practice of "phrenology" championed the idea of localizationism, but, without any means of examining the living human brain directly, it relied on "reading" bumps on the skull and trying to associate different mental attributes to the shape of different parts of the skull. Phrenology enjoyed a brief period of fame in the early 1800s, but growing skepticism produced an overcorrection that led to the scientific community accepting that globalism must be correct. Specifically, Jean Pierre Flourens presented evidence from his studies on lesioning the brains of small animals (e.g., pigeons and rabbits) that suggested that the whole cortex acted together to achieve complex cognitive functions. Flourens did show some degree of localizationism in finding that the cerebellum and brain stem supported coordination and vital life functions (e.g., respiration, breathing), respectively. However, although cognitive functions were associated with the cerebral hemispheres, Flourens was unable to find regions of the cortex that were responsible for specific cognitive functions. Rather, his results suggested that the *amount* of cortical tissue, rather than any specific location, was the important attribute for supporting cognitive functions. Subsequent research would reveal limitations in Flourens' methods that invalidated some of his findings regarding cognitive functions, but at the time his results swayed the scientific community toward accepting the globalist perspective. A few years later, however, Paul Broca published evidence that finally resolved the debate in favor of the localizationist viewpoint and that showed the potential and promise of the field of study now known as **neuropsychology**.

3.2   Neuropsychology

In 1861, Paul Broca published a case study of a patient referred to simply as "Tan." The patient's name was Louis Victor Leborgne, but he was referred to as Tan because that was the only word he was capable of speaking when he was brought to Broca's attention. Tan previously had full and normal use of language, but after a progressive illness he lost almost all ability to produce speech. Critically, Tan's deficit was largely restricted to *speech production*, sparing other motor and cognitive abilities. When Tan passed away, Broca was able to extract and examine his brain, which revealed a large lesion in the left frontal lobe. This was an important first piece of evidence that a specific part of the cortex might support a specific cognitive function (Broca, 1861). Broca then searched for more patients suffering from a similar type of selective impairment of speech production, and in a follow-up paper (Broca, 1865) he showed that 12 patients with a selective deficit in speech production all showed damage to the left frontal lobe. This provided important evidence against the globalist view that all parts of the cortex worked together, and it suggested that different parts of the cortex support different cognitive functions. The localizationism viewpoint was further strengthened when Carl Wernicke (1874) published results regarding a different type of language impairment (speech *comprehension*) that was associated with a *different* area of brain damage (left temporal lobe damage). These early patient studies established the importance of neuropsychology research investigating the cognitive and mental functions of patients with brain damage, and they established neuropsychology as a core method for understanding the human mind.

3.2.1   "Case Studies" versus "Group Studies" in Neuropsychology

As compelling and informative as a sudden loss of a specific mental function can be, an issue with any study of an individual patient is that people are different. A "case study" refers to a description of one patient, and it can vary in the amount of additional information that is collected or known about the individual. Case studies are often the first clue we have that a particular mental phenomenon is composed of specific, partially separable processes. These studies provide sometimes surprising evidence that a particular cognitive process can be impaired while much of the rest of cognition remains intact. Case studies are thus highly useful in helping to identify component processes of cognition, such as Broca's patient Tan revealing that speech *production* is a process that can be separated from the brain's other language processes. It is important in case studies to examine carefully whether the observed deficit is an isolated problem or is simply one of many cognitive deficits. Therefore, a case study typically reports results from testing multiple cognitive functions. In addition, a history of the patient's health and development can also be important for understanding whether the deficit is part of a developmental or progressive disorder affecting multiple mental functions or can be more tightly linked to a deficit following a specific brain injury. Neuropsychology studies typically focus on patients with focal brain damage resulting from a transient event – this allows for stronger conclusions to be drawn about the brain area(s) supporting a particular cognitive function because the event provides a clear demarcation of the patient's abilities before versus after the injury. Case studies can be important for revealing the subcomponent processes of cognition, sometimes in highly surprising ways that provide new insights into how cognition works; however, there are important limitations as well. While much of cognitive neuroscience relies on the assumption that at least most of the linkages between the brain and behavior are highly replicable across people, it must be acknowledged that people are different; genetic differences can cause some people's brains to be different at birth, experiences during development can lead to further differences between people, and recovery from injury can be different across people in ways we have yet to understand. Thus, while the existence of a single patient with a specific cognitive deficit can provide new insights regarding *possible* subcomponent processes of a mental phenomenon, something more is needed to allow us to conclude that it is actually a *separate* component process in *all* people.

This is where "group studies" are critically important. A group study reports on the deficits of multiple patients with similar lesions or with similar behavioral deficits, and these studies are important for understanding how generalizable the results of a single case study may be. By combining and analyzing data from multiple patients, group studies can address the question of whether the deficits seen in a case study are truly indicative of how the human brain typically functions or whether a case study may be due to other factors – besides the lesion of interest – that may be unique to that individual person. Depending on the rarity of the type of deficit or the type of brain injury, group studies may consist of only a few additional patients. In those situations, the goal may simply be to see if the deficit replicates beyond the initial case study. In situations in which the deficit or lesion is not as rare, however, larger group studies can be performed that allow for greater confidence that specific cognitive functions can be linked to a specific region or set of regions in the brain.

3.2.2   Single versus Double Dissociations in Neuropsychology

A goal of neuropsychological research is to link cognitive and mental functions to regions of the brain by carefully assessing those functions after brain injury. In a **single dissociation**, a particular patient or patient group is found to have a specific cognitive deficit. For example, in Broca's studies described earlier, patients with damage to the left frontal lobe were found to have a deficit in speech production. Through assessment of other types of language and cognitive tests, it could be shown that the deficit is primarily restricted to the production of speech. While this single dissociation suggests that the injured brain area helps support that specific cognitive function, such conclusions are limited. Specifically, it could be that the deficit is observed in only one type of test because that test is simply harder overall than the tests that the patient did well on. In that situation, the injured brain area may not specifically support that specific task but rather may just be needed more whenever task difficulty reaches a certain threshold. A related limitation in using a single dissociation to conclude that a brain region supports a specific process is whether enough tasks have been used to allow one to isolate a cognitive process.

A **double dissociation** provides greater confidence that particular brain areas are associated with separate mental processes. A double dissociation consists of two different types of patient groups that have two different types of behavioral deficits. For example, whereas Broca's initial findings alone consisted of a single dissociation compared against the intact functioning of healthy controls, Wernicke's findings of a different patient group with a different pattern of deficient versus spared processing provided critical evidence for a double dissociation. Broca's patients with left *frontal* damage had a deficit in production with relatively spared comprehension, whereas Wernicke's patients with left *temporal* damage had a deficit in speech comprehension with relatively spared production. Note that the spared processes are critical in this double dissociation because those spared processes address the problem described in the previous paragraph about not knowing if a deficit is due simply to one task simply being harder than another task. The finding of a double dissociation provides stronger evidence that different brain areas are responsible for different deficits. Although double dissociations provide the strongest evidence from neuropsychology to link cognitive processes to brain areas, it is difficult to find two patient groups with exactly opposite patterns of deficits versus spared processing; therefore, a good deal of neuropsychology research is still focused on the reporting of single dissociations. Furthermore, these results still rely on only correlational evidence. Due to obvious ethical reasons, human patients are not randomly assigned to different lesion groups, and naturally occurring lesions vary widely in location and extent. This latter limitation in human neuropsychology research can be addressed through the use of *transcranial neurostimulation* and *nonhuman animal studies*, as described later in this chapter. First, though, we turn to a method that addresses a critical limitation in early neuropsychology studies: that damage to the brain couldn't be assessed until the patient had passed away.

3.3   Neuroimaging

The ability to acquire images of the living human brain represents a relatively recent advance in the study of the mind, one that has greatly accelerated the rate at which we can gain greater understanding of human cognition. Early neuropsychology research on human patients was severely limited by not being able to identify the exact sites of brain injury until postmortem extraction of the brain. This meant that the brain could sustain further injuries between the initial identification of the deficit and the patient passing away, making it difficult to know which injury led to the deficit of interest. In addition, doctors might lose track of patients whose deficits had been noted years earlier and have no opportunity to examine their brains after death. For most of medical history, postmortem analysis was the only way to see the extent and locations of lesions to the brain. The advent of neuroimaging methods allowed the living brain to be examined at the time of injury and was thus a tremendous advance in studying the link between the brain and cognitive function.

3.3.1 Structural Neuroimaging: CT, MRI, DTI, and DWI

3.3.1.1 CT

In the late 1960s, Godfrey Hounsfield at Electric and Music Industries (EMI) Central Research laboratories developed a method of using X-rays to provide images of the living brain, which would come to be called computerized axial tomography (CAT), often referred to as simply computerized tomography, or CT scanning. CT scanning involves taking a series of X-rays through an object from successive positions circling 360 degrees around the object. The "computerized" part of CT then comes into play, as computation-intensive algorithms are run to calculate the relative intensity of each point within the brain based on the X-rays that traveled through that particular point from multiple different directions. The ability to obtain images of the structure of the living brain was quickly realized to represent a major advance in medicine, and by the early 1970s CT scanners had been installed in thousands of hospitals worldwide. CT scanning greatly enhanced neuropsychology, as it became possible to identify the location of a patient's lesion shortly after an injury. This meant that the injury could be localized before any other lesions or changes due to brain plasticity could occur, significantly enhancing the ability to link behavioral deficits to specific brain regions. In addition, the widespread availability of CT scanners in hospitals around the world led to brain injuries being localized with a speed and precision never before possible. Thus, CT scanning provided a critical tool for neuropsychology and made larger-sample-size group studies possible, greatly improving the ability to link focal brain lesions to specific cognitive deficits. It is interesting to note that the invention of the CT method has links to both ancient Egypt and to the musical group The Beatles. The funding from EMI for the original project to develop such a scanner was made possible in part by the large windfall the company enjoyed from the sales of The Beatles' albums, and Godfrey Hounsfield's idea for using probing rays from multiple directions to examine the hidden insides of an object was reportedly inspired by his contemplating how one might look for secret chambers inside the great pyramids!

Since the initial development of CT scanning in the 1960s, further developments allowed for faster acquisition times and increasingly high spatial resolutions of the images obtained. Hounsfield's first scanner took multiple hours to acquire a single scan and multiple days to reconstruct the images. Today's CT scanners (Figure 3.3) can acquire and fully process whole-brain images in just minutes. Such advances in speed not only made the technique feasible for large-scale studies, but the increased speed of imaging meant that motion in the scanner (from simple respiration, blood flow, or the patient moving) could be mitigated. The spatial resolution of CT scanning has also vastly improved: A typical modern-day scan will have a resolution of 0.5 mm or better. Despite these improvements, there remain two major limitations to CT scanning as a tool for cognitive neuroscience research. First, it uses X-rays and therefore

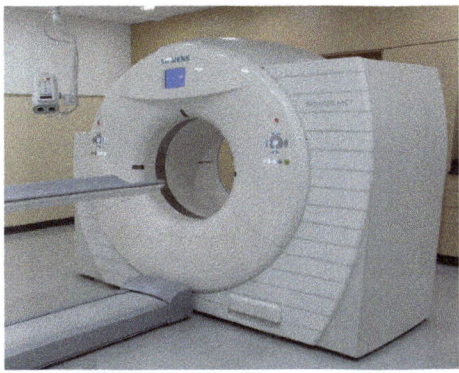 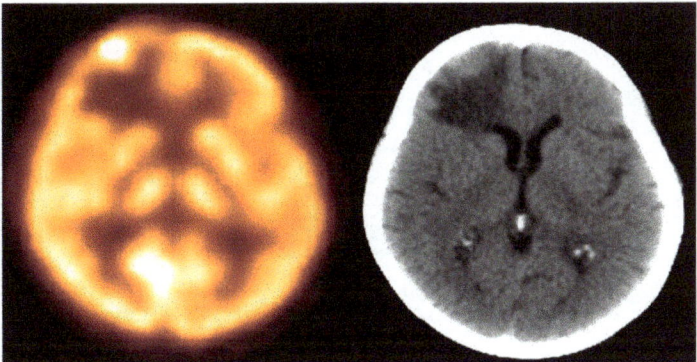

Figure 3.3 CT/PET scanner. Left: A scanner that combines CT and PET in one machine. Photograph courtesy of Brian Strickland Photography and the Biomedical Research Imaging Center at the University of North Carolina at Chapel Hill. Middle and right: PET image (middle) and CT image (right) of an axial slice through the brain. Reprinted from Pietrzak et al. (2021), *Scientific Reports*, Springer Nature, Creative Commons CC BY.

involves the transmission of radiation. Therefore, it's not a good method for longitudinal research studies or any research in which multiple scanning sessions would be required. The second major limitation of CT scanning is that it has a lower spatial resolution than the subsequently developed **magnetic resonance imaging (MRI)** method described in the next subsection. The development and proliferation of MRI scanners has largely obviated the need for CT scanning in much of cognitive neuroscience research.

3.3.1.2 MRI

MRI provides high-resolution images of structures throughout the body, and its ability to differentiate gray matter, white matter, and cerebral spinal fluid (CSF) makes it one of the most important methods in cognitive neuroscience (Figure 3.4). The initial principles leading to MRI were developed in the 1950s (Carr, 1952; Hahn, 1950), and by the early 1970s the first MRI images of a living (nonhuman) animal were published (Lauterbur, 1974). The development of the echo-planar imaging technique (Mansfield & Grannell, 1975), which is still used today, significantly enhanced the promise of MRI by reducing scanning times from hours to seconds. This quickly led to the adoption of the technique as a clinical tool, and by the late 1980s MRI scanners had become commonplace in hospitals. The ability of MRI scanners to provide high-resolution images of the living brain without any exposure to radiation enhanced neuropsychological patient studies and provided a safe and powerful technique for studying developmental changes in the brain as well.

The MRI method is based on a few fundamental principles: (1) Protons (hydrogen atoms, such as in water molecules) are prevalent throughout the body; (2) protons will align with a very strong magnetic field, such as that created by the MRI scanner; (3) when aligned, protons can be "excited" into a higher-energy state through application of a radiofrequency pulse; (4) the exact frequency of the pulse needed to excite a proton (i.e., the "resonant frequency") is determined

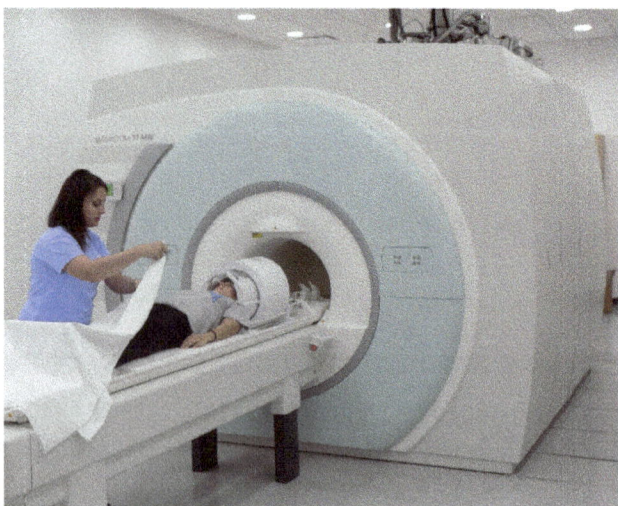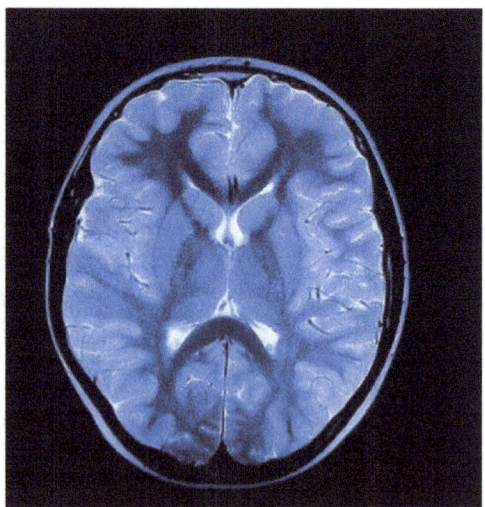

Figure 3.4 MRI machine and structural brain image. Left: 7 Tesla MRI machine. Right: Structural MRI image from a 3 Tesla MRI machine. Photographs courtesy of Brian Strickland Photography and the Biomedical Research Imaging Center at the University of North Carolina at Chapel Hill.

by precisely where that proton is within a magnetic field; (5) in addition to the constant, very large magnetic field gradient running the length of the scanner, smaller magnetic gradients are produced in orthogonal directions by gradient coils within the MRI machine, so each location throughout the brain will have a unique resonant frequency; (6) MRI machines generate a sequence of radiofrequency pulses to "excite" protons in successive slices through the brain; (7) protons do not remain in this excited energy state for long, and when they "relax" back to their state of being aligned with the strong magnetic field, this process releases energy at the exact same frequency as that which excited those protons; (8) radiofrequency-receiving coils can then measure the energy being released, which indicates the precise location of the source; (9) the rate at which this "relaxation" occurs differs across tissue types, which allows MRI to differentiate gray matter, white matter, and CSF; and (10) different "sequences" of the timing, duration, and frequency of the excitation pulses result in different amounts of contrast between tissue types (e.g., the sequence that provides the clearest difference between gray and white matter is different than the sequence that provides the clearest differentiation between white matter and CSF).

The overall "strength" of the MRI magnet refers to the size of the main magnetic field generated by the machine, and this strength is associated with the degree of spatial resolution that can be achieved. The most common MRI machines in use in hospitals are 1.5 T (Telsa), and these provide high-quality structural images of the brain that satisfy most clinical needs. Many research centers, however, utilize 3 T scanners because these provide stronger signals for the type of imaging sequences needed to look at function (described in Section 3.3.2). Scanners with strengths of 7.0, 9.4, and 11.7 T have been used in a few research centers for human research, and even higher-strength magnets (up to 21.1 T) have been used in non-human animal imaging work.

3.3.1.3 DTI and DWI

In addition to providing high-resolution structural images that differentiate between tissue types (e.g., gray matter vs. white matter), MRI scanners can also provide images that reveal the connections between brain areas. **Diffusion-weighted imaging (DWI)** is a different type of MRI sequence, in which the intensity from each voxel (or "volume element," the 3D version of a pixel in neuroimaging data) is related to the diffusion of water molecules, which makes it ideal for highlighting white matter tracts between brain areas. After acquiring MRI data with a DWI sequence, **diffusion-tensor imaging (DTI)** analysis procedures produce detailed images of the white matter connections between brain regions. This method can be important when comparing groups that are hypothesized to have differences in brain connectivity. Most MRI research studies do not typically include DWI scans, however, as they add significant time to the scanning session, and the DTI procedures require substantially more postprocessing to be completed after data collection compared to the basic structural scans.

Finally, MRI scanners can also provide images of the neurochemistry of selected regions of the brain by means of a type of imaging referred to as **magnetic resonance spectroscopy (MRS)**. This method can provide detailed information on the concentration of different nuclei in the brain, including sodium, fluorine, phosphorus, and carbon nuclei, and it has been used to measure a variety of biochemicals such as glucose, creatine, lactate, and N-acetylaspartate. Because of its ability to measure metabolites and other biochemicals, MRS can provide critical information for the diagnosis of cancers, and it is also being used to study Parkinson's disease, Alzheimer's disease, and epilepsy. This type of scanning is relatively slow and is typically only done for a small, predefined region of the brain where a possible abnormality is being investigated. Therefore, most cognitive neuroscience research doesn't utilize this type of scanning. Recently, however, whole-brain spectroscopy methods have been developed (Li, Strasser, Jafari-Khouzani, Thapa, Small, Cahill, Dietrich, Batchelor, & Andronesi, 2020), which may expand the application of this type of scanning.

3.3.2 Functional Neuroimaging: PET, fMRI, and fNIRS

One of the most influential methods of cognitive neuroscience is *functional neuroimaging*. This method seeks to identify the location(s) of brain activity in the living human brain that can be associated with specific cognitive processes. Although the goal is to track brain activity, none of the methods discussed in this section can measure neural activity directly. Rather, these methods rely on blood flow changes that are correlated with neural activity. Specifically, there is a localized increase in blood flow to brain regions where neural activity has recently occurred. This enhanced blood flow is quite slow, peaking ~6–8 seconds after the neural activity and returning to baseline ~12–16 seconds later. Therefore, these methods are quite limited in their temporal resolution. The exact timing between neural activity and the subsequent peak of the blood flow response can vary between regions (as well as between people), so these methods are not sufficient to track the temporal order of fast-acting cognitive mechanisms. However, although the temporal resolution of these methods is quite low, the strength of these methods is their excellent spatial resolution. The increase in blood flow that follows neural activity extends to only a couple of millimeters from the center of neural activity. Combined with a powerful scanner, the localized blood flow response means that these methods provide a level of localization of brain activity that is unmatched by any other noninvasive human neuroscience technique.

3.3.2.1 PET

The first method of functional neuroimaging to be widely used in cognitive neuroscience research was **positron emission tomography (PET)**. PET scanners are used for a range of medical diagnostic purposes because they can be used to measure a variety of different chemicals in the brain. PET scanning involves the injection into the blood of a radioactive isotope, and, depending on the tracer chemical being used, the PET results can show areas of high metabolism, chemical absorption, or even the amyloid protein implicated in Alzheimer's disease. For the purposes of cognitive neuroscience research, however, using the isotope oxygen-15 (O^{15}) is most effective for tracking cognitive processes. O^{15} can be incorporated into water (H_2O), and when injected into the bloodstream it will provide an excellent measure of where blood is going in the brain. As explained earlier, there is a localized increase in blood flow shortly following neural activity, so comparing the blood flow responses across two or more conditions provides a measure of relative increases and decreases in brain activity.

The PET scanner itself is similar in shape to a CT scanner (see Figure 3.3), and the participant lies on a table that is surrounded by a circular array of "coincidence detectors" that measure the emission of gamma rays. When a radioisotope decays, it releases a positron that travels only a short distance before being annihilated by interacting with an electron. This annihilation destroys both particles and produces a pair of photons (gamma rays) that travel in 180-degree opposite directions. The PET system is able to calculate, from the activity measured by its coincidence detectors, where annihilation events occurred with high spatial resolution. In order to achieve sufficient signal to noise, however, PET scanning usually involves collecting data for at least 30 seconds or more and summing all of the measured activity over that time period together to form a single image. Therefore, each experimental condition has to be performed continuously for at least that long, and each condition requires its own injection. All of this constrains the types of experimental designs that can be used in PET studies. In addition, even an isotope with a relatively short "half-life" (the amount of time for half of the isotopes to decay) will require ~10 minutes to decay enough before another dose can be injected. This is the major reason why cognitive neuroscience research has used O^{15} – it has a half-life of 2 minutes, whereas the other isotopes used in PET have half-lives of anywhere from 10 minutes up to 5 days. The temporal resolution of PET scanning is therefore very poor, as experimental conditions must be separated by at least 10 minutes, and within any single condition the activity is summed over the entire 30–60 seconds of the task being performed. Nonetheless, many important and foundational research findings in cognitive neuroscience came from PET studies because from the early 1980s until the early 1990s PET was the best method for localizing neural activity while acquiring whole-brain data in a living human subject. In the early 1990s, however, a new method was developed that produced even better localization and had the added advantage that there was no need to inject the subject with radioactive isotopes. Plus, this new technique utilized equipment that was already in most hospitals around the world.

3.3.2.2 fMRI

Equipped with the knowledge that blood flow could be used to track neural activity and seeing the important advances that PET research made possible, MRI researchers worked to develop methods to track blood flow using MRI scanners. In the early 1990s, researchers presented new scanning sequences that would allow MRI scanners to image a correlate of neural activity. Since these new imaging techniques were measuring the functioning of the brain instead of simply its structure as previous MRI sequences had done, the new technique became known as **functional magnetic resonance imaging (fMRI)** (Figure 3.5). Similar to PET, fMRI doesn't measure neural activity directly but is sensitive to changes in blood flow that follow neural activity. Briefly, as mentioned in the PET section above, there is a localized increase in blood flow that occurs after a burst of neural activity. The increase in blood flow peaks 6–8 seconds after the neural activity and returns to baseline ~12–16 seconds after the initial activity. The signal that the fMRI sequence measures is sensitive to the proportion of oxygenated to deoxygenated hemoglobin. Oxygenated hemoglobin in the bloodstream becomes deoxygenated when its oxygen is extracted by active neural regions. Deoxygenated hemoglobin is more magnetic than oxygenated hemoglobin, so the immediate effect of a brain area using more oxygen is that the local environment becomes more magnetic. The more highly magnetic environment results in more disruption of the signal that fMRI measures, so there is an "initial dip" in the fMRI signal (within ~2 seconds of the neural activity). However, this initial dip is small and hard to detect; therefore, this is rarely reported. Luckily for fMRI research, the subsequent blood flow response, in which blood flow increases much more than is needed to simply resupply that level of oxygen, produces a longer-lasting and more robust change in the local environment in which the ratio of oxygenated hemoglobin to deoxygenated hemoglobin is greatly increased for a short time. This temporarily increased ratio results in an increase in the fMRI signal, which corresponds to the change in blood flow

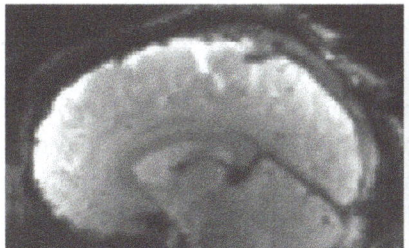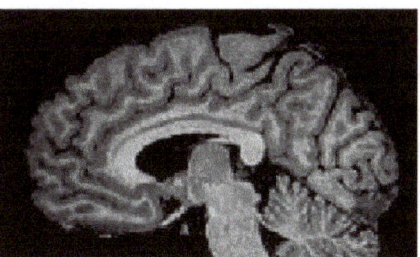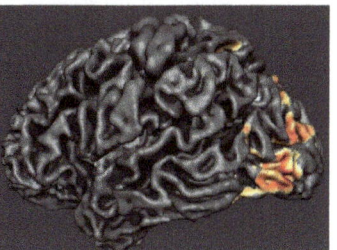

Figure 3.5 fMRI image and overlay of active regions on a 3D brain surface. Left: fMRI image, medial sagittal slice. Middle: Structural MRI image of a medial sagittal slice. Right: Areas of activity from series of fMRI scans displayed on a 3D-rendered cortical surface, generated from structural MRI data.

described earlier. Because the signal is a result of the ratio of oxygenated to deoxygenated molecules, this type of fMRI imaging is referred to as blood oxygenation level-dependent (**BOLD**) imaging.

Block Design versus Event-Related Design in fMRI

Most fMRI experiments have used the so-called boxcar or block design, in which "on-task" and "off-task" periods (or simply two different experimental conditions) are alternated. In this way, there are continuous "blocks" of time, usually ~30 seconds, in which one task is performed, alternating with equally long periods during which the task isn't performed. This type of design is sometimes called a "boxcar" design because when the activity is plotted over time the alternating on-task versus off-task periods produce a pattern similar to the boxcars on a train. This type of design was based on the idea that the best way to measure mental activity was to maintain a task for long enough to ensure the BOLD response would have time to peak, and that the maintenance of that level of activity over an extended time would allow task-evoked activity to be differentiated from more transient sources of noise in the data. This proved to be an efficient and powerful design for fMRI studies; however, the need to continue one task for an extended period of time limits what cognitive processes can be investigated. Block designs cannot address some of the most interesting cognitive questions because if a participant knows that every trial (or at least most of them) will be the same for an entire block of trials, the way they perform the task may be quite different. For example, the cognitive processes evoked by any sort of unexpected stimulus (e.g., a cue pointing to the wrong location; an incongruent target; a novel stimulus in a sequence of familiar ones) can't be measured if one must have an entire block of only that type of stimulus, since the "unexpected" quickly becomes expected. Once the timing and shape of the blood flow response was more completely understood, however, it became possible to use "event-related fMRI" designs. In these types of designs, the order of different trial types can be randomized and many different trial types can be included in each run, thereby greatly increasing the types of cognitive questions that can be addressed in fMRI studies. These techniques typically model the hemodynamic response as peaking ~6–8 seconds after the event, with the response then undershooting baseline a few seconds later, and finally returning to baseline by ~16 seconds. It should be noted, however, that event-related fMRI designs still have limitations because of the sluggishness of the blood flow response; specifically, these designs must include at least some trials with longer periods between successive trials in order to link the slow hemodynamic response to the event of interest.

Regardless of whether an event-related or a block design is chosen, fMRI can't determine the temporal order of activity when multiple areas are involved in a task. This is due to the variability of the hemodynamic response across brain regions and across people, combined with its overall sluggish response to neural activity. Therefore, although fMRI provides excellent spatial resolution for localizing the brain regions involved in a task, its effective temporal resolution is too poor to resolve the sequences of brain activities that underlie cognitive processes. The methods described in Section 3.4, however, address this limitation and have complementary strengths and weaknesses compared to hemodynamic neuroimaging. Finally, some authors have raised concerns that relying only on the ability of fMRI to localize brain activity is reminiscent of previous attempts to associate complex mental functioning to bumps on the head (Box 3.1).

Box 3.1 Is neuroimaging "modern-day phrenology?"

In the late 1800s, with no methods available to directly measure the living human brain, Franz Gall suggested that details of brain structure could be assessed by measuring the contours and deformations of the skull. He proposed that individual differences in skull topography reflected underlying brain differences and therefore should be associated with distinct psychological attributes and cognitive abilities. This method became known as "phrenology," and although it enjoyed popularity in the early 1800s, scientific studies soon revealed fatal flaws in its principles. For one, differences in the shape of the skull are not caused by differences in the underlying brain structure. Furthermore, the psychological functions being attributed to different areas were often highly complex and not easily defined (e.g., personal attributes such as constructiveness, ideality, adhesiveness, or wonder). Indeed, the maps of different phrenologists could vary widely in the number of areas and the attributes associated with each. Today, the use of neuroimaging methods (fMRI, PET) in cognitive neuroscience is sometimes referred to, by critics, as "modern-day phrenology." The concern is that the aim of localizing mental events to distinct areas of the brain is essentially what Gall was trying to do, and that MRI is just a more technologically advanced way to do that. They argue that creating a map of where functions can be localized in the brain doesn't actually explain mental processes or how we experience them. On one hand, the comparison to actual phrenology is exaggerated, at least insofar as fMRI is able to image the organ of interest itself, not the bony structure encasing it. On the other hand, the criticism should be heeded as we continue to develop and use neuroimaging to ensure that we don't make the mistake from the nineteenth century of ascribing very complex psychological phenomena to discrete and independent areas of the brain. Research must seek to better understand the different component processes that make up any psychological phenomenon, and theories about what a region's activity represents must be carefully tested. These concerns highlight the importance of combining knowledge gained from different methods to ensure that conclusions about the function of an area converge across methods that provide information at different spatial and temporal scales. The concern that any complex mental process cannot be accounted for by the activity of just one region also highlights the increased use of analysis procedures that seek to understand the connections and interactions between brain areas. These connectivity-related analyses are now used across a range of methods (e.g., MRI, DWI, fMRI, EEG, MEG) and provide critical information when interpreting the results of stimulation methods (e.g., TMS, tDCS) that can affect areas far from the site of stimulation. In these ways, cognitive neuroscience methods have hopefully moved well beyond the mistakes and errors of a popular fad from a previous century.

3.3.2.3 fNIRS

One other method that uses differences in blood oxygenation to localize regions of neural activity is **functional near-infrared spectroscopy (fNIRS)**. fNIRS is a noninvasive method that involves placing light sources and detectors on the scalp (Figure 3.6). This method involves

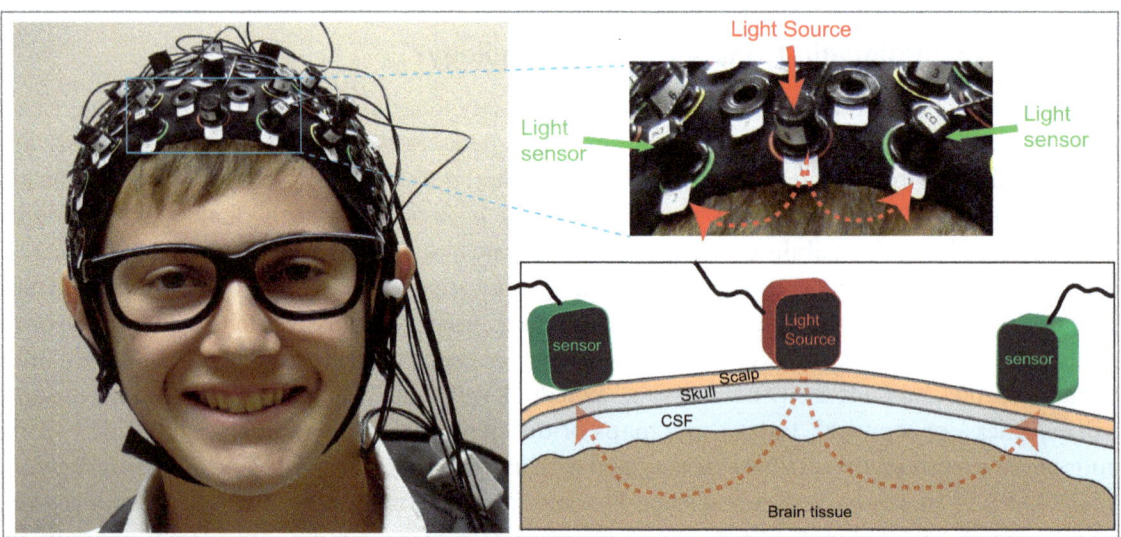

Figure 3.6 Functional near-infrared spectroscopy (fNIRS). Left: fNIRS "cap," which includes light sources and sensors. Picture courtesy of Carol Cheatham Laboratory, University of North Carolina at Chapel Hill and the UNC Nutrition Research Institute in Kannapolis. Right: Illustration of the pathway of light and underlying brain regions that would be detected from a given light source and optical sensor.

shining intense light into the brain and measuring the reflectance. The near-infrared wavelengths (~700–900 nm) used in this method largely pass through skin, tissue, and bone and therefore can reach inside the head. Hemoglobin, however, absorbs light at these wavelengths, and the amount of light reflected from the blood provides a measure of the concentration of hemoglobin. Furthermore, the absorption spectra are different for deoxygenated versus oxygenated hemoglobin, so the relative concentrations of each can be tracked across an experiment by using two (or more) different wavelengths. As described earlier, the ratio of oxygenated to deoxygenation hemoglobin can be used to measure neural activity. Since fNIRS is measured at the scalp, its ability to localize activity is significantly worse than fMRI; however, fNIRS has the unique advantage, among hemodynamic methods, of being portable. Once the light sources and detectors are attached to the scalp, the subject can move about freely, and the equipment can be carried in a small backpack. The other hemodynamic methods (PET and fMRI) require subjects to lie flat on their backs and remain as still as possible throughout the duration of a scan because even small movements of the head can significantly degrade the data quality. Therefore, those methods can be very challenging to use in populations such as young children, for whom remaining still is very difficult. For these populations, fNIRS can be a good solution. And since fNIRS can be measured while the subject is freely moving, it opens up the possibility of tracking hemodynamics in real-life situations.

Unlike the whole-brain coverage that is standard in PET and fMRI studies, fNIRS is limited to measuring superficial cortical activity. Most of the reflectance being measured comes from within a few millimeters of the light source, and the diffuse dispersion of the light within the tissue means that the localization of the source of the reflected light is less precise than what other neuroimaging methods achieve. Although fNIRS relies on the same hemodynamic mechanisms that fMRI depends upon, the temporal sampling rate of fNIRS (1 sample/~100 milliseconds) is better than typical whole-brain fMRI (1 volume/~1–2 seconds). This increased sampling rate can improve the ability to measure the *onset* of the hemodynamic event of interest and provides a more precise estimation of the shape of the hemodynamic response, which helps differentiate it from physiological and motion-related artifacts. However, as mentioned in Section 3.3.1.2, any method that measures blood flow will lag behind the neural activity that is of interest. Therefore, in order to measure brain activity in real time, one must turn to methods that directly measure the activity of active neurons. In cognitive neuroscience research in humans this is usually done by recording the electrical potentials and magnetic fields generated by active neurons, as described in Section 3.4. However, before leaving these optical imaging methods, it's important to point out that a different version of the technique used in fNIRS, called **event-related optical signals (EROS)** or "fast optical imaging," has a high temporal resolution. Instead of relying on the slow hemodynamic response that typical fNIRS measures, EROS measures the scattering of infrared light that occurs when neurons become active (reviewed in Gratton, Chiarelli, & Fabiani, 2017). The temporal resolution of EROS is 8–10 *milliseconds* – much faster than the many *seconds* it takes for the hemodynamic response that typical fNIRS (and fMRI) measures. However, the scattering of light that EROS detects is a much smaller signal and is more difficult to detect than the subsequent hemodynamic response, so this newer technique has not yet become widely used in cognitive neuroscience.

3.4 Electroencephalography and Magnetoencephalography

3.4.1 EEG and the ERP

In 1924, Hans Berger placed electrodes on a subject's scalp and recorded human brain activity for the first time using the method we now refer to as **electroencephalography** (**EEG**; Figure 3.7). Although it subsequently became one of the most important research methods for studying the brain, Berger was hesitant at the time to present such a revolutionarily new method of science and did not publish his work until 5 years later (Berger, 1929). Although initially met with skepticism and even derision, by the late 1930s the technique had gained widespread acceptance, and its use for clinical diagnosis of such things as epilepsy, sleep disorders, and coma increased rapidly.

EEG doesn't measure action potentials directly but rather the postsynaptic potentials that are generated in a population of neurons that are receiving action potentials. When a sizable population of neurons are aligned, such as pyramidal cells in the cortical ribbon, and their postsynaptic potentials are responding in synchrony to incoming action potentials, the net electrical change over the population can be measured at some distance from the source. The electric fields that are generated propagate through the brain tissue, CSF, skull, and scalp before being recorded by electrodes on the scalp. The EEG method was the first to provide researchers with a completely noninvasive way to measure neural activity with millisecond precision.

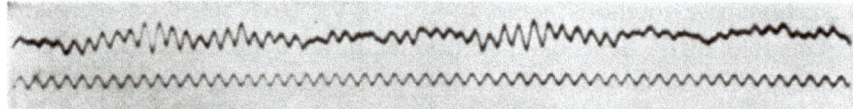

Figure 3.7 Original EEG recording from Berger (1929). The top trace shows Berger's recording from an electrode on the human scalp. The bottom trace shows a perfect 10 Hz mechanically produced waveform for comparison. Note that there are prominent periods of 10 Hz activity within the scalp recording. Reprinted from Berger H., Über das Elektrenkephalogramm des Menschen. *Archives für Psychiatrie*. 1929; 87:527–70 (public domain image).

EEG recordings are sometimes colloquially referred to as "brainwaves," dating at least as far back as Herbert Jasper's 1948 article titled "Charting the Sea of Brainwaves." In that article, Jasper discussed some of the different patterns observed in human EEG recordings, and he described five "frequency bands," which are still in use today. These bands refer to the frequency of the oscillations being recorded, and the relative intensities of these different bands have been associated with different mental states. For instance, Berger (1929) identified a frequency of brainwave that occurred at ~10 Hz frequency (10 cycles per second) and that was enhanced when the subject closed their eyes and relaxed. Berger labeled this oscillation the "Alpha" wave, presumably because it was the first wave to ever be identified and labeled, and he compared it to a slightly faster frequency of activity that he labeled the "Beta" wave. Subsequent research identified and labeled additional frequency bands, and the major divisions still used today are Delta (1–3 Hz), Theta (4–7 Hz), Alpha (8–12 Hz), Beta (13–30 Hz), and Gamma (30–100 Hz). These waves have been associated with generalized brain states: Delta waves with slow-wave sleep and periods of deep relaxation; Theta waves with learning processes; Alpha waves with a calm, relaxed state; Beta waves with alert, active processing; and Gamma waves with some higher-order cognitive processes. The ongoing EEG can thus provide insight into general states of brain functioning, but it doesn't provide much information on specific cognitive processes. The real power of EEG in cognitive neuroscience is therefore not in simply looking at the continuous, ongoing activity but rather in extracting and averaging together portions of the EEG that can be related to specific cognitive or mental events. This averaging procedure is what allows one to go from the continuous EEG to the **event-related potential (ERP)**.

The ERP is created by extracting portions of the ongoing EEG that are time-locked to a specific mental event of interest and averaging those sections together (Figure 3.8). The result is the electrical *potentials* that are *related* to a specific *event* – the *ERP*. Whereas the EEG indicates general brain states, ERPs provide more precise information about the neural processing of specific cognitive events. By extracting and averaging together the portions of the EEG whenever a specific perceptual or cognitive event occurs, it is possible to observe the neural processes related to just that event. The neural activity generated by any one event is typically much smaller in size than the ongoing EEG recorded at the scalp, so it is usually necessary to average many instances of the event of interest in order to obtain an ERP signal that is not contaminated by the large, ongoing EEG. The exact number of trials needed to obtain a clean ERP signal depends on the size and consistency of the component being

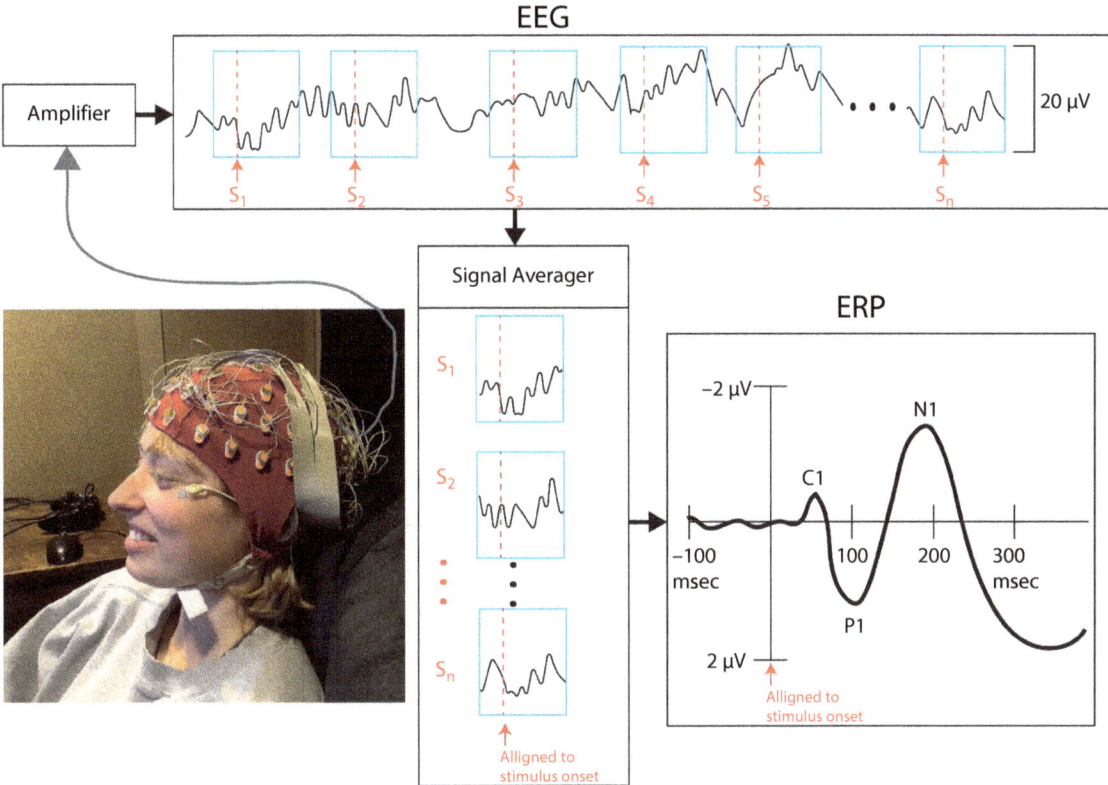

Figure 3.8 Creation of ERPs from an EEG waveform. Lower left image shows a participant in an EEG study, wearing an electrode cap. Signals from each electrode on the head are amplified and recorded separately; the EEG box at the top of the figure shows a section of raw EEG from a single electrode on the head. The red arrows and dashed vertical lines on the EEG show each occurrence of a specific stimulus event of interest (e.g., a brief appearance of a visual stimulus at one location). By extracting a section of the EEG for each occurrence of the stimulus, aligning each section by the precise onset of the stimulus, and averaging these with a signal averaging program, the resulting average response is the event-related potential (ERP) – the voltage *potential* across time *related* to the *event* of interest. The ERP box shows a cartoon waveform made to show multiple visually evoked components on a single electrode. In reality, these components are maximal at different locations across the scalp. Most sensory-evoked ERP components are labeled according to their polarity ("P" for positive, "N" for negative) and relative timing (P1 is the first exclusively positive component, P2 the second, etc.). The visual "C1" is different in being labeled based on its scalp location ("C" for central), because its polarity depends on where in the visual field the stimulus was presented. Note that positive is plotted downward, in accordance with the classical ERP convention.

measured. Large, longer-latency components that index higher-order cognitive processes may require only tens of trials, whereas components related to perceptual processes may require hundreds of trials of the event. A simple measure of the signal-to-noise ratio of an ERP can be observed by comparing the size of the component(s) of interest to the prestimulus baseline period (the time period just before the event occurred). Since activity evoked by any mental event should occur only *after* that event occurs, the prestimulus ERP activity should be zero.

As shown in Figure 3.8, the extracted signal from any single trial of an event will be dominated by the large, ongoing EEG, and therefore the prestimulus period may show a large amount of activity because the ERP of interest is only a relatively small activity on top of the large, ongoing activity recorded in the EEG. However, over many trials, the ongoing EEG will have been different at each occurrence of the event – sometimes high, sometimes low – whereas the signal generated by the event of interest will have been the same for each event. Over enough trials, the EEG activity that is *unrelated* to the event of interest will therefore average to zero, leaving only the activity that is present on every trial: the ERP. The process by which random noise is averaged to zero occurs at all time points of the ERP but is most easily seen in the prestimulus baseline period – since there should be no ERP activity before the event of interest, the activity in the prestimulus period provides an indication of the "noise" remaining in the ERP average. Although ERPs require a large number of trials to obtain a clean and strong signal, uncontaminated by noise, this method can track neural activity in near real time, which is critical for understanding the fast-acting dynamics of human cognitive processes. This millisecond temporal resolution and the ability to acquire these data in an inexpensive and noninvasive manner are why the ERP represents one of the most important methods in cognitive neuroscience.

Furthermore, EEG/ERP systems can be made to be completely portable. Battery-operated systems that transmit signals via wireless connections allow researchers to record ongoing neural activity while the subjects perform tasks naturally in the real world. There are, however, some disadvantages with portable EEG systems, including: Portable systems do not have the high-density electrode arrays available in nonmobile systems; throughput may be more reliable with wired connections; movement with the portable systems is likely to produce artifacts and lower signal quality; and sources of environmental electrical noise can contaminate the data when an electrically shielded recording booth isn't being used.

Spatial Localization of EEG/ERPs   Although EEG and ERPs have excellent temporal resolution, their spatial resolution is limited. In part, this is because the electrical fields generated by neural activity can be attenuated and partially distorted by the brain, CSF, skull, and scalp before reaching electrodes on the outer surface of the scalp. In addition, since air is a very poor conductor, EEG and ERP studies typically use a saline gel solution to essentially create a salt bridge between the scalp surface and the recording electrode. Various methods can be used to localize the intracranial source of scalp-recorded potentials. Having a greater number of electrodes on the scalp provides more detail on the distribution of charge across the scalp, which helps to differentiate ERP components and enhances the accuracy of source estimation. Using modeling techniques that include separate layers (for the brain, CSF, skull, and scalp) can also provide more accuracy by better accounting for the different conductivities of those layers. When possible, having a detailed structural MRI of each subject can be especially helpful, because detailed "boundary-element models" (BEMs) or "finite-element models" (FEMs) can be constructed in which each subject's unique anatomy can be used to build a more accurate model of how electrical charge would propagate from brain to scalp. However, even with the most detailed anatomical models and the highest-density electrode arrays, any method that relies on modeling intracranial sources from activity outside the scalp suffers from the "**inverse problem**." The inverse problem refers to the fact that there is no one unique solution to the source configuration that would produce a given pattern of activity at the scalp. The "**forward solution**" of calculating what activity will be on the scalp if a given source in the head is active can produce a single unique solution because this is simply a matter of calculating how the fields will propagate through to the scalp. But the reverse process is a problem, without a single unique solution, because any pattern of activity on the scalp could be accounted for by multiple different configurations of sources in the brain. Some solutions may be much more plausible than others, but it remains true that any pattern of activity recorded on the scalp cannot definitively establish the source of that activity. For this reason, localizations of EEG/ERP results are usually interpreted with reference to converging results from studies using methods with better spatial resolution. This is also why some researchers have done studies that combine ERP and neuroimaging methods within a single experiment: Neuroimaging provides the precise location of the sources, and ERPs provide the timing of activity in those sources. Box 3.2 provides more discussion of the issues involved in combining these methods.

Box 3.2　The best of both worlds? Combining fMRI and EEG methods

Since fMRI and EEG have complementary strengths, in terms of temporal and spatial resolution, why not simply record both at the same time? This has been done in a few research studies, but there are practical and methodological reasons why this isn't a simple solution (see the discussion by Mangun, Hopfinger, & Heinze, 1998). Because standard EEG equipment uses metal electrodes and wires, it would be hazardous to place it in or near an MRI scanner, with its powerful magnetic fields. Special EEG equipment can be made with materials that are safe to go inside an MRI scanner, but those systems are significantly more costly and can be harder to maintain because of the rarity of the equipment. Even with these MRI-safe EEG systems, data cannot be acquired simultaneously from the EEG and fMRI systems. The fields generated when MRI data are being collected would greatly overwhelm the relatively tiny electrical signal from the brain (and would impact the filters set during EEG data collection), so EEG data can only be collected when the MRI machine is not actively sending pulses and collecting data. Given the sluggishness of the hemodynamic response, EEG could collect its signal when the trial starts, and the fMRI signal could then be collected a few seconds later when the hemodynamic response finally peaks. However, turning the EEG and MRI equipment on and off has consequences as well: The amplifiers and filters in the EEG system take some time to recover back to baseline, and MRI systems can require precise adjustments of the magnetic fields to achieve optimal signal acquisition. Therefore, longer lags between the collection of EEG and fMRI data are usually necessary. Another critical limitation in combining these methods is that the optimal experimental design for an ERP study is different than the optimal design for an fMRI study. For ERPs, it is critical to obtain data over many (often 100 or more) trials, and short intertrial intervals are optimal in terms of both efficiency and ensuring that the participants remain in a consistent alert state. For fMRI, the intertrial interval must be slowed down (on average) to allow the sluggish hemodynamic

Box 3.2 (cont.)

response to peak and return to baseline. Since fMRI doesn't typically require as many trials as ERPs to achieve acceptable levels of signal to noise, the slowed pace of trials isn't an issue when only fMRI data are being collected in an experiment. But when trying to combine ERP and fMRI in a single session, the experimental paradigm will usually be suboptimal for at least one – and usually both – of the methods. Combined with the added costs of a specialized EEG system, these constraints make the combination of fMRI and ERPs in a single session problematic. Therefore, a number of studies that have desired to combine the excellent spatial resolution of fMRI with the excellent temporal resolution of ERPs have instead done so using separate sessions for each method. This method has its own limitations, in that the sessions are separated, sometimes by days, and so it wouldn't be appropriate for such things as tracking mechanisms of learning that may occur in one session more than the other. For other cognitive functions that are more easily replicable over multiple sessions, however, this way of combining methods can be highly effective.

Event-Related Oscillations   As described earlier, ERPs are created by averaging together extracted sections of the EEG that are time-locked to an event of interest. This simple procedure highlights the sequence of neural processes that are consistently active when that cognitive process is engaged. However, other information can be lost during this averaging procedure. Specifically, if there is a burst of oscillatory activity at a particular frequency on every trial, it may not produce any activity in the averaged ERP if the phase is not tightly synchronized across trials. However, the occurrence of such a burst of oscillatory activity could be observed if, instead of a simple averaging procedure, the amount of activity within each frequency band is measured for a series of small time windows following the event of interest. Through this procedure, it is possible to measure whether a cognitive event is consistently associated with bursts of oscillatory activity in different frequency bands. Given that this procedure requires that the activity (within a frequency band) is assessed over a wider window of time, the temporal resolution is not as precise as EEG and ERPs, but it still provides resolution on the scale of tens of milliseconds. These methods of analysis are sometimes referred to as "frequency analyses" or "event-related synchronization/desynchronization," and the results are providing important new evidence linking transient bursts of oscillatory neural activity to different stages of cognitive processing.

Electrocorticography   A variation of EEG with higher spatial resolution is **electrocorticography (ECoG)**. ECoG involves placing a grid or strip of electrodes directly on the surface of the brain, and therefore it provides better localization of the potentials being measured. The electrodes may still pick up activity from other sources, but the measured activity will be dominated by the activity underneath the electrode, and the issues described earlier relating to the skull and scalp attenuating and distorting the signal are mitigated. ECoG thus has the same excellent temporal resolution as other EEG methods but significantly better spatial resolution because of its intracranial placement. However, since ECoG is a very invasive procedure that requires surgical removal of part of the skull and scalp, it is only performed on patients who may require neurosurgery to treat a serious brain condition. For example, patients with intractable epilepsy may have an ECoG grid inserted in order to find the precise initiation point of their seizures, in the hope that finding and later surgically removing that area may eliminate – or at least greatly reduce the frequency of – their seizures. Since the ECoG grid will usually need to stay in place for at least a few days, the patient may also choose to take part in research studies during that time. ECoG studies have been able to provide important findings regarding the precise temporal and spatial precision of some cognitive processes. However, one caveat regarding this method is that it can only be performed in patients who have a serious brain condition; this severely limits experimental sample sizes, and it also means that the brain region being recorded may have previously undergone some degree of damage or compensatory reorganization. This, combined with the caveat of all case studies that the findings may be idiosyncratic to the individual patient, means that the results of ECoG studies may not always be representative of typical brain organization.

3.4.2   MEG and the ERF

Magnetoencephalography (MEG) involves recording the magnetic fields that are generated by populations of active neurons. The MEG signal is produced by the same postsynaptic dendritic potentials that produce the EEG signal. Specifically, an electric current induces a magnetic field in the orthogonal direction; thus, there are always magnetic fields corresponding to EEG signals. In the same way that time-locking to events and averaging across multiple sections of the EEG can produce ERPs, the **event-related magnetic field (ERF)** is produced from extracting and averaging sections of the ongoing MEG data linked to specific events. ERFs are thus used in the same way as ERPs: to track the sequence of activities that are related to cognitive and perceptual events of interest. A significant advantage of MEG over EEG is its spatial localization. Magnetic fields are much less attenuated and distorted by the brain, skull, and scalp compared to the electrical potentials measured by EEG. Therefore, MEG can, at least in some situations, provide much more accurate source localizations of the underlying activities.

As with EEG, there must be a sizable quantity (tens of thousands) of neurons active and aligned in the same direction to produce fields large enough to record outside the head. The MEG signal comes predominantly from pyramidal cells in the cortical ribbon, and because MEG is most sensitive to magnetic fields oriented tangentially to the sensor, the measured signal usually come from activity within sulci. Although not attenuated or distorted by the brain and skull to the degree of electrical fields, magnetic fields do decay more rapidly with distance than electric fields. MEG is most sensitive to superficial cortical activity within sulci, and it does not detect deep brain sources as well. Therefore, MEG is insensitive to some sources of brain activity, but the activity it does measure can be localized more precisely because of the known constraints on what it's able to measure. In particular, when the neural activity originates from sulci near the cortical surface, the localization of MEG can be highly precise, on par with some neuroimaging techniques.

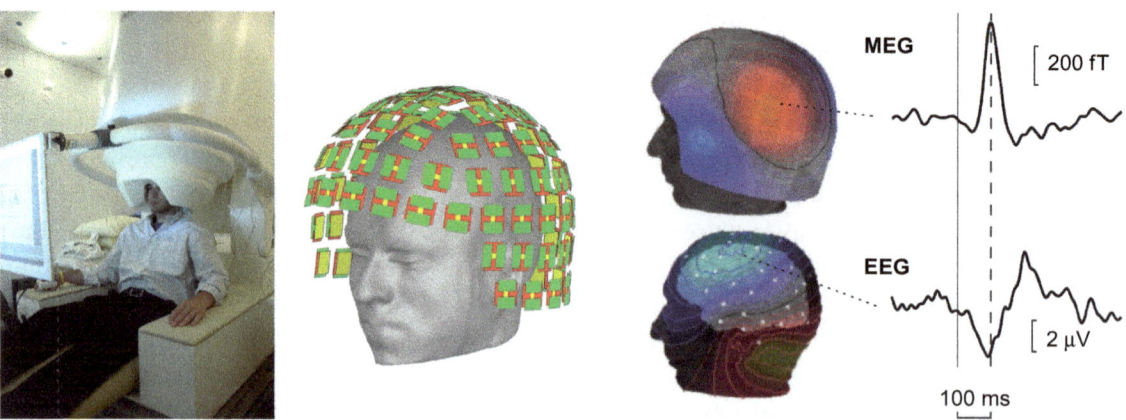

Figure 3.9 MEG machine, helmet of sensors, and recordings. Left: A participant viewing visual stimuli on a special monitor while their head is inside the helmet section of the MEG machine. Reprinted from file: NIMH MEG.jpg (2020, October 18); Wikimedia Commons, the free media repository. Middle: Image of the locations of the sensors inside the MEG helmet relative to the participant's head. Reprinted from Pantazis & Adler (2021), *Sensors*, with permission from MDPI. Right: Topographies and waveforms for MEG and EEG data, illustrating the orthogonal magnetic and electric fields produced by neural activity and the high temporal resolution of both methods. Reprinted from Ahlfors & Mody (2019), *Organizational Research Methods*, with permission from SAGE.

However, as with EEG, MEG still suffers from the "inverse problem" because there cannot be a single unique solution to the pattern of magnetic or electrical fields recorded outside the head. The MEG sensors are permanently installed inside the MEG machine, and the participant sits upright within the lower "helmet" section of the MEG machine (Figure 3.9). MEG sensors do not need to be attached to the scalp as EEG electrodes are because magnetic fields pass freely through air. However, since the signal is still being acquired at some distance from the neural sources, there can still be multiple possible solutions to account for the MEG measurements. Converging evidence from other methods and the plausibility of the different possible solutions can help constrain the number and location of likely sources – in this way, MEG can provide highly accurate localization of activity, along with the same excellent temporal resolution as electrophysiological methods.

Despite its significant strengths, MEG is not as widely used as other methods because of the significant monetary cost. MEG equipment and installation costs are closer to those for MRI and PET, but there is a smaller market because MEG doesn't have as many clinical applications. The reason that MEG is so much more expensive than EEG is that the MEG machine has to create one of the most magnetically quiet environments on the planet. We are constantly surrounded in our everyday lives by magnetic fields that are many orders of magnitude greater than the tiny fields our brain activity produces. The ERFs generated by cortical activity have the approximate strength of a computer transistor die (~10^{-13} Tesla), whereas typical urban noise is in the range of 10^{-8} Tesla. Therefore, MEG machines must create an environment around the head that is shielded from the much larger magnetic fields typically surrounding us and be able to detect the small fields coming from inside the brain.

Modern MEG machines do this through the use of active (noise cancellation) and passive (magnetically shielded rooms) systems and by using specialized superconducting quantum interference devices (SQUIDs) to measure the tiny magnetic fields produced by neuronal activity. The high cost of this equipment and setup has greatly limited the use of this otherwise very powerful technique in cognitive neuroscience studies.

3.5 Neurostimulation

Functional neuroimaging and magneto/electrophysiology have provided a wealth of evidence that activity in different regions of the brain is *associated* with different cognitive functions. Although these findings have greatly advanced our understanding of the neural mechanisms of cognition, those methods cannot assess *causality* – whether activity in that brain region *causes* the mental event. Despite all their strengths, fMRI/PET/MEG/EEG cannot assess whether a certain brain activity is responsible for the cognitive process of interest, nor whether that brain area is necessary for that mental function. Neurostimulation methods fill this critical need. By stimulating a part of the brain and observing the effects on cognition and behavior, we can better understand the neural basis of cognitive processes.

Direct brain stimulation began to be used in the 1930s, when Wilder Penfield developed neurosurgical techniques to treat epilepsy and some brain cancers. In the procedure, Penfield would stimulate small areas of the brain to better localize where seizures may have been originating, and he would test areas that were being considered for surgery to lessen the side effects of these procedures. These stimulations could be done while the patient was conscious, and thus they provided a wealth of evidence regarding the phenomena experienced by the patients during stimulation. The maps of primary somatosensory and motor areas of the brain that Penfield was able to generate from these studies are still useful today. These techniques can also be used before surgical extraction of cancers to try to ensure that the most critical brain regions are spared. As described earlier, however, the results of any neuropsychological case study can be due to idiosyncrasies of that person. Furthermore, patients who have epilepsy or tumors may have experienced some plasticity or reorganization of neural processes. Finally, use of this highly invasive method is only ethical for the small portions of the brain where surgery is required to treat a serious brain condition, and different patients will undergo stimulation in different areas. Therefore, although highly informative, this method also has severe constraints. As described in the next two subsections, the recent development of stimulation methods that can be safely performed in a noninvasive manner and replicated precisely across subjects has been an important advance in cognitive neuroscience.

3.5.1 Magnetic Stimulation

Transcranial magnetic stimulation (TMS) stimulates the brain by creating a brief magnetic field close to the head. As discussed in Section 3.4.2, magnetic fields are not attenuated or distorted much by the air, scalp, or skull. Therefore, generating a strong magnetic field near the head causes this field to propagate into the brain. The transient magnetic field induces a change in electrical potential of the underlying tissue, and a strong enough field can trigger activity in the neurons in a select area. Since magnetic fields fall off rapidly with distance, the technique is most useful for stimulating superficial cortical areas (an area of ~2 cm at depths of ~2–3 cm; Deng, Lisanby, & Peterchev, 2013). When TMS is applied over the back of the head, near the early visual processing regions of the brain, subjects may report perceiving "phosphenes," or brief flashes of light, in the absence of any actual visual stimuli. Similarly, TMS over the motor cortex can produce involuntary muscle twitches in distinct body locations. Indeed, a first step in some TMS studies had been to locate the thumb region of the motor cortex by eliciting a twitch of the thumb or a motor-evoked potential via electromyography on the forearm. This procedure is done in order to determine what strength of stimulation to use and to locate the brain area to be stimulated in the experiment. Regarding the first point, determining the threshold for triggering a motor response must be done individually for each subject, since there is variability across people in what strength of stimulation is required to trigger neural activity from a coil placed outside the head. Once that threshold is determined, the strength of the field to be used in the experimental session is set to be a smaller percentage of what was able to trigger motor activity. Regarding the second point, the location of the brain area to be stimulated in the experiment was estimated, in early studies, by measuring how far that area would be expected to be from the thumb area of the motor cortex. Much more precise localization of the target area can be achieved now by using structural and functional MRI data from each individual subject, when available.

The mechanism by which TMS works is based on the principle of induction. The TMS stimulator is composed of a capacitor and coils of wire (encased in a plastic case), through which a rapid electric current is sent. This changing *electric* field produces a brief *magnetic* field in a perpendicular orientation (Figure 3.10). That magnetic field passes freely through the scalp

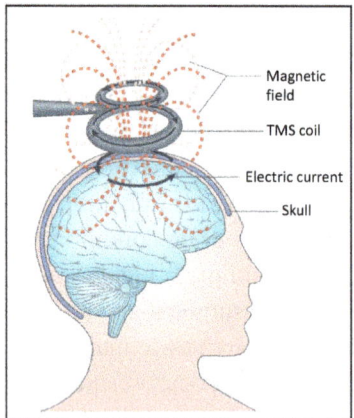 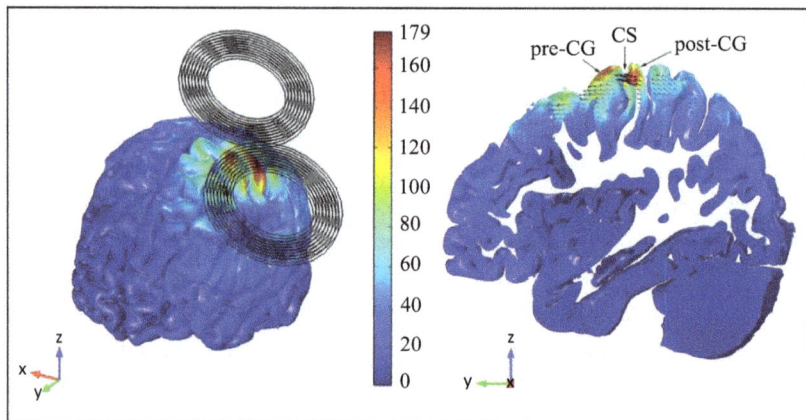

Figure 3.10 TMS method and area of induced current. Left: Illustration of the electric current and magnetic field induced by a figure-eight TMS coil. Reprinted from Ridding & Rothwell (2007), *Nature Reviews Neuroscience*, with permission from Springer Nature. Right: Example of using 3D models of the brain to estimate the cortical regions that would be maximally affected by a given set of stimulation parameters. The position of the figure-eight coil is represented by the sets of black circles over the brain. CG = central gyrus; CS = central sulcus. Reprinted from Salvador et al. (2015), *Frontiers in Cellular Neuroscience*, Creative Commons CC BY 4.0.

and skull (decaying only because of distance) and in turn induces an electric current in the underlying neural tissue. Early TMS equipment used a single circular coil, but most current systems use a "figure-eight" design in which there are two adjacent coils. Whereas the single-coil systems were able to stimulate a large, circular area corresponding to the entire circumference of the single coil, the figure-eight system can achieve much better spatial resolution. By controlling the direction of electric current through the two coils, figure-eight systems can create a much smaller area of maximal stimulation precisely where the two coils meet (Figure 3.10). The spatial extent of the stimulation and its effects on neural firing depend on many parameters, including the size and shape of the coils, the positioning of the coils, the intensity of the current, and the frequency of electric pulses.

There are two main types of TMS studies: repetitive TMS (rTMS) and single-pulse TMS. rTMS is often used to test whether or not an area is involved at all in a cognitive process, whereas single-pulse TMS allows one also to determine *when* that area is involved in that process. rTMS studies are sometimes referred to as "virtual lesion" studies because the targeted area of the brain is essentially made to go offline for a short period of time, and therefore this method can be compared to actual lesion studies. The recovery period depends upon the parameters and duration of stimulation. In most research experiments, the recovery period will be set to be a few minutes up to an hour. In addition, whereas stimulation at lower frequencies (e.g., 1 Hz) results in reduced cortical excitability, stimulation at higher frequencies (e.g., 10 Hz) increases excitability. During the recovery period when the targeted brain region is not working normally, researchers can test the participant on cognitive tasks to investigate if that region affects those cognitive processes. Neurostimulation methods are unique in being able to test whether the brain region is *necessary* for the cognitive process of interest. The significant advantages this sort of "virtual lesion" method holds over neuropsychological studies of patients with actual lesions are: (1) Baseline task performance can be measured for every participant before and after the "lesion," allowing for better accounting for individual and idiosyncratic differences in task performance; (2) the temporary duration of the "virtual lesion" is too short to allow time for plastic changes to occur in the brain; and (3) the experiments include many participants, and thus the results can be assumed to be more representative of the general population. The effect produced through rTMS is temporary in cognitive experiments, and participants return to their prestimulus baseline shortly after stimulation ends. However, it should be pointed out that rTMS can be used in a different way to treat clinical disorders such as severe depression. When used to treat a disorder, the rTMS procedure is different in critical ways: The TMS pulses are performed repeatedly over longer periods and over many sessions, spanning days or often weeks. Given the reports of TMS successfully treating some forms of depression, it appears that stimulation through rTMS can change the brain in more lasting ways, at least when used in that way. However, when rTMS is used for cognitive neuroscience experiments, it is used in much briefer durations and without the repetition over many days or weeks that is critical for clinical treatments.

The second major way in which TMS is used in cognitive studies is "single-pulse" TMS, which allows researchers to make use of the excellent temporal resolution afforded by this method. The TMS pulse can be very rapid (in the order of microseconds), and a single pulse produces an effect on neurons that may last for only a brief duration (as little as 10–20 milliseconds). Therefore, using single-pulse TMS can allow researchers to investigate *when* a brain region is involved in a cognitive process. The researcher is able to selectively impair functioning of an area at different time points (across different trials) in order to test precisely when that brain area is involved with processing that stimulus. For example, in one classic study, the primary visual cortex was stimulated at different latencies after a letter was briefly presented on screen (Corthout, Uttl, Ziemann, Cowey, & Hallett, 1999). The results revealed that participants' ability to perceive the letter was significantly impaired if the TMS stimulation occurred between ~70 and 130 milliseconds after the letter appeared. However, stimulation shortly before or any time after that time range had little to no effect on the participants' perception. This study provided evidence that the primary visual cortex was necessary for visual perception and, furthermore, that it was necessary precisely 70–130 msec after the stimulus appeared.

Despite its strengths, there are notable limitations to TMS as a method of research. Since magnetic fields decay with distance, they are not very useful for testing deep structures. The field strength that would be required to reach deep structures would stimulate the intervening areas to an even greater degree, impairing the ability of TMS to isolate the brain area of interest. One solution to this problem involves using two TMS stimulators oriented such that their fields converge on a deeper structure. This can be effective, but, given the size of typical TMS stimulators, they have to be quite far apart on the head, limiting the areas that can be targeted in this way. In addition, targeting some deep structures would entail risk because of their closer proximity to brain stem structures that are necessary for maintaining critical body functions. Another limitation of TMS, as it relates to experimental design, is that it creates a sudden loud "click" every time the electrical current is sent rapidly through the coil. This loud click can provide the participant with knowledge that they are being stimulated every time the sound occurs. Therefore, care must be taken to also include "sham" trials in any study so that results in the main stimulation condition cannot be attributed to subjects performing differently simply because they think their brain should be affected at those times. Sham trials do not stimulate the brain but include the same type of loud click, either by using specialized equipment that can produce the click without generating a magnetic field or by orienting the stimulator away from the head during sham trials so that the field is not reaching any part of the brain. These sham trials provide an important way of checking that any observed effects of stimulation are truly due to the stimulation and not to subjects' expectations alone. Another experimental limitation of TMS is that it can stimulate muscles, and this can be especially uncomfortable and unnerving when it happens to muscles in the face. Therefore, stimulation of temporal lobe and lateral frontal areas near the face are usually avoided in TMS research. Finally, another issue that must be accounted for in experimental designs and in interpretation of results is that every TMS experiment is actually multisensory because of the loud auditory clicks and the possible stimulation of muscles and somatosensory receptors.

3.5.2 Electrical Stimulation

Transcranial direct current stimulation (tDCS) is a method for stimulating the brain using electrodes placed on the scalp. The amount of current used in tDCS studies (1–4 mV) is not strong enough to trigger action potentials, but it does modulate the excitability of the

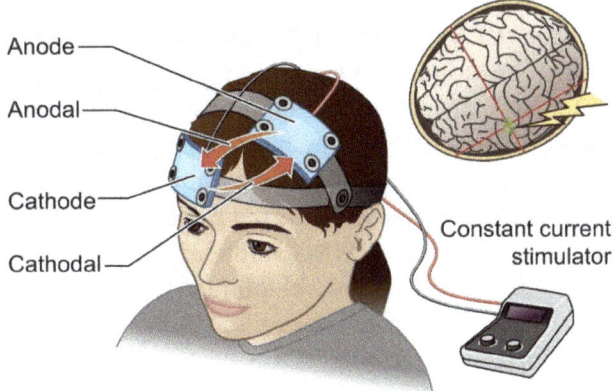

Anode
Anodal
Cathode
Cathodal

Constant current
stimulator

Figure 3.11 Equipment and setup for tDCS studies. Illustration of a standard tDCS method, with two sponge electrodes attached to the head. The mild current is generated and controlled by a small stimulator device, with the area under the anode becoming more depolarized and the area under the cathode becoming more hyperpolarized. Reprinted from Banich & Compton (2023), *Cognitive Neuroscience* (5th ed.), with permission from Cambridge University Press.

underlying brain regions. Given the modest current that is applied to the scalp, it is considered a very safe and noninvasive method. In addition, the cost of tDCS equipment is significantly less than that for TMS, and electrical stimulation techniques are increasing in popularity. tDCS can enhance or suppress excitability depending on where the electrodes are placed. In a typical setup, there are two electrodes: the anode and the cathode. The brain region under the anode becomes depolarized and therefore more likely to fire, while the area under the cathode becomes hyperpolarized and is inhibited from firing. The area of stimulation is quite broad compared to TMS, especially when using the typical setup of two relatively large sponge electrodes (Figure 3.11). In order to provide greater precision of a targeted region, newer systems have been developed that use smaller and multiple anode or multiple cathode electrodes. For example, it is possible to restrict the area that becomes more excitable under the anode by surrounding it by four nearby cathodes. While this placement will distribute the regions of inhibition more widely, the area of enhanced excitability under the single anode will be more precisely targeted. The duration of the induced hyperpolarized and depolarized states depends on the stimulation parameters. However, given that the stimulation is quite mild and often not even perceived by the subjects, most tDCS experiments collect data during the stimulation period itself. As opposed to the loud clicking noise associated with TMS pulses, tDCS is completely silent. Some stimulation can be felt as a tingling sensation on the scalp, but this usually occurs only in the period when the current is first being turned up. Therefore, experiments typically use a control condition in which the current is initially turned on for only a few seconds to mimic the initial sensation on the scalp but is then turned off for the remainder of that condition to serve as a no-stimulation comparison.

Transcranial alternating current stimulation (tACS) uses the same hardware as tDCS but allows researchers to investigate the influence of specific *frequencies* of neural activity (e.g., Alpha, Gamma, Theta). With tACS, oscillating currents are generated instead of the constant current used in tDCS. The researcher can control the rate of the oscillations through the electrodes and thereby induce oscillatory activity in the underlying regions. tACS can be used to test how different frequencies of neural oscillatory activity affect cognitive processes. In addition, the coupling between areas of the brain can be studied by using multiple stimulators and adjusting whether the oscillatory patterns are in phase or out of phase with one another. This technique offers the opportunity to test the causal influences of the oscillatory activities measured as correlations in EEG and MEG studies.

Note that the term transcranial electrical stimulation (tES) is sometimes used to refer to tDCS and tACS, as well as to transcranial pulsed current stimulation (tPCS) and transcranial random noise stimulation (tRNS). tDCS and tACS are the most common electrical stimulation methods used in cognitive neuroscience studies, but tRNS and tPCS are being explored as well. Whereas tDCS uses a constant current intensity, tRNS uses a range of current intensities and frequencies that are randomly distributed around a set mean intensity. Similar to tDCS, tRNS results in a subsequent period of excitability or suppression that extends longer than the stimulation itself. tRNS has been suggested to be more comfortable for subjects and therefore may be useful in clinical treatment settings. It has also been suggested that the random stimulation currents may help in obscuring from subjects which condition they are participating in. tPCS delivers the same sort of constant-intensity stimulation as tDCS but for shorter periods of time, alternating with equal periods of no stimulation. This alternation means that tPCS creates phasic stimulation in addition to tonic effects, and this combination could be useful in treating some disorders (reviewed in Ganguly, Murgai, Sharma, Aur, & Jog, 2020).

3.6   Nonhuman Animal Studies

This chapter focuses on methods that allow us to measure brain activity in humans because most of the research described in this book utilizes those methods. However, some of the methods that can be used only in nonhuman animals provide the highest levels of experimental control, temporal resolution, and spatial resolution. The methods described in this section allow for the direct measurement and manipulation of neural activity in real time and with a spatial resolution not possible in human studies. These methods cannot be used in human subjects because they are invasive and can result in permanent damage to the brain. Nonetheless, these methods have provided some important insights into the neural mechanisms of attention that would not have been possible using human methods alone.

3.6.1   Cellular Recording

Neurophysiologists can measure brain activity in real time with precise spatial resolution using depth electrodes inserted directly into the brain through invasive surgical operations. The placement of the electrode can be done with guidance from neuroimaging scans and can be verified through postmortem dissection. It is possible to record from within a single neuron, but such "intracellular" recordings require a more difficult operation procedure and the cell can be damaged in the process. Thus, most cognitive neuroscience studies use "extracellular recordings," in which the depth electrode is placed outside the cell and records activity from a small set of nearby neurons. This is sometimes referred to as "single-cell recording" because, by using computational algorithms to further process the data, this method can assess the contributions from a single cell. The unmatched spatial and temporal precision of this technique make it an important method for studying brain function and processes. Since it is used only in nonhuman animals, however, many complex cognitive functions cannot be assessed using this method. In addition, animals typically require extensive training on tasks, often involving multiple stages of training before recording activity during the task of interest. Issues can thus arise when trying to compare results from nonhuman experiments that involved days or weeks of training to human experiments that were completed in a single brief experimental session.

3.6.2 Lesions

Although human neuropsychology studies provide critical insights into the workings of the mind, the methods used in these studies are limited because the damage to the brain is naturally occurring and not under experimental control. In nonhuman animals, however, lesions can be made precisely and solely to the area of interest, and the animal can be tested on the task(s) before and after the lesion to better understand the effects of the lesion. A number of methods can be used in nonhuman animals, from nonreversible surgical operations that permanently remove an area of the brain, to reversible methods (e.g., cooling, chemical injections) that cause an area of the brain to be inactive only temporarily. These latter methods are less spatially precise, in that surrounding areas may also be affected, but they have the advantage of allowing the animal to be tested before, during, and after the area is temporarily disabled. All of these lesioning methods are valuable in assessing whether a brain area is necessary for a mental function of interest.

3.6.3 Stimulation Studies

Stimulating an area of the brain can greatly advance our understanding of its function(s). As noted earlier, TMS, tACS, and tDCS can be used in humans to noninvasively stimulate the brain, but in nonhuman studies invasive stimulation procedures can provide better precision and control. Electrodes surgically inserted into the brain can be used to directly stimulate the brain, with the spatial extent of the stimulation determined by the strength of the current. Optogenetics is a recent method that provides incredible precision but it is highly invasive. Optogenetics involves inserting DNA into the cells of interest to create light-sensitive ion channels. Cells with these channels can then be triggered into activity by shining a specific wavelength of light near that region. This method of neurostimulation is unmatched in its precision, and its level of spatial and temporal specificity makes this an exciting method for investigating brain functioning; however, it is used only in nonhuman animal studies.

This method may be developed further to treat some brain diseases, but, due to its highly invasive nature, it cannot be used as a method for testing cognition in healthy humans. Recent research in nonhuman animals has shown that focused ultrasound stimulation (FUS; see Yang, Phipps, Newton, Chaplin, Gore, Caskey, & Chen, 2018), guided by MRI, may also be an effective method for brain stimulation. Although lacking the spatial precision of the methods described earlier, FUS is done noninvasively from outside the head, and therefore it holds promise as another method to safely and transiently modulate brain activity.

3.7   Comparing Methods

As noted throughout this chapter, all of the cognitive neuroscience methods have advantages and limitations. Figure 3.12 presents a summary of the methods used in human subjects compared on their spatial and temporal resolution as well as their portability. It is noteworthy that portability is typically correlated with cost as well: The methods that are completely immobile, such as fMRI, PET, and MEG, are significantly more expensive because they include large machines and dedicated facilities. In contrast, some recent EEG and fNIRS systems have been made completely portable, allowing the subject to move about freely while recordings are transmitted wirelessly to a nearby computer or wired into equipment that can be worn in a backpack. This portability allows for exciting new research to be conducted in real-world environments. As described in a recent review (Stangl, Maoz, & Suthana, 2023), the ability to

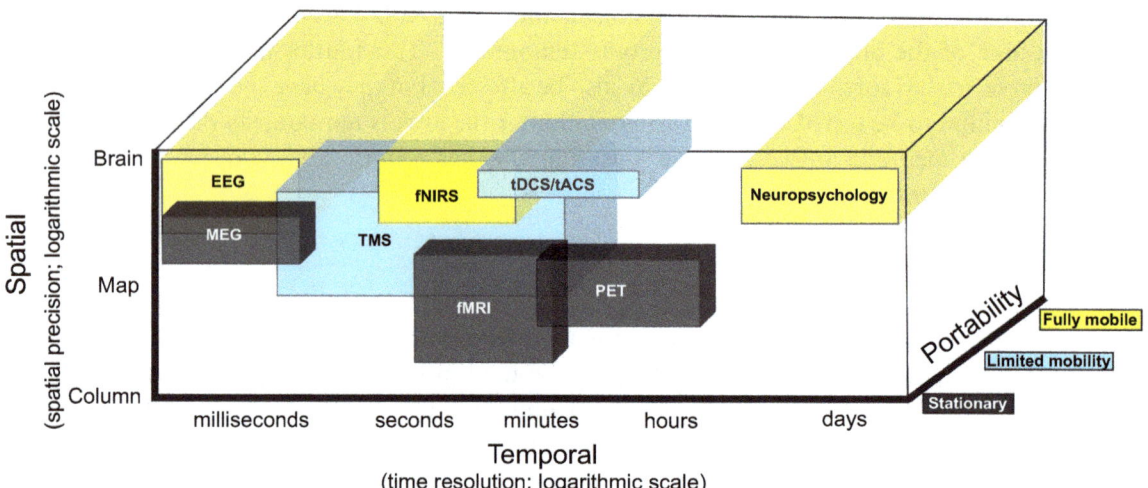

Figure 3.12 Comparison of methods used in human cognitive neuroscience studies. Spatial resolution (*y*-axis) represents how well the technique can localize *where* the neural activity was generated. Temporal resolution (*x*-axis) represents how precisely the technique can identity *when* the neural activity was generated. Portability (*z*-axis) indicates how flexible the technique can be in terms of mobility; this also relates to cost, since completely stationary equipment is typically much more expensive and often requires specially built rooms.

record neural activity in ecologically valid situations may be a game changer in terms of allowing us to understand the neural mechanisms of natural social interactions. Furthermore, the ability to record data in this way may provide new insights into the real-world effects of therapies and treatments for neurological and psychiatric disorders. However, most neuroscience equipment, including most EEG/ERP and fNIRS systems, is still built to be used exclusively in a laboratory and therefore is not fully mobile. These laboratory systems generally produce higher-quality data, in part because the laboratory environment can be better controlled to reduce noise in the data. TMS and tDCS systems have limited mobility, as they can be moved to different locations; however, because the brain is being stimulated, it is usually preferable to have the subject remain seated throughout the experiment.

The main comparison of interest for most researchers, however, is the spatial and temporal resolution of these methods. For research questions that aim to resolve the precise brain *regions* involved in a process, methods such as fMRI and PET are the best choices for human subjects. In contrast, MEG and EEG are the methods of choice when the critical test of a hypothesis requires the precise *timing* of neural activity to be known. Neurostimulation methods (i.e., TMS, tDCS, tACS) have neither the spatial resolution of fMRI nor the temporal resolution of MEG/EEG, but these methods allow causality to be assessed. Within the neurostimulation methods, TMS is significantly more expensive but also has greater precision than tDCS and tACS. It should be noted that the variation in resolution for some of the methods is due to variations in how the method is used. For example, the spatial resolution of TMS depends on how the targeting of the brain region is performed – by using an individual's own MRI and fMRI results, the spatial precision of TMS is greatly enhanced compared to when relying only on standardized scalp locations. The temporal resolution of TMS also varies depending on how it is used – single-pulse TMS affects a region for only a few milliseconds, whereas repetitive TMS may affect a region for minutes to hours.

Researchers are continuing to develop ways to combine these methods to achieve the best possible spatial and temporal resolutions. As described in Box 3.2, researchers have combined ERP and fMRI/PET neuroimaging methods to investigate the effects of attention on visual processing. ERPs can also be combined with fNIRS in specialized caps that allow data from both methods to be acquired simultaneously. Neurostimulation techniques can be combined with other methods, such as by recording ERP, MEG, or fMRI activity immediately following the stimulation. In this way, the effects of the stimulation on neural activity can be assessed and compared to both prestimulation and poststimulation recovery periods. The future of cognitive neuroscience continues to evolve as we assess brain function with both high spatial and temporal precision and extend the recording of neural activity beyond the laboratory into real-world environments.

CHAPTER SUMMARY

- Neuropsychology has been critical for providing new insights into the processes of cognition and their neural bases, but it is limited because natural lesions are variable and plasticity and learning may change how a lesioned brain performs a cognitive task.

- Neuroimaging methods (fMRI, PET, fNIRS) provide excellent spatial resolution of the anatomy of lesions and the areas active during cognitive tasks; however, functional neuroimaging cannot assess the temporal order of activity across areas because of the sluggishness of the hemodynamic response that it measures.
- EEG/ERP and MEG/ERF can track neural activity noninvasively in real time and are thus critical for understanding the sequence of mechanisms involved in complex cognitive process; however, their spatial precision is limited because they are recorded from outside the head.
- Whereas most methods of cognitive neuroscience can only measure whether there is a correlation between activity in a brain region and the performance of a task, neurostimulation methods (TMS, tDCS, tACS) can assess causality by testing whether and when a brain area is necessary for a particular function.
- Nonhuman animal methods, including single-unit recording, permanent or temporary lesions, and stimulation studies, can provide the highest levels of spatial and temporal precision and experimental control. However, since these methods are not ethical for human research, they are limited to mental processes that these animals can be trained to demonstrate.

REVIEW QUESTIONS

Before the advent of modern cognitive neuroscience methods, where were mental and cognitive processes believed to be localized?

Explain the limitations with "case studies" in neuropsychology and what can be done to address those limitations.

Describe the relation between neural activity and what each of the different neuroimaging methods measures/manipulates.

What are the different types of scans (e.g., structural, functional, DTI) that an MRI can provide, and what is each used for?

Compare and contrast electrical (EEG, ERP, tDCS, tACS) versus magnetic (MEG, TMS) neuroimaging methods.

FURTHER READINGS

Engel, S. A., Glover, G. H., & Wandell, B. A. (1997). Retinotopic organization in human visual cortex and the spatial precision of functional MRI. *Cerebral Cortex, 7*(2), 181–192.
  - This article was one of the first to illustrate the high *spatial precision* of fMRI and its ability to map out functionally defined regions of cortex.

Luck, S. J. (2014). *An Introduction to the Event-Related Potential Technique*, 2nd edition. MIT Press.
  - This book explains the neural mechanisms, recording, and analysis of ERPs.

Wassermann, E., Epstein, C., Ziemann, U., Walsh, V., Paus, T., & Lisanby, S. H. (2012). *Oxford Handbook of Transcranial Stimulation*. Oxford University Press.
  - This book explains the methods, neural mechanisms, and analysis of TMS.

4   Deficits in Attention

Learning Objectives

- Compare and contrast the multiple subtypes of attentional neglect syndrome observed in neurological patients
- Identify the mechanisms of attention that are linked to subcortical structures based on studies of patients with brain damage
- Describe how the balance of top-down and bottom-up attention mechanisms relates to developmental disorders of attention
- Compare the processes of attention that are affected in autism and schizophrenia

We often don't realize just how important something is until it stops working. When we fail to pay attention to someone who's talking to us and then have to face the uncomfortable consequences of not having any idea what's been said to us, this momentary lack of attention causes us embarrassment and impairs our ability to respond appropriately. When we get distracted from our work by the notification of a new email or text message, the involuntary capture of our attention undermines our ability to get our work done efficiently. When we lose ourselves in our own internal thoughts and realize we've "read" paragraphs of text without actually processing any of the information contained therein, our lack of attention forces us to spend more time rereading the text again (sometimes again and again). While momentary lapses of attention such as these can be annoying and frustrating, some individuals suffer from more profound and longer-lasting deficits in attentional processes. Primary among these are developmental disorders (e.g., attention deficit hyperactivity disorder [ADHD]) and brain injury (e.g., unilateral neglect syndrome). In this chapter, we will review different developmental and neuropsychological populations that suffer from dysfunctional attention. These populations and individuals are important for both advancing our understanding of the component processes and neural mechanisms of attention and guiding further research into which functions of attention might be most critical to understand in order to better help those who suffer from attention deficits. This chapter will first describe neuropsychological patient groups that have helped identify core processes of attention. We will then turn to developmental disorders that reveal deficits in multiple, sometimes co-occurring processes of attention and discuss how these fit into and extend classic models of attention. Finally, we will review some research directions that may lead to new treatments and therapy options for those suffering from attentional disorders.

4.1   Deficits in Attention Following Brain Injury

4.1.1   Hemispatial Neglect Syndrome

The most common neuropsychological patients used to illustrate the functions of attention are patients with "unilateral neglect," sometimes referred to as "hemispatial neglect," or more simply as just "**neglect**." After suffering a brain injury, an individual with this syndrome would ignore, or neglect, a large portion of space. They may fail to eat food on just one side of their plate or leave half of their face unshaven. They may fail to use the hand on their affected side to steady a sheet of paper they're writing on. These failures persist despite the fact that the person's vision is fine – they have the ability to see objects across their whole visual field. The neglect of their own body persists despite the fact that they have the ability to fully use all their limbs. That these deficits occur despite these patients' sensory and motor-control systems being intact is why the neglect syndrome is so fascinating, and why it's considered a deficit in attention. The patient has the ability to, and in some situations does, perceive visual objects in their affected visual field, but they often simply don't pay attention to anything within an entire half of the visual world in front of them. The patient *can* use their hand on their affected side but oftentimes will simply neglect putting it into action. This syndrome thus illustrates that attention is something beyond perceptual and motor processes.

Neglect is most commonly seen following unilateral damage to parietal and temporal–parietal regions and is more common following damage to the right hemisphere (reviewed in Karnath & Rorden, 2012). A typical test for neglect that can be performed at the patient's bedside in the hospital is the **extinction task**. In this task, the doctor stands with their arms slightly outstretched and with the index finger on each hand pointing upward. The doctor asks the patient to look them straight in the face while detecting which of their fingers in the patient's peripheral vision moves: the one in the patient's left visual field, the one in the patient's right visual field, or both. When the doctor moves their fingers in both visual fields simultaneously, a patient with neglect following damage to the right hemisphere of their brain will typically miss seeing the doctor's finger move in the patient's left visual field. However, when the doctor only moves one finger at a time, the patient can sometimes perceive the single moving finger, even when it is in their left visual field. This provides evidence that the patient is not blind to stimuli in that field, but rather that they ignore, or neglect, the stimulus in their affected field when another stimulus is present in their intact field. In other words, the stimulus in their affected field is "extinguished" when there is a competing stimulus in their good visual field. This pattern also accounts for the reports of patients only shaving half of their face – the unaffected side of the face competes with the affected side for their attention, and the unaffected side continually wins this competition for attention. Another simple task that can reveal neglect is the line bisection task. In a version of this task, patients are given a sheet of paper with many short horizontal lines on it, and they are asked to use a pen to make a mark through the middle of each line (i.e., to bisect each line). The results show that patients with neglect often miss many lines in their affected visual field (i.e., they don't mark most items on the left half of the paper). In addition to neglecting half of space, this task also often reveals that patients may sometimes neglect half of each object as well – specifically, the marks drawn by patients to bisect the lines are often well within the right half of each line, as if the patient

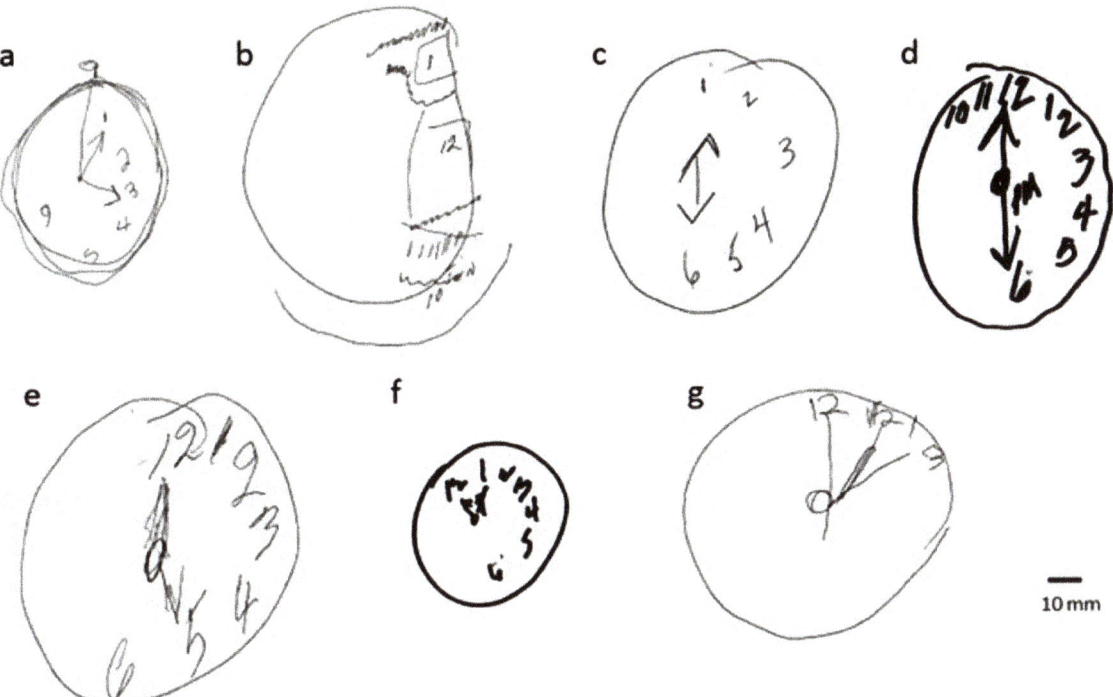

10 mm

Figure 4.1 Results from the clock-drawing task in patients with neglect syndrome. Results from seven patients (labeled a–g) with unilateral hemispatial neglect following right hemisphere stroke injury. In this task, the patients are asked to complete a drawing of a clock within a circle drawn on the page. As shown here, patients perform poorly on this task, typically failing to put any numbers into the left side of the clock. Reprinted from Chen & Goedert (2012), *Journal of Neuropsychology*, with permission from John Wiley & Sons, Inc.

doesn't perceive the left half of each line. In Section 4.1.1.2 we return to this interesting pattern to further describe how attention acts at multiple different levels.

Another simple test of neglect is the clock-drawing task. In this task, a patient is given a sheet of paper with only a large circle drawn on it and is asked to use that circle to draw a clock. As shown in Figure 4.1, patients with left-sided hemineglect will typically put all the numbers into just the right half of the clock, ignoring all the space in the left visual field. The patients know what a clock is, and they know all the numbers that it must contain, but their representation of space is severely impaired. Such a patient may start by putting the number 12 at the top of the clock and draw the right side of the clock without a problem, but then pause when they get to the left side of the clock; at this point, they might simply stop and not include the other half of the numbers on the clock at all, or they might try to go back and squeeze the remaining numbers into the same right half of the clock as where they had already placed numbers 12 through 6. What remains consistent is that most of the left side of the clock is not drawn correctly.

4.1.1.1 Lateralization of Attention Processes in Neglect

The most typical pattern of neglect is the one described earlier, in which damage to the right hemisphere of the brain results in the patient neglecting the contralateral (left) region of space (Karnath & Rorden, 2012). Studies of patients with neglect have thus suggested that the right hemisphere of the brain may be more critical than the left hemisphere for processes of attention. However, some studies have reported patients that exhibit neglect of the right visual field following unilateral damage to the left hemisphere (Ogden, 1985; Stone, Halligan, & Greenwood, 1993). If mechanisms of attention were exactly the same in both hemispheres, we might expect leftward and rightward neglect to be equally frequent. However, although there are reports of right hemineglect following unilateral left hemisphere damage, it is much more common to observe left hemineglect following right hemisphere damage. To account for this lateralization, some accounts of attention have posited that the left hemisphere monitors only the right side of space, whereas the right hemisphere represents both the right and left visual fields. According to this theory, the lack of unilateral neglect following left hemisphere damage may be due to the intact right hemisphere still being able to monitor both visual fields. When the right hemisphere is damaged, however, the left visual space is no longer represented well anywhere in the patient's brain, and therefore they neglect stimuli in that side of space. Other theories have posited that, instead of the brain hemispheres differing in how much of space they represent, there is a difference in the orienting bias. According to this view, when the right hemisphere is damaged, the strong bias of the left hemisphere to orient toward the right side of space is no longer balanced by the right hemisphere, causing a hyperorienting to the right that results in the left visual space being largely ignored. The right hemisphere, according to this view, doesn't bias orienting as dominantly in only one direction, and therefore damage to the left hemisphere doesn't have as severe an effect. There are also theories that the reason for the higher prevalence of neglect following right hemisphere damage could be due to the types of attention being tested in most clinical evaluations for neglect (e.g., Kleinman, Newhart, Davis, Heidler-Gary, Gottesman, & Hillis, 2007). As introduced in the next subsection, there are different varieties of neglect, and only some of these may be associated more strongly with the right hemisphere. Relevant to this possible lateralization of attention functions, some theories suggest that the lateralization of language processes predominantly to the left hemisphere may have led to some mechanisms of attention becoming mostly lateralized to the nonlanguage right hemisphere. Whereas some of the mechanisms of attention may be the same in each hemisphere (e.g., the ability to selectively enhance versus inhibit sensory inputs; the ability to move attention endogenously), other processes of attention, such as the ability to disengage attention from its current focus to orient elsewhere, may be more exclusively lateralized to the right hemisphere. This could be especially the case in brain regions such as the posterior superior temporal cortex, which in the left hemisphere supports critical language functions and in the right hemisphere supports attentional reorienting. Another reason for the rarity of neglect being reported in patients with left hemisphere damage could be that many such patients have profound language deficits that overshadow more subtle deficits in attention.

Whereas neuropsychological evidence has suggested at least some degree of lateralization of attention function to the right hemisphere, neuroimaging studies have consistently found a *bilateral* pattern, with both hemispheres producing activity in dorsal regions of the frontoparietal attention control network (e.g., Corbetta & Shulman, 2002). Many of these neuroimaging studies have found no difference between the hemispheres in those dorsal areas, and others have even found stronger activity in the left hemisphere than the right. We will return to this issue in Chapter 7 when we discuss the control of attention with a focus on neuroimaging results. The discrepancy between the neuropsychological findings and the neuroimaging results could be due to a variety of factors. For example, neuroimaging results can show activity in areas that may be only tangentially involved in a task and aren't necessarily critical for performing that task, whereas neuropsychology deficits indicate that a brain region is critical for that task. On the other hand, results from patients could also be influenced by injury-induced compensation or plasticity following brain injury, and therefore these results might not always best reflect how the healthy brain performs these functions. A consistent take-home message from studies of neglect patients, however, is that when there is damage to the attention system some processes of attention can be seen to be lateralized to the right hemisphere within the human brain.

4.1.1.2 Subtypes of Neglect (Space-Based versus Object-Based)

Careful observation of patients suffering from neglect has revealed different subtypes of the disorder. A major distinction is between patients suffering from primarily space-based neglect and patients suffering from primarily object-based neglect. This distinction highlights how our brains attend to and represent the world. Patients with primarily space-based neglect show the typical pattern described earlier in which they ignore everything in their affected visual field – for example, a patient with right hemisphere damage who neglects objects in the left half of space. Such a patient, however, typically perceives and attends to all objects, in their entirety, in their "good" visual field (in this case the right visual field). Patients with object-based neglect, on the other hand, ignore the left half of every object regardless of where it occurs in space. These patients can perceive and attend to objects in both visual fields but ignore the left half of every object. The distinction between these subtypes of neglect is illustrated nicely by having subjects copy line-drawing pictures. As shown in Figure 4.2 (left), a patient with primarily space-based neglect fails to draw the entire left half of the picture. In contrast, a patient with object-based neglect fails to draw the left half of each object but copies all the objects within each hemifield (Figure 4.2 [right]). These subtypes of neglect also show stark differences on the "gap task," a task in which they are given a page full of circles and partial circles (with a gap on either the left or right side). The subjects' task is to draw an X inside all of the partial circles (the ones with a gap anywhere) and to draw a circle around all of the complete circles. Patients with left spatial neglect ignore everything in the left visual field, not making any marks on anything on the left side of the page, but they do quite well on all the items on the right side of the page. In contrast, patients with object-based neglect make marks on objects across the whole page, but they ignore the left half of each object and therefore mistake partial circles for complete circles when the gap is on the left of the object. The comparison of these subtypes of neglect has revealed that different brain regions support space-based versus object-based attention. Patients exhibiting space-based neglect typically have damage to posterior parietal regions, including the intraparietal sulcus (IPS) and inferior parietal lobule (IPL). Patients with object-based neglect, in contrast, typically have damage to more ventrally located brain regions in the temporal lobe.

Figure 4.2 Results from copying task, illustrating space-based versus object-based neglect. Patients are given the line drawing shown at the top of each image and are asked to copy it. The left image shows the result from a patient with space-based neglect, who ignores everything on the left side of the page but draws both sides of the tree presented on the right of the drawing. In contrast, the right image shows the result from a patient with object-based neglect, who can attend to items across the whole page and attempts to draw each one but neglects the left half of each object. Reprinted from Hillis (2006), *Neurobiology of Unilateral Spatial Neglect* with permission from SAGE.

Other authors have used the term **allocentric neglect** to encapsulate two different ways in which an object could be neglected. Allocentric neglect, sometimes called stimulus-centered neglect, refers to the situation in which one side of a currently viewed stimulus is ignored. There are also cases of *object-centered* allocentric neglect in which one side of a *canonical* orientation of an object is ignored. For example, certain stimuli such as words or maps often have a standard orientation, and we tend to align the object to its canonical representation regardless of the alignment of the current stimulus. For such stimuli, there have been some reports of neglect aligning with the canonical orientation of the object instead of the current physical orientation that is being viewed. Allocentric neglect can be contrasted to **egocentric neglect**, which refers to the space-based form of neglect described earlier (e.g., neglecting everything in the left half of space relative to one's position). Relating this back to the topic of hemispheric differences in neglect, Kleinman and colleagues found that whereas right hemisphere damage led to more cases of egocentric neglect than allocentric neglect, the opposite was true for left hemisphere damage. In their study, patients with left hemisphere lesions were much more likely to exhibit allocentric neglect than egocentric neglect (Kleinman, Newhart, Davis, Heidler-Gary, Gottesman, & Hillis, 2007). This again highlights possible differences in the attention functions supported by each hemisphere. As described in Chapter 3, however, a limitation of neuropsychological studies is that no two patients are alike because the deficits result from naturally occurring brain lesions that vary widely in extent and may include lesions to multiple brain areas. Therefore, many patients with neglect may exhibit features of both egocentric and allocentric neglect. Nonetheless, the existence of cases that have distinct and different primary deficits in terms of neglect provides evidence that these types of attentional allocation rely upon at least somewhat different brain regions.

4.1.1.3 Cortical and Subcortical Connections Underlying Neglect

The classic account of the neglect syndrome held that the right posterior parietal lobe was the critical region supporting attention functions and that it was damage to this region that led to the neglect. However, higher-resolution imaging and lesion analyses have revealed that the most critical regions appear to lie ventral to the superior parietal regions originally thought to be critical for neglect. Karnath, Ferber, and Himmelbach (2001) provided evidence that patients who suffer from neglect most often have sustained damage to the superior temporal gyrus (STG). They make the case that while traditional models have associated neglect with the parietal lobe, those findings often confound neglect with hemianopia (a deficit in vision in one visual field). When specifically isolating neglect *without any hemianopia*, these researchers found the STG to be the critical region. In addition, a number of studies have found cases of spatial neglect associated with damage to ventral regions of the frontal lobe (reviewed in Lunven & Bartolomeo, 2017). Overall, lesion studies, as well as electrical stimulation studies, have converged on the finding that ventral portions of the frontal cortex and the parietal cortex extending into posterior portions of the superior temporal lobe are critical for attention (Figure 4.3).

In addition to these cortical areas, subcortical regions may also be involved in these processes of attention. Neglect has been reported in patients whose damage is mostly

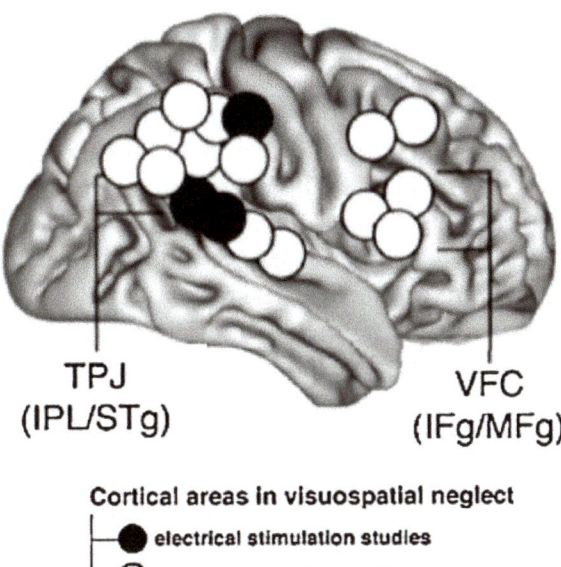

Figure 4.3 Cortical brain regions implicated in hemispatial neglect. Results from several studies of neglect are plotted onto a canonical brain, showing overlap in naturally occurring lesion studies and brain stimulation experiments in which visual-spatial neglect was observed. Two main areas of overlap across studies are in the temporal–parietal junction (TPJ), including the inferior parietal lobule (IPL) and superior temporal gyrus (STg), and the ventral frontal cortex (VFC), including the inferior frontal gyrus (IFg) and the medial frontal gyrus (MFg). Reprinted from Lunven & Bartolomeo (2017), *Annals of Physical and Rehabilitation Medicine*, with permission from Elsevier.

restricted to areas of the putamen and the pulvinar nucleus of the thalamus, and to a lesser degree to the caudate nucleus (reviewed in Karnath, Himmelbach, & Rorden, 2002). Across the many studies of patients with neglect, it is becoming clear that the intact functioning of attention relies on multiple areas across the brain. The connections between these brain regions are therefore critical to healthy functioning. Indeed, research has shown that subcortical regions linked to neglect are strongly connected to the STG, providing evidence for a critical attention network centered on the STG and its links to the pulvinar and putamen. Interestingly, the subcortical lesions associated with neglect have been found to be right lateralized in neglect patients (Karnath et al., 2002), further supporting the right lateralization of some mechanisms of attention. In terms of the importance of the communication between these areas, Lunven and Bartolomeo (2017) have suggested that neglect should be viewed as a disconnection syndrome. They argue that disruptions anywhere along the superior longitudinal fasciculus result in a disconnection between the frontal and parietal areas involved in attentional control and lead to symptoms of neglect. In addition to these connections within a hemisphere, they also suggest that the degree of disconnection *between* the hemispheres may be a critical factor in whether neglect is chronic or is only transiently experienced shortly after the brain injury.

4.1.2 Simultanagnosia

Patients who have suffered from bilateral lesions of the parieto-occipital area may be diagnosed as having **simultanagnosia**, another disorder of attention. Simultanagnosia is sometimes referred to as "Bálint's syndrome" (Bálint, 1909), but the latter includes the additional deficits of oculomotor apraxia (an inability or difficulty in making voluntary saccades to fixate an object) and optic ataxia (an inability or difficulty in manually reaching to and grabbing objects as guided by vision). Simultanagnosia refers specifically to a severe and striking deficit in attention, in which the patient can perceive only one object at a time – the patient gets locked onto a single object and seems unable to attend to anything else. A patient with this disorder will have difficulty disengaging from an attended object in order to perceive anything else in the visual world. In describing his patient, Bálint (1909; Bálint & Harvey, 1995) noted that objects at the center of gaze captured attention so strongly in his patient that they seemed unable to break free from attending to that object. In a complex scene, such patients may only attend to a single item or to only a very few items even after extended viewing, rendering them incapable of describing the gist of the scene as a whole (Duncan, Bundesen, Olson, Humphreys, Ward, Kyllingsbæk, van Raamsdonk, Rorden, & Chavda, 2003). A bedside test of this that can be conducted in the hospital involves the doctor holding up two objects – for example, a comb and a pen – overlapping with each other directly in front of the patient. When the patient is asked to report what they are seeing, they will typically report seeing only one of the two objects, even though they are overlapping in space. Similarly, when presented with line drawings of overlapping objects, the patient will typically identify only one object, even when asked to identify all the items on the page. These patients thus provide compelling evidence that attention can be strongly object-centered, as the patient's attention seems to spread across the attended object but absolutely no further. In addition, these patients provide support for the notion that

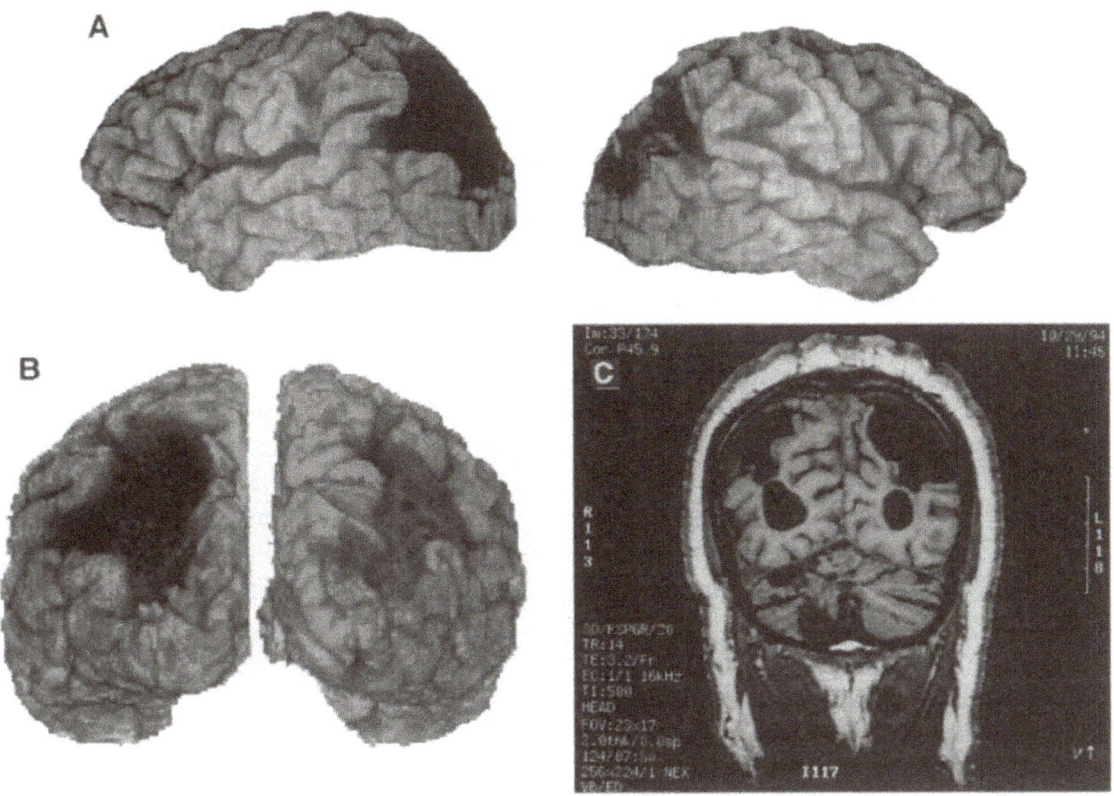

Figure 4.4 Lesion locations of a patient with Bálint's syndrome. (A) and (B) show the extent of the lesion (in dark gray and black) drawn on a 3D reconstruction of the patient's brain created from their structural magnetic resonance imaging (MRI) scan. Note that the primary visual cortex is spared. (C) shows a coronal slice of their MRI, with the lesions in black. Reprinted from Friedman-Hill, et al. (1995), *Science*, with permission from the American Association for the Advancement of Science.

a process of *disengaging* attention from its focus and *reorienting* it to another object is a critical aspect of a fully functioning attention system. When this function is lost, the result is a debilitating hyperfocus on individual objects, sometimes rendering the patient effectively blind to much of the visual world. As noted earlier, this syndrome is only observed in rare cases when there is bilateral damage to parietal–occipital regions in both hemispheres (Figure 4.4).

Luria (1959) described a remarkable patient of this type who couldn't mark the center of a simple cross drawn in the middle of a piece of paper because as soon as their pencil came close to the cross, the patient become so focused on the pencil that they could no longer "see" the cross. This patient also could perceive a Star of David in its entirety when drawn in a single color, but if the two component triangles were drawn in different colors the patient would perceive only one of the two triangles at a time. These results highlight another key aspect of attention: its influence in biasing competition. A number of theories of attention stress that a core component of processing in the brain involves lateral suppression between neural populations coding stimuli competing for resources. Attention is thought to play a key role in biasing this competition, so that the desired stimuli can be processed clearly without undue interference from nearby or related stimuli (e.g., Desimone & Duncan, 1995). Cases of simultanagnosia could be thought of as this system of selection running on overdrive. Without an intact system to allow attention to disengage and explore, the extreme focus and biasing toward the one item continue unchecked. In the case of the overlapping triangles, when the triangles are in the same color they can be viewed as a single object. But when the triangles are in different colors they can be viewed as two objects, and in these patients there is such a strong and prolonged focus on just one of the two objects that they can't "see" the other object overlapping with the object they're attending.

Although it is a rare disorder, some researchers have suggested that there are different subtypes of simultanagnosia. Duncan et al. (2003) provide evidence for distinct effects of lesions to either dorsal areas (i.e., bilateral parietal–occipital regions) or ventral areas (i.e., left occipital regions). Whereas patients with dorsal lesions exhibited the typical and profound deficit of not being able to attend to more than one object at a time, patients with ventral lesions had a less severe restriction of awareness. The patients with ventral damage were able to shift attention across objects somewhat better than patients with dorsal simultanagnosia, but they still had difficulty understanding complex scenes because they could only attend to a few objects, and usually only one at a time. In addition, ventral simultanagnosia patients showed an interesting hyperfocus of attention even with words, as evidenced by their attempts to read by identifying words letter by letter instead of using the typical whole-word analysis employed by healthy readers. In cases of full-blown Balint's syndrome, reading may be severely affected by oculo-motor apraxia, or the inability to move the eyes voluntarily. Of note, the ventral simultanagnosia patients in that study didn't have the typical bilateral pattern of damage, at least according to the neuroimaging scans that were available for those subjects. In neuropsychological studies, the absence of a clear lesion observable in computerized tomography (CT) or magnetic resonance imaging (MRI) is not proof that an area is undamaged – neuroimaging may miss subtle areas of damage that can only be observed in careful postmortem analyses. In the study by Duncan et al. (2003), they also provided evidence that the attentional deficit seen in simultanagnosia goes beyond just the inability to attend to more than one object at a time. They reported on a patient with dorsal simultanagnosia who suffered from impaired processing speed, even within a single-element display. Although the patient could keep their attention on the single item, they were significantly slowed in assessing fundamental aspects of the item. This suggests that even the one object within their awareness suffered from a lack of attentional resources, as it might not have been fully processed. This also relates to the finding that items at the center of gaze are more likely to be the focus of attention in these patients. Objects at the center of gaze undergo stronger visual processing because of the greater processing that occurs within the fovea, and thus they are less likely to be extinguished from consciousness compared to objects in the periphery. Under this view, it may be that a general impairment of visual processing is partially responsible for the signature deficit of seeming to be hyperfocused on just one item. Instead of this simply being due to a broken mechanism of attention, it could be that impaired visual processing results in an even greater disparity of foveal versus peripheral processing, resulting in anything outside the center of gaze being processed so poorly as to have little chance of rising into consciousness. Further research topics in this area involve assessing the extent to which temporal attention (how quickly one can change focus) and spatial mapping (moving attention to the locations of objects) may play a role in the deficits of patients with simultanagnosia.

4.1.3 Progressive Supranuclear Palsy

Neglect patients and patients with simultanagnosia have revealed regions of the cortex that are critical for processes of attention, but studies of other patients have revealed subcortical brain structures that are critical for attention as well. Because strokes to subcortical brain regions can often be fatal, there is relatively little evidence from patients with focal brain lesions in these areas. However, there are degenerative disorders that disproportionately affect subcortical regions, and these can provide important insight into the mechanisms of attention that are supported by these regions.

Progressive supranuclear palsy (PSP) is a progressive, degenerative disease that affects widespread brain areas. PSP is in some ways similar to Parkinson's disease (PD) in that both may involve deterioration in the basal ganglia and both can include symptoms such as loss of balance and slowness of movements. A distinguishing feature of PSP, however, is its effects on eye movements. One of the regions affected by PSP is the *superior colliculus*, and these patients often have difficulty making eye movements, particularly in the vertical direction. Patients may complain of difficulty reading, in part because they have difficulty in moving their eyes downward over the page. This problem with moving the eyes in the vertical direction often extends to upward as well as downward movements. Given the link between eye movements and attention, researchers have been interested in whether these patients may also have deficits in particular functions of attention. Robert Rafal and colleagues provided some of the earliest evidence that damage from PSP affects the orienting of attention (Rafal, Posner, Friedman, Inhoff, & Bernstein, 1988). They compared PSP to PD patients and suggested that PSP patients suffer specifically from a deficit in "moving" attention. As opposed to neglect patients, who show a specific deficit in their ability to *disengage* attention from stimuli in their good field, the main deficit in PSP patients is that they are slow to *move* attention voluntarily, particularly in the vertical direction in which they have problems moving their eyes. In attention cuing tasks in which subjects were asked to keep their eyes fixated on a central spot and simply to detect target stimuli in the periphery that were preceded by cue stimuli at either the same or opposite field location, PSP patients had difficulty shifting attention in the vertical dimension regardless of whether they had been validly or invalidly cued to the correct location. The PSP patients showed much better ability to move attention in the horizontal direction (right or left) than in the vertical direction (up or down), providing evidence that their specific type of eye movement problems were observed in covert attention as well. Given the heterogeneity and widespread damage that can result from PSP, it remains difficult to link specific sites of damage to specific cognitive impairments, but by comparing PSP patients to PD patients Rafal and colleagues attempted to control for some of the common areas affected by these degenerative diseases. More recent work has extended these findings, demonstrating that patients with PSP also show deficits in *visual search* tasks when the targets require attention to be oriented in the affected (vertical) direction (Smith & Archibald, 2019).

4.1.4   Focal Lesions to Subcortical Structures (Superior Colliculus and Thalamus)

As mentioned earlier, naturally occurring lesions in subcortical regions are often fatal, but there have been case studies of patients who survived damage to select subcortical structures. Sapir and colleagues reported on a very rare patient who lost a portion of their superior colliculus in only one hemisphere (Sapir, Soroker, Berger, & Henik, 1999). One of the unique deficits reported in this case study was that the patient didn't generate inhibition of return (IOR) in the affected visual field. IOR, first described by Posner and Cohen (1984), refers to the slowed perception and reaction time to detect stimuli at a location where attention has recently been captured. Specifically, in exogenous/involuntary cuing paradigms, a brief nonpredictive stimulus event (e.g., a flash) at one location in the periphery triggers a rapid orienting of attention to that location. This initial capture of attention results in faster manual response times to detect target stimuli that occur at the location of the flash versus at other locations. This initial capture orients attention rapidly (within 50 msec), but it is short-lived. When subjects know the flash is nonpredictive of target location, there is no voluntary effort to keep attention at the flashed location, and after about 300 ms the subjects move attention back to the central fixation. If the target appears more than about 300 ms after the cue, subjects are significantly slower to respond to stimuli at the location of the previous flash. The interpretation of this result is that after attention had been initially captured at the location of the nonpredictive flash for a brief period, subjects are *inhibited to return* their attention back to that same location. This IOR typically lasts longer then the initial capture period before returning to baseline, and theories have suggested that IOR is an important mechanisms for facilitating foraging behavior, ensuring that the same locations are not repeatedly searched (reviewed in Klein, 2000). In the study by Sapir et al. (1999), the patient exhibited an initial capture of attention followed by IOR in the regions of space represented by their *intact* superior colliculus. However, the patient showed *no* evidence of IOR in the region of space represented by their damaged superior colliculus, despite exhibiting a robust initial capture of attention in that visual field. This study provided unique evidence that IOR is at least partially dependent on processing in the superior colliculus, and furthermore that IOR is a separate mechanism from the initial capture of attention. As discussed in Chapter 3, single-case studies can provide incredibly unique evidence, but because all strokes are unique and individuals can very somewhat in brain activity patterns as well as what compensatory mechanisms may be at play after brain damage occurs, it is important to investigate the issue more widely. Unfortunately, neurostimulation studies in humans cannot provide converging evidence of the role of the superior colliculus due to the difficulty and potential danger of attempting to stimulate subcortical regions that are so close to brain regions critical for maintaining life. There has been some converging evidence from patients with more extensive damage due to degenerative disease: Posner and colleagues found that a group of PSP patients exhibited IOR in the horizontal direction but showed no evidence of IOR in the vertical direction – the direction in which they also showed impairment in moving their eyes (Posner, Rafal, Choate, & Vaughan, 1985). As with the case study of the patient described earlier, these PSP patients showed a typical initial orienting of attention to the peripheral cue, confirming that the initial capture and subsequent IOR are separate mechanisms.

Another rare patient type has provided evidence for a hypothesized "engage" operation of attention. Rafal and Posner studied three patients who had suffered damage to the thalamus due to stroke (Rafal & Posner, 1987). The extent of damage varied across the subjects, but all three patients had damage to the thalamus in only one hemisphere. Therefore, attention could be assessed for both an intact visual field (ipsilateral to the lesion) and an affected visual field (contralateral to the lesion) in each patient. This study used a cuing paradigm, but unlike that described earlier, the cue in this study was *predictive* of the location of the target (80% valid, 20% invalid). Therefore, unlike the studies presented earlier, this study was *not* intended to measure the exogenous attention effect of IOR. The results of this study showed that patients with damage in the thalamus were impaired on this attention task, but the pattern was different than in other patients. As a reminder, PSP patients show impairments in *moving* attention in the affected dimension, and neglect patients have extreme difficulty *disengaging* attention from a cue in their good visual field to attend to a target in their bad field. The patients with damage to the thalamus, in contrast, showed validity effects (faster responses to validly cued location targets versus invalidly cued location targets) of similar magnitude in both the ipsilesional and contralesional visual fields. However, they were much slower overall to respond to any stimulus in the contralesional field. This led to the conclusion that the thalamus is important for *engaging* attention in the contralateral visual field (Rafal & Posner, 1987). These patients showed effects of orienting attention in both visual fields, since the effect of cuing validity was equally strong in both visual fields, but they were delayed in *engaging* attention to the space contralateral to their lesion.

4.2 Deficits in Attention in Developmental Disorders

During the normal course of development over the first months of life, changes in infants' perceptual and motor systems reveal the influence of different neural systems on processes of attention. Although our visual system is functioning at birth, it matures rapidly over the next few months. Careful study of infants' eye movements has revealed that distinct patterns of overt attention coincide with the development of different visual pathways (reviewed in Johnson, 2019). The pathway from the eyes to the superior colliculus is relatively mature at birth, and infants' eye movements in the first few weeks of life appear dominated by reflexive saccades to salient stimuli in the periphery. During the next few months, as connections to cortical areas such as the frontal eye fields develop, an infant's gaze is less controlled by highly salient peripheral stimuli, and they may fixate on interesting objects for longer and engage in movements that are more exploratory than simply stimulus-driven (Bronson, 1994). Related to the process of disengaging attention discussed in earlier sections, the literature on infant gaze research refers to the concept of "sticky fixations." This refers to the findings that infants from 1 to 2 months of age exhibit much longer fixation durations than do younger (newborns) or older (4 months and older) infants, and this has been linked to the development of connections that act to inhibit the superior colliculus (Hood, Murray, King, Hooper, Atkinson, & Braddick, 1996; Johnson, 1990). In addition, sensitivity to visual features such as motion and the ability to execute smooth-pursuit eye movements (i.e., maintaining fixation smoothly on a moving stimulus instead of a series of discrete saccades) come online as connections to higher cortical visual processing areas (such as motion processing areas) develop more fully (Banks & Salapatek, 1983; Johnson, 1990). Overall, such findings support the idea that separate neural systems may underlie the sensory-driven, bottom-up capture of attention versus the more goal-directed, top-down orienting of attention. Higher-order processes of attention, such as selection, inhibition, and executive control, may be linked to the development of connections between and within frontal and parietal structures. In the following subsections, we will discuss disorders in which the attention systems in the brain seem to have developed in an atypical manner, resulting in deficits in one or more processes of attention, starting during childhood.

Before discussing the attention deficits seen in childhood disorders, however, it can be informative to briefly touch upon the other end of the lifespan as well. A recent focus of research has been the investigation of whether different processes of attention may be uniquely affected by healthy aging in older adults. Some studies have provided evidence that the "alerting" aspect of attention may become impaired with normal aging, with the onset of decline starting in the 60–70 years of age range (e.g., Mahoney, Verghese, Goldin, Lipton, & Holtzer, 2010). Regarding executive attention, or the ability to inhibit prepotent responses and filter out irrelevant stimuli, some studies have found that this function is relatively well preserved in the earlier stages of aging but may decline in more advanced age (into the 80s; e.g., Lu, Fung, Chan, & Lam, 2016). Research on the effects of aging on the functioning of attention has not been entirely consistent, however, as other studies have found that older adults' measures of alerting, orienting, and executive attention are all within the typical range of younger adults (Gamble, Howard, & Howard, 2014; Young-Bernier, Tanguay, Tremblay, & Davidson, 2015). A few important factors must be considered whenever assessing age-related differences in cognitive performance (reviewed in McDonough, 2019). The overall slowing of response times in older adults can result in some measures appearing different compared to those of younger adults, but if the overall slowing of manual responses is accounted for in the analyses, attention effects are sometimes found to be quite similar across the groups. Another factor to consider when interpreting any differences between older and younger adults is that some studies of "normal aging" may unintentionally include a percentage of individuals who are in preclinical stages of dementia, which could confound the ability to assess the effects of healthy aging on attention. The investigation of the precise effects of normal aging on the different processes and mechanisms of attention represents an important direction for future research.

4.2.1   Attention Deficit Hyperactivity Disorder

The developmental disorder in children that is most closely associated with attention is, of course, the one that has "attention" in its name: ADHD, sometimes referred to as attention deficit disorder (ADD). Although the attention deficits observed in individuals with ADHD don't perfectly match up with all of the different processes of attention proposed by philosophers and cognitive scientists, it is informative to consider how the deficits in this disorder can inform theories of attention, and how results from basic attention research may improve treatment efficacy for those with ADHD.

The history of doctors making note of patients with attention-related issues dates back at least to the 1700s. According to Barkley and Peters' (2012) investigation, the earliest description in the medical literature of a disorder that we would now label as ADHD was made in 1775, when physician Melchior Adam Weikard described a pattern of behaviors he observed in his *inattentive* patients, including distractibility, impatience, and shallow study habits. The Scottish physician Alexander Crichton formally described a *disease of attention* in 1798, and he further noted that the attention problems became apparent at an early age (Crichton, 1798/ 2008). Over 100 years later, the pediatrician George Still published a lecture about children who show a "quite abnormal incapacity for sustained attention," and he noted that there may be hereditary links (Still, 1902). In 1932, the German physicians Franz Kramer and Hans Pollnow reported cases of children who couldn't stay still, were easily distracted, and were unable to concentrate (described in Lange, Reichl, Lange, Tucha, & Tucha, 2010). Such reports, over many decades, led to the inclusion of a disorder termed "minimal brain dysfunction" (MBD) in the first version of the Diagnostic and Statistical Manual of Mental Disorders (DSM-1) in 1952. The inclusion criteria for MBD included many of the symptoms that today would be associated with ADHD and specified that it did not affect overall intelligence but rather was specific to deficits in impulse control and ability to focus attention. In the second version of the DSM (DSM-2), published in 1968, MBD was replaced by a disorder called "hyperkinetic reaction of childhood." This new disorder highlighted the hyperactivity aspect of ADHD and made it explicitly a childhood disorder. It later became clear, however, that many children with deficits in attention did not exhibit hyperactivity. This in turn led to the DSM-3, in 1980, replacing that disorder with "attention deficit disorder" (ADD), which could be with or without hyperactivity. In the revised version of DSM-3 (DSM3-R), published in 1987, ADD was replaced with "attention deficit/hyperactivity disorder" (ADHD). In 1994, the DSM-4 formally defined three subtypes of ADHD: mainly inattentive; mainly hyperactive; and combined. In the most recent version of the DSM (DSM-5), three subtypes of ADHD are identified: (1) inattentive; (2) hyperactive-impulsive; and (3) combined. DSM-5 also expanded the diagnosis to *adults*, no longer restricting the disorder to just children. Table 4.1 lists the symptoms for the inattentive and hyperactive-impulsive subtypes, according to the US Centers for Disease Control and Prevention (CDC) website. The DSM-5 criteria for the *inattentive* subtype are the presence of six or more symptoms from the inattentive list, present for at least 6 months. The criteria for the *hyperactivity* subtype are six or more symptoms from the hyperactivity and impulsivity list, present for at least 6 months. The *combined* subtype is diagnosed for children with six or more symptoms from both lists. Note that those criteria are for children up to 16 years old; for those aged 17 years or older and for adults, the criteria stipulate five or more symptoms. There are additional conditions that must be met before a diagnosis can be made, such as (but not limited to): several symptoms being present in two or more settings; several symptoms being present before age 12; evidence that the symptoms interfere with or reduce the quality of social, school, or work functioning; and that symptoms are not better explained by another mental disorder.

Table 4.1 **ADHD symptom criteria. Abbreviated descriptions of symptoms; full descriptions are in the DSM-5. Table adapted from www.cdc.gov/ncbddd/adhd/diagnosis.html.**

Inattention symptoms	Hyperactivity and impulsivity symptoms
Is often easily distracted	Often fidgets with or taps hands or feet
Is often forgetful in daily activities	Often leaves seat when should remain seated
Often has trouble holding attention on tasks	Often talks excessively
Often does not seem to listen when spoken to	Often has trouble waiting their turn
Often has trouble organizing tasks	Often interrupts or intrudes on others
Often loses things necessary for tasks	Is often "on the go," acting as if "driven by a motor"
Often avoids or dislikes tasks that require sustained mental effort	Often unable to play or take part in leisure activities quietly
Often does not follow through on instructions and fails to finish tasks	Often runs about or climbs in situations when it is not appropriate
Often fails to pay close attention to details or make careless mistakes in schoolwork	Often blurts out an answer before a question has been completed

There have been conflicting reports in the literature, across a range of attention tasks, regarding whether ADHD individuals show impairment in selective attention tasks (Carter, Krener, Chaderjian, Northcutt, & Wolfe, 1995; Mason, Humphreys, & Kent, 2003) or not (Dalebout, Nelson, Hletko, & Frentheway, 1991; Hooks, Milich, & Pugzles Lorch, 1994). Several studies have used the attention network task (ANT) to investigate multiple processes of attention, including alerting, orienting, and executive control. Across studies using the ANT, there has been a consistent finding that executive control is impaired in ADHD relative to age-matched controls (e.g., Johnson, Robertson, Barry, Mulligan, Dáibhis, Daly, Watchorn, Gill, & Bellgrove, 2008; Konrad, Neufang, Hanisch, Fink, & Herpertz-Dahlmann, 2006; Oberlin, Alford, & Marrocco, 2005). The executive control measure in the ANT is calculated as the difference between trials with response-incompatible flanking elements around the central target versus trials with response-compatible flanking elements. Although it is a relatively simple display of elements and responses, this measure appears to tap into processes of inhibition and cognitive control in the real world that are deficient in some individuals with ADHD. Indeed, there has been consistency across tasks in finding that the specific process of being able to *filter out* irrelevant distracting information is impaired in ADHD relative to control children (Carter et al., 1995; Mason et al., 2003; Prior, Sanson, Freethy, & Geffen, 1985). There have been inconsistent findings, however, regarding whether ADHD individuals show a deficit in the *alerting* mechanism of attention. The ANT measures alerting as the enhancement of responses on trials when subjects are given a warning cue before the target appears compared to trials with no warning cue. Despite some earlier studies finding no differences in this alerting score, a subsequent large-sample study of adolescents with ADHD found a significant deficit in this network as well (Johnson et al., 2008). Regarding the orienting measure, studies consistently report no differences between ADHD and control groups on this measure in the ANT. Specifically, both groups show a similar ability to *move* attention and detect target stimuli faster after a spatially directive cue compared to trials when the target was preceded by only a warning cue that didn't indicate the location of the upcoming target. This suggests that the basic orienting mechanism is intact in individuals with ADHD. However, the standard ANT paradigm only includes 100% valid cues. In studies that have used variations of the ANT that also include a proportion of invalid-cue trials (in which the cue is wrong in its prediction of the target location), results reveal significant differences between ADHD and control subjects on invalid trials. Those differences are sometimes measured as worse behavioral performance for invalidly cued targets (Swanson, Posner, Potkin, Bonforte, Youpa, Fiore, Cantwell, & Crinella, 1991) or, in functional MRI (fMRI) studies, by increased activity in frontal regions, which suggests more effort was needed to complete the task on invalid trials (Konrad et al., 2006). Such results provide evidence that although individuals with ADHD may be able to *orient* attention well, they have more difficulty than do control subjects in *disengaging* their attention from an invalidly cued location in order to move it rapidly to the location of the target.

In addition to selective attention processes, studies of ADHD have reported deficits in other mechanisms of attention as well, including the maintenance of attention (Egeland, Johansen, & Ueland, 2009), temporal perception processes (Toplak, Dockstader, & Tannock, 2006), and response selection and inhibition (Barkley, 1997). Deficits in response-related mechanisms have been linked specifically to the *precision* of responses in both the spatial and temporal domains. Specifically, studies have found that, compared to non-ADHD controls, individuals with ADHD exhibit more variability in their response times to detect targets in attention tasks (reviewed in Kofler, Rapport, Sarver, Raiker, Orban, Friedman, & Kolomeyer, 2013), possibly due to a less consistent ability to shift attention at the required time. In addition, other studies have found more variability in ADHD in the execution of motor movements to spatial locations (Papadopoulos, Rinehart, Bradshaw, Taffe, & McGinley, 2015), and the ability to achieve high spatial precision has been linked to prefrontal cortices (Stuss, Murphy, Binns, & Alexander, 2003). Together, such results suggest that some problems for ADHD individuals could arise because of a relative deficit in the precision of when and where attention is engaged. Finally, connections with another prefrontal region, the anterior cingulate cortex (ACC), have been implicated in the interaction of motor execution and reward processing (Hayden & Platt, 2010), with some theories suggesting that ADHD could be viewed as a disorder of motivation and reward (Luman, Oosterlaan, & Sergeant, 2005).

Finally, other lines of research have suggested that it may be the *balance* between top-down and bottom-up attention that is critical in many of the difficulties experienced by children with ADHD (Mueller, Hong, Shepard, & Moore, 2017). Specifically, a reduced ability to endogenously focus attention on the goals of a task combined with an unusually strong exogenous orienting to irrelevant salient sensory events could result in attention being consistently pulled away from what should be attended in order to complete a task efficiently. This also relates to some of the neurochemical systems implicated in ADHD. Dopamine signaling has been linked to top-down attention, and abnormal dopamine functioning has been linked to ADHD (Swanson, Kinsbourne, Nigg, Lanphear, Stefanatos, Volkow, Taylor, Casey, Castellanos, & Wadhwa, 2007). However, not all studies find deficits in the dopamine system (Gonon, 2009), and other research suggests that the influence of the cholinergic system may be more critical, as it has been linked to bottom-up attention processes (Knudsen, 2011). Additionally, the catecholamine system has also been linked to ADHD symptoms (Arnsten, 2006). Overall, there is unlikely to be a single neurochemical treatment for ADHD given the heterogeneity of subtypes of ADHD and the multiple different brain systems that are involved in the mechanisms of attention and the range of ADHD symptoms. However, research into the different systems of the brain that support these functions of attention holds promise for improving our understanding and treatment of ADHD in the future. Researchers have also investigated potential sex differences in ADHD to try to better understand the disorder (Box 4.1).

4.2.2 Autism

Another developmental disorder that has sometimes been linked to atypical attention functioning is autism spectrum disorder (ASD). A full description of ASD is beyond the scope of this book, so this subsection will focus only on research related to the functioning of attention processes in individuals with ASD. One of the more well-studied aspects of ASD concerns the processing of social information, and experiments have revealed that both children and adults with ASD exhibit atypical orienting toward faces and social-emotional stimuli. In studies examining patterns of eye movements across pictures of scenes, individuals with ASD have sometimes been found to fixate on faces less often than do control subjects (Klin, Jones, Schultz, Volkmar, & Cohen, 2002; Pierce, Conant, Hazin, Stoner, & Desmond, 2011). Similarly, it has been found that individuals with ASD attend less to emotional faces and socially relevant stimuli (Dawson, Meltzoff, Osterling, Rinaldi, & Brown, 1998; Kliemann, Dziobek, Hatri, Baudewig, & Heekeren, 2012). Researchers have used the ANT to investigate whether this atypical orienting is specific to social stimuli or whether individuals with ASD may have dysfunction in core mechanisms of attention. These studies tend to show that the orienting process is most impaired in individuals with ASD, even for nonsocial stimuli, while alerting and executive control processes are relatively less impaired or even are found to be intact in some studies (e.g., Keehn, Lincoln, Müller, & Townsend, 2010). In other types of attention paradigms in which visual search is tested, studies have found that individuals with ASD show *enhanced* abilities compared to control subjects in finding and detecting simple physical features (Joseph, Keehn, Connolly, Wolfe, & Horowitz, 2009). This has led to theories proposing that there may be enhanced bottom-up attention mechanisms, driven by physical salience, that could lead individuals with ASD to show a preference for salient physical features over social stimuli, possibly resulting in impaired learning of social skills (Amso, Haas, Tenenbaum, Markant, & Sheinkopf, 2014). In an fMRI study using only nonsocial stimuli, Murphy and colleagues (Murphy, Norr, Strang, Kenworthy, Gaillard, & Vaidya, 2017) found that ASD and control subjects showed activity in temporoparietal and frontal attentional control regions during a simple attention task; however, the enhanced activity was in different *conditions* for the two groups. Control subjects showed greater activity in these regions when the stimuli contained a highly salient visual item that would trigger an orienting response (e.g., when one of the stimuli was a complex colorful pattern as opposed to simple solid-colored squares). However, the individuals with ASD showed activity in those same brain regions even for displays without a highly salient, uniquely patterned stimulus (i.e., when the display only contained squares that were of the same solid color); control subjects did not show enhanced

Box 4.1 Sex differences in attention deficits?

A controversial topic in psychology is whether there are sex differences in cognitive and mental processes. Authors who have argued that it is important to take sex differences into account when trying to understand cognitive processes point to a number of studies that have found evidence for differences between males and females (reviewed in Cahill, 2006; Spets & Slotnick, 2021). Other authors, however, argue that there are no reliable sex differences (Haut & Barch, 2006; Joel, 2011) and suggest that the studies reporting differences could be due to small sample sizes, confounds in their designs or analyses, chance occurrence, or publication bias (David, Naudet, Laude, Radua, Fusar-Poli, Chu, Steganick, & Ioannidis, 2018). While more work will surely be done to examine sex differences in cognitive processes in neurotypical populations, it may also be interesting to consider such differences found in developmental disorders that affect mechanisms of attention. According to a 2019 analysis from the CDC (www.cdc.gov/nchs/fastats/adhd.htm), 8.8% of children aged 3–17 had at some point been diagnosed with ADHD (or ADD). Critically, the percentage of young males diagnosed (11.7%) was substantially higher than the percentage of young females diagnosed (5.7%). Previous analyses in the literature have suggested an even larger sex difference, with Willcutt (2012) finding an approximately 3:1 male:female ratio in community-based samples. In addition to sex differences in the diagnosis of ADHD, researchers have also debated whether there are differences in symptom severity between the sexes. There are, of course, multiple potential reasons as to why such sex differences might be found. Selection bias, missing symptoms, and a lack of measurement invariance are just a few of the ways in which sex differences could be observed in incidence rates even if no real differences are present (Arnett, Pennington, Willcutt, DeFries, & Olson, 2015). "Selection bias" is particularly important to consider when examining the severity of symptoms, as it could be the case that those with more severe symptoms would be more likely to be referred to clinical studies (Gershon, 2002). "Missing symptoms" refers to the potential that the standard criteria for diagnosis of ADHD may include symptoms most likely to occur in males whereas symptoms that are more likely to occur in females might not be part of the standard criteria. "Lack of measurement invariance" refers to a related concept that the rating scales may not be psychometrically equivalent for males and females; this could be due to an overdiagnosis of males as having ADHD in early studies, leading to diagnostic criteria being produced that are indicative of those samples. However, despite these potential explanations for why a sex difference might be observed even if none existed, some research that has tried to control for these factors has nevertheless found such sex differences (as reviewed in Arnett et al., 2015). In regards to the other developmental and clinical disorders covered in this chapter, it is noteworthy that there have been recent studies reporting neurobiological sex differences in dyslexia (Krafnick & Evans, 2019), autism (Werling & Geschwind, 2013), and anxiety disorders (Bangasser & Cuarenta, 2021), with there being some links to deficits in attention functions in these disorders. Future studies will need to specifically investigate the attentional deficits within those groups to better determine whether there are significant sex differences in those attention mechanisms as well. Much work remains to be done before firm conclusions can be drawn, but the investigation of sex differences could be important for improving our understanding of the disorders and the neurobiological and genetic factors that underlie them and for tailoring treatments to individuals.

activity in attentional control regions during this condition. This has been taken as evidence that individuals with ASD may be overly influenced by unusually strong bottom-up processing, with their attention being biased by the physical features in the environment, even when only relatively low-salience stimuli are present. Regarding the alerting and arousal systems, several studies have reported *elevated* levels of arousal and alerting during attention tasks in ASD compared to control subjects (Geva, Zivan, Warsha, & Olchik, 2013; Keehn, Müller, & Townsend, 2013; Liss, Saulnier, Fein, & Kinsbourne, 2006). This has led to the idea that individuals with ASD may experience a relative hyperarousal to even relatively nonsalient stimuli, and some studies have found this to be especially apparent for stimuli in the periphery that could trigger exogenous attention (e.g., Zivan, Morag, Yarmolovsky, & Geva, 2021). This may also relate to findings of hyperorienting in some studies of ASD (Greene, Colich, Iacoboni, Zaidel, Bookheimer, & Dapretto, 2011), which has been suggested to account for poor executive decisions about what stimuli should receive attention (Mutreja, Craig, & O'Boyle, 2016). Key regions supporting executive attention/decision/control processes in the brain include the ACC, prefrontal cortex, and midfrontal regions. These regions have been found to be affected in ASD, as children with ASD have been reported to have enlarged frontal lobes (Carper & Courchesne, 2005) and hyperactivity in the ACC (May & Kana, 2020) relative to typically developing children. It has been suggested that these effects may be due to inefficient pruning of frontal lobe connectivity during early development.

It has been proposed that a critical cause of dysfunction in ASD may be the connectivity and interactions between the different networks of attention (Sabag & Geva, 2022). Within the dorsal attention network, critical for the orienting of attention, some research has found that the connections between visual processing regions and the frontal eye fields may be enhanced in ASD, whereas connectivity between those visual regions and the IPS is reduced (e.g., Fitzgerald, Johnson, Kehoe, Bokde, Garavan, & Gallagher, 2015). A more consistent finding across the literature, however, is atypical connectivity with the ventral attention network, which is involved more in the disengagement and reorienting of attention. Studies of resting-state activity (i.e., fMRI scanning with no task) have revealed hyperconnectivity between the temporal–parietal junction (TPJ) and ventral frontal cortex (Chien, Lin, Lai, Gau, & Tseng, 2015; Farrant & Uddin, 2015). It has also been noted that parts of the TPJ are involved in processing social information (Blakemore, 2008), and therefore altered connectivity between the TPJ and frontal parts of the mirror neuron system could relate to diminished social orienting in individuals with ASD (Chan & Han, 2020; Fitzgerald et al., 2015).

Sabag and Geva (2022) suggest that a key interaction in ASD involves the connectivity between the executive control network and the ventral orienting network. In typical development, these two networks have a moderating influence on each other (Figure 4.5A). In ASD, however, the executive control network has been found to have less of a moderating influence over ventral attention network areas (Fitzgerald et al., 2015). These networks are more strongly and positively connected in ASD (Figure 4.5B), possibly due to incomplete neural pruning during development. This could result in atypical patterns of attention orienting in ASD, as executive control regions are less able to inhibit the reorienting of attention to even mildly salient stimuli in the environment, even in the presence of social stimuli that may be deemed

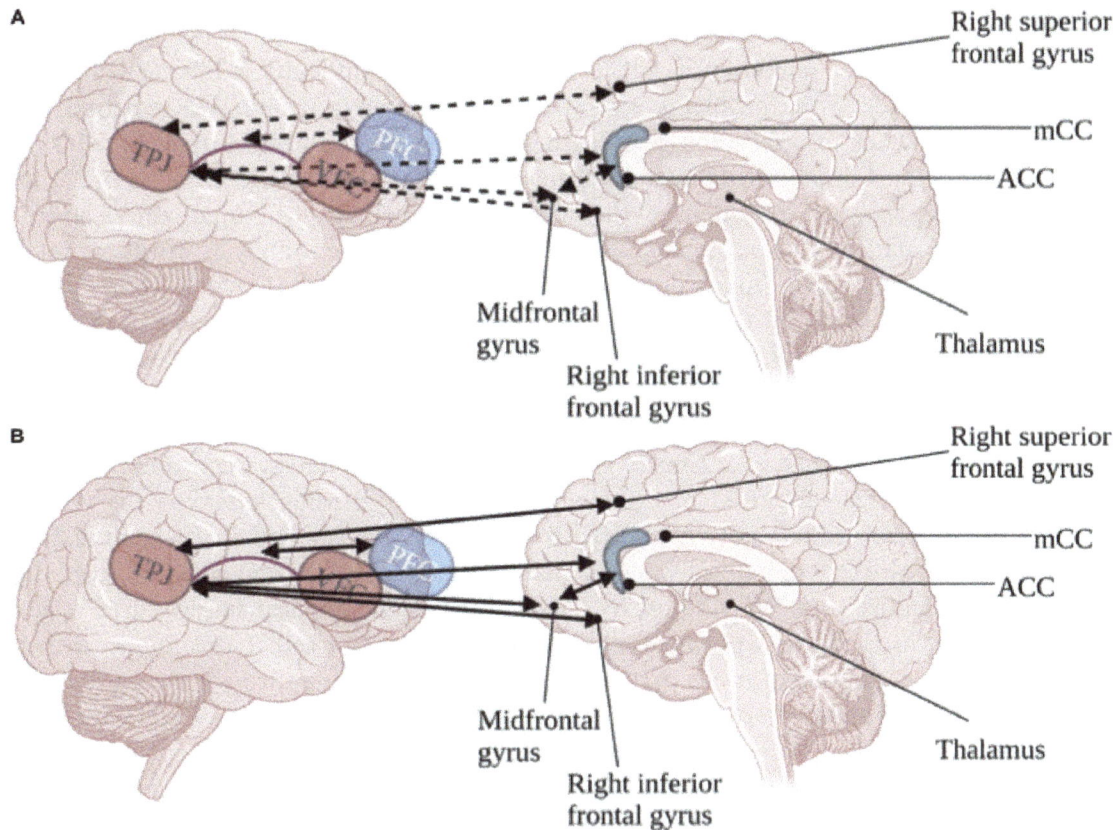

Figure 4.5 Model of attention network connectivity in ASD. Models of the pattern of connectivity in (A) typical development and (B) ASD. Areas in the executive attention network are shown in blue and areas of the ventral orienting network are shown in red. Solid lines represent positive correlations; negative correlations are shown with dashed lines. Note that a key difference between the groups is that the typical inhibitory influences of these networks on each other (shown in A) are replaced in the ASD group with facilitatory effects. ACC = anterior cingulate cortex; mCC = medial cingulate cortex; PFC = prefrontal cortex; TPJ = temporal–parietal junction; VFC = ventral frontal cortex. Reprinted from Sabag & Geva (2022), *Frontiers in Human Neuroscience*, Creative Commons CC BY 4.0.

more important by non-ASD individuals. The authors also suggest that enhanced activity in arousal networks, triggered by stimuli that may not be salient enough to trigger attention orienting in typical development, and the atypical connectivity between the arousal and dorsal attention networks may result in inefficient and relatively less selective engagement of the orienting response. Finally, the progression of ASD over development is an active area of research, and some studies have provided evidence that some of the patterns of hyperconnectivity between attention networks observed in children with ASD become patterns of hypoconnectivity in adults with ASD (Farrant & Uddin, 2015). Further research is needed to better understand how the interactions between attention networks and the changes in these interactions over the lifespan contribute to typical and atypical attention-related behavior.

4.2.3 Dyslexia

Dyslexia refers to a condition in which individuals have difficulty reading. A diagnosis of dyslexia is made if the problems with reading cannot be attributed to impaired general vision or to a lack of opportunities to learn to read. Most accounts of dyslexia focus on the language-related impairments in translating the written word to the meaning of the word, and techniques used to improve the impairment include structured and step-by-step methods of matching letters to sounds, learning to recognize sounds in words, or integrating multiple senses (e.g., touch or sound) when learning to read. However, other work has suggested that some of the difficulties in dyslexia may be due to impaired attention (Vidyasagar & Pammer, 2010). Shallice and Warrington (1977) described two patients who developed a rare form of dyslexia following damage to the parietal lobe. The patients could read isolated letters well but had difficulty when a letter was surrounded by other letters. Note that this is a similar task to the Eriksen flanker task and the "flanker interference" effect discussed in Chapter 1, but here it is using letters as the target and flankers. Critically, the ability to identify the target letters was impaired when the flanking stimuli were also letters but not when the flankers were of a different category (e.g., digits). This led Shallice and Warrington to conclude that the impairment was not a general impairment in vision or a problem with linguistic access but rather was an impairment in the *selective processing* of just a part of the information coming through the senses – a core process of attention. Further, because the deficit in Shallice and Warrington's study occurred only when there were competing letter stimuli in very close spatial proximity, the authors labeled this rare form of dyslexia "attentional dyslexia." Patients with this rare form of acquired dyslexia may also show "letter migration," in which flanking letters (sometimes from within nearby words) are inserted into the word being read (Saffran & Coslett, 1996). Such letter migration can also be observed in typical readers under more extreme experimental parameters, such as when words are presented for only very brief durations and are quickly masked by other stimuli (Mozer, 1983; Treisman & Souther, 1986). Together, these results point to the critical importance of attention during reading (Davis & Coltheart, 2002). When the selection mechanisms of attention fail to filter out competing stimuli, reading becomes much more difficult, as letters from nearby words are being processed to the same extent as the letters of the words that the individual is trying to read.

Two competing accounts have been offered for the role of attention in reading deficits. One account suggests that attention acts early, before word identification takes place, to focus the attentional window on just the letters or word being identified. According to this model, attention helps to suppress the competing information from nearby (in space and time) stimuli. When this suppression cannot take place, the letters or words are processed to a higher level, where they interfere with correct identification. Another model suggests that attention acts at a later, postperceptual level. According to this model, all stimuli may proceed to the level of perceptual identification, but attention then selects the information coming from the intended location and inhibits any further processing of nearby stimuli that could interfere. In a case study reported by Mayall and Humphreys (2002), a patient with attentional dyslexia was presented with a pair of words with normal spacing between the words, such as:

bit led

On over 50% of trials using this spacing with words that contained letters that could be rearranged into different words (e.g., "bed" and "lit" in this example), the individual reported the rearranged words instead of what was actually presented to them. However, if the words were separated with more space between them, such as:

bit   led

the individual performed significantly better – they reported rearranged words on only 39% of trials when there was more space between the words. Therefore, *spatial* attention may be playing an important role. This patten of results suggests that the deficit in reading in some individuals may be due, at least in part, to the inability to restrict the attentional window to just the space occupied by one word. Without an ability to restrict the attentional window to a small set of letters, more letters are processed, leading to confusion about the letters constituting a given word. Another intriguing finding from this study was that individuals also performed better if the constituent letters of a word could be easily segregated from the letters of a nearby word by other features. For example, they performed significantly better if the case of the two words differed, with one being entirely in lower case and the other in upper case, such as:

bit LED

It should be noted that the participants were given explicit instruction to report only the word in lower case (or, in other trials, only the word in upper case). Under those conditions, migration errors were significantly reduced, suggesting that after letters pass through the attentional window other features can still be used to reject mismatched combinations. For example, a mismatched word such as "bed" may be less likely to be created because its constituent letters would be a mix of upper- and lower-case fonts (e.g., bED).

Although these accounts are of acquired dyslexia occurring in previously typical readers after suffering damage to the brain, these findings have led to the proposal that at least some forms of developmental dyslexia may also be due to a deficit in attention (Vidyasagar & Pammer, 2010). The theory that attention may be a factor in the development of dyslexia relates to how a high-fidelity mechanism of attention may be critical to the act of reading. As discussed previously, attention is strongly linked to eye movements. Although covert shifts of attention show that attention is not simply tied to where we're looking, in most situations in our daily lives attention goes with the eyes. When viewing a scene, the frequency, speed, distance, and number of eye movements can vary widely depending on the goals of the observer (Yarbus, 1967), but in general initial saccade lengths are rather large in order to rapidly assess the full extent of the environment, and fixation durations may be rather lengthy depending on the salience of the objects. In contrast, reading requires a much more rapid series of eye movements and requires the fixations to occur much closer together than when viewing other things in our environment. If a child is having difficulty with this level of fidelity in rapidly moving and focusing attention at locations in very close proximity, the ability to read at a typical pace may be impaired. In turn, difficulty keeping pace with age-appropriate reading materials may lead to frustration and even less time spent reading, or to the development of coping strategies (e.g., guessing based on context instead of sounding out new words). The idea that problems with attention might contribute to dyslexia has led to the development of different training techniques to try to ameliorate the disorder. For example, Lorusso and colleagues trained children to make rapid attention shifts into the periphery, specifically to the right or left visual field (Lorusso, Facoetti, Paganoni, Pezzani, & Molteni, 2006). They were able to demonstrate that, compared to a control group of students, the subjects who took part in the training significantly improved their reading accuracy and reading speed. While much further work needs to be done, this finding suggests the possibility that attention training could potentially help some children with dyslexia.

The possibility that attention plays a core role in what is often thought of as a purely language-related disorder highlights how fundamental attention can be to the development of other aspects of cognition. From perception to consciousness to memory to language, the intact functioning of attention systems in the brain is essential for many other cognitive processes. In the case of dyslexia, this highlights why it is important for us to understand how attention is oriented, the speed with which it is deployed, and the limits of its spatial precision. The neural mechanisms supporting those different aspects of attention will be covered in more detail in the following chapters.

4.3   Deficits in Attention in Clinical Disorders

4.3.1   Anxiety Disorders

Studies have found that in various clinical disorders attention may be involuntarily captured by stimuli that are relevant to the particular disorder (e.g., pictures of drug paraphernalia capturing the attention of individuals with substance use disorder). Mogg and Bradley (2018) reviewed several studies suggesting that people suffering from *anxiety disorders* may have an atypical pull on their attention toward threatening stimuli. Some theories suggest that this bias to orient attention to threat-related stimuli may cause even more heightened levels of anxiety due to this continued involuntary focus on threatening stimuli.

One task that has been used to test how attention may be captured by particular types of task-irrelevant images is the "dot-probe paradigm" (Figure 4.6). In this paradigm, the subject must quickly detect a small dot presented very briefly on one side of the computer screen. What's critical here is that the dot to be detected is immediately preceded by two pictures, one at each of the two locations where the dot could appear (typically, one location to the left of fixation and the other to the right). The location of the target dot is completely randomized in terms of the side of the display and the type of picture it will appear behind. For example, the target may occur equally often at the location of a threatening stimulus as at the location of a nonthreatening stimulus, so subjects have no reason to orient to either picture voluntarily. A number of studies using this type of paradigm have found that anxious individuals are faster at detecting the dot when it occurs at the location where a threatening face just occurred

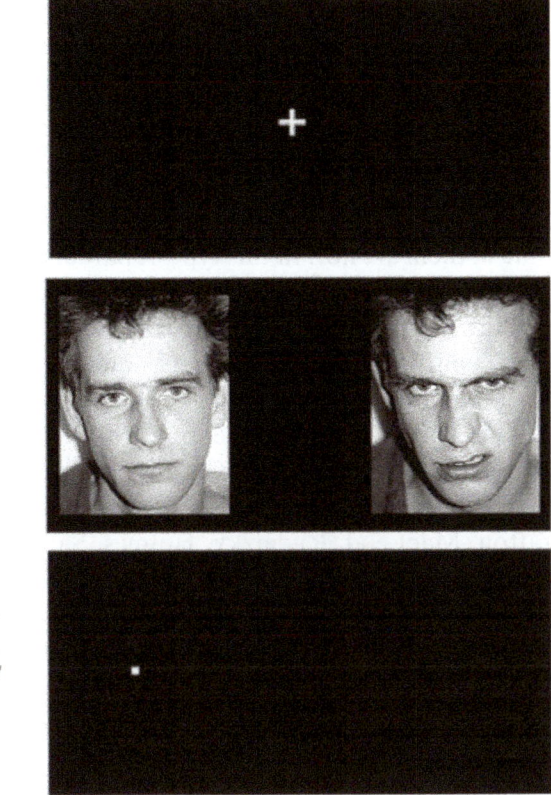

Figure 4.6 Dot-probe task with threatening versus neutral stimuli. A single trial in a dot-probe task is illustrated here. The trial begins with only a fixation cross on the screen (shown at top). A short time later, two pictures appear very briefly and simultaneously in the periphery – one to the left, the other to the right. These stimuli are completely irrelevant to the subject's task and are uninformative regarding where the target will occur. When the pictures disappear, the target – a small dot – appears briefly at one location and requires a speeded manual button press. If subjects are faster when the dot appears behind one particular type of stimulus (e.g., a threatening face), this provides evidence that their attention was involuntarily captured by that stimulus. Reprinted from Mogg & Bradley (2018), *Trends in Cognitive Sciences*, with permission from Elsevier.

compared to the location of a neutral face, suggesting that these individuals are orienting toward the threatening stimulus (reviewed in Bantin, Stevens, Gerlach, & Hermann, 2016). However, other studies have found no evidence for a capture of attention toward the threating stimuli, and yet others find that individuals behave in a "threat-avoidant" manner, essentially suppressing orienting to threatening faces relative to neutral objects (Chen, Ehlers, Clark, & Mansell, 2002; Mansell, Clark, Ehlers, & Chen, 1999). According to theories that suggest that attention bias to threat *causes* further anxiety, a proposed treatment for anxiety disorders is to train anxious individuals to better avoid attending to the threatening stimuli. This has led to the adoption of attention bias modification (ABM) training, in which anxious individuals are trained to explicitly direct their attention away from threat. ABM training can utilize the type of dot-probe paradigm described earlier, but such training would have the dot occur only rarely behind the threatening stimulus; in most trials the target dot would occur behind the picture that was *nonthreatening*. In this way, the ABM training would give subjects repeated practice at attending away from the threatening stimulus because the target usually occurs at the opposite location. This repeated practice and the implicit reward of doing well on the task is thought to reinforce the orienting of attention *away* from the threat.

Another type of ABM training uses simple visual search tasks, where some of the stimuli are threatening and some of them are not. In this type of paradigm, the subject might see a screen full of a few different pictures, some of which are threatening and a few of which are either positive, neutral, or calm. Participants are told specifically to look for the positive pictures (or calm pictures in other conditions) and to respond as quickly as possible when they find those pictures. The point of this training is to try to get these individuals to specifically search out and find positive stimuli and to suppress orienting toward the threatening stimuli. There is evidence that these types of ABM training may be effective in anxious individuals, as some studies have shown a reduction in anxiety as well as in attentional bias toward threat in these types of tasks (Grafton, MacLeod, Rudaizky, Holmes, Salemink, Fox, & Notebaert, 2017; MacLeod & Clarke, 2015). However, other studies have failed to find a correlation between a change in attentional bias following attentional bias training and a reduction in anxiety symptoms (Mogg, Waters, & Bradley, 2017). A potential confounding factor is that many of the assessments of attentional bias use the exact same tasks that were used in the attention bias training. Therefore, it's possible that the improvements in the attentional bias measures might be very task specific and may not translate well beyond that task.

Further work is needed to understand in what situations – or for which individuals – the capture by threatening stimuli occurs (reviewed in Bantin, Stevens, Gerlach, & Hermann, 2016). There's growing evidence that anxious individuals do not consistently show attentional bias toward threat (Van Bockstaele, Verschuere, Tibboel, De Houwer, Crombez, & Koster, 2014; Waters, Bradley, & Mogg, 2014). This has led to some models of anxiety being proposed that suggest that cognitive processes beyond simply the orienting of attention need to be considered, such as how bottom-up, stimulus-driven orienting interacts with top-down executive attention and cognitive control processes (e.g., Yiend, Mathews, Burns, Dutton, Fernández-Martín, Georgiou, Luckie, Rose, Russo, & Fox, 2015). Some of these executive control processes include the allocation of resources to evaluating the stimulus, the balancing of goals, and working memory. For example, if threatening stimuli remain active in working memory for a long period of time, those stimuli are more likely to influence the person's attention and affect.

This sort of attentional bias toward meaningful stimuli has also been found in other clinical disorders. For example, studies have found that patients with depression have a bias toward loss-relevant cues (Gibb, McGeary, & Beevers, 2016; Gibb, Pollak, Hajcak, & Owens, 2016; MacLeod, Mathews, & Tata, 1986), and patients with substance use disorders (e.g., for alcohol, nicotine) have been shown to have an attention bias toward cues related to the addiction (reviewed in Field & Cox, 2008). Overall, these studies provide evidence that the attentional landscape can be tilted, in an automatic and involuntary way, toward stimuli that have special meaning to an individual, and this can be critically exacerbated in clinical disorders.

4.3.2 Schizophrenia

Although not considered as a core or defining deficit in the disorder, a number of studies have reported attentional deficits in patients with schizophrenia. These have included slower search rates in visual search tasks (Fuller, Luck, Braun, Robinson, McMahon, & Gold, 2006), impaired accuracy in visual working memory tasks (Stäblein, Sieprath, Knöchel, Landertinger, Schmied, Ghinea, Mayer, Bittner, Reif, & Oertel-Knöchel, 2016), and a slower recovery in attentional blink tasks (Mathis, Wynn, Breitmeyer, Nuechterlein, & Green, 2011). In a recent fMRI study, Hahn and colleagues designed a task to investigate the focusing of attention and selective higher-order visual processing under different levels of difficulty (Hahn, Robinson, Kiat, Geng, Bansal, Luck, & Gold, 2022). At the start of each block of trials, subjects were shown an instruction screen that consisted of a picture of either a face or a house that would serve as the target in that block. In each trial, subjects would see a single picture of either a face or house, and they were required to press one button if the picture was the target for that block or a second button if the picture was any other face or house (Figure 4.7). There were two main conditions: the *sequential* and the *overlay* conditions. The *sequential* condition was designed to be a relatively easy condition, in terms of the need for attention, because the two different types of stimuli were presented one at a time, with plenty of time between stimuli to detect what stimulus had just occurred before the next stimulus appeared. In terms of attention, this condition would not require much selection and biasing because there would be no direct competition between the clearly separated stimuli. The *overlay* condition, on the other hand, would require attentional focus and selection because the two stimuli would be overlapping in space and occur simultaneously. Thus, attention would be needed in this condition to select the type of stimulus that must be discriminated in each block. For example, when the target stimulus was a particular face, then for all trials in that block faces had to be attended and houses ignored within each face/house overlay pair; when the target was a particular house image, then houses had to be attended and faces suppressed in each face/house overlay pair.

This fMRI study used a region-of-interest (ROI) approach, which focused on just the two different areas of higher-order visual object processing that are most selective for processing faces and places. Specifically, before the attention trials began, subjects participated in scans that identified the fusiform face area (FFA; selective for images of faces) and the parahippocampal place area (PPA; selective for images of places) individually in each subject. These subject-specific ROI areas were then used to extract that set of voxels from the fMRI scans acquired during the attention task. The ROI approach allows for more sensitive analyses in those specific regions, avoiding the need to implement the multiple-comparison statistical corrections that need to be used in fMRI studies when testing for any active voxels across the whole brain.

Activity in each of the ROIs in this study was dominated by the type of stimulus present on the screen when only one stimulus was present at a time. In the sequential condition, the FFA was more active to faces than houses and the PPA was more active to houses than faces. The critical comparison was whether this activity would be modulated by attention (i.e., when subjects were instructed to detect the presence of a specific face or a specific house). In the easy task (sequential condition), activity in the FFA and PPA showed no modulation by

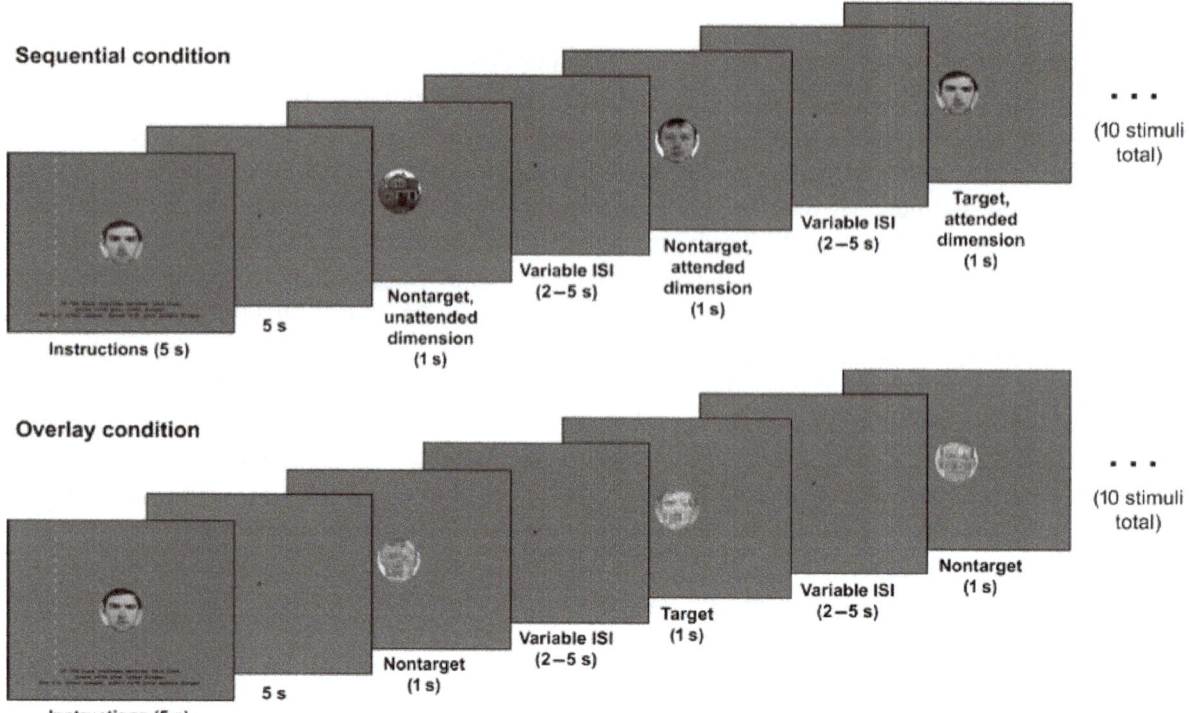

Figure 4.7 Task used during fMRI to assess possible attention deficits in schizophrenia. In each condition, the first screen of each block of trials (shown at far left of each row here) presented the instruction, which was to press one button for any upcoming images that matched the stimulus shown on that instruction screen and to press a different button for any other images when they appear. In every block, pictures of both faces and houses would appear, with only one specific image being the target. In the "sequential condition" (top row), each image was of a single object (a face or a house). In the "overlay condition" (bottom row), each image was composed of two superimposed semitransparent pictures (one face, one house). Reprinted from Hahn et al. (2022), *Cerebral Cortex*, with permission from Oxford University Press.

attention in healthy control subjects – whether the subject was attending to faces or to places made no difference to the amount of activity in those regions in those subjects. The lack of attention effects in this condition can be attributed to the lack of competition between the stimuli, which were separated by multiple seconds. Therefore, attentional selection mechanisms weren't required to perform the task well. However, individuals with schizophrenia showed a significant attention effect in the FFA in this easy (sequential) condition. Specifically, when a face appeared on the screen, activity in the FFA was significantly greater when the individuals with schizophrenia were paying attention to faces versus when they were attending to houses, even though there was no direct competition with the current stimulus. This biasing of attention in the easy condition has been interpreted as showing a *hyperfocusing* of attention in schizophrenia. This is in line with other research that has found that individuals with schizophrenia may exhibit unusually strong attentional selection, measured as the filtering of peripheral distractors, because of a more narrow and enhanced focus on the items at fixation (Kreither,

Lopez-Calderon, Leonard, Robinson, Ruffle, Hahn, Gold, & Luck, 2017). This intense focus at the center of gaze can be observed in individuals with schizophrenia even when this is detrimental to task performance because the task requires a broader distribution of attention (Elahipanah, Christensen, & Reingold 2010).

In the more difficult (overlay) condition, in which faces and houses were always overlapping in space and time, healthy control subjects now showed a significant attention effect. Specifically, activity in the PPA was significantly greater for the overlay images when the subject was attending to houses versus when attending to faces. This suggests that control subjects exerted attentional control to better focus on the target stimulus within the overlapping images by boosting the processing of the house when places needed to be attended and suppressing the processing of house stimuli when faces needed to be attended. Individuals with schizophrenia, on the other hand, showed no attentional modulation in either ROI during this difficult overlay condition. Together, this pattern of results across conditions suggests that individuals with schizophrenia show a *deficit* in being able to focus attention when selective processing of competing stimuli is most needed, but they also exhibit a *hyperfocusing* of attention under circumstances when there isn't as much of a need to engage attentional selection.

Other lines of research have used a "go/no-go" paradigm to investigate executive control and processes of inhibition. In a typical go/no-go task, the participant is presented with one of two types of stimuli, one of which requires a behavioral response (the "go" stimulus) and another type that requires that no response be made (the "no-go" stimulus). In most versions of the task, the stimuli are presented one at a time in the middle of the screen, and the go stimulus is presented frequently, whereas the no-go stimulus is presented more rarely. This sets up a need to engage cognitive control when the no-go stimulus is presented in order to inhibit the prepotent response that builds up over many consecutive trials of making the same repeated response to the go stimulus. Event-related potential (ERP) studies using this paradigm typically find that go stimuli elicit a robust P3 component over posterior parietal sites, in line with the standard appearance of that component when subjects need to process and respond to a stimulus (Kropotov, Ponomarev, Hollup, & Mueller, 2011). In response to the rare no-go stimuli, however, two other components are observed at a shorter latency and distributed over more anterior central sites. The earlier of these two components is the N2, and subsequent research has revealed that it is evoked not only by the rare no-go stimuli but also by rare go stimuli in some experiments, suggesting that it indexes a process of *conflict detection*, not the inhibition of a response (Randall & Smith, 2011). However, the second component elicited by no-go stimuli, the anterior P3, has been linked more directly to the inhibition of a prepotent action plan (Enriquez-Geppert, Konrad, Pantev, & Huster 2010).

Kropotov and colleagues (Kropotov, Pronina, Ponomarev, Poliakov, Plotnikova, & Mueller, 2019) used ERPs in a large-sample-size study to investigate the cognitive control abilities of individuals with schizophrenia and ADHD in a go/no-go task. The behavioral results confirmed previous findings that both patient groups, compared to healthy control participants, were impaired on the task, producing significantly more errors and much greater variance in response times (indicating inconsistent attention). The schizophrenia group in this study exhibited larger impairments on both of these overt response measures compared to the

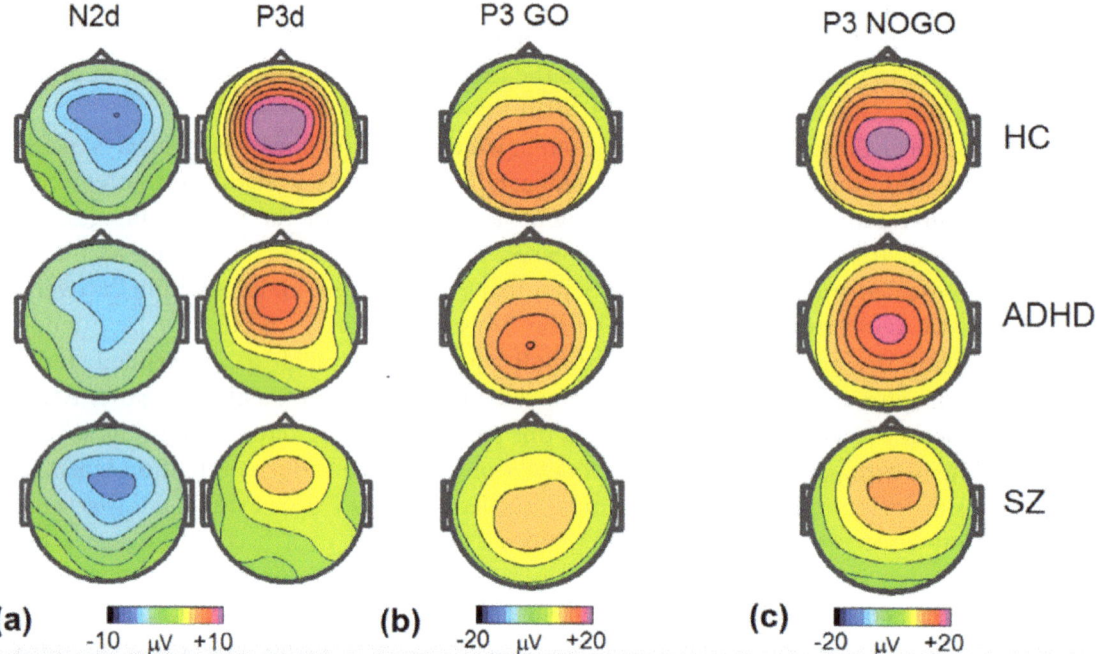

Figure 4.8 ERP results from a go/no-go task in individuals with schizophrenia or ADHD. Scalp topographies for healthy control (HC) subjects (top row), ADHD subjects (middle row) and individuals with schizophrenia (SZ; bottom row). Topographies of the *difference* between no-go and go trials are presented in the left-hand columns labeled N2d (~240 ms) and P3d (~400 ms). The right two columns present the separate scalp topographies during the P3 range for go trials (a posterior peak occurring at ~320 ms) and no-go trials (an anterior peak occurring at ~370 ms). Reprinted from Kropotov et al. (2019). *Clinical Neurophysiology*, with permission from Elsevier.

ADHD group. The ERP results showed that the N2 effect (calculated as no-go trials > go trials) was similarly robust for all groups, suggesting that the process of conflict detection was intact in both the schizophrenia and ADHD groups (Figure 4.8, left columns). However, the P3 was significantly *reduced* in individuals with schizophrenia compared to individuals in the other two groups, suggesting a selective impairment in the mechanism of inhibiting responses that was unique to the schizophrenia group (Figure 4.8, middle and right columns).

Although this study provided evidence from standard ERP component analyses for an impairment in only the mechanism of response inhibition, the authors went further and used advanced analysis techniques to extract latent components from these EEG data. Using blind-source separation algorithms, they were able to isolate brain processes that are not easily observed in the standard ERP methods. This method utilized an unmixing matrix defined from the ERPs of the large group of healthy control participants (132 individuals) in their study and included both temporal dynamics and scalp topography. These new analyses revealed a latent component localized to frontal regions that was present during the time range of the N2. Critically, this latent component was significantly reduced in the individuals with schizophrenia relative to the other two groups. This provides critical new information that the behavioral impairments observed in such tasks in individuals with schizophrenia may begin at an earlier stage of processing than previously thought, and these results suggest that the impairment includes the initial stage of *conflict detection* as well as the later stage when prepotent action plans must be suppressed. Interestingly, the latent component analyses also revealed frontal activity in the latency range of the P3 that showed a reduction both in individuals with schizophrenia and in those with ADHD relative to controls. This raises the possibility that new analysis methods such as modeling latent components in ERP data may reveal more subtle dysfunctions of individuals with attention deficits. As discussed throughout this chapter, neuroscience studies investigating the brain basis of impaired and atypical attention in neurological and clinical patients provide critical insights into the mechanisms of attention, and this knowledge can, in turn, lead to improved strategies for addressing attention impairments.

CHAPTER SUMMARY

- Neuropsychological patients suffering from the neglect syndrome have revealed a network of temporoparietal and ventral frontal regions – lateralized largely to the right hemisphere – that is critical for disengaging and reorienting attention.
- Neglect patients provide evidence for different neural structures supporting space-based attention versus object-based attention.
- Damage to the thalamus and to the superior colliculus is associated with deficits in engaging and moving attention, respectively.
- ADHD is a complex developmental disorder that may affect multiple processes of attention, including executive control, filtering of irrelevant distractors, and the balancing of top-down and bottom-up influences on allocating attention.
- Attention tasks can be adapted to help treat attention problems, such as the development of attention bias training to help individuals with anxiety disorders.
- Even in clinical disorders that are not associated primarily with attention, atypical attention processes are sometimes observed and could be targeted to enhance treatment and therapy plans.

REVIEW QUESTIONS

Describe evidence from different groups of patients suffering from the neglect syndrome that implicates multiple cortical and subcortical areas in the functions of attention.

Compare the attention deficits observed in simultanagnosia versus neglect versus progressive supranuclear palsy.

Explain how the study of developmental disorders has provided insights into the interactions of multiple attentional processes and networks.

Describe the different attention tasks that have been most useful in identifying deficits in attention and that have provided critical insights into the neural mechanisms of attention processes.

FURTHER READINGS

Mogg, K., & Bradley, B. P. (2018). Anxiety and threat-related attention: Cognitive-motivational frame-work and treatment. *Trends in Cognitive Sciences*, *22*(3), 225–240.
  – This article reviews evidence and new treatment strategies for the attentional bias to threat seen in some individuals with anxiety disorders.

Moore, M. J., Milosevich, E., Mattingley, J. B., & Demeyere, N. (2023). The neuroanatomy of visuo-spatial neglect: A systematic review and analysis of lesion-mapping methodology. *Neuropsychologia*, *180*, 108470.
  – This article provides an excellent and recent review of the regions underlying multiple subtypes of the attentional neglect syndrome.

Mueller, A., Hong, D. S., Shepard, S., & Moore, T. (2017). Linking ADHD to the neural circuitry of attention. *Trends in Cognitive Sciences*, *21*(6), 474–488.
  – This article summarizes neuroscience results investigating the brain networks of attention that have been found to be affected in ADHD.

5   The Effects of Attention on Neural Processing

Learning Objectives

- Compare theories of when, where, and how attention modulates information processing
- Describe the evidence, from multiple methods of cognitive neuroscience, investigating the "early selection" versus "late selection" debate
- Identify controversies and current issues regarding how attention is distributed and the mechanisms by which it modulates processing
- Compare and contrast the neural effects of spatial attention versus feature attention
- Describe the effects of cross-modal attention and whether visual, auditory, and tactile attention should be considered separate or integrated systems

5.1   The Focus and Distribution of Attention: Theory and Models

Before describing the cognitive neuroscience evidence for what attention does, it will be useful to briefly revisit the theories of how attention may be distributed. As introduced in Chapter 1, Helmholtz (1896) demonstrated that attention isn't simply where the eyes are looking. Rather, attention can be moved and focused independently from the eyes – what we now call **covert attention**, in contrast to **overt attention**, which involves moving the head or eyes to fixate the region being attended. Much of the research in cognitive neuroscience employs paradigms that test covert attention, because in this way the differences in visual acuity across the retina are not confounding factors – any difference in the brain's processing of the input can be ascribed to mental functions such as attention, not what part of the retina received the input. As described in Chapter 1, a popular analogy for attention is that it works in a similar fashion to a spotlight. According to this spotlight model, attention can be thought of as highlighting a small portion of the visual world, similar to a spotlight in a theater illuminating an actor onstage. While the rest of the stage may still be visible, the area within the spotlight is much clearer and easily perceived. Similarly, attention can be thought of as providing improved clarity and focus on a selected region of the world. The zoom-lens model of attention (Eriksen & St. James, 1986) stressed that the *size* of the attentional window wasn't fixed. Similarly to a flashlight that can be adjusted to make the beam wider or more narrowly focused, the zoom-lens model of attention holds that the size of the attentional window can be adjusted depending on current conditions and goals. Furthermore, the zoom-lens model included the idea that attention was a limited "mental resource," and that the effects of attention would be lessened as the size of the attentional window was widened, similarly to the intensity of a flashlight beam being reduced as the beam is widened. Both the spotlight and zoom-lens models attempt to explain how attention is allocated across space, and Section 5.2 describes what is known about how, when, and where (in the brain) *spatial* attention affects visual processing.

Although spatial location does play a strong role in attention, we can appreciate from our everyday lives that attention can be allocated in other ways as well. When looking for a car that will be picking us up outside a busy airport, it is the color or the type of vehicle that we try to attend to rather than just spatial location. Behavioral research has confirmed the existence of this type of feature-based attention, with studies showing faster and more accurate responses to target stimuli that match the features being attended versus ignored (Yantis, 2000). Much of the research on feature-based attention has built upon work in the visual system that identified core features that are processed preferentially in different visual areas or pathways. Indeed, research on feature-based attention has shown that attention can be effectively allocated to such visual features as color (Sàenz, Buraĉas, & Boynton, 2003), motion (Liu & Mance, 2011), and orientation (Baldassi & Burr, 2000). These studies were able to show that behavioral responses were faster and more accurate when the target included the features that were being attended; in the sections to follow, we will discuss the neural regions and processes that are involved in this type of attention.

In addition to attention being flexible enough to be allocated to either space or features, it is also important to understand how these different varieties of attention may work together. Specifically, as we attend to an object such as a face, our attention may be more selectively focused on different parts of the face (e.g., the eyes, the hair, or the movements of the mouth when the person is speaking). In such a situation, is our attention distributed only across the area corresponding to the specific feature we're trying to attend, or is the entire object still receiving a boost from attention? Clever work by Egly and colleagues (Egly, Driver, & Rafal, 1994) showed that once attention is allocated to any part of an object, the rest of that object enjoys some attentional benefit as well. These studies of object-based attention have provided evidence that attention automatically spreads across an object. Using a variation of the Posner cuing paradigm, Egly et al. replicated the standard finding that the attended *location* showed the greatest benefit of attention (i.e., faster response times [RTs] to targets at that location), but they also found that unattended regions of space that were *within* the same object enjoyed some degree of attentional enhancement as well (Figure 5.1). Targets that appeared at an unattended location within the *same* object as where spatial attention was focused were responded to more quickly than targets that were equally distant from the attended location but within a *different* object. In this design, the cue stimulus accurately predicted the location of the target with 75% accuracy. Therefore, subjects voluntarily attended to that one spatial location, and responses were fastest overall to that specific location. However, the two other possible target locations were equally likely; note that targets never occurred at the diametrically opposite location from the cue in this experiment. Specifically, the unattended location in a *different* object was just as likely to contain the target as was the unattended location within the *same* object as where spatial attention was focused. Despite this, subjects were consistently faster at detecting the target at the unattended *within-object* location compared to the unattended *different-object*

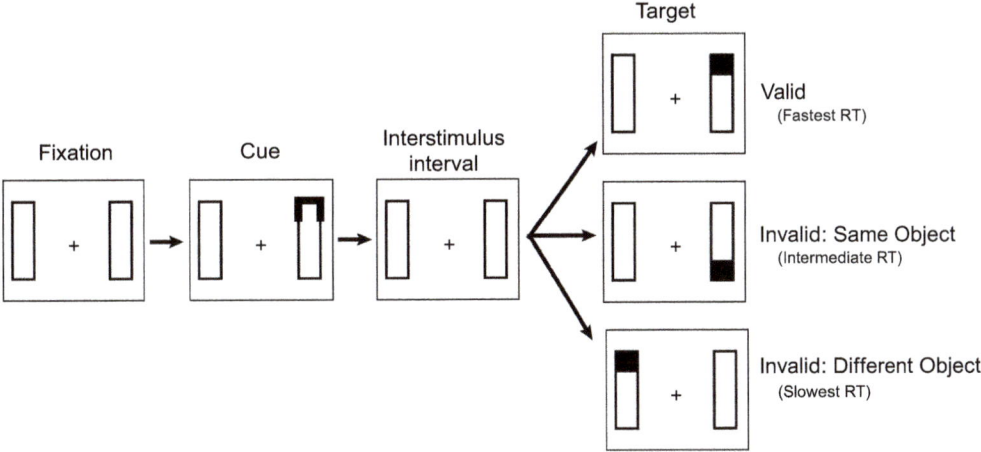

Figure 5.1 Object-based attention task. The cue, here shown as a thickening of the end of one of the rectangles, was predictive of the target location: 75% of trials were "valid" trials. The remaining 25% of trials were divided evenly between the other two locations equally distant from the cued corner: 12.5% of trials were "Invalid: Same Object" and 12.5% were "Invalid: Different Object." Subjects made a simple, speeded button press as soon as they detected the target. Note that the background rectangles, shown as vertical rectangles here, were oriented horizontally, with one rectangle centered above fixation and the other below on half the trials. Figure based on design used by Egly et al. (1994). Note that the typical Posner cuing paradigm uses a very similar design, but the boxes marking the possible target locations would be smaller and the targets would only occur in the center of the location marker boxes.

location. Thus, attention appears to automatically spread across an object when any part of it is attended. It should be noted, though, that spatial attention still had the largest effect, as the attended *spatial location* within the attended object received the overall greatest attentional enhancement.

The importance of object-based attention has also been highlighted in studies that found that it is quicker to detect and discriminate the presence of two different types of features (e.g., color and form) when those features are part of the *same object* as opposed to split across objects (e.g., Goldsmith, 1998). This effect occurs even when both features overlap within the same space, suggesting that object attention may bind the features together more efficiently than spatial location. This is in contrast to Treisman's feature integration theory, which had suggested that spatial location and a master map are key to binding features. Similarly, Rodríguez, Valdés-Sosa, and Freiwald (2002) used transparent planes in which objects could overlap in terms of spatial location in order to test object attention independent from spatial location. They found that visual search was more efficient (in this experiment for the features of form and motion) when they were part of the same "transparent plane" as opposed to being on separate planes. To bring this section full circle and return to the more real-world situation in which we move our eyes to what we are attending, Theeuwes, Mathôt, and Kingstone (2010) tested the effects of objects on overt attention. Using a modified version of the Egly et al. (1994) task shown earlier, Theeuwes et al. found that voluntary eye movements were more likely to be made within object boundaries, before moving to locations within other objects. In their study, small letters occurred at the ends of each rectangle, and subjects needed to fixate the letters in order to be able to discriminate the small letters and detect the target letter. As in the original study, the rectangles were oriented horizontally on half the trials and vertically on the other half. The target letter was most likely to be at the cued location (66% valid) but was equally likely to be at the two other equidistant locations (17% invalid: same-object; 17% invalid: different-object). As would be expected, subjects almost always looked at the cued location first (93% of trials), but the critical finding was that the next saccade was significantly more likely to go to another location within the *same object* (56%) compared to the equidistant location in the *other object* (40%). Overall, behavioral studies of attention have shown that spatial location, feature characteristics, and objects influence how attention is focused and distributed. The following sections present cognitive neuroscience studies that provide evidence for how, where, and when brain processes are modified by attention. Since visual-spatial attention is the most well-studied type of attention, the next section will focus specifically on this type of attention. Feature- and object-based attention, in the visual modality, will be covered in subsequent sections, and this chapter will wrap up with the effects of attention in the other sensory modalities and how attention affects the linking of information across sensory modalities.

5.2   The "Sites" of Spatial Attention Effects

Whereas behavioral research has provided important insights into the different varieties of attention, cognitive neuroscience studies have been critical for investigating exactly *where* in the brain and exactly *when*, during the multiple stages of perceptual processing, attention can modulate processing in the brain. The current chapter will focus on the *effects* of attention (i.e., the "sites" in the brain where attention modulates processing and the nature of those effects). The following two chapters will discuss the brain regions and mechanisms involved in *controlling* the focus of attention (i.e., the "sources" of attentional control).

Some of the earliest theories of attention focused upon when, during the multiple stages of perceptual and higher-order processing, attention affects processing. As introduced in Chapter 1, this "level-of-processing" debate concerned how soon processing in the brain was changed by attention. Figure 5.2 (top) shows a simplified model of a few stages of processing between the initial inputs to the sensory organs and the final overt response. The inputs here are just a small representation of all the information coming in from the senses at every moment. The totality of all the inputs is obviously enormous, and since we are consciously focused on only a tiny fraction of those inputs, understanding how attention selects what information gets through to our conscious awareness has been an important topic since the beginnings of research into cognitive psychology and the mind. Figure 5.2 (bottom left) presents an extreme version of an "early filter" model of attention, based on Broadbent's (1958) theory, in which all stimulus processing except that small part being actively attended is filtered out at early levels of perceptual processing. Figure 5.2 (bottom middle) presents an

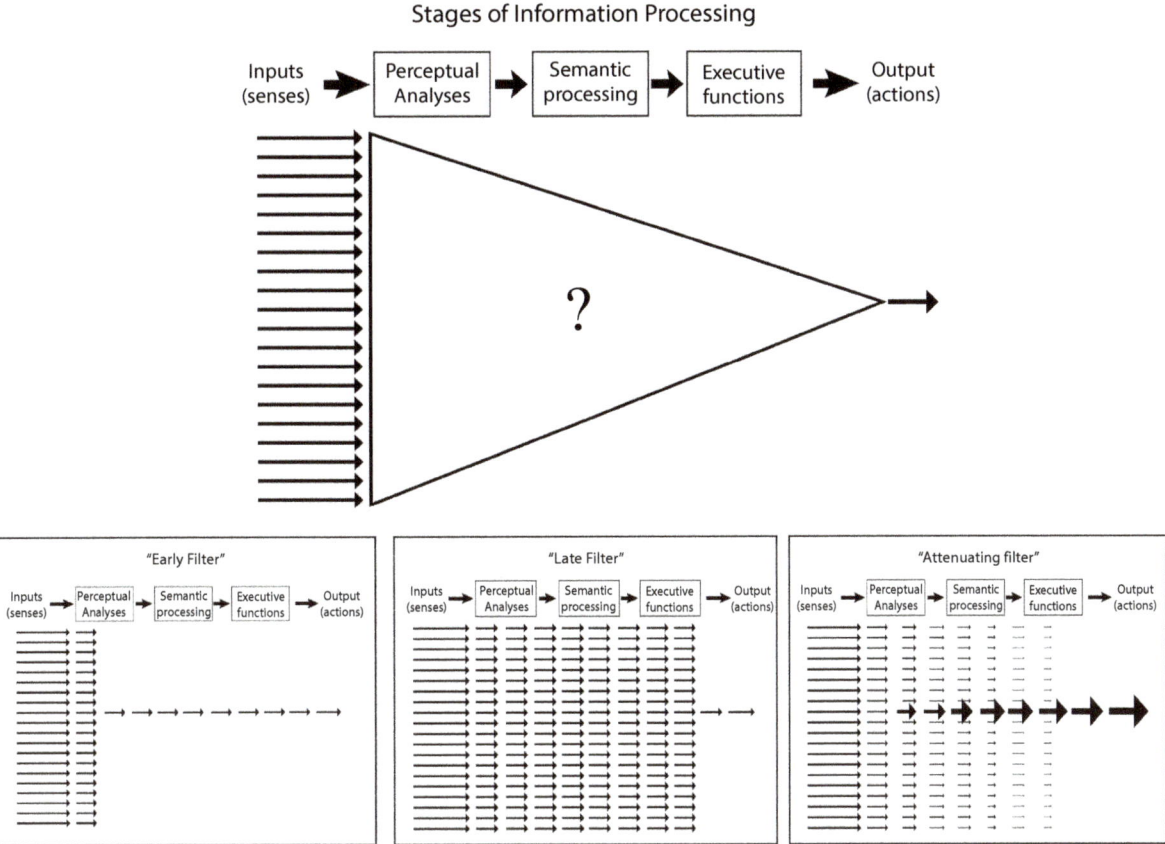

Figure 5.2 Predictions of different attention models on stages of information processing. Top shows a few hypothesized stages of processing, from the multiple sensory inputs to the final action and overt response to just a small portion of those inputs. The question mark indicates the unknown of how attention selects what gets processed and to what level. The bottom three panels illustrate extreme versions of the early filter, late filter, and attenuating filter hypotheses.

extreme version of a "late filter" model, based on that proposed by Deutsch and Deutsch (1963), in which everything coming into the senses proceeds through multiple levels of processing (including full semantic analysis) before attention finally acts to allow the highly processed information to rise to conscious awareness. As can be seen in Figure 5.2, these two different models of attention differ greatly in the overall amount of processing that would need to be done, and each has potential advantages and disadvantages. Early filter models are obviously much more efficient in terms of requiring much less neural processing because most of the information from the senses is filtered out early in processing. In contrast, late filter models require a great deal of processing of sensory inputs that may be irrelevant to us at the time. However, late filter models have the significant advantage of ensuring that anything critically important in an unattended channel would be processed to a high level of meaning, allowing important stimuli to rise to awareness. As noted in Chapter 1, Anne Treisman (1960) provided evidence that, instead of a strict filter that completely eliminates processing of the unattended information, attentional selection appears to work more as an "attenuating filter" to suppress unattended information relative to attended information. Figure 5.2 (bottom right) presents an attenuating filter model in which attended information is boosted while unattended information is relatively suppressed. Across these different models of attention, questions remain as to exactly when, where, and how such a hypothesized "filter" actually works in the brain. To answer those questions, we can turn to evidence from cognitive neuroscience studies.

5.3   When Attention Affects Processing: The Timing of Spatial Attention Effects

In order to test exactly how the brain accomplishes the selection process(es) of attention, a method is required that provides a highly precise measure of exactly when in time a neural process is being modulated. As described in Chapter 4, electroencephalography (EEG) and magnetoencephalography (MEG) are ideal methods for the noninvasive recording of neural activity in healthy human subjects with millisecond resolution, enabling us to address these sorts of questions. Figure 5.3 (top) shows an idealized event-related potential (ERP) waveform along with the simplified model of the stages of processing from Figure 5.2. Since visual attention has been the subject of more cognitive neuroscience studies than the other sensory modalities, this section focuses on visual-spatial attention, and the idealized ERP in Figure 5.3 shows visually evoked components. It must be noted that this idealized ERP is showing all of the components here as being observable on the same single-electrode/ERP waveform, when in reality the different components would be observed at different electrodes. The purpose here, however, is to indicate that there are a series of ERP components that can be associated with different stages of processing. Thus, ERPs are well suited to addressing the questions of when and how attention affects information processing in the brain. Figure 5.3 shows hypothetical results that would be predicted by each of the three major filter models. An early filter model of attention would be supported if ERP evidence were to show that unattended sensory information was completely cut off very early in the processing stream (Figure 5.3, bottom left). In contrast, if ERP evidence were to show that attended and unattended information generated the same size and strength of ERP components throughout most stages of processing, with the unattended processing only being stopped at a relatively late stage, that would support a late filter theory (Figure 5.3, bottom middle). Note that Figure 5.3 shows extreme versions of the predicted results from these theories, for illustration purposes, to make the differences between models clearer. Finally, ERP evidence could provide support for "attenuating filter" theories in a couple of different ways. One manner in which an attenuating filter could be instantiated would be by changing the speed at which processing advances through the stages. The ERP method, because of its excellent temporal resolution, is able to detect latency shifts in the components of interest. As shown in the hypothetical results (Figure 5.3, top right), one possible means of attenuation would be to slow down the processing of unattended information

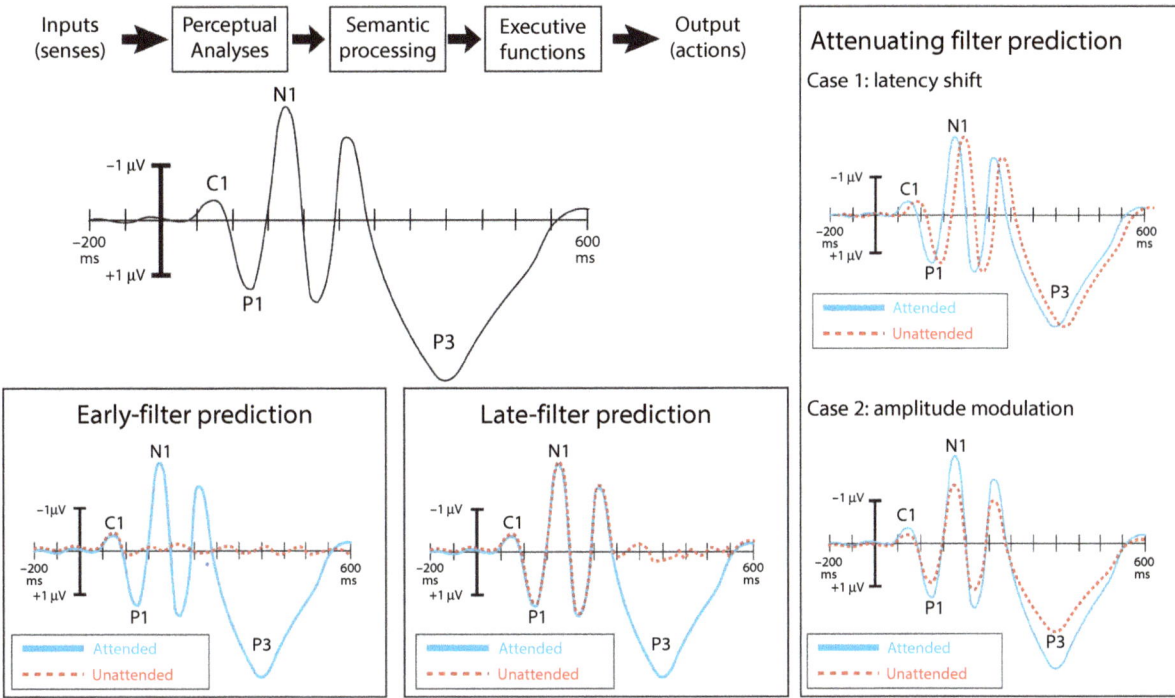

Figure 5.3 Predicted ERP results from levels-of-processing filter models. The top left image shows a cartoon ERP waveform made to show multiple components, evoked by a visual stimulus in the upper lateral visual field, on a single electrode. In reality, these components would be observed at different electrode site locations across the scalp, but they are shown on one waveform here for simplicity. The "C1" is found at central ("C") posterior scalp locations, and it is typically the earliest component observed on the scalp evoked by a visual stimulus; the C1 is a negative-going wave (as shown here) for upper visual field stimuli but is a positive-going wave for lower visual field stimuli. The "P1" is the first visual component that is always positive (regardless of where the visual stimulus appears), and the "N1" is the first visual component that is always negative. The "P3" is sometimes called the P300, to indicate that it typically begins at around 300 ms. The boxes present the hypothetical predictions of the different attention filter theories, as described in the text.

at multiple levels. According to this prediction, faster reaction times to attended information would be due to the quicker processing of the attended information at these stages. Another possible result of an attenuating filter, however, would be that the amplitude – or strength of processing – for each ERP component is modulated by attention, without affecting the latency of each component at all (Figure 5.3, bottom right). This type of effect could explain the behavioral patterns as being functions of the stronger processing at multiple stages, leading the attended information to be responded to more accurately and quickly because it has been processed more robustly, enabling a quicker and more confident response. Another option (not shown) is that both latency and amplitude could be affected by attention. In all of these models, the question remains as to exactly how early in the stages of processing attention can begin to modulate the neural processing.

5.3.1   ERP Studies of Visual-Spatial Attention

Haider, Lindsley, and Spong (1965) provided some of the first evidence that attention can be allocated to one sensory modality over another. They presented subjects with alternating streams of visual flashes and auditory clicks and instructed them to attend to one sensory modality and ignore the other in different blocks of trials. They found that components measured at an occipital site (over visual cortex) were larger when subjects were instructed to attend to the visual stimuli, and that components over temporal areas (auditory cortex) were larger when subjects were attending to the auditory stimuli. Although this study showed that attention could be allocated to different types of sensory input and affect processing within the first couple of hundred milliseconds, the effects were not specific to spatial attention, and the use of alternating stimuli could have led to phasic differences in arousal. Eason, Harter, and White (1969) provided the first evidence that *spatial* attention affects processing at a relatively *early* level. In their study, visual flashes occurring on the attended side produced enhanced processing, relative to visual flashes at the unattended side, as early as ~160 ms, corresponding to what is now called the visual N1 component. However, in that study, attention was sustained at one location for a prolonged time, and the location of the flashes wasn't completely random, leading Näätänen (1975) to suggest that the modulation of processing may have been due to arousal, different anticipatory states, or the learning of the task sequences rather than being due to an orienting of spatial attention. Further, that study didn't incorporate an objective measure of eye movements, leaving open the possibility that the ERP differences could have been due to eye position and attended versus unattended stimuli falling on different parts of the retina. Van Voorhis and Hillyard (1977) addressed these concerns in their study, which carefully monitored eye position and ensured that the location of the stimuli was random and unpredictable to subjects. In that study, Van Voorhis and Hillyard provided the first evidence from a well-controlled paradigm that sustained visual-spatial covert attention modulates processing at an early stage. They replicated the earlier finding of an N1 attention effect (~160 ms in their experiment) and provided the first evidence that spatial attention can act even earlier. Although not as robust as the N1 attention effect, the P1 component (~100 ms) was larger for stimuli in the attended versus unattended hemifield.

Although those studies had shown that early components could be affected by spatial attention, those experiments utilized paradigms in which attention was sustained at a spatial location over many trials before switching to the other location for a block of trials. This block design could potentially have led to learning effects over time related to only one location being responded to for many trials in a row. In order to better assess the immediate effects of a rapid shift of covert spatial attention, researchers began to use variants of the Posner cuing paradigm (e.g., Posner, 1980) described in Chapter 1. As opposed to the earlier ERP studies that instructed subjects where to attend before an entire block of trials, in the cuing paradigm a cue stimulus (e.g., an arrow pointing left or right) is presented briefly at fixation at the beginning of each trial (see Figure 1.7), which informs the subject of the likely location of the upcoming target stimulus. As described in Chapter 1, behavioral results from this paradigm have shown that participants are faster and more

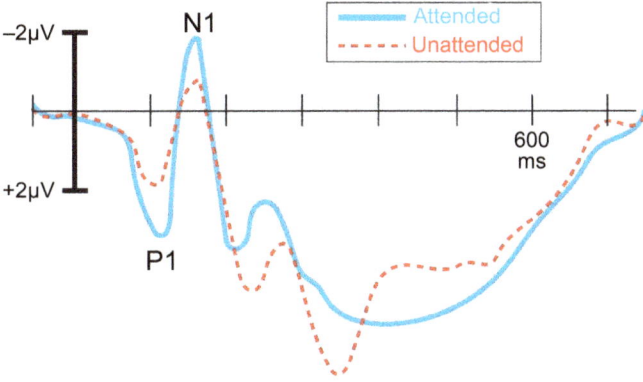

Figure 5.4 ERP effects of spatial attention. Results from Mangun and Hillyard (1991) showing the effects of spatial attention on early visual processing. This pattern of results supports an "attenuating filter model" in which spatial attention acts relatively early to modulate sensory-level stages of processing. Note that the scalp location shown here is a lateral occipital location where the P1 and N1 are robustly observed, but other components are maximal at different scalp locations. Figure adapted from Mangun and Hillyard (1991), *Journal of Experimental Psychology. Human Perception and Performance*, with permission from American Psychological Association.

accurate at responding to targets at the location where they had been instructed to pay attention (the "attended" target) relative to a target at a different location (the "unattended target"). ERP studies using this paradigm provided new evidence that covert visual-spatial attention modulates processing as early as 90–100 ms, as indexed by the P1 component (e.g., Mangun & Hillyard, 1991). As shown in Figure 5.4, Mangun and Hillyard (1991) found that spatial attention, cued on a trial-to-trial basis, modulated the amplitude of the P1 and N1 components without much change to the latency or waveshape of these components. Thus, these data, and many replications of this pattern (reviewed in Hillyard & Anllo-Vento, 1998), support an attenuating filter-type model, in which attention modulates (but does not completely filter out) sensory processing at a relatively early stage of processing in the brain.

Further support for the theory that attention acts *early* and on the basis of *spatial location* is that P1 attention effects are observed for the processing of both target stimuli (e.g., Mangun & Hillyard, 1991) and irrelevant "probe" stimuli (Heinze, Luck, Mangun, & Hillyard, 1990). In Heinze et al.'s (1990) study, subjects attended to one spatial location and responded to only a certain class of "target" stimuli. Nontarget stimuli, however, could also occur at the attended location or at unattended locations. The results showed a significantly larger P1 to the probe stimuli when they were at the attended location relative to the unattended location. If attention was modulating processing based upon the identity of the stimulus itself (e.g., as a target), then the irrelevant probe stimuli should not have been affected. That the P1 enhancement was as large for probe stimuli as for target stimuli provides evidence that the attention effects at this early level are based on where in space attention is focused, with all stimuli at that location enjoying the boost of spatial attention.

5.3.2 Dissociating Early ERP Attention Effects

In each of the hypothetical "attenuating filter" predictions illustrated in Figure 5.3, all the ERP components are affected by attention in the same way. This type of modulation could happen *if* attention worked at just one early stage, with that speeded-up or enhanced processing simply propagating along to each successive stage in turn. Or such a pattern could result from attention acting at multiple stages but with the same exact type of modulation happening at each step. However, ERP studies have revealed that this is not the case. These experiments have found that even components that are adjacent in time and thought to index a similar type of processing can show distinct effects of attention. For example, a number of studies have dissociated the effects of the P1 and N1. As opposed to the N1 simply carrying on the modulated processing begun at the P1 stage, studies have shown these components to be sensitive to different attentional and task demands. These dissociations provide critical insights into the processes of attention, revealing that it's not the case that a simple filter or sensory gain modulation acts at one stage but that instead there are multiple, somewhat independent and unique processes of attention.

Although many studies have found spatial attention to affect both the P1 and N1, research has also revealed dissociations between attention effects on these two components (e.g., Luck, Heinze, Mangun, & Hillyard, 1990; Vogel & Luck, 2000). Indeed, even in the Mangun and Hillyard (1991) study that produced the effects shown in Figure 5.4, the authors found differences between the P1 and N1 attention effects in other experiments. Specifically, they found that spatial attention modulated both the P1 and N1 during a difficult discrimination task (judging the height of a bar stimulus), but that it only affected the P1 when the task was a simple detection task. The authors suggested that the P1 may reflect the spatial focus of attention regardless of task, but that the N1 may only be modulated by attention when the task requires a deeper level of engagement. Other studies have found N1 attention effects without P1 attention effects, suggesting that the N1 may index a higher-level discrimination process (e.g., Vogel & Luck, 2000). Note that the N1, like the C1 and P1, will be generated by any visual stimulus, but that the modulation of that process by attention depends on the subject's goals.

It should be noted that even referring to the N1 as a single component can be misleading because there are multiple activities occurring during the time of the N1 component, and researchers have defined a number of subcomponents of the N1 (Clark, Fan, & Hillyard, 1994; Di Russo, Martínez, & Hillyard, 2003). Therefore, it is clear that multiple different brain activities are occurring shortly after the P1 in the latency range associated with N1 components (~140–200 ms). Here, we are focusing only on the early components that have been found to be affected by spatial attention. In addition to the theories presented earlier, Luck and colleagues have proposed that a core difference between the attention effects on the P1 and N1 is that they reflect the costs and benefits of attention, respectively (Luck, Hillyard, Mouloua, Woldorff, Clark, & Hawkins, 1994). Luck et al. (1994) had subjects perform a typical attention cuing task in which one location was cued on most trials, but they also included a neutral condition in which all four possible target locations were cued. They found that the P1 was reduced in the invalid (unattended location) condition relative to both the valid (attended) and neutral conditions, whereas the N1 was enhanced in the valid condition relative to both the invalid and neutral conditions. The P1 was thus showing the costs of attention, whereas the N1 was showing the benefits. The authors proposed that the P1 attention effect indexes a process of suppressing information from unattended locations, which would help reduce interference from potential distractions, and that the N1 attention effect indexes the engagement of a higher-level discriminatory process at only the attended location and only when required by the task. Box 5.1 discusses the various ways in which "neutral" can be defined and how that choice affects the interpretation of costs and benefits.

5.3.3   Attention Effects on "Late" Components

In addition to the dissociation between the P1 and N1 components, further dissociations can be seen between the attention effects on these early components and the effects on later higher-order components. Unlike what's shown in the cartoon predictions of Figure 5.3, an enhancement of the P1 and N1 does not necessarily lead to changes in later components such as the P3. This is important for understanding that attention can act at multiple "sites" in the brain to modulate different aspects of information processing. The C1, P1, and N1 are sometimes referred to as "externally evoked components" because they are sensitive to physical properties of the stimulus (e.g., brightness, intensity). The P3, and most components occurring after 300 ms, are sometimes referred to as "internally generated components" because they depend on the subject's evaluation of the stimulus, and they are not typically affected by simple physical characteristics. The P3 was originally labeled the P300, in reference to its onset latency of ~300 ms. Chapman and Bragdon (1964) first identified this large positivity and suggested that it reflected whether the stimulus was meaningful to the subject. In their study, subjects saw intermixed sequences of visual flashes and numbers and had to respond to questions about each pair of numbers; the visual flashes were to be ignored. They found that the P300 was generated only in response to the number stimuli; the visual flashes did not evoke this component at all. An earlier component (e.g., what would later be called the N1) was evoked similarly by both types of stimuli, and this component was sensitive to the physical properties of the stimuli (e.g., brightness). This pattern of findings led the authors to suggest that the earlier component represented early perceptual-level processing, whereas the P300 represented a deeper level of processing related to the meaningfulness of the stimulus. Sutton, Braren, Zubin, and John (1965) extended this work, finding that the P300 was also related to expected likelihood; specifically, this positivity was larger for task-relevant stimuli that were less expected/more rare. These early findings have been replicated and extended, and this component, now referred to as the P3b, has been shown to be a useful measure of the timing and resource allocation of cognitive evaluation mechanisms (for a review, see Polich, 2007).

Another late positivity component occurring within a very similar latency range has been labeled the P3a. Whereas the P3b is located more posteriorly on the scalp, over parietal regions, the P3a is found over *frontal* scalp sites, with a slightly *earlier* peak latency, usually ~250–280 ms. The P3a is thought to index a different cognitive process than the P3b. Specifically, the P3a is often referred to as the "novelty P3" because it is evoked by unexpected and rare events. In "oddball" studies that include a few rare stimuli within streams of standard stimuli, the rare

Box 5.1 What's "neutral"? Assessing the costs versus benefits of attention

Studies of attention often compare "attended stimuli" to "unattended stimuli," showing that attended stimuli are responded to faster and more accurately and with greater neural processing. However, this difference can be interpreted as an *enhancement* of processing for the attended stimuli (benefit of attention), or a *suppression* of the unattended stimuli (cost of attention), or both. In order to determine whether attention works through enhancement or suppression, a "neutral" or "baseline" condition must also be included. In Posner's original papers using the cuing paradigm (Posner, 1980; Posner, Nissen, & Ogden, 1978), it was found that reaction time and accuracy were better for the items at the attended location relative to a neutral condition (i.e., a *benefit* of attention) and worse for the unattended location items relative to the neutral condition (i.e., a *cost* of attention being allocated away from that location). However, due in part to the complexity of defining what attention is and how it is allocated, there are different possible neutral conditions, and they may not all produce the same results. This means that findings of enhancement or suppression may depend on the type of neutral condition that is used. In Posner's original study, for example, the directional cue was an arrow presented at central fixation that indicated the likely location of the upcoming target, whereas the neutral cue was a plus sign ("+") presented at fixation that gave no information about the likely location of the target. While it is clear that the arrow cue should result in orienting and maintaining attention to the indicated location, it is not as clear what subjects are doing with the neutral cue. Although the assumption is that they simply are not orienting attention on neutral trials, they could be doing various other things instead, such as: focusing attention on the central fixation spot; widening their "zoom-lens" and trying to attend to both locations at once; rapidly orienting back and forth between the two possible target locations; or simply guessing on some trials and picking one location to attend. Of course, subjects could also be trying each of these different strategies at different times throughout the experiment. Thus, it is difficult to assess absolute costs versus benefits of attention because of the ambiguity of the neutral condition. Indeed, Wright, Richard, and McDonald (1995) found that the presence of costs and benefits depended on the type of neutral cue that was used. When the neutral condition was two simultaneous cues, one at each of the two possible target locations, they found a *cost* of attention (slower responses to invalid than neutral cues) but *no benefit* (responses to valid and neutral cues were equally fast). When the neutral cue was a brief flash of the entire screen, however, both costs *and* benefits were found. In addition, only the least bright of the whole-screen flash intensities resulted in this pattern; for more intense screen flashes, only costs were found. Thus, it is important to carefully consider the neutral condition in any study that proposes to show absolute costs or benefits of attention.

oddball events trigger a P3a (for a review, see Donchin, 1981). Unlike the P3b, the P3a can be triggered by task-irrelevant stimuli. As long as the "oddness" of a stimulus can be discerned, a P3a may be evoked by it, and it has been suggested that this may index the triggering of attention orienting. The P3a will be discussed more in Chapters 6 and 7 when the exogenous control of attention is discussed, because this component has been linked to the involuntary orienting of attention triggered by surprising events. The effects of attention on memory and higher-order cognitive processes will also be discussed in subsequent chapters.

5.4   Where Attention Acts in the Brain

Before describing where attention affects information processing, it is useful to briefly review the visual system and describe the pathways from the eye to higher-order visual areas. The majority of output from the retina goes to the lateral geniculate nucleus of the thalamus (LGN) and then on to the cortex, specifically to the "striate cortex" located in the medial posterior occipital lobe. The striate cortex (also known as Brodmann area 17) is the region of cortex surrounding the calcarine sulcus, and its name comes from its striped appearance compared to the surrounding cortex. Francesco Gennari first described this part of the cortex in 1782 and noted that its striped appearance was different from the surrounding cortex (Gennari, 1782). This identification of a part of the brain that was different from surrounding areas was a critical step in moving scientists beyond the notion that all parts of the brain contributed equally to all functions. By showing that part of the brain was structured differently, this led to the idea that different parts of the brain might do different things, which in turn led to efforts to localize specific mental functions to different brain areas. We know now that the stripe identified by Gennari is a band of myelinated axons from the LGN terminating in layer IV of the striate cortex. Approximately 90% of outgoing fibers from the retina are along the retinogeniculate pathway that projects to the striate cortex. Most of the remaining output from the retina goes along the retinocollicular pathway, which connects the retina to the superior colliculus and on to the pulvinar nucleus of the thalamus. Given that the vast majority of visual processing happens along the retinogeniculate pathway through to multiple cortical visual processing areas, most attention research has focused on this pathway. We will revisit the retinocollicular pathway in Chapter 7 when discussing involuntary influences on the control and orienting of attention.

Studies of visual processing going back to Hubel and Wiesel (1959, 1962) have shown that the mammalian brain contains a series of visual processing "areas," typically referred to as V1, V2, V3, V4, etc. (Felleman & Van Essen, 1991). The numbers refer to the sequence by which visual information from the eyes enters and progresses through the cortex (Figure 5.5). Progressing from the retina to the LGN to V1 and on to the higher areas, the receptive fields of the individual cells become larger and the cells are responsible for processing increasingly more complex stimuli. Although much of the work defining the different visual areas came from cellular recordings from nonhuman animals, functional magnetic resonance imaging (fMRI) studies are now able to map out the borders of visual areas in humans (Engel, Glover, & Wandell, 1997; Sereno, Dale, Reppas, Kwong, Belliveau, Brady, Rosen, & Tootell, 1995). This mapping of human visual areas is possible because of the knowledge gained from animal studies that each visual area contains a complete map of one quadrant of visual space (i.e., upper left, upper right, lower right, lower left) and that each successive visual area in the brain codes that space in a mirrored direction from the previous area. In other words, if one imagines looking at

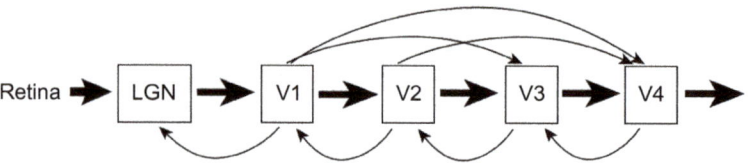

Figure 5.5 Visual processing regions. A very simplified representation of the first few stages in the primate visual system; LGN = lateral geniculate nucleus of the thalamus; V1 = first region in the cortex to receive visual information; V2 = second, etc. Many connections and areas are not shown in this diagram, but these first few areas have been of particular interest to attention researchers. Note that some higher-order areas beyond V4 can receive information directly from lower areas, bypassing intermediary areas, and the retina also projects to the superior colliculus (not shown here).

the center of a clock, one visual area in the left hemisphere would map the upper right visual field staring at the 12:00 position and progressing through that space in a clockwise manner until reaching the 3:00 position. The next visual area would start at the position of 3:00 and progress though space in a counterclockwise direction back up to the 12:00 position; the next visual area would then progress clockwise from 12:00 to 3:00, etc. Therefore, borders between visual areas can be identified by marking where the mapping reverses. This method of defining the visual areas of the cortex is referred to as "retinotopic mapping" because each visual area maintains the type of smooth and continuous representation across space that is encoded by the retina. In fMRI studies, the identification of visual areas is done by having subjects view a flashing stimulus (typically a thin wedge, with its point at a central fixation location and its wide edge at the periphery) that rotates progressively through 360 degrees around the central point, covering both visual hemifields. V1 in each brain hemisphere contains a complete map of the contralateral visual space. After V1, there is a split into dorsal and ventral visual areas for areas V2 and V3, in which upper visual field stimuli are processed in the ventral stream and lower visual field stimuli are processed in the dorsal stream. What's critical for the mapping, however, is that each subsequent area has a map that is mirror-reversed from the preceding area. Therefore, fMRI studies can map out the borders between visual areas (for detailed instructions, see Warnking, Dojat, Guérin-Dugué, Delon-Martin, Olympieff, Richard, Chéhikian, & Segebarth, 2002).

5.4.1   Neuroimaging Studies of Spatial Attention Effects

Whereas ERP studies have been instrumental in revealing the time course of *when* attention can first affect stimulus processing in the human brain, that method has limited ability to localize the precise neural sources of that activity. On the other hand, neuroimaging methods such as fMRI and positron emission tomography (PET) have excellent spatial precision but are very limited in identifying the timing of activity because they measure blood flow or a derivative of it. The earliest neuroimaging studies of the effects of spatial attention on visual processing used PET. Heinze and colleagues found that spatial attention resulted in robust effects in the fusiform gyrus, contralateral to the direction of attention (attention to a location in the left visual field resulted in enhanced processing in the right fusiform gyrus, and attention to the right visual field boosted processing in the left fusiform gyrus; Heinze, Mangun, Burchert, Hinrichs, Scholz, Münte, Gös, Scherg, Johannes, Hundeshagen, Gazzaniga, & Hillyard, 1994). The exact visual area corresponding to the fusiform gyrus attention effect could not be determined in that study because retinotopic mapping isn't feasible with PET. The reason for this is that only one condition can be run every 10 minutes or so when using PET, and therefore it would take exceedingly long – and require too many injections – to acquire all the conditions necessary to map out the visual areas. Although that study was unable to determine the exact visual area corresponding to the attention effect in the fusiform gyrus, that gyrus extends into the temporal lobe, and it is clearly separate from the striate cortex. Thus, this study was able to conclude that the attention effect was occurring in extrastriate cortex and outside of V1. Subsequent PET studies replicated and extended these results, finding that, in addition to the fusiform gyrus, the middle occipital gyrus – another extrastriate visual processing area – also showed effects of spatial attention (Mangun, Hopfinger, Kussmaul, Fletcher, & Heinze, 1997). fMRI studies of spatial attention further confirmed these results, showing that the attention effects in the fusiform and middle occipital gyri were robust and identifiable in individual subject data (Mangun, Buonocore, Girelli, & Jha, 1998). These studies all used stimuli in the upper visual field, and thus the extrastriate attention effects were observed predominantly in the more ventral visual processing regions. A PET study using lower visual field stimuli found extra-striate attention effects in more dorsal regions, corresponding to the split for lower- versus upper-field stimuli (Woldorff, Fox, Matzke, Lancaster, Veeraswamy, Zamarripa, Seabolt, Glass, Gao, Martin, & Jerabek, 1997). In all of these studies, attention effects were found in extrastriate cortices but not in the striate cortex, in line with conclusions from ERP studies that attention doesn't typically modulate processing in V1. In order to localize the visual area(s) showing these attention effects, fMRI can be used to perform retinotopic mapping for each of the subjects. Using these techniques, Hopfinger, Buonocore, and Mangun (2000) were able to localize the spatial attention effect in the fusiform gyrus to area V3 (Figure 5.6). In that study, an additional attention effect was also observed in a more dorsal area, but this was outside of the visual areas (V1–V4) that could be identified with the retinotopic mapping methods utilized.

Another cognitive neuroscience technique that has been utilized in the study of spatial attention is **event-related optical signals (EROS)**, a method related to functional near-infrared spectroscopy (fNIRS); both are sometimes referred to as "optical imaging." As described in Chapter 4, fNIRS involves shining near-infrared light into the brain and observing the reflection on nearby sensors on the scalp. This technique relies on differences in reflectance between oxygenated and deoxygenated hemoglobin, which varies depending on the activity of nearby neurons. When an area becomes more active, glucose utilization goes up, which involves taking more oxygen from the bloodstream, which results in an increase in deoxygenated hemoglobin relative to conditions involving less brain activity in that region. fNIRS is thus dependent on the sluggish blood oxygenation-level dependent (**BOLD**) signal that fMRI measures and that peaks ~6 seconds after the neural activity. In contrast, EROS is able to measure neural activity with high temporal resolution because it measures the scattering of infrared light that occurs directly when neurons are active (reviewed in Gratton, Chiarelli, & Fabiani, 2017). Although optical imaging techniques cannot localize activity precisely and they are not able to detect deep

Selective attention (targets)
(single subject)

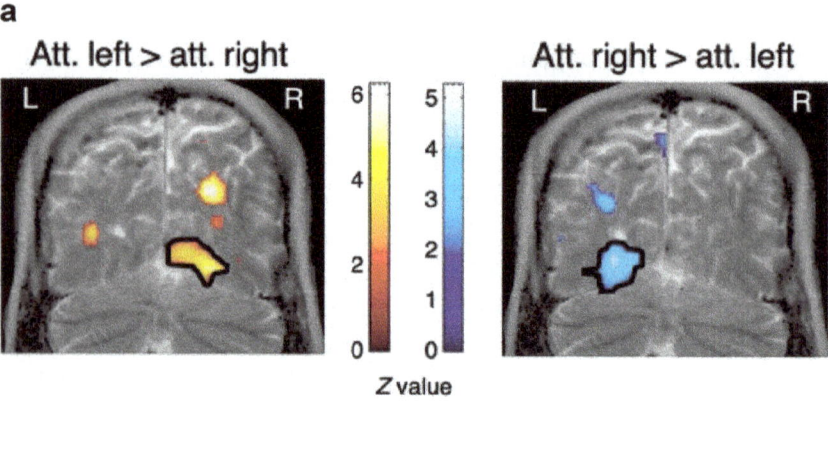

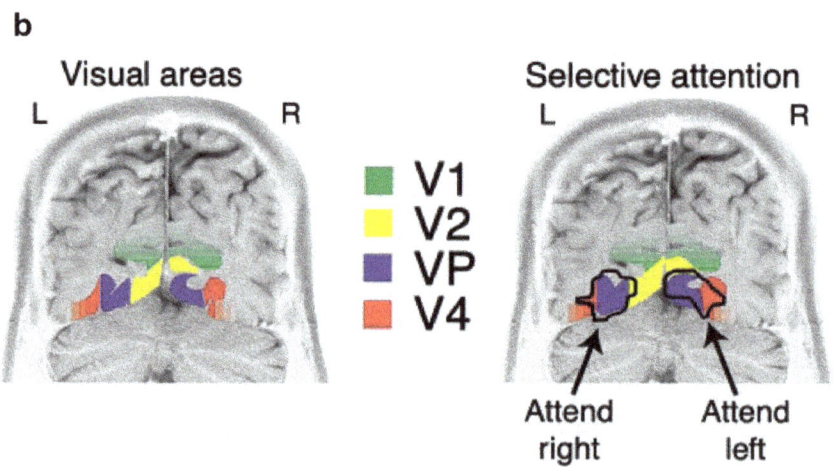

Figure 5.6 Attention effects in V3/VP. Top (a) shows the locations of spatial attention effects from an fMRI study, with a thick black outline around the attention effects in early visual processing regions. Bottom left (b) shows the location of visual areas V1–V4 defined from separate retinotopic mapping scans. Bottom right (b) shows attention effects overlaid onto visual areas, illustrating that the attention effects occurred in VP (ventral-posterior). Note that VP in this figure corresponds to what is now commonly called ventral V3. Reprinted from Hopfinger et al. (2000), *Nature Neuroscience*, with permission from Springer Nature.

sources well, the temporal resolution of EROS is much greater than fMRI, and it can measure activity from neural sources near the optical sensors quite well. In one of the first studies using EROS to study attention, Gratton (1997) positioned sensor probes on the back of the head to measure activity near the occipital pole (corresponding to activity from V1) and more laterally (corresponding to extrastriate processing). This study found no effects of attention in early time windows at the occipital pole (likely corresponding to activity reflected in the C1 ERP component), but it did find activity over extrastriate regions, in a similar time window to P1 attention effects in ERP research. Thus, across multiple methods, there was general consistency in the finding of strong attention effects in early extrastriate visual processing areas without attention effects in the striate cortex or earlier (but see Box 5.2).

Using large full-hemifield checkerboard stimuli that could evoke robust BOLD signals in subcortical (as well as cortical) regions, O'Connor and colleagues (O'Connor, Fukui, Pinsk, & Kastner, 2002) presented fMRI evidence that spatial attention significantly boosted activity in

---

Box 5.2 Can attention affect the earliest stage of cortical processing?

The C1 ERP component is considered the most reliable marker of the first cortical processing stage for incoming visual information. Source modeling of ERP and event-related field (ERF) data localize this component to the striate cortex, and combined neuroimaging studies have identified its source as V1. Most studies have found that the C1 is not modulated by spatial attention, but a few EEG experiments have reported attention effects coming from the striate cortex (reviewed in Slotnick, 2012). There has been debate, however, about whether those can be taken as evidence for attention affecting the earliest cortical processing. In one of those studies (Slotnick, Hopfinger, Klein, & Sutter, 2002) the latency of the effect localized to the striate cortex was later than the standard C1 latency, and in another (Rauss, Pourtois, Vuilleumier, & Schwartz, 2009) the topography of the early effect suggests it may not have been evoked in the striate cortex. A third study found a significant effect of spatial attention on the C1, and with the typical latency and scalp distribution for that component (Kelly, Gomez-Ramirez, & Foxe, 2008). These authors suggested that the reason they were able to find C1 effects is that they accounted for individual difference in anatomy of the calcarine sulcus when determining the placement of the stimuli to ensure that a robust C1 could be recorded from each subject. This, they argued, was key to obtaining a measure of the C1 that was sensitive enough to reveal attention effects. However, when Baumgartner and colleagues (Baumgartner, Graulty, Hillyard, & Pitts, 2018) attempted to replicate that study, they failed to find any attention effect on the C1 despite replicating the longer latency attention effects (e.g., P1, N1). Thus, further research is needed to understand the specific conditions that may lead to C1 attention effects in ERP studies. Adding to the debate, a recent MEG study found spatial attention effects at 60–120 ms (Kurki, Hyvärinen, & Henriksson, 2022). The 60-ms latency would normally be considered within the range of the first volley of activity in V1. However, the authors note the standard grand-average waveform analysis only found an early attention effect to stimuli in one of the two hemifields. When using multivariate analyses that combine data from across sensors and model differences between conditions, the more robust effect at 60 ms was revealed. Interestingly, the multifocal stimuli and orthogonal sequences of flashing patterns used in that study are quite similar to those of the Slotnick et al. (2002) study that also localized attention effects to the striate cortex. Together, these results suggest that the very dense field of stimuli and the complex orthogonal presentation sequences used in those studies could be important factors in understanding when attention might act on the earliest processing in the striate cortex. Overall, the majority of ERP and MEG studies find that spatial attention affects processing only *after* the initial processing in V1, but the site of the earliest possible stage of attention selection remains an open question.

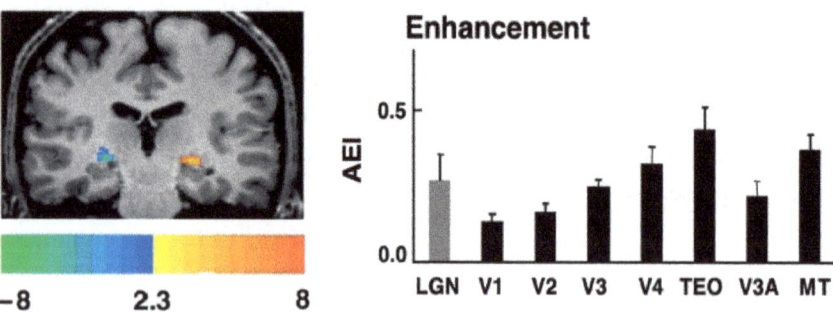

Figure 5.7 Attention effects in the LGN. The left panel shows the areas of the LGN activated by the large checkerboard stimuli used in this study. This image is not showing attention effects but the subtraction of the effects of left visual field stimuli from right visual field stimuli (note that this image is in radiological convention, so the left side of the brain image is the right hemisphere). The right panel shows the Attention Enhancement Index (AEI) for the LGN and a number of the visual areas defined by separate retinotopic mapping scans. MT = middle temporal; TEO = temporal occipital; V3A = V3 accessory; Adapted from O'Connor et al. (2002), *Nature Neuroscience*, with permission from Springer Nature.

the LGN (Figure 5.7). In addition, that study included separate retinotopic mapping fMRI scans to identity the first few visual areas. As shown in Figure 5.7 (right panel), attention effects were found throughout the visual areas. This study was important for showing that spatial attention could modulate processing in V1 and, furthermore, could even affect subcortical visual processing in the thalamus. Regarding activity in V1, Kelley, Rees, and Lavie (2013) reported evidence that areas as early as V1 show effects of attention and competition. Specifically, V1 activity to a target stimulus was enhanced by attention when there was a response-incongruent item also present, but there was no change in V1 when a neutral distractor (a stimulus that never required a response) was present. This suggests that it is not simply the presence of another stimulus but rather the presence of a competing stimulus that engaged the attention effects at this level. Overall, these findings of attention effects in the LGN and V1 could be taken as evidence *against* the conclusions from ERP and EROS studies that the earliest effects of attention occur *after* V1. In order to understand how these seemingly contradictory findings can be resolved, we turn to studies that have combined methods to provide measures of attentional selection with high spatial and temporal resolution.

5.4.2 Combining Methods for an Enhanced Spatiotemporal Localization of Attention Effects

As described in Chapter 3, PET and fMRI have poor temporal resolution and therefore cannot determine the timing or order of when different brain areas are active. In regards to attention studies, this means that these methods can't assess which attention effects occurred first, nor whether the change in activity in those regions was during the initial feed-forward pass through those areas or was instead a modulation of activity after feedback from higher areas had arrived. This is an important point as it relates directly to the early versus late filter theories of attention discussed earlier. Note that these are often referred to more generally as early versus late *selection* theories, in relation to the level of processing at which attention selects information to be modulated. Since the same area in the brain can be active at multiple points in time, in order to fully understand the processes and mechanisms of selective attention, the timing of the attention effect(s) must also be known. In order to obtain the temporal and spatial resolution necessary to understand exactly when and where attention effects are happening in the brain, researchers have combined ERPs with neuroimaging (Box 3.2 in Chapter 3 explained the general benefits and limitations of combining neuroimaging and ERPs).

In one of the very first studies to combine these methods, Heinze and colleagues had subjects perform a visual-spatial attention task while either EEG or PET data were collected in separate sessions (Heinze et al., 1994). Using source-modeling techniques to help localize the neural regions generating the scalp-recorded ERPs and seeding these models with the PET results, the authors determined that the earliest effect of spatial attention was an enhancement of activity in the fusiform gyrus beginning at 80 ms, as indexed by the P1 component. A subsequent study combining ERPs and PET replicated and extended these findings, demonstrating that spatial attention enhanced the P1 and activity in the fusiform gyrus, but also that a significant attention effect occurred in the middle occipital gyrus (Mangun et al., 1997). In that study, the P1 component and the PET activity in the fusiform gyrus showed the same main effects and interactions between attention and task difficulty. This covariation across tasks and attention conditions further strengthens the conclusion that the brain activity recorded during the P1 is generated in the extrastriate fusiform gyrus. Furthermore, using source-modeling procedures, the PET attention effect in the middle occipital gyrus was associated with ERP activity ~200 ms after the P1 (Mangun et al., 1997). As described in Mangun, Hopfinger, & Heinze (1998), data from different methods can be combined in a variety of ways. Conducting source analyses, especially with MRI-informed, subject-specific realistic head models (as done in Mangun et al., 1997), can be very effective, but it is also important to look for covariations of effects measured by the different methods across task and stimulus conditions to ensure that the effects across methods are in fact tightly linked.

Although there is strong evidence that the P1 attention effect at ~90 ms can be localized to extrastriate processing, fMRI studies have found attention effects in the striate cortex (e.g., Brefczynski & DeYoe, 1999; Wordon & Schnider, 1996) and even subcortically, in the LGN (O'Connor et al., 2002). These findings could be taken as evidence that attention acts earlier than the extrastriate locus of attention effects found with the combined PET/ERP studies. In addition, as described in Box 5.2, a few ERP experiments found evidence that spatial attention may modulate processing as early as the C1 component, which has been localized to area V1. If the fMRI findings of attention effects in the LGN and V1 index the initial feed-forward activity from the retina, then the C1 component should show attention effects. But since the majority of the ERP evidence suggests that the C1 is not affected by spatial attention, how can we explain attention effects as early as the LGN and V1 without there being any effects on the C1? In a series of studies by Hillyard and colleagues that combined fMRI with ERPs, these authors provided critical evidence that fMRI effects in the striate cortex are related to ERP effects occurring well after the C1 component. Firstly, Martinez and colleagues performed retinotopic mapping with fMRI and found attention effects in multiple visual areas from V1 through V4

(Martinez, DiRusso, Anllo-Vento, Sereno, Buxton, & Hillyard, 2001). In separate ERP sessions, there were significant attention effects on the P1 and N1 but there was no attentional modulation of the C1. The authors suggested that the V1 attention effect must therefore be generated during reentrant processing to V1 after the initial bottom-up processing has progressed through the higher visual areas.

Then, Noesselt and colleagues extended these results by including MEG in the combination of fMRI and ERPs. The results confirmed that the earliest activity from the striate cortex (60–90 ms) wasn't modulated by attention, but that a later attention effect (140–250 ms) could be localized to that same striate region (Noesselt, Hillyard, Woldorff, Schoenfeld, Hagner, Jäncke, Tempelmann, Hinrichs, & Heinze, 2002). Di Russo et al. (2003) furthered these results, showing that a later attention effect (150–225 ms) was localized to V1, and that this activity showed properties of striate-evoked activity (e.g., the polarity of the ERP activity is flipped for upper- versus lower-field stimuli due to the cruciform organization of the striate cortex surrounding the calcarine sulcus). These studies have provided strong evidence that V1 attention effects found with fMRI experiments are associated with later reentrant processing in that region, not with the initial processing (within the first 100 ms) that evokes the C1 component. This is also in line with nonhuman primate electrophysiology evidence that attention effects in the striate cortex occur at latencies (~200 ms) that are much later than the initial feedforward activity in that area (Roelfsema, Lamme, & Spekreijse, 1998; Vidyasagar, 1998). Overall, these results make sense when considering the extensive feedback connections in the brain from "higher" to "lower" stages of processing in the brain. Relating back to the O'Connor et al. (2002) results, this suggests that attention effects in the V1 and LGN are likely due to feedback to those areas, after attention has already modulated processing in higher visual areas.

5.4.3   Single-Unit Recordings

Another method that has provided critical insights into the mechanisms of visual-spatial attention is nonhuman electrophysiology. By implanting microelectrodes directly into the brain, activity from a single neuron can be recorded – this method is often referred to as **single-unit recording**. Using this method, researchers can record the frequency and firing rate (also known as the spiking rate) of individual neurons. This method can also be used to measure the ionic changes associated with neural activity in populations of neurons, often referred to as local field potentials. Due to the highly invasive nature of this method, however, this type of research is mostly limited to nonhuman animals. Intracranial depth electrodes are sometimes used in human patients for the purposes of monitoring epileptic seizures or to stimulate the brain (e.g., deep-brain stimulation treatments for Parkinson's disease), so some such research has been conducted with these patients. However, the electrodes are placed according to medical needs, not to test theories of attention, and there are limitations in interpreting data from these patient groups. Therefore, single-unit recording as a cognitive research tool is used mainly in nonhuman animal studies, and research in macaques has proven especially important for the purposes of studying visual attention because of the high similarity between the human and nonhuman primate visual systems.

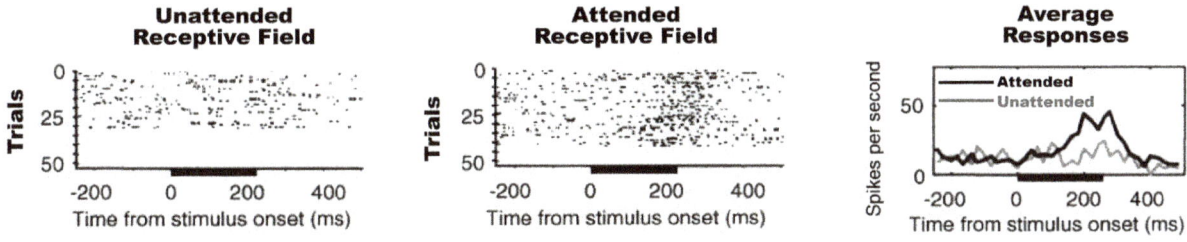

Figure 5.8 Single-unit recording during a spatial attention task. The plots show the activity of an individual neuron from area V4 in a macaque brain. The left panel shows spiking activity when a visual stimulus was presented at an unattended location; the middle panel shows when the stimulus was at the attended location. Each row in the plots represents a separate trial, and each black mark indicates when the neuron fired. The black rectangle along the x-axis of each plot shows when the stimulus was on the screen. Attention resulted in a higher firing rate, especially shortly after the stimulus appeared. The right panel shows the average response of the cell over all trials, separated by attended versus unattended condition; an attention effect is clearly seen beginning around 100 ms. Figure adapted from Reynolds et al. (2000), *Neuron*, with permission from Elsevier.

Since this method provides excellent spatial and temporal precision, it has provided important insights into the mechanisms of how spatial attention modulates visual processing in the brain. Studies using this method have shown that stimuli in attended regions of space produce enhanced firing rates compared to when those objects occur in ignored locations (e.g., Moran & Desimone, 1985; Spitzer, Desimone, & Moran, 1988). As shown in Figure 5.8, the spiking activity in V4 is enhanced for attended versus unattended stimuli (Reynolds, Pasternak, & Desimone, 2000), in agreement with the human studies described earlier showing robust effects of attention in extrastriate visual processing areas. Some single-unit research has reported attention effects in V1 (e.g., McAdams & Reid, 2005; Motter, 1993), but studies typically find more robust attention effects in higher visual areas, such as V4 (Luck, Chelazzi, Hillyard, & Desimone, 1997). In addition, some of the V1 attention effects are consistent with feedback from higher regions (e.g., McAdams & Reid, 2005). Thus, overall, there has been agreement between nonhuman single-unit studies and human neuroscience studies regarding the level at which spatial attention affects visual processing.

However, in addition to simply showing that attention modulates processing in these visual regions, single-unit studies have helped reveal the processes and mechanisms by which attention modulates processing. Two of the most important of these mechanisms are: (1) attention functioning to bias and resolve competition and (2) synchrony as a way of enhancing information processing. In regards to the former, single-unit studies were instrumental in the development of the biased-competition model of attention (Desimone & Duncan, 1995). This model stresses that attention is needed to resolve competition and that the level at which attention effects are observed depends on the level at which that competition is present. To understand this model, it is helpful to review receptive fields. The receptive field of a neuron is the region of space to which a neuron responds, and the size of the receptive field increases across the visual hierarchy from the retina to the LGN and through at least the first few visual areas (in areas beyond V4 receptive fields already encompass most of the whole visual field). In addition to a spatial location, a neuron will also be optimally sensitive to a particular feature (e.g., a specific orientation; a specific color) depending on the visual area. For example, a neuron may be highly active to a certain "preferred" stimulus (e.g., horizontal lines) in a specific region of space but may not be very active to other "nonpreferred" stimuli (e.g., vertical lines) in that same region of space. (Note that other neurons in the same visual area would have different preferences, so over an entire visual region there would be equal numbers of neurons coding the different stimuli.) Research on attention has revealed that when both preferred and nonpreferred stimuli are presented at the same time within a neuron's receptive field, the neuron's response depends on attention. Specifically, if the animal is not attending to that region of space at all, the neuron's response will be approximately an average of the response it shows to the preferred and nonpreferred stimuli when presented alone. But if the animal is attending to that region of space, the neuron's response is biased strongly by the attended stimuli. In this example, if the animal is attending to the preferred stimulus for the cell being recorded, that neuron's response to the two stimuli will be close to the firing rate it shows when only the preferred stimulus is present; if the animal is attending to the other stimulus, however, the neuron's response to those same two stimuli will be greatly reduced, with a firing rate close to that when the preferred stimuli is not even present. Luck et al. (1997) showed that these attention effects were observed in V2 and V4 when the competing stimuli were presented simultaneously and when both stimuli fell within the receptive field of the cell being recorded. If the stimuli were not both within the receptive field or if the stimuli were not presented at the same time there were much weaker or completely absent effects of attention. The researchers argued that this shows that attention is engaged at the level(s) of processing where it is needed to resolve competition.

A second major concept to come from single-unit research is that the modulation of the synchrony of firing, both within and across areas, is a critical mechanism of attention. An increase in the synchrony of firing among cells within a region can result in a more effective transmission of information to other areas because the spikes of activity will arrive at the subsequent area at the same time. That stronger input could in turn increase the likelihood of triggering activity in the subsequent region, boosting communication throughout the system. It has also been suggested that the synchrony of firing across different areas could be important for the binding of different types of features together as part of an object. In addition, different rates of firing have been associated with different functions, and therefore synchronizing firing patterns within a region could ensure that the proper signal is efficiently transmitted. For example, Fiebelkorn and Kastner (2019) have proposed a **rhythmic theory of attention**, which posits that low-frequency oscillations organize regions into alternating states to avoid conflicts between sensory acquisition and oculomotor movements. Behavioral patterns of responses provide evidence for this rhythmic sampling of the environment, and this sampling has been linked to theta-band activity (3–8 Hz) in regions associated with the control of attention (e.g., frontal eye fields, lateral intraparietal area, and the pulvinar; these regions and the control processes of attention will be covered in more detail in Chapter 7). The effect of this control is observed as alternating states of increased gamma-band activity (30–100 Hz), which has been associated with attentional enhancements of visual processing, versus increased alpha-band activity (8–12 Hz), which has been associated with reduced sensory processing. Regarding activity in area V4, gamma-band activity (~30–80 Hz) shows a significant *increase* in spike-field coherence in this area for attended versus ignored stimuli, while alpha-band activity (~9–11 Hz) in V4 is associated with a significant *decrease* in coherence for attended versus ignored stimuli (Bichot, Rossi, & Desimone, 2005; Fries, Womelsdorf, Oostenveld, & Desimone, 2008; Taylor, Mandon, Freiwald, & Kreiter, 2005).

Information from single-unit recording has thus greatly advanced our understanding of the mechanisms by which attention affects visual processing. However, for understanding mental processes that involve multiple areas, these techniques can be limited in that they are not able to sample from areas across the whole brain at the same time. Advanced methods can provide single-unit data from multiple areas, but those still can only sample from a very small set of regions at the same time. Therefore, techniques that sample the entire brain simultaneously (e.g., fMRI, EEG, MEG) remain critical for understanding the full picture of what attention does across the brain. Regarding neural firing rates and the effects of attention, studies in humans measuring the neural activity of regions with high temporal precision (i.e., EEG and MEG) have also found effects of spatial attention within different frequency bands and on the synchrony of activity across regions. Although EEG and MEG studies cannot resolve the activity patterns of individual neurons, these studies have found evidence at the level of regions of the cortex that focused spatial attention is associated with reduced alpha activity (Thut, Nietzel, Brandt, & Pascual-Leone, 2006; Worden, Foxe, Wang, & Simpson, 2000) and enhanced gamma activity (Gruber, Müller, Keil, & Elbert, 1999; Müller, Gruber, & Keil, 2000).

5.5 Advanced Questions on the Mechanisms of Visual-Spatial Attention

The previous sections in this chapter described cognitive neuroscience studies that have provided critical insights into when and where in the brain spatial attention can affect visual processing. Results across these investigations, however, raised further questions about the mechanisms of how attention works in the brain. The robustness and replicability of the spatial attention effects described in the previous sections of this chapter made it possible to use these indices of attention to test further theories of the mechanisms of attention. A number of these theories originated from behavioral studies, as described in Chapter 1, such as whether the spotlight of attention can be split, if attention is allocated in an all-or-none manner, and the possibility of an inhibitory surround. Each of these is reviewed briefly in the following subsections.

5.5.1 Splitting the Spotlight: Can Attention Be Divided Across Spatial Locations?

As described in Chapter 1, a common analogy for spatial attention is the spotlight. The idea of a spotlight confirms our intuition of a unitary focus of attention. Early attention research suggested that a zoom-lens metaphor better accounted for our ability to spread attention out over different-sized regions of space. However, other theories suggested that attention doesn't need to be a unitary focus but rather could be split across multiple locations. The key difference between these theories is how intervening locations are processed. According to the spotlight and zoom-lens models, any locations between to-be-attended locations must also be processed as being attended. According to the theory that attention can be split, however, intervening locations would not be attended at all. The first ERP studies that investigated the possibility of splitting the spotlight of attention reported evidence that the spotlight could *not* be split. Heinze and colleagues performed a study in which subjects were asked to pay attention to two locations and detect targets there, while probe items that did not require any response (but that generated a strong ERP signal) could be presented at one of the cued locations or at the location between the two locations that needed to be attended (Heinze, Luck, Munte, Gös, Mangun, & Hillyard, 1994). The results showed that the intervening location produced the same P1 attention effect as the two to-be-attended locations. Thus, attention wasn't able to split across two noncontiguous locations but rather was widely focused across all three locations. This study also provided support for a zoom-lens model, in that the P1 attention effect was larger when attention was allocated across two adjacent positions as opposed to across three locations.

In a series of studies, Eimer (1999, 2000) used ERPs to investigate the flexibility of the spotlight of attention by having subjects attend to ring-shaped regions around fixation. In those studies stimuli could appear at different eccentricities around the central fixation point, and subjects were instructed in different conditions to attend to one or more of these rings. Critically, when subjects were instructed to attend to the more eccentric rings, stimuli at the inner rings should be ignored. However, the results showed that in those conditions stimuli at the inner rings produced the same attentional effects on the P1 and N1 as did stimuli at the locations that were supposed to be attended. This result provided additional support to Heinze et al.'s (1994) conclusion that the attentional spotlight could be spread out like a zoom-lens but not split to exclude intervening locations. However, although Eimer's results on early attention effects supported the unitary focus of attention, there were effects on postperceptual processing stages (negativities at ~200 ms) that suggested that at later levels of analysis attention could be split.

A subsequent study using EEG and steady-state visual evoked potentials (SSVEPs), however, found that an intervening location did *not* receive the attentional boost that the two noncontiguous to-be-attended locations showed (Müller, Malinowski, Gruber, & Hillyard, 2003). The SSVEP is the response recorded over visual processing regions in response to a sequence of stimuli presented at given rate. For example, a sequence of visual stimuli presented at 10 times per second will evoke a 10-Hz SSVEP; stimuli presented at 20 times per second will evoke a 20-Hz SSVEP. These can be recorded simultaneously from multiple streams of stimuli presented at different rates, and postprocessing selection of the relevant frequency can pull out the brain response to a particular stream. Previous studies had shown that the amplitude of the SSVEP was larger for attended versus ignored stimuli. Müller et al. (2003) presented four streams of stimuli across the horizontal midline (each at a slightly different rate of presentation), and subjects were instructed which two streams to attend to in each block. Whichever two locations were attended showed the typical attention effect of an enhanced-amplitude SSVEP. However, when the locations were nonadjacent, the intervening location did not show this amplitude enhancement; this location was being ignored, at least as measured by the SSVEP response. This was taken as strong evidence that the focus of attention could indeed be split between noncontiguous locations, without the intervening area being attended. These studies were followed by fMRI studies that also found evidence for a split focus of attention (McMains & Somers, 2004, 2005). In those studies, retinotopic mapping and localizer scans allowed the researchers to locate the borders of visual areas and to demarcate the location within each visual area that represented the regions of space where different stimuli could occur. Using these methods, they found the typical attention effects of enhanced processing in early visual areas, but they also found that intervening locations between two noncontiguous attended locations did not show attention effects. Thus, these data provided evidence that even in early visual areas (e.g., V1 and V2) the spotlight can be split.

Although these studies provide evidence that the spotlight of attention can be split, subsequent behavioral studies have raised questions regarding this point. For example, Dubois, Hamker, and VanRullen (2009) showed that the ability to divide attention across noncontiguous locations is very short-lived. In a behavioral study, they found that at short intervals (150 ms or less) after the cue, attention could be evenly split to two nonadjacent locations, excluding intervening spaces. However, at longer intervals after the cue, attention was found to be focused on just one of the locations. Thus, a very short-lived split of attention quickly became a unitary focus. The authors suggest that some previous support for a split of attention focus could have been due to sampling responses in the brief period when attention is split, or due to a rapid switching of attention from one attended location to the other. Jans, Peters, and de Weerd (2010) evaluated the many studies on this topic spanning psychophysics, electrophysiology, and neuroimaging. On the basis of four criteria that should be met to rule out alternative explanations (specifically: sufficiently difficult task, brief stimulus presentation, appropriate cue–target interval, and dense probing of nontarget locations), they concluded that no single experiment has been able to conclusively show a true splitting of attention. Furthermore, the authors argue that the ability to split the spotlight of spatial attention is not an easy or natural function, but rather is a specific skill that would have to be learned with training. This also implies that evidence for a splitting of attention may be quite specific to the particular experimental setup and task design being used in the tests. Therefore, although there is support for the idea that the spotlight of attention *can* be split, there is continued debate as to whether that ability is a general mode of allocating attention that can be used whenever needed in the real world or whether it is more of a laboratory phenomenon that only occurs after training and practice.

5.5.2 Does the Spotlight Have an Inhibitory Surround?

The spotlight and zoom-lens analogies for spatial attention typically represent attention as illuminating the region within the spotlight while everything outside of it is unaffected. However, a number of studies have reported that the area immediately surrounding the attended region is actually affected, specifically by being actively suppressed. The general concept of an inhibitory surround is well established in basic visual processing, as Hubel and Wiesel's original studies showed that the receptive fields of visual cortical neurons are highly responsive to stimuli within a specific area and are inhibited by stimuli in the immediate surrounding area (Hubel & Wiesel, 1959). The original accounts of attention hadn't presumed that the mechanisms of attention could also include an area of suppression surrounding the attended region. Mounts (2000) provided one of the first direct tests of a possible inhibitory region surrounding the spotlight of attention. In visual search experiments, participants responded most slowly to the items that surrounded the item being attended, with items at farther distances not showing this suppression relative to a control condition in which attention wasn't focused on one location. The exact stage of processing of this inhibitory surround wasn't clear from this initial behavioral study, but subsequent studies used cognitive neuroscience methods to address this question. Slotnick, Hopfinger, Klein, and Sutter (2002) used EEG to record visual cortical activity while subjects performed a visual detection task amid an ongoing sequence of checkerboard patterns flickering in orthogonal sequences at 60 locations distributed across the central 15 degrees of the visual field. Unlike standard ERP analysis procedures, the high degree of temporal overlap in this study required specialized procedures involving cross-correlating each electrode's response with the specific sequence of each checkerboard location. Dipole modeling was then performed with these responses to estimate the size of neural response from the cortical generators. In this study, the dominant response was localized to the striate cortex. For the task, subjects were instructed to attend to one specific patch of checkerboard in the upper right visual field and count the number of times a small dot in the center of that patch changed between red and green. In the control condition, subjects performed the color-change counting task for a dot overlying the central fixation point. The results revealed that the attention to the peripheral checkerboard boosted evoked activity at that location and suppressed activity at the patches immediately surrounding that location. At greater distance from the attended region there were no effects of attention.

In a subsequent study of spatial attention, Slotnick, Schwarzbach, and Yantis (2003) used fMRI and subject-specific retinotopic mapping to precisely identify the borders of the first few visual areas of cortex. They results showed that the inhibitory surround can be detected within multiple visual regions, from V1 through V3. Furthermore, they found that the inhibitory surround in V1 started at closer spatial locations than the inhibitory areas of V2 and V3. This aligns with the receptive field sizes of neurons in those areas and suggests that the inhibitory surround of visual space may be optimized separately to best help reduce uncertainty in each visual area. Although this experiment was able to show that V1 exhibits an inhibitory surround, the fMRI data do not have the temporal resolution to determine whether that V1 activity represents the initial bottom-up volley of activity evoked by the stimulus or if it instead reflects feedback from higher visual areas.

In order to provide greater temporal precision, Hopf, Boehler, Luck, Tsotsos, Heinze, and Schoenfeld (2006) conducted a spatial attention study using MEG. In that study, they replicated the finding of an inhibitory surround, in that visually evoked activity was significantly reduced to stimuli occurring at a location adjacent to the attended location (Figure 5.9). The MEG data were broadly localized to posterior occipital areas, so it wasn't clear which visual area or areas were contributing most to the suppression effect, but the time course results revealed that the suppression was occurring at between 130 and 150 ms. This latency is later than the latency at which the first bottom-up wave of activity would be processed in V1, suggesting that at least for the effect that could be observed in this study, the inhibitory surround was being generated at some point after visual information had first entered higher visual areas.

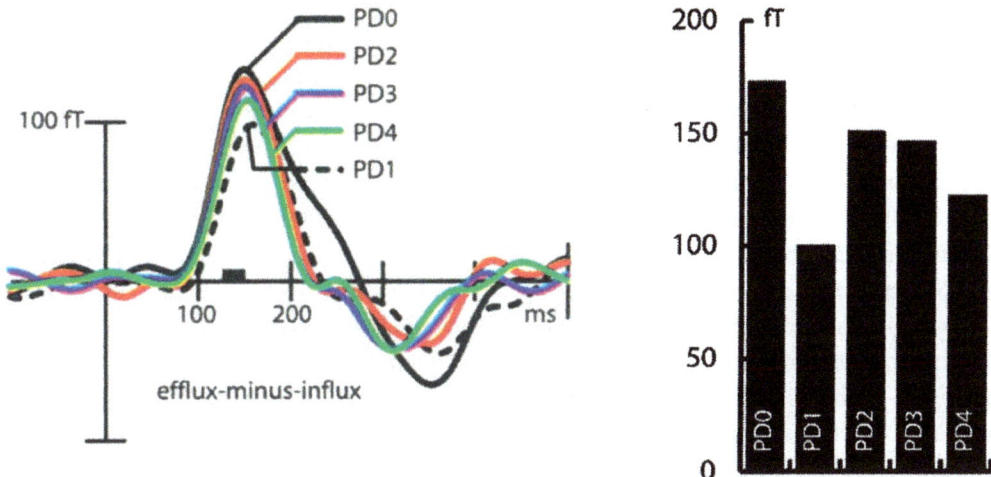

Figure 5.9 MEG investigation of an inhibitory surround. Event-related magnetic fields (ERFs) evoked by nontarget visual "probe" stimuli and generated in posterior occipital region. PD0 (Probe Distance 0) refers to a probe that occurred at the attended (target) location, PD1 (Probe Distance 1) is when the probe occurred one location away from the attended location, PD2 is when the probe occurred two locations away from the attended location, etc. The small black rectangle on the x-axis represents the time period (130–150 ms) that is plotted in the bar graph at the right, showing the mean activity level for each probe location. As can be seen, the MEG response is largest at the attended location and significantly suppressed at the immediately adjacent location. Probe locations farther away from the attended target location show an intermediate level of activity. Reprinted with permission from Hopf et al. (2006), *Proceedings of the National Academy of Sciences*. Copyright (2006) National Academy of Science, USA.

5.5.3 An Inhibitory "Surround" for Feature Attention?

In addition to spatial attention, feature-based attention also has also been found to have an inhibitory surround. Störmer and Alvarez (2014) investigated attention to *color* and found that colors that were close to the attended color (e.g., orange when attending to red) were relatively suppressed compared to colors that more distinct from the attended color (e.g., yellow when attending to red). In addition to the behavioral effect of reduced accuracy, the authors found evidence of neural suppression from SSVEPs – oscillatory potentials from visual cortex that have the same frequency as the driving stimulus and are modulated by spatial attention. Störmer and Alvarez found that the SSVEP was largest (in amplitude) for flickering colors that matched the attended color, suppressed for colors that were close to the attended color, and back to baseline for colors more easily distinguished from the attended color. Recent work by Liu and colleagues used a challenging dual-task paradigm to investigate attention to the feature of *orientation* (Liu, Fang, & Saba-Sadiya, 2023). Their task consisted of three streams of rapidly presented circular-shaped stimuli with gratings at different orientations presented centrally and to the left and right of fixation. Subjects had to detect the possible presence of a target of predefined grating orientation in the central stream while also monitoring for a change in luminance in either of the two peripheral streams. The critical manipulation in this study was that the orientation of the gratings in the peripheral stimuli, which was irrelevant to both tasks, could exactly match the orientation that subjects were attending to in the central stream (0 degrees offset) or could vary from it by 22.5, 45, 67.5, or 90 degrees offset. The results showed that subjects performed the peripheral dimness-detection task best when the (irrelevant) orientation of the peripheral stimuli exactly matched the attended orientation for the center task, as would be expected based on feature-based attention. Surround inhibition was also found, as performance was worst when the orientation was close to the attended orientation (22.5 and 45 degrees offset), became progressively better with increasing offset (67.5 degrees), and was almost completely recovered from any suppression at 90 degrees offset. This study also included an easier version of the task in which the nontarget stimuli presented in the central stream were always clearly distinct from the target in terms of orientation of the gratings. In that simpler task in which there wasn't a need to finely discriminate the orientations to detect the target because it was very distinct from the other stimuli presented in the center stream, there was no evidence of an inhibitory surround. In that easier condition, performance was still best at 0 degrees offset but gradually decreased as the peripheral orientation was progressively farther from the attended orientation. In a second experiment, surround suppression was also not found when the target orientation for the central stream was made to be a wide range (across 40 degrees) of orientations. In that condition, subjects were presumably not attending to one specific orientation, so a fine-grained surround inhibition of similar orientations would not be helpful. Thus, across conditions and experiments, this study was able to show that feature-based surround suppression exists, but that it isn't hardwired; rather, it appears to be flexible and depends on task demands and stimulus context. Future studies using cognitive neuroscience methods are needed to better understand the neural mechanisms underlying this type of feature-based inhibitory surround for attention to orientation. ERP, fMRI, and MEG studies of typical feature-based attentional *enhancements* are discussed in Section 5.6.

5.5.4   Is Attention All-or-None?

Whereas the design of most studies assumes that attention is either allocated to a location or not, researchers have found that it's possible to allocate attentional resources proportionately. Mangun and Hillyard (1990) instructed participants to direct their attention to one of two possible target locations in varying amounts: 100%, 75%, 50%, 25%, or 0%. They found evidence that participants were able to do this, as reaction times and discrimination accuracy showed steady improvements across increasing levels of attentional engagement. Furthermore, visual P1 and N1 ERP components also showed a graded effect of attention, with increasing amplitudes occurring as progressively more attention was paid to a specific spatial location. These effects were observed for both target and nontarget stimuli, providing additional evidence that spatial attention works at these levels to boost the processing of all stimuli at that location. As opposed to the graded effects on the P1 and N1, showing effects that increased with the amounts of attention, the P300 showed more of an all-or-none type of pattern. As noted earlier, this component is only generated to potential target stimuli, so there was no P300 at all to stimuli at the 0% attended location because those stimuli would never require a response.

The P300 was significantly larger in the 100% condition compared to the other partial attention conditions (75%, 50%, 25%), but there wasn't a clear distinction between those three partial conditions. This result provides further evidence that attention effects for higher-order stages of processing may be gated in a way that is distinct from the graded effects of attention on earlier perceptual-level processes.

The allocation of attention doesn't depend only on what percentage we're trying to allocate, however, but also on the difficulty of the task that we're performing. Lavie and Tsal (1994) suggested that the amount of resources at any one location is a function of the goals of the observer and the difficulties of the task. Furthermore, they argue that attentional resources are always utilized in full – if 100% isn't allocated to one location, the remaining percentage is not simply unused but is instead allocated to some other location or object. Therefore, if a task is easy and doesn't require much attention, the unused attentional resources will get used up doing something else, potentially processing irrelevant information and leading to distraction. In regards to the early versus late selection debate, Lavie and colleagues have provided behavioral evidence that attentional selection acts at an early level to filter out distractors when the perceptual task is difficult, but that it acts at a late level when the task is perceptually easier (e.g., Lavie, 1995; Lavie & Tsal, 1994). Specifically, they found that nearby distractors produced more interference (slowed reaction times and impaired accuracy) when the perceptual task was easy. Overall, these findings suggest that we automatically allocate all of our attentional resources, and any unused resources will be allocated to processing something other than the task we're trying to complete.

Handy and Mangun (2000) tested these hypotheses with ERP experiments, investigating whether perceptual load affects the early stages of processing at which spatial attention effects are observed (i.e., the P1 and N1 components). Subjects performed a typical cuing paradigm in which a central arrow cue indicted the likely location of the upcoming target. However, the difficulty of the target discrimination was manipulated across conditions. In the "low-load" condition, subjects discriminated between two clearly different letters: a capital "A" and a capital "H." In the "high-load" conditions, the letters were changed to be more similar (i.e., the vertical lines in the "A" and "H" shapes were manipulated to be similar) or a dense visual mask obscured the target stimuli almost immediately after their presentation. Their results showed that attention effects on the P1 and N1 components were affected by perceptual load, with larger attention effects occurring in the high-load conditions. Furthermore, across experiments, the effects of load on the P1 and N1 were dissociated. The P1 was sensitive to even moderate levels of perceptual load that didn't modulate the N1, whereas the N1 was more sensitive to higher levels of perceptual load.

Lavie and colleagues have also made a distinction between "perceptual load" and "cognitive load" (de Fockert, Rees, Frith, & Lavie, 2001; Lavie, 1995). As described earlier, "perceptual load" refers to the difficulty of the perceptual aspect of the task being performed (e.g., detecting a certain stimulus within a complex, crowded, or masked display). However, "cognitive load" refers to the overall expenditure of resources across mental tasks. Whereas attentional focus becomes more narrowly focused on a region of space as the perceptual difficulty rises, the opposite happens for cognitive load. For instance, in a dual-task situation, the difficulty level of a primary task can affect the amount of resources that will be left over for a secondary task.

De Fockert et al. (2001) investigated this with an fMRI study in which participants performed a working memory task while also performing an attention task in which they had to classify written names of famous people as "politician" or "musician" while ignoring pictures of faces (that were also of politicians or musicians). The results showed that irrelevant faces triggered activity in the **fusiform face area (FFA)**, and that the processing of these irrelevant stimuli was stronger when the working memory task was more difficult (i.e., more items needed to be held in memory). This showed that the ability to inhibit the processing of irrelevant information depends on having enough cognitive resources available to implement this suppression; when resources are not available because of a mentally taxing main task, unattended (irrelevant) stimuli in a secondary task may not get filtered efficiently.

5.6   Nonspatial Visual Attention

5.6.1   Feature-Based Visual Attention

Based in part on the spatially defined layout of the retina and the retinotopic mapping in early visual regions, as well as on the metaphor of a spotlight theory of attention, there was initially an assumption in attention research that "space is special." While it was known that attention could be directed based on visual features (e.g., color, motion, orientation, size), early research provided evidence that feature-based attention acted only at later levels of processing, after the early stages at which spatial attention affected processing. Whereas experiments consistently find that visual-*spatial* attention robustly modulates the early P1 component, early ERP studies found that the effects of feature-based attention occurred later, at around 150–300 ms.

Feature-based attention studies utilized tasks that were similar to the spatial cuing studies described earlier, except that instead of being directed to a location, subjects were cued to attend to one feature. For example, Harter and Guido (1980) cued subjects to attend to either vertical or horizontal gratings within centrally viewed patterns. They found that the effect of paying attention to orientation resulted in enhanced processing starting at around 150 ms. This attentional modulation was observed as a negative-going wave that was sustained for about 100 ms. Many ERP studies testing attention to features (e.g., color, motion, orientation) have found this pattern of results, and this enhanced sustained negativity over posterior visual cortex for attended versus unattended features has been termed a "selection negativity" (SN; Eimer, 1995; Hillyard & Münte, 1984; Kenemans, Kok, & Smulders, 1993; Wijers, Mulder, Okita, Mulder, & Scheffers, 1989). Figure 5.10 shows an example of the SN component. In addition to the SN, some studies have found that feature attention also produces a "selection positivity" (SP) – sometimes called an anterior P2 attention effect – during a similar latency range over anterior scalp sites (Anllo-Vento, Luck, & Hillyard, 1998).

Hillyard and Münte (1984) directly compared the effects of feature and spatial attention in the same experiment by having subjects attend to a particular color and a specific location. They found the typical effects of enhanced P1 and N1 components for attended versus unattended spatial locations, and these effects of spatial attention occurred regardless of

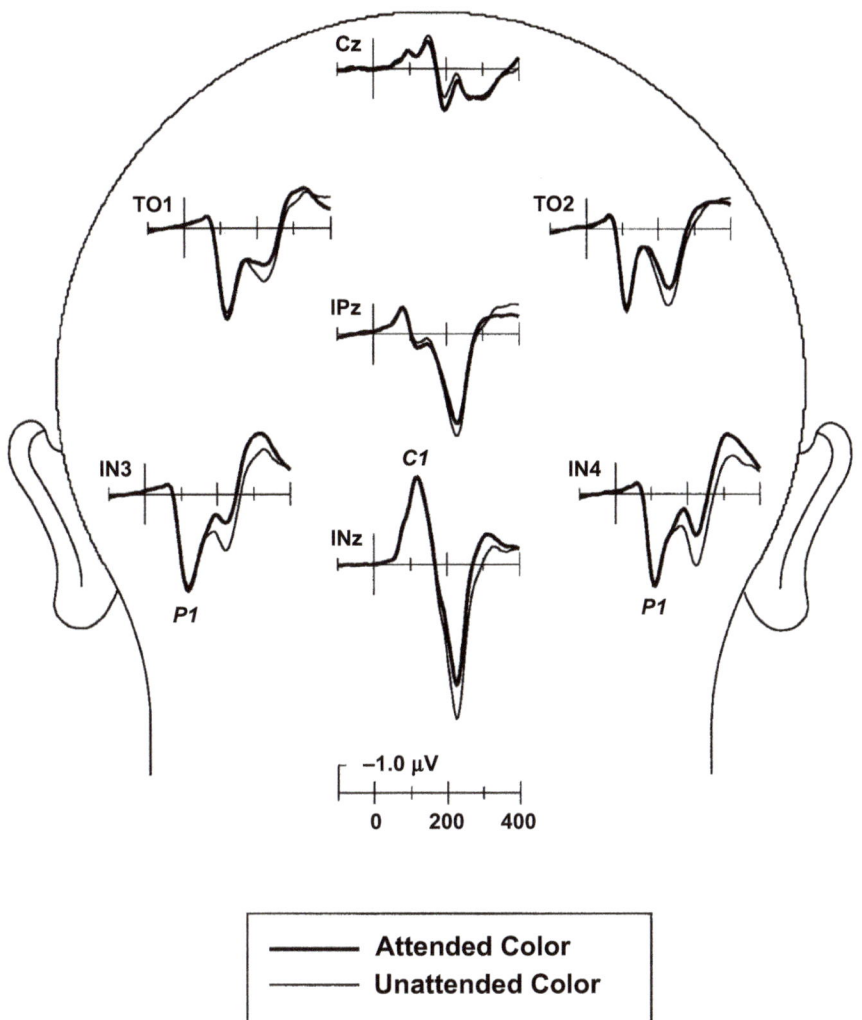

Cz

TO1

TO2

IPz

IN3   C1   IN4

INz

P1   P1

−1.0 µV

0   200   400

Attended Color
Unattended Color

Figure 5.10 Feature attention and the SN. ERP results from posterior scalp electrodes showing visually evoked components (labeled C1 and P1) and the SN (not labeled here). The SN is the sustained negativity for the attended compared to unattended color at between 160 and 400 ms, observed most strongly at the most posterior electrodes IN3, INz, and IN4. Electrodes are labeled according to the International 10–20 system. Note that the main effect of feature attention here is the SN, with little to no change in the early sensory components. Reprinted from Anllo-Vento et al. (1998), *Human Brain Mapping*, with permission from John Wiley & Sons, Inc.

what feature was being attended. The earliest feature-based attention effect was the SN, starting at ~150 ms, and this attention effect was only found when the stimuli were at the cued location. Together with previous results showing the SN occurring well after the earliest effects of spatial attention, this supported the idea that space was special, in terms of spatial attention modulating visual processing earlier than feature attention and the effects of feature attention being dependent on the location of spatial attention.

Subsequent work combining ERPs with MEG revealed that attention to features can produce earlier effects than initial studies had found (Schoenfeld, Hopf, Martinez, Mai, Sattler, Gasde, Heinze, & Hillyard, 2007). Schoenfeld and colleagues found that attention to *motion* elicited a sustained negativity that began at around 110 ms, and that attention to *color* produced a sustained positivity starting at around 90 ms. These feature-based attention effects were substantially earlier than the typical SN and SP found in most other studies of feature attention. The authors suggested that the reason as to why they found earlier effects was that subjects were cued to pay attention to an entire feature dimension (e.g., attend motion versus attend color) as opposed to most other studies in which attention was cued to a within-dimension feature (e.g., attend to red versus blue; or attend to upward versus downward motion). The authors argued that attentional filters may be set at an earlier stage when attention can be focused on discriminating between different types of features. Subsequent ERP work has extended this result, finding a larger and earlier SN for stimuli that match the attended dimension (i.e., color versus motion) compared to the SN for stimuli matching the attended feature attribute within an attended dimension (i.e., specific direction of motion; Gledhill, Grimsen, Fahle, & Wegener, 2015). These studies revealed that mechanisms of feature-based attention can be engaged earlier when attending to one whole feature dimension instead of attributes within a dimension, although these attention effects were still later than those seen with spatial attention.

Some studies have found that feature-based attention can affect processing at even earlier stages under the appropriate conditions. Biased competition models of attention hold that attentional selection occurs when it's necessary to resolve conflict (Desimone & Duncan, 1995). Accordingly, the lack of early attention effects in most feature-based attention studies may have been due to the nature of the tasks, which often consisted of stimuli presented one item at a time, thereby reducing competition between stimuli. When studies have tested feature attention under conditions in which attended and unattended features were presented simultaneously, thereby increasing competition and the need for early selection, feature-based attention effects were found as early as the P1 (e.g., Valdes-Sosa, Bobes, Rodriguez, & Pinilla, 1998). Zhang and Luck (2009) found that visual items that were in an attended color produced an enhancement of the P1 component relative to when those items were in an unattended color, and this P1 effect began at 80 ms, in line with the P1 effects of spatial attention. Critically, Zhang and Luck showed that this feature-based enhancement occurred *only* under conditions in which both attended and unattended colors were *simultaneously* present. In one of their experiments in which the attended and unattended colors were never on the screen at the same time, there was no modulation of the P1 component, in agreement with previous feature-based attention studies in which attended and unattended features never appeared simultaneously. Together, these results suggest that feature attention *can* modulate processing as early as the P1 component, but only when there is immediate competition between attended and unattended features. As the authors noted, spatial attention is still unique among the varieties of attention for being able to modulate this early stage of processing regardless of whether or not there is simultaneous competition (e.g., attended and unattended *location* stimuli present at the same time).

The most robust feature-based attention effect measured with ERPs remains the SN. The neural generator of this ERP effect, however, is more difficult to localize than the visually

evoked sensory components that show effects of spatial attention. This difficulty is due in part to the broader scalp distribution of the SN and its sustained duration, meaning that it overlaps with other concurrent neural activity. However, neuroimaging studies have been able to localize multiple brain regions that show effects of feature-based attention. In one of the first studies to use PET to investigate attention, Corbetta, Miezin, Dobmeyer, Shulman, and Petersen (1990) found that attending to color, motion, or shape resulted in different sites of attentional effects. In that experiment, each trial consisted of a pair of displays made up of a number of small colored bars moving across the screen; within each display, all the bars had the same size, color, and motion. Subjects were instructed to detect whether anything changed from the first to the second display. In different conditions, subjects had to pay attention and respond to only changes in color, shape, or motion. Attention to motion resulted in enhanced activity in an inferior parietal region, attention to color enhanced activity in lateral occipital gyri, and attention to shape enhanced activity in multiple ventromedial occipital regions along the fusiform and parahippocampal gyri. Subsequent studies with fMRI have replicated and extended these results, showing that attention to color and motion produce attention effects in the extrastriate areas (V4 and V5, respectively) that are specialized for processing those features (e.g., Chawla, Rees, & Friston, 1999; Liu, Slotnick, Serences, & Yantis, 2003; O'Craven, Rosen, Kwong, Treisman, & Savoy, 1997). A recent fMRI study investigating the combination of spatial attention and attention to orientation has provided evidence for feature-based attention effects in the early visual processing areas V1, V2, and V3 (Foster & Ling, 2022), although the limited temporal resolution of fMRI doesn't permit conclusions to be drawn regarding the relative timing of those activities. By combining fMRI with MEG and EEG, Schoenfeld et al. (2007) were able to provide evidence for the timing of feature-based attention effects. They found that attention-to-motion effects in V5 were associated with activity from 110 to 280 ms, and that attention-to-color effects in V4 were associated with activity from 105 to 210 ms. These results would suggest that effects in earlier visual areas may be due to feedback from these higher areas.

Beyond locating the spatial location and timing of feature-based attention effects, recent studies have also provided insights into the mechanisms of this type of attention. Bartsch, Loewe, Merkel, Heinze, Schoenfeld, Tsotsos, and Hopf (2017) conducted an MEG study in which an irrelevant probe stimulus was presented simultaneously with a target stimulus. The target consisted of a circle with one half colored red and the other half colored green, and the subjects had to quickly determine whether the left or right half was red. On each trial, one irrelevant probe was presented in the opposite visual field, and the probe color was one of seven different shades of color varying from dark red (matching the target color exactly) through light pink. The results provided evidence that feature attention entails a progressive sharpening of tuning curves (how responsive a neuron is to a specific stimulus versus similar ones) through feedback from higher-order visual areas. Specifically, early effects of feature-based attention were found to occur at ~200 ms, were coarsely selective to color (e.g., differentiating the deep red target only versus pinks), and were generated in higher visual processing regions (anterior ventral extrastriate regions past V4). About 100 ms later, the feature-based attention effects were more finely selective (e.g., the deep red target color was processed better than other shades of red as well as pinks), and these effects were localized to lower visual processing regions

(V4 and nearby retinotopically organized visual regions). The authors propose that feature-based attention entails back-propagation from higher-order visual areas that sharpens selectivity by attenuating the responses to similar but nonmatching features in earlier visual processing regions. The back-propagating influence of higher processing regions will be discussed again when we introduce predictive coding models in Chapter 9.

5.6.2 Object-Based Visual Attention

O'Craven, Downing, and Kanwisher (1999) provided strong evidence from an fMRI study that attention to object categories results in robust modulations of activity in visual areas specializing in processing those types of objects. On each trial in their study, subjects viewed a centrally presented image that consisted of two semitransparent overlapping images – one of a face and the other of a place – that could be moving or static. Subjects were instructed to detect, in separate blocks, whenever the face, place, or motion was identical in two trials in a row. This attention to faces, places, or motion resulted in enhanced processing in specialized extrastriate areas: the FFA when attending to faces, the **parahippocampal place area (PPA)** when attending to places, and area V5 (also known as area MT) when attending to motion. In a second experiment, the authors tested how object-based attention affects processing when the object is irrelevant. In that experiment, one of the two overlapping images was in motion and one was static (randomly assigned to the face or place), and subjects were instructed to attended to either the motion direction or to the exact location of the static image and detect any change (in motion or location) in successive trials. Despite the object category (face versus place) being completely irrelevant to this task, the results showed a robust effect of which object belonged to the target category (moving or static). Specifically, even though both a face and a place were presented on each trial, and both categories were always irrelevant, the activity in the FFA was significantly greater when subjects were attending to the motion or location of the stimulus that was a face, and vice versa for activity in the PPA. Although enhancing the processing of the identity of the face or place was not necessary for this task, it appears that when any aspect of that stimulus was being attended (motion or location), attention automatically modulated processing in specialized extrastriate areas coding the entire object.

This automatic spreading of attention to objects relates back to the object-based attention study shown in Figure 5.1, in which attention was found to automatically spread from a single attended location across an entire object (Egly et al., 1994). In this task, there are two rectangles on screen on every trial, oriented either vertically or horizontally, and subjects are cued to a spatial location at the end of one of these rectangles. The target occurs at that location on most trials (i.e., cued-location trials). Critically, on uncued-location trials, the target is equally likely to occur within the same object (at the opposite corner of the rectangle that was cued) or at an equidistant location in a corner of the other rectangle. Müller and Kleinschmidt (2003) had subjects perform this task during fMRI scanning, and they replicated the behavioral results, showing that subjects were faster at the cued location than anywhere else (i.e., the standard spatial cuing effect) and significantly faster at the same-object uncued location compared to the different-object uncued location (i.e., the effect of object attention). The fMRI results from retinotopically defined visual areas revealed location- and object-based attention effects in V1

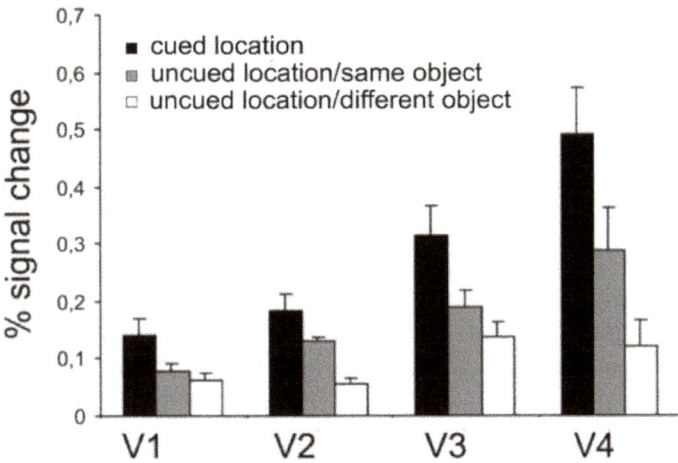

Figure 5.11 Object-based attention effects in early visual areas. Results of an fMRI study (Müller & Kleinschmidt, 2003) using a version of the Egly et al. (1994) spatial and object attention paradigm. The fMRI results from regions of interest in the first four visual areas (V1, V2, V3, and V4) show that spatial attention (cued location versus all uncued locations) has the largest effect in each area, but that the automatic spreading of attention throughout an object also enhances activity (same versus different objects for unattended locations) in early visual processing regions. Adapted from Müller and Kleinschmidt (2003), *Journal of Neuroscience*, with permission from the Society of Neuroscience (Copyright, 2003, Society for Neuroscience).

through V4 (Figure 5.11). The effect of spatial attention was larger than the effect of object attention, but critically this study showed that attention to one location in an object automatically spreads across that entire object. Furthermore, this study provided evidence that this effect of spreading of attention across objects is present in early visual processing areas. As with all fMRI studies, it isn't possible to accurately assess the relative timing of the activity in these visual areas, but it is noteworthy that the largest effects of object attention occurred in the highest area tested here (V4).

In another fMRI study that used a similar overlapping face/place task to the O'Craven et al. (1999) study described earlier, Cohen and Tong (2015) proposed that object-based attention effects in the FFA and PPA project back to earlier visual areas and make those areas respond more selectively to the attributes of the attended object. The results showed that attention to faces versus places modulated the activity patterns in early visual areas (V1–V4), such that those patterns across each area could be used to predict which of the overlapping objects (face versus place) the subject was attending. The authors argue that this feedback from higher areas is a core process by which attention to a broad category can result in attention being allocated to specific patterns of that stimulus even when it's overlapping in space with other stimuli. Baldauf and Desimone (2014) used MEG in a similar face/place paradigm to show that even higher-order centers may control the initial object-based attention effects in the FFA and PPA. They found increased synchrony in the gamma band (60–100 Hz in this study) between the inferior frontal junction and the specialized visual area (FFA or PPA) corresponding to the object being attended (face or place, respectively). Importantly, the gamma phases were 20 ms earlier in the inferior frontal gyrus (IFG) compared to either the FFA or PPA, suggesting that the IFG was initiating the synchrony with the visual processing areas.

5.7   Neural Effects of Auditory Attention

Some of the very first cognitive neuroscience studies of attention investigated auditory attention. Using paradigms similar to the dichotic listening experiments that investigated the cocktail party phenomenon, Hillyard and colleagues (Hillyard, Hink, Schwent, & Picton, 1973) had participants listen to tones presented separately to the right and left ear. Subjects were instructed to attend to the input from one ear and detect target tones only in that ear. In this way, the experimenters were able to isolate attention, because the exact same stimuli could be analyzed as a function of whether they were in the attended or the unattended ear. They found that the auditory N1 component was relatively enhanced for attended tones relative to when the same tone was unattended. Note that the *auditory N1* is distinct from the *visual N1* discussed in previous sections. The auditory N1 is the first, large, reliable negative component evoked by auditory stimuli. Unlike the visual N1, which is generated by neural sources well past the first visual processing area of cortex (and more than 150 ms after the presentation of a visual stimulus), the auditory N1 is generated within the first 100 ms, and it is thought to come from an early auditory processing region. Note that there are believed to be multiple processes and neural generators active during the latency range of the auditory N1 (Näätänen, 1975), but part of that processing has since been localized by combined MEG/ MRI studies to auditory-evoked sensory processing activity within **Heschl's gyrus** (Woldorff, Gallen, Hampson, Hillyard, Pantev, Sobel, & Bloom, 1993). Heschl's gyrus, anatomically defined as the transverse temporal gyrus, lies in the superior temporal lobe and includes the first auditory processing region of the cortex (i.e., primary auditory cortex, "A1"). The finding of attention effects at a latency of less than 100 ms on a component generated in this part of the cortex suggests that auditory attention can act at a slightly earlier stage than visual processing. Both the visual and auditory ERP results, however, show that spatial attention modulates the *strength* of processing at relatively early perceptual stages, before higher-order semantic analyses.

Importantly, the enhancement of the auditory N1 has been found for both target and nontarget stimuli within the attended channel (i.e., attended ear). This tells us that auditory spatial attention is working at a stage of processing before identification of the target. The finding of the same degree of N1 enhancement to target and nontarget stimuli provides additional support for an early selection hypothesis, in that all stimuli in the attended channel are enhanced regardless of their identity. It is noteworthy that in these experiments later, higher-order components such as the P3 were dependent on whether or not the stimulus was a target. Specifically, Hillyard et al. (1973) found an enhanced-amplitude N1 for all stimuli in the attended versus unattended ear, but a P3 was generated only to the rare target stimuli. Thus, the N1 attention effect behaves in a manner predicted by early-stage attenuation models of attention, in which all stimuli from the attended channel receive enhanced processing, whereas the P3 shows an effect more similar to an all-or-none "filter" model of attention, in which processing at that level is only evoked if the stimulus is determined to be relevant and requiring of a cognitive or behavioral response.

In subsequent studies, the earliest stage of processing affected by auditory attention was shown to occur even earlier than the N1. Woldorff and Hillyard (1991) found that auditory attention affects processing as early as 20–50 ms after stimulus presentation! This very early ERP effect (labeled the P20–50 attention effect) was replicated in a study that combined the functional time course from MEG data with anatomical data from structural MRI to model the early attention effect as arising from Heschl's gyrus (Woldorff et al., 1993). The investigation of such early attention effects is challenging due to the very rapid and small amplitude of activity, and the effects may only be observable under certain conditions. Therefore, most studies of auditory attention report the larger and more robust N1 attention effects. However, given appropriate electrode/sensor placement, a high sampling rate, robust stimuli, and appropriate analysis procedures, these early potentials can be measured reliably. Finding attention modulations of these early components still depends, however, on the difficulty of the task and the amount of attentional resources allocated to the task. As mentioned in previous sections, theories of attentional load suggest that attention affects processing at different stages depending on the difficulty of the task and the engagement level of the subject (Lavie, 1995).

5.7.1   Localizing Auditory Attention Effects with fMRI

Given the inherent limitations of localizing the sources of ERPs and the need to rely on inverse modeling to estimate the sources of MEG results, researchers turned to neuroimaging methods when these became available. Grady and colleagues were among the first researchers to utilize fMRI to study the effects of attention on auditory processing (Grady, Van Meter, Maisog, Pietrini, Krasuski, & Rauschecker, 1997). The fMRI results from that study provided spatially precise evidence that attention to auditory stimuli produces enhanced processing in regions around Heschl's gyrus. In that study, the largest attention effects were observed in the auditory areas surrounding primary auditory cortex, suggesting that, like visual attention, auditory attention may act mainly after the initial stage of cortical processing. However, the stimuli in that experiment were complex (words), and attention was directed to specific words, not to spatial locations or primary features. Therefore, this left open the possibility of earlier attention effects if attention was focused on more elementary auditory characteristics. In addition to the task not orienting attention to the sort of basic features that could require attentional focusing at the earliest stages, there is a specific limitation of fMRI when it comes to studying auditory processing: the noises of the MRI scanner. Specifically, typical scanning sequences produce a series of loud noises throughout their entire run time. Furthermore, these noises are not easy to habituate to because they are complex and varied. Thus, localizing auditory attention effects within the noisy environment of an active MRI scanner is quite challenging, and the lack of robust attention effects in primary auditory cortex in early fMRI studies may have been due in part to this limitation.

Hall and colleagues (Hall, Haggard, Akeroyd, Summerfield, Palmer, Elliott, & Bowtell, 2000) addressed some of these limitations in an fMRI study in which subjects had to detect specific rare tones in an active attention condition versus a passive listening condition. In addition, they addressed the issue of ongoing scanner noise by using a novel "temporally sparse imaging" sequence, in which the scanner is silent during each 10-second block of stimulus presentation and then a full brain volume is collected immediately after each stimulus block ends. Although this greatly reduces the number of brain volumes collected, it has the advantage of ensuring that scanning noises aren't interfering with perception of the auditory stimuli of interest. And since the hemodynamic response peaks 6–8 seconds after neural activity and takes 12–18 seconds to fully return to baseline, the fMRI signal can be measured well after the auditory stimuli have ended. This study found robust auditory evoked activity in Heschl's gyrus and surrounding cortex, but attention effects were still only found in secondary auditory processing regions in that study.

Other research, however, has found evidence for attentional modulation in the primary auditory cortex, in line with what the timing of attention effects in ERP and MEG studies had suggested. Jäncke, Mirzazade, and Shah (1999) found that attention modulated activity in primary as well as secondary auditory cortex. In that fMRI study, subjects listened to consonant–vowel syllables (e.g., /da/, /pa/, /ta/) while either attending to the stimuli in order to detect a target syllable, passively listening to the stimuli, or ignoring the auditory stimuli and concentrating on their hands and feet. Activity in the region of Heschl's gyrus was found to be significantly greater in the active attention condition. The surrounding region, assumed to be secondary auditory cortex, also showed significant attention effects. One possible reason as to why this study found attention effects in primary auditory cortex whereas some other studies did not is that the task was difficult and required attention to fine details of the stimuli.

5.7.2   Auditory Attention Effects in "What" versus "Where" Regions

Using high-resolution anatomical MRI images and surface mapping techniques to flatten the cortex and investigate fMRI-measured activity across Heschl's gyrus and the surrounding planum temporale region, Petkov and colleagues identified two different subregions of auditory cortex: one that was affected by attention and another that was sensitive to physical differences but not attention (Petkov, Kang, Alho, Bertrand, Yund, & Woods, 2004). Specifically, they found that more medial regions were sensitive to physical features of the auditory stimuli, such as location and frequency (pitch), but were unaffected by attention. More lateral regions, however, showed enhanced activity when subjects were attending to auditory or visual stimuli. This study suggested that even within early auditory processing regions there may be subregions that more faithfully transmit sensory information and others that are modulated by the current focus of attention.

Other studies have compared how attention affects spatial versus feature coding in subregions of the auditory cortex. Warren and Griffiths (2003) found that different regions of the planum temporale responded preferentially to pitch versus the spatial location of sounds. Specifically, a posteromedial region was sensitive to changes in the location of stimuli, whereas an anterior region was sensitive to changing pitch. Building upon this distinction between "where" and "what" regions of auditory cortices, Ahveninen and colleagues used MEG and

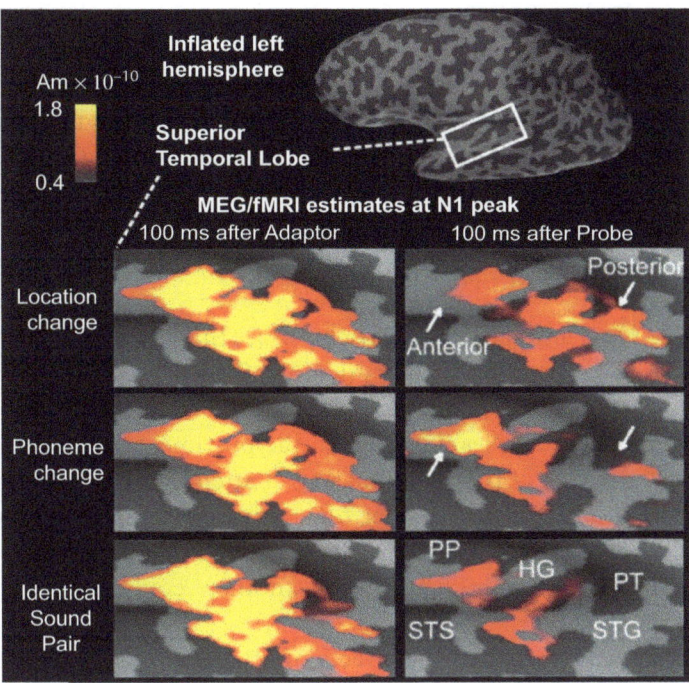

Figure 5.12 Areas of auditory cortex sensitive to location versus feature attention. Results from a combined MEG and fMRI study showing activity in the superior temporal lobe during the latency range of the auditory N1. The left column shows activity in response to the initial sound ("adaptor") and the right column shows activity in response to the second sound of the pair ("probe"). The top row shows that on trials in which attention was focused on location changes of the sounds, posterior superior temporal gyrus (STG) regions were most active, whereas attention to feature changes (middle row) resulted in more activity in more anterolateral regions, including Heschl's gyrus (HG). PP = planum polare; PT = planum temporale; STS = superior temporal sulcus. Reprinted with permission from Ahveninen et al. (2006), *Proceedings of the National Academy of Sciences.* Copyright (2006) National Academy of Science, USA.

fMRI, in separate sessions, to investigate the effects of auditory attention with high temporal and spatial resolution (Ahveninen, Jaaskelainen, Raij, Bonmassar, Devore, Hämäläinen, Levanen, Lin, Sams, Shinn-Cunningham, Witzel, & Belliveau, 2006). Subjects were presented with pairs of vowel sounds that could be phonetically different, spatially different (coming from different locations), or identical on both dimensions. In different blocks, subjects attended to location or phonetic similarity and had to determine whether the pairs were the same or different on the attended dimension. The results provided evidence for separate regions – outside of the primary auditory cortex – that are sensitive to different types of attention and auditory features. As shown in Figure 5.12, attention to location modulated processing in more posterior regions (specifically, the planum temporale and posterior superior temporal gyrus), whereas attention to phonetic features modulated processing in more anterior regions (specifically, anterolateral Heschl's gyrus and anterior superior temporal gyrus). The MEG activity localized to these regions showed that each of these attention effects occurred ~100 ms after stimulus presentation, in the time range of the N1 component. As noted earlier, it is thought that there are multiple sources of activity occurring within the latency range of the auditory N1, and in this study the *location-sensitive* aspect of the MEG signal was found to occur during the earlier part of the N1 (~90 ms), whereas the *phonetic-sensitive* component was best fit to a later part of the N1 (~120 ms). The authors suggested that this temporal ordering may imply a hierarchy in terms of spatial information being used in subsequent stages of auditory feature perception.

5.7.3   Remapping of Auditory Cortex by Attention

Studies of auditory attention often purposely ensure that, from trial to trial, the subject cannot know what spatial location or auditory feature they will be asked to attend. In this way, researchers can ensure that any observed effects are due to the orienting of attention in response to the cue stimulus. The studies discussed in the previous subsection utilized this method to isolate and test the effects of transient focusing of attention. However, in discussing the effects of attention, it is also useful to consider how attention being focused consistently over time could have additional effects. For example, when subjects are trained over a set of trials to attend to one type of stimulus and ignore another, could this change the way auditory information is mapped on the cortex?

Researchers have investigated the effects of attention on the short-term *plasticity* of the auditory cortex. Capitalizing on the enhanced localization abilities of MEG, Ozaki and colleagues (Ozaki, Jin, Suzuki, Baba, Matsunaga, & Hashimoto, 2004) provided evidence that attention to a specific feature (pitch) resulted in a change in the tonotopic mapping of primary auditory cortex. This study focused on the auditory N100m (the MEG component corresponding to the N1 auditory ERP) and built upon the known tonotopic organization of the auditory cortex (i.e., laid out from lower to higher frequencies). Using MEG allowed the researchers to localize the neural generators of different tones to different regions of primary auditory cortex. The distance between the areas representing 400 and 4,000 Hz was mapped before and after subjects were trained, in two sessions, to respond to one of two different tone pips that they would hear (400 or 4,000 Hz). The researchers found that the distance between the dipole estimates of the cortical generators of the N100m was *expanded* following training. This increase in distance between the regions representing the attended and ignored stimuli was taken as evidence that the tonotopic mapping of the auditory cortex had been modulated by short-term attention training. Research using fMRI has also found that training subjects to attend to a specific auditory tone can trigger a reorganization of the cortical mapping in primary auditory cortex (Jäncke, Gaab, Wüstenberg, Scheich, & Heinze, 2001).

5.7.4   Subcortical Auditory Attention Effects and the Cocktail Party Revisited

Rinne and colleagues used fMRI to show that auditory spatial attention can also affect subcortical processing. In addition to showing the effects of auditory spatial attention in primary auditory cortex, they found similar attention effects in the inferior colliculus (Rinne, Balk, Koistinen, Autti, Alho, & Sams, 2008). In terms of the bottom-up stream of auditory processing from the cochlea, the inferior colliculus is the region that projects to the medial geniculate nucleus of the thalamus, which in turn sends the information to primary auditory cortex. Thus, this finding is important for showing that top-down attention to location can modulate subcortical auditory processing at a very early level; however, as noted in previous sections, fMRI is very limited in its temporal resolution because of its reliance on the relatively sluggish hemodynamic response. Therefore, fMRI findings of inferior colliculus attention effects cannot prove that attention acts during the initial bottom-up processing in that region.

In order to provide a more temporally precise measure, Forte, Octave, and Reichenbach (2017) developed a mathematical technique to estimate brain stem responses from EEG recordings. Furthermore, they used this method to investigate the attentional effects of attending to one stream of speech amid other speech, providing a link between laboratory studies using only simple tones and an actual cocktail party situation. Subjects were presented with two streams of speech diotically (i.e., in both ears) and asked to attend to one of the two streams and answer comprehension questions about that speech after each block. The authors report that attention enhanced the processing of speech in the brain as soon as 10 ms after presentation. This early timing and the analysis techniques suggest that the activity arose from subcortical regions. Additional studies are needed to replicate this finding and to further validate the novel decomposition technique and assumptions underlying the localization to brain stem responses, but these results would push the site of attentional enhancement in the auditory domain even earlier than the P20–50 attention effect described above. Another interesting aspect of this study was that it didn't separate the speech streams by different ears but rather depended on the subjects being able to attend to one speaker based on features of that speaker's voice (in this case, one male and one female speaker). This highlights that early attention effects on auditory processing are not limited to spatial attention but can occur due to feature-based attention as well. Relating back to Section 2.3 of Chapter 2, a recent study found that comprehension of speech was impaired by the presentation of concurrent background music, and more so for music that contained vocals compared to instrumental music (Brown & Bidelman, 2022). Furthermore, the impairment from vocal music was significantly worse for unfamiliar music. Familiar music was easier to ignore than unfamiliar music. This study recorded EEG and used temporal response functions (TRFs) – a type of linear stimulus–response model that calculates the brain's response to complex continuous stimuli (such as speech) – in its analysis. The TRF deflection at a latency of ~100 ms corresponds to the typical auditory N1 to isolated auditory stimuli. The authors found that background music slowed the latency of this activity, suggesting that the music was delaying the processing of the target speech, in line with studies of perceptual masking on the auditory N1. Furthermore, the slowing of target speech processing by background music was significantly worse if the music contained vocals (versus instrumental music), but only for *unfamiliar* music. Familiar music did not show this effect. The influences of memory and expectancy on perception and attention will be discussed again in Chapters 6 and 9.

5.8   Attention Effects Across Sensory Modalities

The vast majority of cognitive neuroscience research on the effects of attention on sensory processing has investigated the visual and auditory modalities. Investigating the effects of attention on *olfactory* (smell) processing has been hampered by difficulties in being able to precisely control the stimulus input. However, in a study that developed specialized equipment to be able to study olfaction in the MRI environment, the researchers found that sniffing (as opposed to passive smelling) resulted in a transient increase in activity in primary and secondary olfactory areas in the temporal and frontal lobes (Sobel, Prabhakaran, Desmond, Glover, Goode, Sullivan, & Gabrieli, 1998). Although not the usual way to manipulate attention, the authors suggested that sniffing may be related to an attentional shift, in terms of preparing the olfactory regions for an upcoming odor. In behavioral studies that have used a more typical and explicit manipulation of attention, it has been found that paying attention to smell (versus paying attention to vision) leads to faster responses to olfactory stimuli (e.g., Spence, McGlone, Kettenmann, & Kobal, 2001). In the realm of *gustatory* (taste) processing, there is also evidence that paying attention to a specific taste can lower detection thresholds for gustatory stimuli (Marks & Wheeler, 1998).

Behavioral studies of *tactile* attention have shown similar enhancements of response speeds to attended tactile stimuli, and ERP studies have been able to reveal the timing of these attention effects. Forster and colleagues (Forster, Sambo, & Pavone, 2009) found that spatial attention in a purely tactile condition produced enhanced tactile components starting at 130 ms after tactile stimulus onset. Interestingly, in an intermodal condition in which the subjects kept their eyes open and fixated on a central point, the effects of tactile attention were seen even earlier, at around 90 ms, indexed by the somatosensory P100 component. Overall, these findings provide evidence that attention in the gustatory, olfactory, and tactile domains boosts relatively early perceptual processing in a similar way as for visual and auditory spatial attention.

Karns and Knight (2009) investigated how paying attention to one sensory modality versus another affects perceptual-level processing. In their study, subjects performed a speeded detection task in which they were presented with auditory, tactile, or visual stimuli while EEG was recorded. Critically, the possible stimuli all came from the same location: The subject placed their hand at the location of small speaker that was placed just behind a translucent screen onto which the visual stimuli could be projected. This was to control for spatial attention effects that may have advantaged visual attention in previous studies comparing attention to the different sensory modalities. In Karns and Knight's design, spatial attention would be allocated to the same location in all conditions. The conditions consisted of subjects needing to respond to just one of three types of stimuli: a brief auditory broadband sound (a duck quack), the appearance of a blue square, or a brief vibration delivered to the hand. The stimuli could be presented alone or simultaneously with stimuli in the other modalities. The ERP findings showed that attending to any one sensory modality while ignoring the other modalities resulted in modulations of sensory processing, although this was noticeably later for the tactile modality. Specifically, attending to the somatosensory domain resulted in enhanced processing of the tactile stimulus (relative to how that stimulus was processed when attending to one of the other modalities), but not until 165 ms, as a modulation of the tactile N160 component. In contrast, auditory and visual attention affected processing in those modalities within the first 100 ms. Attending to auditory stimuli produced modulations of auditory processing as early as 29 ms, generally in line with the auditory spatial attention effects described earlier. Attending to the visual modality resulted in enhanced visual processing at the stages of the visual P1 and N1 components, as described earlier, but also a modulation of the C1 component at ~70 ms. This latter effect has only rarely been observed in studies of visual-spatial attention (see Box 5.2). One possible reason for the attention effect on the C1 in this study is that the comparison was between attention to vision versus a condition of completely ignoring visual stimuli. Most of the studies that did not find a C1 attention effect compared conditions of attended versus unattended *locations*, but always in the visual domain, so not all visual processing would be suppressed.

Although most of these studies investigated the effects of attending to one sensory modality while explicitly ignoring other modalities, in the real world we often want to pay attention to all features, across modalities, at the attended spatial location. Indeed, theories of attention have long assumed that there may be a master map or shared attention focus that benefits all senses and that could be used to bind features across modalities into multimodal object perceptions (reviewed in Spence, 2021). Again, the most well-studied links have been between vision, hearing, and touch. Spence and colleagues (Spence, Nicholls, Gillespie, & Driver, 1998) conducted a series of behavioral experiments showing that stimuli capturing attention in one sensory modality resulted in enhanced processing (i.e., faster RTs) for stimuli in other sensory modalities occurring at that same location. Specifically, they found that orienting attention to a tactile stimulus enhanced visual and auditory processing at that location. They also found that orienting to a visual or auditory stimulus enhanced tactile processing. Other studies have shown that orienting to an auditory stimulus enhances visual processing at that location (Schmitt, Postma, & de Haan, 2000; Spence & Driver, 1997). However, the link between visual cues and auditory targets had been unclear, as some initial studies failed to find enhanced auditory processing at the location of a visual cue (Schmitt, Postma, & de Haan, 2000; Spence & Driver, 1997), whereas other studies did find such a link (Ward, 1994). A subsequent study by McDonald and colleagues controlled for factors (e.g., the decision criterion) that may have contributed to the inconsistency in the previous results, and they found robust evidence that orienting of spatial attention by a peripheral visual stimulus did result in enhanced auditory processing at that location (McDonald, Teder-Salejarvi, Heraldez, & Hillyard, 2001). Furthermore, that study included ERPs, which showed that spatial orienting to the visual cue produced enhanced perceptual processing of the subsequent auditory stimulus as early as a latency of 100 ms. This attention effect – an enhanced negativity that peaked at around 120–140 ms – is later than the purely auditory spatial attention effects described earlier in this chapter, but it is still relatively early in the processing stream, and before levels of semantic analysis. Overall, the results from cross-modal attention studies provide support for the theory that there may be a common supramodal mechanism of attention, possibly most aligned with *spatial* attention, that links our senses together and may help bind features from across these modalities.

CHAPTER SUMMARY

- Integrating results from methods that have high temporal precision with results from those with high spatial precision is critical for understanding the effects of spatial attention, as some effects of attention occur during the initial feedforward processing and others occur as a result of feedback processing.
- Attention can modulate processing at multiple stages depending on the amount and type of attention being allocated, the task being performed, and the level at which competition needs to be resolved.
- The effects of attention include boosting activity, suppressing activity, synchronizing activity, and tuning cells to respond more precisely and selectively to the target.
- Attention can affect relatively early stages of processing in the visual, auditory, and tactile domains, but differences exist in the timing and nature of those effects.

REVIEW QUESTIONS

What is the evidence from ERP, MEG, and EROS methods that spatial attention works through early selection, and how do those pieces of evidence converge with neuroimaging and single-unit results?

Compare and contrast the effects of spatial attention versus feature-based attention and object-based attention.

Discuss the roles of enhancement versus suppression in the effects of attention.

How have cognitive neuroscience studies advanced our understanding of the "cocktail party phenomenon"?

FURTHER READINGS

Baumgartner, H. M., Graulty, C. J., Hillyard, S. A., & Pitts, M. A. (2018). Does spatial attention modulate the earliest component of the visual evoked potential? *Cognitive Neuroscience, 9*(1–2), 4–19.
- This discussion article and the series of commentaries following it illustrate the high *temporal precision* of ERPs and why that precision is critical for understanding the brain mechanisms of visual-spatial attention.

Fiebelkorn, I. C., & Kastner, S. (2020). Functional specialization in the attention network. *Annual Review of Psychology, 71*(1), 221–249.
- This review article covers recent fMRI and single-unit studies of attention and discusses classic and recent models of attention.

Hopf, J. M., Boehler, C. N., Luck, S. J., Tsotsos, J. K., Heinze, H. J., & Schoenfeld, M. A. (2006). Direct neurophysiological evidence for spatial suppression surrounding the focus of attention in vision. *Proceedings of the National Academy of Sciences of the United States of America, 103*, 1053–1058.
- This empirical article presents an MEG study of attention that confirms that an area of inhibition surrounds the focus of attention.

Karns, C. M., & Knight, R. T. (2009). Intermodal auditory, visual, and tactile attention modulates early stages of neural processing. *Journal of Cognitive Neuroscience, 21*(4), 669–683.
- This article presents experiments testing the effects of paying attention to different sensory modalities.

6   Voluntary versus Involuntary Attention

Learning Objectives

- Identify differences between endogenous and exogenous influences on visual processing.
- Compare and contrast the neural effects of top-down and bottom-up attention.
- List and describe event-related potential components indexing attention effects on higher-order processing.
- Compare the attention effects of "special" classes of stimuli (e.g., faces, new objects, rewarded stimuli) to physically salient stimuli.
- Describe the effects of different types of memory on attentional allocation.

6.1   Voluntary versus Involuntary Attention

The previous chapter discussed the effects of attention after it had been allocated to a spatial location, feature, or object. In the studies described in that chapter, subjects were instructed where or what to attend and had to *purposely* orient and maintain attention on that location, feature, or object. Either through trial-by-trial cuing or through instructions at the beginning of each block of trials, subjects were to orient their attention *voluntarily*. In some experiments, this was done by asking subjects to only respond to one type of item (based on location or feature), whereas in other experiments a cue stimulus presented before the target display was predictive of the likely location or features of the target. These informative, predictive cues were sometimes wrong, but they usually correctly predicted the target's location/feature/identity. Although we clearly have the ability to orient our attention in that way – voluntarily and on purpose – our attention can also be captured to stimuli *involuntarily*. As described in earlier chapters, the sound of our name in a nearby conversation at a party or a notification popping up on our phone captures our attention regardless of our intentions. Even decades before Cherry's (1953) first investigation of the cocktail party effect, Titchener had already described this type of attention:

There is, in the first place, an attention that we are compelled to give and are powerless to prevent . . . there are impressions that we cannot help attending to . . . they force their way to the focus of consciousness, whatever the obstacles that they have to overcome. (Titchener, 1916).

This difference in whether attention is oriented on purpose or despite our intentions has led to research investigating whether involuntary attention has the same effect on information processing as does voluntary attention. Before describing that research, it is useful to explain the different terms used to refer to these types of attention. *Voluntary* attention is sometimes referred to as **endogenous attention** because the control of orienting comes from goals and intentions that are internal (endogenous) to the person, whereas *involuntary* attention can be referred to as **exogenous attention** because what drives this orienting of attention is external to the person (i.e., physical stimuli and events in the environment). Along these same lines, voluntary attention is sometimes referred to as "top down," whereas involuntary attention is sometimes referred to as "bottom up." These latter terms refer to the idea that the inputs through our senses that drive exogenous orienting are at the bottom of the hierarchy of processing steps, and that these inputs come from below our brain, whereas our internal goals can be considered at the top of the processing hierarchy, and this type of attention would thus have to reach "down" to affect early sensory-level processing. In addition, "automatic" or "reflexive" is used to refer to involuntary attention, whereas "effortful" or "controlled" is used for voluntary attention. Given the different assumptions and implications behind some of these terms, we will mainly use "exogenous" and "endogenous" in the rest of this chapter since these terms nicely reflect what is thought to be the critical difference that establishes these two distinct types of attention. This chapter will first compare and contrast the effects of exogenous and endogenous attention, as typically defined, and then will present research showing other influences on attention that don't fit simply into those categories. Note that this chapter is focused on the *effects* of exogenous attention on subsequent processing and comparing these to the *effects* of endogenous attention covered in Chapter 5; the brain systems and mechanisms that *control* the orienting and focusing of attention will be covered in Chapter 7.

6.1.1 Behavioral Effects of Exogenous versus Endogenous Attention

The original studies used to compare voluntary and reflexive attention largely focused on how attention was oriented (Jonides, 1981; Posner, 1980; Posner & Cohen, 1984). According to the assumption of there being a single beam of attention, the idea was that this beam could be moved either automatically or on purpose, but that it was the same focus of attention being moved. In early studies comparing the effects of informative (i.e., predictive of subsequent target location) central arrow cues and predictive peripheral cues, exogenous attention triggered by the peripheral cues was found to be oriented more quickly, was less dependent on cognitive resources, and was more resistant to interference than the central arrow cues (e.g., Jonides, 1981; Müller & Rabbitt, 1989). In addition, research using uninformative peripheral cues provided details about the time course of attention, showing that involuntary attention affected processing very shortly after the capturing effect (even within 50 ms), in contrast to voluntary attention, which took much longer to exert its strongest effects (earliest robust effects at ∼ 300 ms; Cheal & Lyon, 1991). These reaction time and accuracy results supported the idea of there being two different mechanisms of orienting but were largely in line with the idea that it was the same spotlight being moved. However, other results began to raise the possibility that what was being oriented was not in fact the same beam.

One of the first results to suggest that exogenous attention was doing something different than endogenous attention came from analyzing the aftereffects of reflexive attentional capture. Whereas the initial involuntary capture to a salient stimulus is obvious, what happens next is not so easy to appreciate. Indeed, philosophers had described the capture of attention for hundreds of years without mentioning the more subtle effect that follows. It was not until Posner and Cohen's (1984) landmark study that we came to understand that shortly after attention has been initially captured at a location there is a period of suppression at the cued location. Posner and Cohen found that a peripheral flash initially captured attention and resulted in quicker reaction times to targets at the cued location versus uncued locations; however, this only occurred when the target occurred shortly after the cue, with stimulus onset asynchronies (SOAs; the time between the onset of the cue and the onset of the target) of 200 ms or less. At SOAs of more than 200 ms, targets at the cued location were responded to significantly slower than targets at uncued locations (Figure 6.1). This effect has been termed inhibition of return (IOR), referring to attention being inhibited from returning to a location that had just recently captured attention. One of the reasons IOR is not easy to appreciate in everyday life is that we typically are interested in what had captured our attention, so the initial involuntary capture of attention is quickly followed by a voluntary maintenance of attention at that location. In the laboratory, however, in tasks that include hundreds of trials and instructions to respond rapidly to brief stimuli, researchers are able to isolate the effects of exogenous attention and track those effects over time. This research has shown that while endogenous attention takes longer than exogenous attention to orient to a location, it can be sustained on

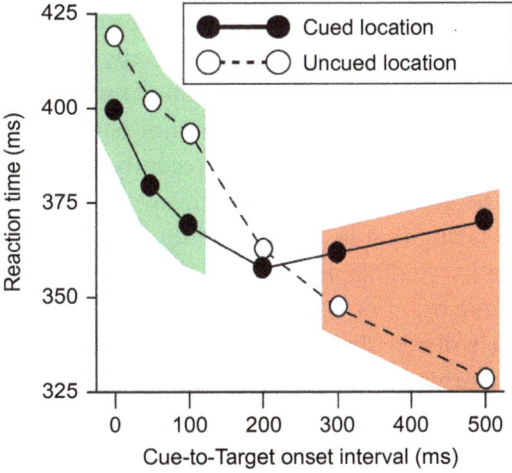

Figure 6.1 Effects of exogenous cues on target reaction times. Reaction time results from Posner and Cohen's (1984) cuing study of exogenous capture. "Cued location" refers to targets occurring at the same location as a preceding nonpredictive peripheral flash; "Uncued location" refers to targets appearing at a different location than the preceding flash. Green shading highlights the cue-to-target timing intervals at which capture resulted in facilitation at the cued location; orange shading highlights timings at which the cued location is responded to more slowly than other locations (i.e., IOR). Reprinted from Klein (2000), *Trends in Cognitive Sciences*, with permission from Elsevier.

a location for as long as the person has the resources and will to do so. The exogenous capture of attention, however, acts quicker but is then inhibited due to IOR.

In order to understand what purpose this inhibitory mechanism may have, Klein and MacInnes (1999) tested whether IOR might be important in facilitating foraging behavior. The idea is that when searching through space it may be helpful to tag locations where attention has already been to ensure efficient searching of unexplored space. In a clever experiment that utilized overt orienting, they tracked participants' eye movements while they performed a "Where's Waldo?"-type search paradigm (looking for a target character in a very cluttered scene in which different features of the target character's appearance occur in many other objects in the scene). At some points during the participants' searching, a black circle could flash on the screen, and the participants had to quickly fixate this circle. The interactive computer program used the participants' recent fixations during their search to place the black circle at the location of their most recent *previous* fixation or at control locations equidistant from the current eye position. The results showed that participants were significantly slower to return to a recently fixated location than to other locations equally far away. This slowing was greatest at the exact location of the previous fixation and became less pronounced with distance, demonstrating progressively less inhibition at farther distances from the previous fixation. As this paradigm used overt shifts of attention, it is relevant to note that Redden and colleagues have proposed that there are two different forms of IOR (Redden, MacInnes, & Klein, 2021). They suggested that an "input" form of IOR suppresses the salience of items at the cued location and that an "output" form biases overt responding toward the cued location. Based on a review and modeling of previous datasets, they suggested that the activation state of the reflexive oculomotor system determines the type of IOR that ensues. When the reflexive oculomotor system is suppressed, as must be done in paradigms that require subjects to maintain fixation throughout trials that include peripheral events that would normally trigger eye movements, the saliency of items at the cued location is reduced. In situations in which the eyes are allowed to move to salient items, however, the salience of items may not be affected as much, but eye movements become biased toward locations that haven't recently been fixated. Although more work is needed to determine the mechanism or mechanisms of IOR, it is clear that IOR represents a form of attentional biasing that is triggered uniquely in an involuntary manner.

In addition to IOR, other findings also support the idea that these two types of attention are not simply orienting the same beam of attention. Briand and Klein (Briand, 1998; Briand & Klein, 1987) found qualitatively distinct behavioral effects of each depending on the type of task being performed. When comparing performance on feature search tasks (e.g., detecting a letter "O" among "T" and "X" distracters) with performance on conjunction search tasks (e.g., finding the blue "O" among green "O" distracters and "T" and "X" distracters that could be green or blue), they found interactions between attention type (exogenous versus endogenous) and search type (conjunction versus feature), suggesting that exogenous attention produces a significantly larger difference in performance between these types of search tasks. In typical cuing studies, Berger, Henik, and Rafal (2005) found that when both types of attention are engaged on the same trials, they can have independent effects on reaction times. Prinzmetal, McCool, and Park (2005) also found difference in the effects of each, finding that exogenous attention affected reaction times but not accuracy, whereas endogenous attention significantly affected both.

More recently, Fernández, Okun, and Carrasco (2022) tested how exogenous and endogenous attention affect how the brain fine-tunes its responses to target orientation and the spatial frequency of visual stimuli. In this experiment, subjects had to detect when the test stimuli contained a target (a 2-cycles-per-degree vertical sinusoidal grating pattern, known as a Gabor patch, amid a visual noise pattern) versus when it was only a visual noise pattern. Subjects were cued with either a predictive central cue pointing at the likely location of the test stimulus or a nonpredictive flash of the outlines at one of the possible target locations. The cue-to-target intervals were set to achieve the greatest facilitatory effect of each type of attention (shorter for the exogenous condition, longer for the endogenous condition). The results showed that both types of attention improved performance (increased sensitivity for cued-location targets versus uncued-location targets) and that both types of attention similarly modulated the tuning of orientation representation by enhancing the gain without affecting the sharpness of the tuning curve; in other words, the range of orientation representation remained the same, but all orientations within that range were boosted similarly following either type of cue. With regards to the tuning of spatial frequencies, however, the two types of attention had qualitatively different effects. Endogenous attention enhanced the gain of spatial frequencies both above and below the target frequency without a change in peak sensitivity (at the target frequency of 2 cycles per degree), whereas exogenous attention most strongly affected target frequencies above the target frequency, and the peak sensitivity was shifted to a higher frequency. Overall, these findings from behavioral studies do not converge on a simple explanation for the differences between voluntary and involuntary attention; however, they provide compelling evidence that endogenous and exogenous attention involve at least somewhat different mechanisms of action.

6.1.2 Neural Effects of Exogenous versus Endogenous Attention

As reviewed in Chapter 5, numerous event-related potential (ERP) studies of endogenous attention have found that voluntary spatial attention enhances early stages of visual cortical processing. Although there remains controversy regarding whether voluntary attention can modulate the initial cortical visual response, indexed by the C1 from the striate cortex (e.g., V1), there is reliable and robust enhancement of the extrastriate cortex-generated P1 at about 100-ms latency. There have been fewer ERP studies of the exogenous attention effects, especially on the short cue-to-target intervals for which this type of attention enhances behavioral responses. The short cue-to-target intervals are necessary to observe the facilitative effects of attentional capture before IOR sets in. However, exogenous attention acts so quickly that it is challenging to isolate the effects on target processing from the sensory processing of the cue itself. Using analysis techniques originally developed by Woldorff (1993) for studying short-latency auditory potentials, Hopfinger and Mangun (1998) studied the exogenous capture of attention at short cue-to-target intervals. In that study, targets were preceded by brief, nonpredictive peripheral cues at short intervals (34–234 ms). In addition to replicating the standard speeding of reaction times at the cued location, the ERPs revealed that exogenous attention modulated target processing at multiple stages. As shown in Figure 6.2A, exogenous attention significantly enhanced the strength of the P1 component.

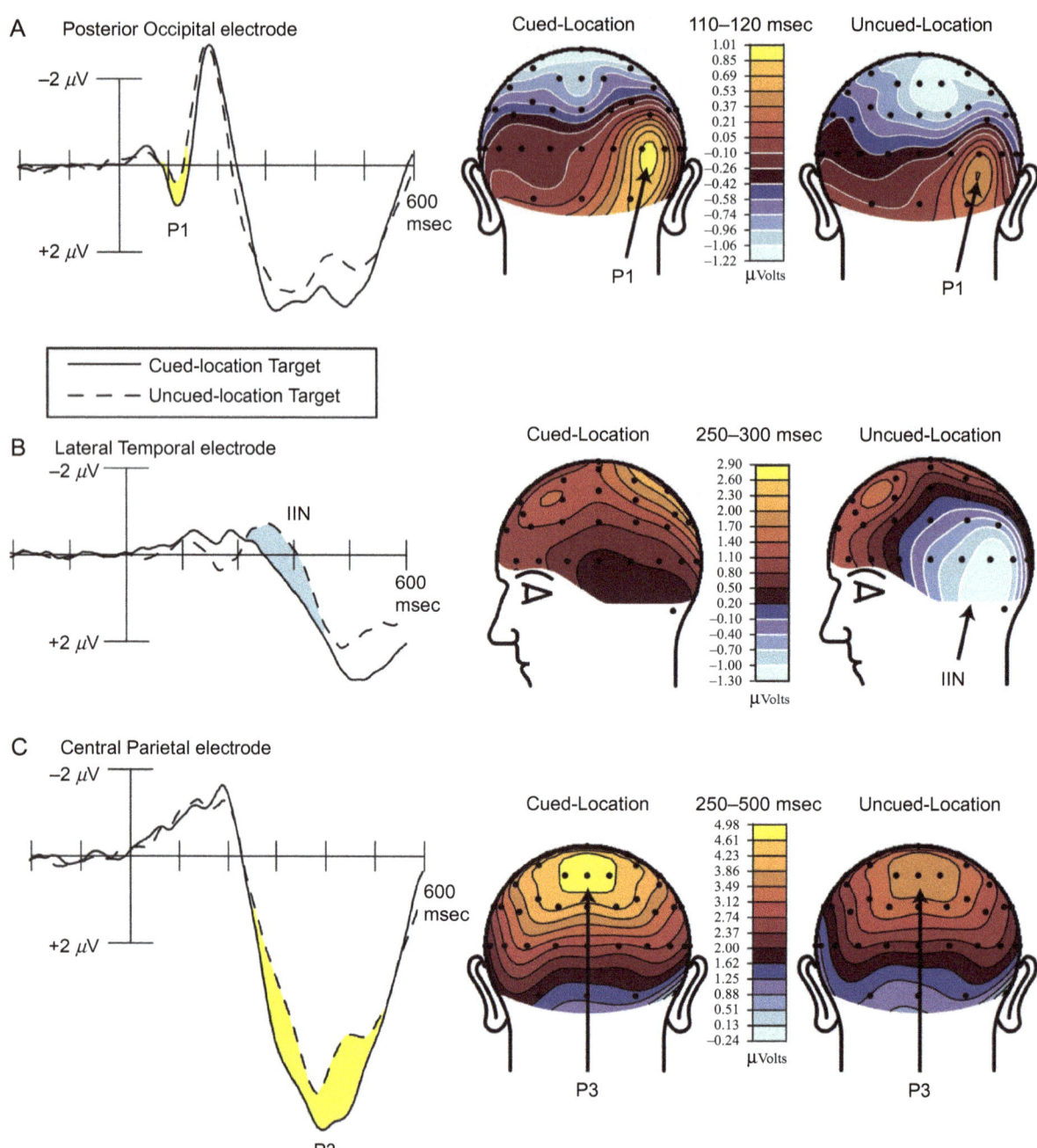

Figure 6.2 ERP effects of exogenous capture of attention on target processing. Shown here are the effects on target processing at *short* cue-to-target intervals (68–268 ms), when exogenous attention produces facilitation in behavior. "Cued-Location" refers to targets that occurred at the same location as the preceding nonpredictive peripheral cue stimulus. "Uncued-Location" refers to targets that were preceded by a cue at a different location. (A) The amplitude of the P1 component was significantly greater for the cued-location targets, reflecting an enhancement in early sensory processing by exogenous attention. The scalp topographic maps show that the distribution of the P1 was the same, but that it was stronger at the cued location. (B) The ERP and topographies show that the "ipsilateral invalid negativity" (IIN) was produced only when involuntary attention had been captured to the wrong location (i.e., uncued-location target) and attention needed to be reoriented. (C) The P3 was produced by both stimuli, but its amplitude was significantly enhanced for cued-location targets. Reprinted from Hopfinger and Parks (2012), with permission from Oxford University Press.

This attention effect of exogenous orienting on the P1 has been replicated across discrimin-ation and detection tasks (Hopfinger & Mangun, 2001; Ries & Hopfinger, 2011), and it has been found even when subjects are just passively viewing the stimuli (Hopfinger & Maxwell, 2005). As with the effects of endogenous attention covered in Chapter 5, the earliest effects of exogenous attention are modulations of this P1 component. The earlier C1 component, gener-ated in V1, is not typically found to be modulated by exogenous spatial attention (e.g., Fu, Greenwood, & Parasuraman, 2005; Hopfinger & Maxell, 2005). Although exogenous and endogenous attention have similar effects on the C1 and P1 components, the subsequent N1 shows a difference. The visual N1 is not usually enhanced by exogenous attention, regardless of task, whereas the N1 can be robustly enhanced by endogenous attention under many task conditions. Most studies of exogenous attention, at short SOAs, have either reported no effect on the N1 or found an enhanced positivity that extends through the P1 and N1 latency ranges. Thus, even relatively early in processing, within the latency range of the N1, there are differ-ences between the effects of endogenous and exogenous attention. Regarding the extension of a positivity through the N1 latency range, some researchers have argued that exogenous attention may inhibit the stage of processing indexed by the N1 and facilitate the processing of the P1 (Fu, Fan, Chen, & Zhuo, 2001; Fu et al., 2005).

In addition to enhancing early visual processing, as indexed by the P1, exogenous attention also produces effects at later stages of processing. These studies of exogenous attention revealed a negativity starting at ~200 ms that is produced only when a target occurs at the uncued, or invalid, location (Figure 6.2B). This effect was labeled the ipsilateral invalid negativity (IIN), and it has been proposed as an index of when attention needs to be reoriented rapidly due to exogenous attention being captured at the wrong location (i.e., invalidly cued). It is observed in discrimination and detection tasks (Hopfinger & Mangun, 1998, 2001) but not in passive viewing conditions when the subject doesn't need to reorient attention in order to respond to a target (Hopfinger & Maxell, 2005). Finally, exogenous attention at short cue-to-target intervals also modulates the P3 component, specifically the P3b, located over central parietal regions and thought to be involved in context-updating and decision-making processes (Figure 6.2C). The effects of exogenous attention on this P3 component may indicate that the cued-location stimuli are briefly treated as more highly relevant to response goals.

ERPs have also been used to investigate exogenous attention at the longer cue-to-target SOAs when IOR is typically observed in behavior. In ERP studies with cue-to-target SOAs of longer than 500 ms, the P1 is significantly *reduced* for cued-location targets relative to uncued-location targets (Hopfinger & Mangun, 1998; McDonald, Ward, & Kiehl, 1999). This reduction of the P1 provides evidence that visual processing at the exogenously cued location is sup-pressed a short time after that location had enjoyed a boost in processing from attention capture. This reduction of the P1 has been replicated across many studies and tasks (Hopfinger & Mangun, 1998; McDonald et al., 1999; Prime & Jolicoeur, 2009; Prime & Ward, 2004; Wascher & Tipper, 2004). Unlike the effects of exogenous attention at short SOAs, the IIN and the enhancement of the P3 are not reliably found at long SOAs. The visual N1 component had not been found to be suppressed in most early studies, but Prime and Jolicueor (2009) found an effect on this component when they ensured that attention was reoriented to fixation before the target appeared. Behavioral studies of IOR had shown that inhibition is most consistent and robust if a "reorienting event" occurs between the cue and the target. The reorienting event is typically a flash of the fixation point to draw attention back to center, and the idea is that this ensures that the subject hasn't decided to dwell on the cued location on some trials. Using this reorienting event between the cue and target in their experiment, Prime and Jolicueor (2009) found that both the P1 and N1 showed IOR-like suppression of the cued versus uncued location. Thus, ERP studies have found that an inhibitory aftereffect of exogenous capture suppresses relatively early levels of visual processing but doesn't appear to affect the very earliest stage of cortical processing (i.e., the C1). Box 6.1 describes the brief inhibitory surround centered at the location of exogenous attentional capture.

Before ERP researchers began trying to isolate and dissociate exogenous and endogenous orienting, many early studies of attention used cues that engaged both systems. In those experiments, cues that would trigger exogenous orienting (e.g., flashes in the periphery) were also predictive of the location of the target, therefore engaging endogenous orienting as well. In most of those studies, the cue-to-target interval time was long enough that the observed effects would not be due to the early effects of exogenous capture, and the effects were usually attributed to the endogenous maintenance of attention at that location (e.g., Mangun & Hillyard, 1990). However,

---

Box 6.1 Inhibition over space and time?

The preceding subsections have discussed how one unique aspect of exogenous attention that distinguishes it from endogenous attention is the subsequent inhibition that follows after attentional capture. This IOR is thought to play an important role in suppressing the salience of objects at locations that have recently captured attention to ensure efficient search and allocation of further attentional resources. However, it is important to clarify that this inhibition at the cued location is different than the inhibitory surround discussed in Chapter 5 regarding the effects of endogenous attention. The inhibitory surround in endogenous attention refers to the suppression of *nearby* locations (or *similar* features in the case of feature-based attention), and this suppression occurs at the same time as the endogenous enhancement of the attended location (or feature). This inhibitory surround is thought to enhance the ability to discriminate more finely between information coming from the attended space versus neighboring locations. The inhibitory surround in the case of endogenous attention thus refers to inhibition *across spatial locations* at the same time as when endogenous attention is focused on one location. In contrast, IOR is an inhibition at the *same spatial location* as where attention had been exogenously captured, but at a later *time*. However, recent research has provided evidence that the exogenous capture of attention is also associated with an inhibitory spatial surround. Baruch and Goldfarb (2020) found that the exogenous capture of attention created an inhibitory surround in space similar to what is seen with endogenous attention. This exogenous version of the inhibitory surround, however, was found to be very short-lived, lasting only about 200 ms after the capturing event. Thus, studies have shown that exogenous attention triggers an initial short-lived inhibition of surrounding spatial locations followed later in time by an inhibition at the captured location. These mechanisms of suppression are critical for the selective processing that attention enacts.

it should be noted that exogenous attention was triggered in the many studies that used peripheral cues, and the lack of IOR in behavior and the absence of an inhibited P1 component in those experiments provided some evidence that IOR can be overcome by voluntarily attending to that location. However, more recent research has found that a reduction of the P1 can be found even with long intervals after a *predictive* peripheral cue (Chica & Lupiáñez, 2009), suggesting that IOR can play a role in perceptual processing even when endogenous attention is focused on that location.

Given the very fast time course of exogenous attention, from the initial rapid enhancement of sensory processing to its replacement a couple of hundred milliseconds later by IOR, most neuroscience research into this type of attention has used methods with excellent temporal resolution. Neuroimaging methods, because of their reliance on the sluggish and temporally imprecise hemodynamic response, are poorly suited to dissociating the initial peripheral cue processing from the target processing that happens shortly thereafter. However, Müller and Kleinschmidt (2007) designed a paradigm that avoided overlapping functional magnetic resonance imaging (fMRI) responses by presenting cues and targets in separate hemifields. By placing cue and target stimuli near to each other in space but on opposite sides of the vertical midline, the major contralateral processing of cues and targets occurred in opposite hemispheres of the brain. Although this design has limitations in understanding the spatial extent of attentional capture and IOR, the two possible target locations were widely separated in the upper versus lower halves of the visual field, and therefore it was possible to see general effects of attention. Müller and Kleinschmidt (2007) found that at short SOAs target processing in early visual areas (V1/V2 and V3/V4) was enhanced at the exogenously cued location ("valid" trials) relative to the uncued location ("invalid trials"). At longer SOAs this pattern was reversed, with a relative suppression at the cued location in these visual areas. These data provide confirmatory evidence that exogenous attention affects early levels of visual processing, boosting processing at short SOAs and suppressing that processing at long SOAs. Although this study found attention effects in V1/V2, it should again be noted that fMRI does not have the temporal resolution to discern whether that modulation is occurring during the initial bottom-up processing or represents later reentrant processing in those areas.

Finally, when comparing the effects of exogenous and endogenous attention, it should be noted that most studies of endogenous, or voluntary, attention have required subjects to attend to a location (or feature or object) that the experiment has told them to attend to. Therefore, this orienting may not be purely voluntary and could instead be at least partially influenced by things outside of the person's own choice. Furthermore, studies of endogenous attention often utilize stimuli that can trigger some aspects of exogenous attention as well. Even studies that use stimuli appearing at a central fixation spot to instruct subjects where to attend in the periphery have limitations in their ability to isolate and test voluntary attention. For instance, the abrupt appearance of a stimulus, even at fixation, can engage the exogenous attention system, and this can result in an additional process of effortfully disengaging from the location of that stimulus in addition to the subsequent shifting of endogenous attention to the periphery. Also, responding in a certain way to a certain cue (e.g., orienting to the right following a red cue) over many trials can lead to that action becoming somewhat automatic, meaning that the cue may be triggering automatic as well as voluntary mechanisms of attention. Some neuroimaging research studies have attempted to better isolate the purely voluntary control of attention, and this type of voluntary shifting of attention has been variously called "internally driven attention" (Taylor, Rushworth, & Nobre, 2008), "self-initiated shifts" of attention (Hopfinger, Camblin, & Parks, 2010), or "willed attention" (Bengson, Kelley, Zhang, Wang, & Mangun, 2014; Bengson, Liu, Khodayari, & Mangun, 2020). These investigations into the brain areas and processes underlying purely voluntary control will be covered in the next chapter on the *control* of attention.

6.1.3   Interactions of Exogenous and Endogenous Attention

The previous subsections focused on studies that attempted to isolate exogenous and endogenous attention in order to see what each type of attention alone can do. But what happens when both attention systems are engaged together? In the real world, there is ongoing competition between our goal to attend to one thing and the capture of our attention by other things. To examine the interactions between endogenous and exogenous attention, researchers have investigated what happens when both systems are engaged concurrently. In an ERP study, Hopfinger and West (2006) made use of the fact that the optimal timing for exogenous capture is shorter than the optimal focusing time for voluntary attention to compare the effects of both systems when at their peak times of focus. On each trial in that study, an instructive central cue first directed subjects where to attend voluntarily; they were to pay attention and only respond to targets (vertical checkerboards) at that one location. Behavioral responses to targets were therefore only at the endogenously cued location, but since this was an ERP study, the neural responses to task-irrelevant probes (horizontal checkboards) could be measured at both attended and unattended locations. The time between the endogenous cue and target was long enough (>800 ms) to permit the orienting and focusing of voluntary attention to be established before a nonpredictive exogenous cue (peripheral blink at one of the locations) occurred. There was only a short SOA (34–234 ms) between the exogenous cue and the target, so that the peak of exogenous attention could be measured before effects of IOR could set in. The results showed that when concurrently engaged at the timing at which both types of attention typically produce their maximal effects, exogenous attention dominated the earliest stage, but interactions were also observed. Specifically, exogenous attention boosted the P1 (i.e., a larger P1 at the exogenously cued location versus the exogenously uncued location), but the size of this involuntary boosting effect was larger at the *endogenously cued* location versus the endogenously *uncued* location (Figure 6.3, left). For endogenous attention, the typical enhancement of the P1 at the endogenously cued location (versus the endogenously uncued location) was not observed when exogenous attention had also been triggered on that trial; however, the N1 was still significantly boosted at the endogenously cued location. Furthermore, the *latency* of when endogenous attention had its largest effect on the N1 was affected by where exogenous attention had been cued (Figure 6.3, right). Thus, overall, endogenous attention affected the *amplitude* of the exogenous attention effect on the P1 and exogenous attention affected the *latency* of the endogenous attention effect on the N1. At later stages, processing was completely dominated by endogenous attention – the P3 was significantly enhanced by endogenous attention without any interaction with exogenous attention, and the IIN and P3 effects typically seen following exogenous shifts of attention were not observed when voluntary attention was concurrently engaged.

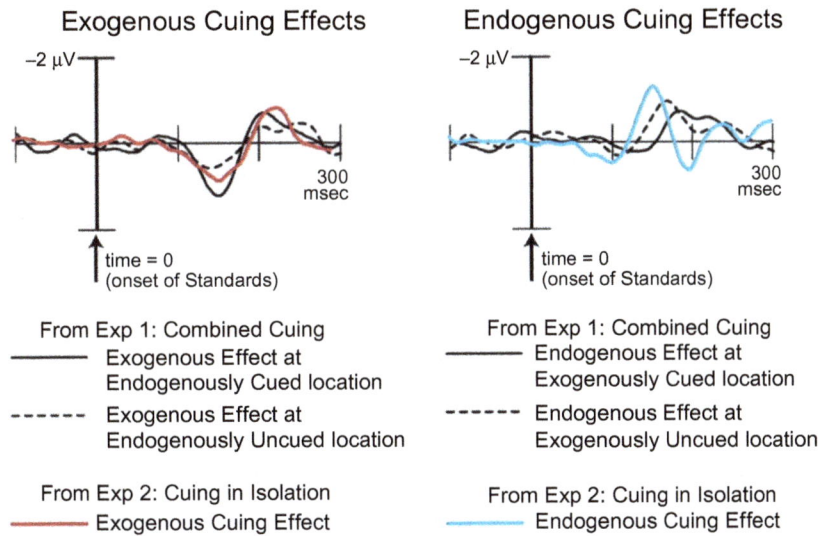

Exogenous Cuing Effects

−2 µV

300 msec

time = 0
(onset of Standards)

From Exp 1: Combined Cuing
—— Exogenous Effect at
Endogenously Cued location
- - - - - Exogenous Effect at
Endogenously Uncued location

From Exp 2: Cuing in Isolation
—— Exogenous Cuing Effect

Endogenous Cuing Effects

−2 µV

300 msec

time = 0
(onset of Standards)

From Exp 1: Combined Cuing
—— Endogenous Effect at
Exogenously Cued location
- - - - - Endogenous Effect at
Exogenously Uncued location

From Exp 2: Cuing in Isolation
—— Endogenous Cuing Effect

Figure 6.3 ERP effects of concurrent endogenous and exogenous attention. Shown here are *difference waveforms* between attended and unattended probe stimuli at a lateral posterior electrode where the P1 and N1 are typically observed. The left column shows the exogenous cuing-effect differences, which were calculated by subtracting the exogenously uncued-location probe from the exogenously cued-location probe. The red line shows the effects of exogenous cuing from a control experiment in which only exogenous cuing was engaged (endogenous attention was never engaged). The solid black line shows the effects of exogenous cuing at the location where endogenous attention was also focused; the dashed line shows the effects of exogenous cuing at the endogenously uncued location. A modulation of the strength of the exogenous P1 attention effect can be seen peaking at ~150 ms. The right column shows the endogenous cuing effect, calculated by subtracting the endogenously uncued-location probe stimuli from the endogenously cued-location probes for the control condition of endogenous attention alone (blue line) and the endogenous attention effect at the exogenously cued location (solid black line) and exogenously uncued location (dashed line). Exogenous attention modulated the timing of the large negativity (~N1) enhanced by endogenous attention. Adapted from Hopfinger and West (2006), *NeuroImage*, with permission from Elsevier.

In another ERP study that investigated the consequences of exogenous and endogenous attention being engaged concurrently, Chica and Lupiáñez (2009) used peripheral cues that were predictive and therefore engaged both exogenous and endogenous attention. As described earlier, that study found an IOR-type reduction of the P1 at the location of the peripheral cue, even though that location should have been voluntarily attended because the target was likely to occur there. This result suggests that exogenous attention dominated this early level of visual processing over endogenous attention. In contrast to the P1 effect, Chica and Lupiáñez found that the P3 component was dominated by endogenous attention. Although these authors used a single cue to trigger both systems and used a cue-to-target SOA when *IOR* would be produced, the findings converge with Hopfinger and West's (2006) study that used different stimuli to trigger each system and an SOA when exogenous *facilitation* would be present. Specifically, both studies found that when both attention systems are active at the same time, *exogenous* attention dominates the *earliest* levels of processing and *endogenous* attention dominates *later* stages of processing. Whereas these studies have attempted to isolate the reflexive versus controlled systems of attention, other research has questioned whether the capture of attention is ever truly automatic, as described in the next section.

6.2   Automatic Capture versus Contingent Capture

Although many studies have attempted to isolate endogenous and exogenous attention, other studies have suggested that top-down control settings are always affecting what is able to capture our attention. In a highly influential series of studies, Folk and colleagues provided evidence that peripheral cues only captured attention when the cue stimuli matched, in some way, the expected features of the target stimuli (Folk, Remington, & Johnston, 1992; Folk, Remington, & Wright, 1994). For example, Folk et al. (1994) developed a cuing paradigm in which target stimuli were of a consistent type within blocks, but cues could either match the critical target feature or not. Figure 6.4 illustrates this type of design: In "onset target" conditions, the target was a single stimulus that would appear ("onset") alone, requiring a simple response based on its identity. In the "color target" condition, the target display contained multiple items, and the target was the uniquely colored item of a predefined color; for example, the one red item presented in a display of mostly black items. Within each of these predefined target blocks of trials, cue displays could be either an "onset" cue (i.e., a single group of dots surrounding just one location) or a "color cue" (one location surrounded by red dots, while simultaneously black dots surrounded all other peripheral locations). The cues were always nonpredictive of the location of the upcoming target and were thus meant to engage only involuntary attentional orienting. According to theories that argue that exogenous attention is automatically oriented to the most salient item in a display, each of the cue types should trigger an automatic shift of exogenous attention to the location of the salient cue (i.e., to the location of the single-onset cue or to the location of the uniquely colored item in the color cue). However, Folk et al. (1994) found that attentional capture, as indexed by faster reaction times to the cued-location target versus uncued-location targets, only occurred when the type of cue matched the expected target type. Color cues only captured attention in blocks in which the targets were always color-defined targets, and onset cues only captured attention in blocks in which the targets were always single-onset items. Based on these results and subsequent replications and extensions (e.g., Folk & Remington, 1999; Lien, Ruthruff, Goodin, & Remington, 2008), it was proposed that attentional capture is contingent upon top-down settings for the types of target displays that were relevant. This **contingent capture** proposal was not suggesting that subjects used endogenous attention to voluntarily orient to the cued location, but rather that the involuntary shift of attention to the cued location was dependent on whether features of the cue matched critical features of the expected target.

The contingent capture theory has been vigorously debated, however. Other researchers have claimed that salient items always trigger exogenous attention, and that the reason it isn't observed in the contingent capture paradigm is because of the time interval between the cue and target displays (Geyer, Müller, & Krummenacher, 2008; Mounts, 2000; Theeuwes, 1991, 1992, 1994, 2004). This argument holds that the differences between cue conditions (matching

Onset Target Conditions   ## Color Target Conditions

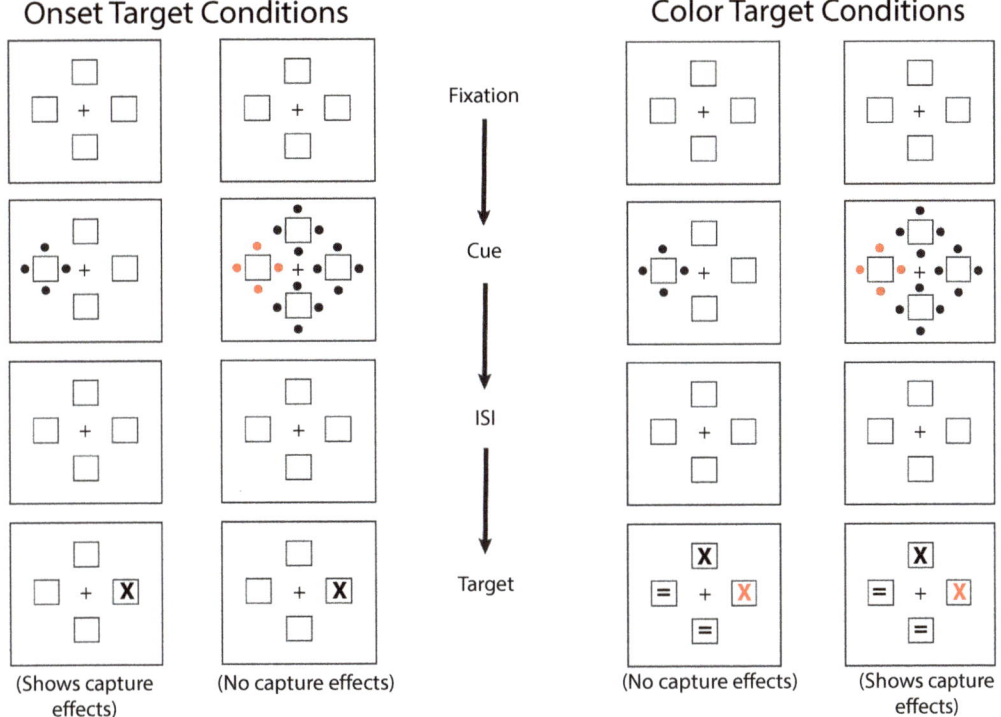

Figure 6.4 Contingent capture paradigm. Shown here is a version of the paradigm developed by Folk et al. (1992). In "onset target" blocks the target screen consists of a single new (onset) stimulus, and in "color target" blocks the target screen consists of stimuli appearing at all four locations with the target being defined by a predefined unique color (here, red). The subjects are informed what type of target display will be presented in each block, but either type of cue can occur in each target block. An "onset cue" display consists of a single set of dots at one location; a "color cue" display consists of sets of dots appearing at each location, but one is uniquely colored (the same color as the color target). For the color cue trials, a "cued-location" target would be a target that appears at the location where the unique color was present in the cue display. All cues are nonpredictive of target location. Shown below each trial sequence in this figure is whether behavioral experiments have typically found attention capture effects (faster reaction times to cued-location versus uncued-location targets) in these conditions. ISI = interstimulus interval.

cues showing capture; nonmatching cues showing no capture) are because of differences in how long it takes to *disengage* attention from the cue: When the cue doesn't match the target type, attention can be rapidly disengaged and attention is no longer at the cued location when the target appears. When the cue matches the target, attention dwells at the cued location longer because of it matching the target properties and capture is observed in reaction times. According to this theory, the initial orienting is automatically triggered by salient events, and any contingencies with top-down target settings occur only after exogenous attention has been captured to that location. However, the evidence that attention is indeed being captured automatically comes from a different type of paradigm; specifically, a visual search-type paradigm in which multiple stimuli appear simultaneously (or near-simultaneously), and evidence for capture is often indexed by comparing the presence or absence of a salient stimulus on target responses across different set sizes. Thus, behavioral studies alone haven't conclusively converged on whether the capture of attention can be driven completely automatically by salient stimuli or whether capture is contingent on the top-down setting for expected target properties (Wang & Theeuwes, 2020; Wu, Remington, & Folk, 2014).

Studies using ERPs have the ability to look at the processing of stimuli at multiple levels before the final overt response. Thus, these studies can potentially observe attention effects on neural processing that may not be observable via overt behavior. For instance, of the multiple ERP effects of exogenous attention described earlier, are all these contingent on top-down settings, as the behavioral results from the contingent capture paradigm would suggest? Or are some or all of these neural effects automatically triggered by salient stimuli? Hopfinger and Ries (2005) conducted an ERP study based on the contingent capture paradigm to address these questions. Across three experiments, the contingent capture pattern was replicated in behavior: Cued-location targets were responded to significantly faster than uncued-location targets, but only when the cue type matched the expected target type. When cues didn't match the target type for that block, there was no attention effect on reaction times. However, the ERP results revealed that some neural effects were contingent on top-down target settings, and others were more automatic. The enhancement of the P1 component by exogenous attention was *not* dependent on top-down settings, but rather was dependent on just the *physical salience* of the cue. When the cue was a single abrupt onset ("onset cue") or when there was a strong luminance difference between the colors in the color cue, the P1 was automatically enhanced at the cued location regardless of the top-down settings for expected target type. The enhancement of the early part of the P1, including the peak of that component, was enhanced *regardless* of the top-down settings, when the cues were highly salient. However, the *duration* of that enhancement depended on top-down settings, as the boost in processing lasted longer when the cues matched the expected target type. In a third experiment in which the salience of the color cue was reduced by making the unique color isoluminant with the other colors in the display, there was no attentional modulation of the target P1 in either contingency condition. Thus, when the salience of the cue was reduced enough, top-down control settings alone couldn't trigger a P1 attention effect. Overall, these experiments suggest that the exogenous enhancement of the P1 is driven by physical salience, although top-down settings can modulate the duration of that enhancement (see also Ries & Hopfinger (2011) for a similar experiment and P1 results with motion cues versus color cues).

These experiments also provided insights into exogenous attention effects at later stages of processing (Hopfinger & Ries, 2005). As described earlier, the IIN component is found 200–300 ms after a target appears at an uncued location, and this component is thought to be related to the disengagement of attention and reorienting of resources to the target location. In these experiments on contingent capture, the generation of an IIN was affected by both salience and top-down settings, depending on the type of cue. Cuing with color cues produced an IIN only to the target when it was congruent with top-down settings (i.e., color cues in the color target blocks); in onset target conditions, there was no IIN on color cue trials. This result would support contingent capture claims. However, an IIN was automatically produced, regardless of top-down settings, when the cue was a unique onset, suggesting that the abrupt appearance of an object may be special in triggering exogenous orienting and the need to reorient on invalid trials. The exogenous attention effect on the P3 component (cued location targets > uncued location targets at ~300–500 ms) wasn't affected by top-down settings in these experiments, although the attentional enhancement of the target P3 by color cues was much smaller than the enhancement by onset cues, suggesting that the appearance of a single new object in the onset cue displays may be special in triggering this effect as well. Section 6.4 discusses in more detail the potential special status of new perceptual objects in triggering exogenous attention. The question of whether attention can be captured to different degrees is discussed in Box 6.2.

Box 6.2 Is involuntary capture all-or-none?

Typically, the exogenous capture of attention is thought to either happen or not happen, with there being no in-between state. We know that we are sometimes captured by an unexpected event, but other times we seem to be "in the zone" and to be able to remain undistracted by whatever is going on around us. The idea of being "partially captured" by something may not be intuitive. However, this assumes a single unitary focus of attention. As described in Chapter 5, studies have shown that voluntary attention can be allocated in various amounts. Mangun and Hillyard (1990) found that as the amount of voluntary attention increased (from 0% to 25% to 50% to 75% to 100%), the attention effects on the early visual P1 and N1 ERP components increased progressively. However, can exogenous attention also be allocated to varying degrees? Anderson and Folk (2010) found evidence from reaction times that capture was indeed graded by how similar a cue color was to the target color: The closer the cue color was to the target color, the greater the attentional capture effect on behavioral responses. Importantly, these authors analyzed the data to see whether the smaller average attention effect to the less-well-matching cues was due to each trial showing a smaller amount of capture, or whether it was simply that attention was captured fully on a smaller percentage of trials and therefore the averaging of these few capture trials along with trials with no capture led to an overall smaller attention effect. These analyses supported the model in which cue–target similarity affected the *amount* of attention captured on *each* trial, not the number of trials on which attention was captured. These findings suggest that exogenous attention may not be all-or-none, but instead can be allocated to varying amounts. The definition of "capture" may need to be refined if one wants to refer specifically to the instances in which attention is fully pulled away from the intended locus entirely to the unexpected salient event. Using a newly developed analytical technique that accounts for changes across different target set sizes, Rigsby, Stillwell, Ruthruff, and Gaspelin (2023) reported evidence that attention isn't captured as often as previous research had suggested based on grand averages across cued and uncued trials. They reported that even cues that typically produce robust attention effects (i.e., match target features) only captured attention on ~30% of their trials. Incidentally, this is a similar percentage to what Wood and Cowan (1995) found in terms of how often our name captures our attention via the cocktail party phenomenon. Combining neuroscience methods with these new techniques that allow for a more precise estimate of when attentional capture has occurred will be critical for assessing what aspects of exogenous attention can be allocated *partially* versus *all-or-none*.

6.3   Visual Search, Orienting, and the "Suppression Hypothesis"

Studies of visual search often involve either cooperation or competition between endogenous and exogenous attention. Both types of attention may work together when targets have a unique feature (e.g., color, motion, form) compared to all other items in the display. In those cases, exogenous attention is captured to the unique stimulus, and endogenous attention maintains focus on that location while the subject makes a decision about the stimulus there. For example, if the task is to discriminate the orientation of a red rectangle presented among multiple green shapes, exogenous attention is likely captured to that red item, and endogenous attention keeps the focus on that item until the subject can discriminate the attribute of the target (e.g., the orientation of the long axis) needed to make the response. However, in conditions in which the most salient item is not the target, exogenous attention works against endogenous attention. For example, if the target in the previous example is not the unique red rectangle but is rather a green circle in a display consisting of that red item plus many green ellipses, then the bias of exogenous attention to orient to the uniquely colored red rectangle must be overcome by endogenous processes trying to focus on the green circle target. Unlike the cuing paradigms described earlier, in which reaction times to a target stimulus can be directly compared for cued-location targets versus uncued-location targets, visual search studies often *infer* attentional capture to nontarget items by comparing trials with a salient distractor to trials without a salient distractor. These studies also compare across trials with different numbers of nonsalient distractors (e.g., green ellipses when searching for a green circle) in the display (i.e., "set sizes"). Attentional capture to a single salient stimulus can be inferred on the basis of a "flat search slope" across set sizes, meaning that as the set size increases (more nonsalient distractors in the display), the time to detect a *salient* unique target stays the same, providing evidence that the salient item is always the first item to be attended. This is also referred to as "parallel search" because it is as if there is an initial stage of search across all items simultaneously that quickly identifies any unique items (in terms of color, luminance, motion, etc.) in the display and moves attention to that spot. Parallel search is contrasted to "serial search," in which the search slope is not flat but steadily increases with each extra distractor item added to the display. In serial search, the idea is that attention must focus on each item one at a time to discern whether it is the target or not, so every additional item adds more time to the search process. Serial search slopes are seen in cases in which the target is defined by a conjunction of features in a display that contains no clear and obvious unique singleton items (e.g., a green circle in a display full of green ovals, with no other-colored stimuli). Studies of visual search often find that highly salient singletons are the first items searched, produce flat search slopes when they are the target, and slow subjects' responses to other targets when the salient singleton is a distractor. These types of visual search paradigms have provided much of the evidence used by some researchers to claim that exogenous attention is automatically captured to the most salient item in the display (e.g., Geyer et al., 2008; Theeuwes, 1992), as described in the previous section on contingent capture.

ERP studies have identified a series of components that are relevant to the debate regarding the automatic versus contingent capture of attention. These include the N2pc, N_T, and P_D components illustrated in Figure 6.5. As opposed to most of the ERP components discussed in

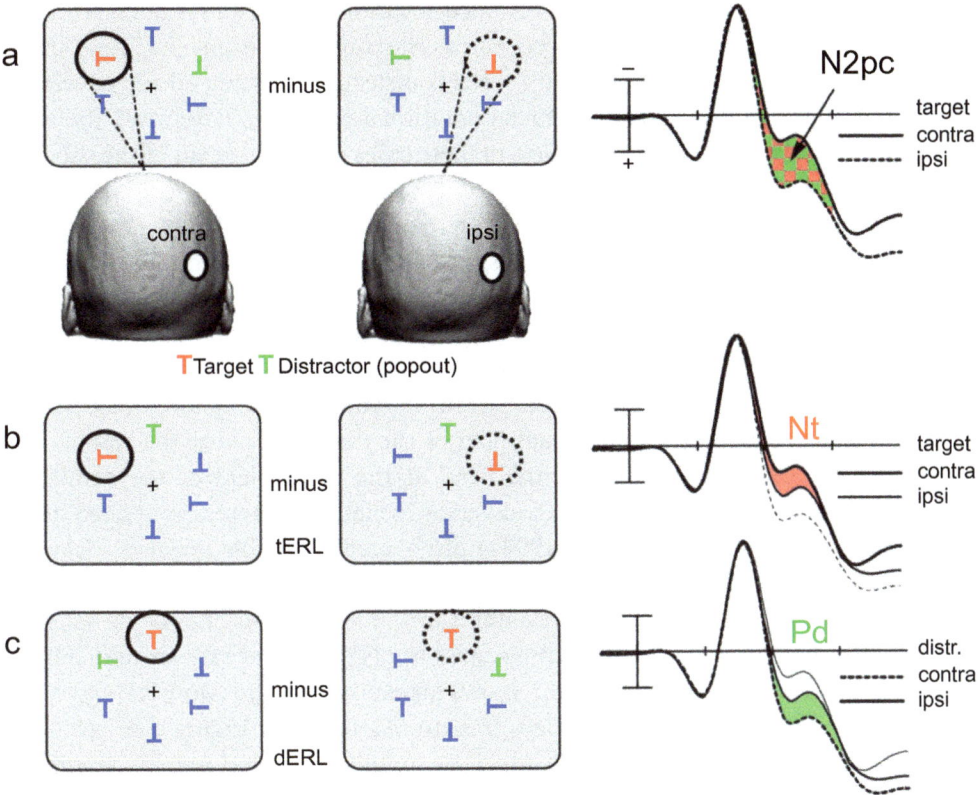

Figure 6.5 ERP components associated with visual search, orienting, and suppression. ERP difference waveforms evoked by the visual search displays shown. In all displays, the target is the red T, most of the distractors are blue T's, and a single green T is a salient distractor. Part (a) shows a display in which the target and salient distractor occur in opposite visual fields. The ERP plot shows electrode sites that are contralateral (contra; solid line) and ipsilateral (ipsi; dashed line) to the location of the target. The N2pc is the negative-going deflection for contralateral versus ipsilateral sites at 200–300 ms, being maximal over contralateral scalp sites. Part (b) shows the ERP response when the target is again lateralized to one visual field, but the salient distractor is always on the midline of the display. In this situation, the difference between contralateral and ipsilateral sites only reflects target processing, and this is indexed by the N_T wave. Part (c) shows the ERP response when the salient distractor is lateralized but the target is on the midline. In this case, the target does not contribute to a lateralized ERP effect, so the lateralized difference wave reflects only the processing of the salient distractor. This ERP difference shows a *positive*-going difference between the contralateral and ipsilateral electrodes, which has been labeled the P_D (distractor positivity) and has been linked to *suppression* of a salient *distractor* stimulus. Note that contralateral and ipsilateral waveforms for the P_D are defined in reference to the location of the salient distractor, not to the location of the target, as in parts (a) and (b). dERL = distractor event-related lateralization; tERL = target event-related lateralization. Reprinted from Donohue et al. (2020), *Communications Biology*, Springer Nature; Creative Commons CC BY 4.0.

previous sections, these are commonly referred to as "difference wave" components. These types of components are defined by the difference between two conditions or two electrode sites, and they are sometimes plotted as a single ERP just showing the difference between conditions.

Figure 6.3 presented difference waves for cued-location targets minus uncued-location targets, to more easily compare the cuing effects across the three conditions (i.e., only three difference waves were needed to compare the cuing effects instead of plotting all six waveforms for all of the cued and uncued locations from each condition separately). Regarding the N2pc, N_T, and P_D components, the difference waveform in these cases is not the result of the subtraction of one *condition* from the other, but rather comes from comparing one scalp *hemisphere* to the other. In visual search tasks, the displays often contain many stimuli distributed across the entire visual field, and the sensory-evoked components discussed in previous sections are evoked by all of those stimuli at once, making it difficult to disentangle which stimulus evoked which of those overlapping responses. However, the activity evoked by stimuli that are lateralized to one visual field can be isolated by subtracting the activity from electrodes on one side of the scalp from that received by electrodes on the other side. "Contralateral electrodes" are those that are over the brain hemisphere opposite to the visual field of the item(s) evoking the response; "ipsilateral electrodes" are on the same side of the head as the visual field of the item(s) evoking the response. By comparing the contralateral versus ipsilateral differences evoked in visual search displays, Luck and Hillyard (1990, 1994a) discovered a reliable negative-polarity deflection around the time of the N2 (~200–300 ms) that was evoked by visual search displays and was maximal at posterior scalp sites contralateral to the visual field of a single salient target that subjects attended. They labeled this component the N2pc (negativity around 200 ms at posterior contralateral electrodes). Figure 6.5 shows an example of the two ERP waveforms that are compared to identify whether an N2pc is present. As shown in Figure 6.5a, an N2pc would be evoked by a bilateral visual display in which the uniquely colored red "T" is *lateralized* to one visual field. The solid line in the ERP plot shows the response from an electrode contralateral to the red target, and the dashed red line shows the response from an electrode over the ipsilateral side of the head. The more negative amplitude for the contralateral electrode compared to the ipsilateral electrode from ~200 to 300 ms is labeled as the N2pc, and this has been taken as evidence of enhanced processing of that stimulus in the contralateral hemisphere. The N2pc has been replicated numerous times (Eimer, 1996; Hickey, Di Lollo, & McDonald, 2009; Luck & Hillyard, 1990, 1994a; Tay, Harms, Hillyard, & McDonald, 2019), and many authors use it as an index of the orienting of attention to a location in space. Although the N2pc can be elicited in visual search tasks in which a salient item "pops out" automatically, it is not thought to be simply an index of an involuntary capture of attention. This is because it is not evoked by salient *nontarget* stimuli that are easily dismissed as nontargets, but it is produced by nontargets that are *similar* to targets; thus, if attention is needed at that location to determine whether the stimulus is a target, it is likely to produce an N2pc. In addition, some authors have found evidence dissociating the N2pc from the actual shifting process of attention, arguing that it reflects the focusing of attention on a specific location to discern what's there in detail (e.g., Kiss, Van Velzen, & Eimer, 2008). This interpretation is similar to Luck and Hillyard's (1994b) original description of the component as indexing a filtering process that is engaged specifically when attention must be highly focused at one location to discriminate attributes of a target stimulus amid nearby distractors. Indeed, the N2pc is also greatly reduced or absent if there are no distractors present in the display (Luck & Hillyard, 1994b), and its amplitude increases with the number of distractors in the display (Mazza, Turatto, & Caramazza, 2009). Finally, the size of the N2pc has been found to be larger when attentional demands are higher, such as when distractors are closer to the target and when targets are defined by a conjunction of features as opposed to a single salient feature (e.g., Luck & Ford, 1998). Overall, the N2pc is a useful index of the timing and location of attentional selection.

Further research has revealed that the N2pc measured in some studies may be composed of two separate processes that can be indexed by different ERP difference wave components: the N_T and the P_D components (e.g., Gaspar & McDonald, 2014; Sawaki, Geng, & Luck, 2012). In a typical study of the N2pc that includes a salient distractor stimulus, the distractor is often in the opposite visual field from where the target occurs. Therefore, the difference between contralateral and ipsilateral sites can reflect an enhanced processing of the target and/or a suppressed processing of the distractor. Further studies sought to disentangle these two processes. The N_T component, referring to the "target negativity," has been linked to processes of *target selection* (e.g., Hickey et al., 2009). As shown in Figure 6.5b, the N_T is the enhanced negativity, starting around 200 ms, at scalp sites that are contralateral versus ipsilateral to a salient target stimulus. The N_T can be observed when the only highly salient *lateralized* stimulus is the target, such as when there is no salient distractor or when the only salient distractor is on the midline of the subject's field of view and therefore not lateralized to either hemisphere (as shown in the example displays in Figure 6.5b). The N_T is thought to index the lateralized processing of the attended target stimulus (this was sometimes reported as simply an N2pc in earlier studies).

The other process that may contribute to an N2pc-like difference wave is the P_D, or "distractor positivity," which has been associated with the *suppression* of a salient lateralized distractor. As shown in Figure 6.5c., the P_D occurs in a similar time window to the N2pc, but it is a positive-going wave that is contralateral to a salient *distractor* item when that item is known to be a nontarget that should be ignored (Hickey et al., 2009). Note that the ERP waveforms in Figure 6.5c are defined as contralateral and ipsilateral to the salient *distractor*, whereas in Figure 6.5a,b the waveforms are defined in terms of the *target* location (and since the distractor in those displays was always in the opposite visual field, the labeling is reversed in terms of what location is contralateral to the *distractor*). The P_D can be more challenging to detect because it is often smaller than the N2pc that is generated when that same stimulus is a target. However, this component has been found consistently across studies that utilize search tasks in which "singleton-search mode" cannot be utilized to locate the target. "Singleton-search mode" refers to being able to locate the target by simply finding the most salient unique item in the display, and under that strategy the most salient items capture attention. When the search is more complex, however, such as if searching for a green, diamond-shaped target among other green shapes with straight edges, then a highly salient, uniquely colored (e.g., red) distractor is found to produce a robust P_D component (e.g., Eimer & Kiss, 2008; Sawaki & Luck, 2010). Since the salient item producing the P_D should be ignored, and since the presence of a P_D has been linked to better task performance, it is thought to index the successful suppression of salient distractors. Further evidence linking the P_D to suppression comes from studies that have found that the *size* of the P_D is correlated with the *magnitude* of suppression measured in behavior (e.g., Feldmann-Wüstefeld et al., 2020; Gaspar et al., 2016; Gaspelin & Luck, 2018; Weaver et al., 2017).

The P_D has recently become a central point in the debate about the automaticity of attentional capture. Sawaki and Luck (2010) proposed the "signal suppression hypothesis," which attempts to resolve the capture debate by suggesting that the ability of a salient item to capture attention can be overridden by suppressing the processing of highly salient nontargets. Importantly, this suppression is not a focusing of voluntary attention at a different location before the display appears, because the locations of the target and salient distractor are random in these experiments. Thus, the P_D and the proposed suppression occur on the basis of the salience of the items and the known features of the target and nontarget stimuli, not on the basis of where in space voluntary attention can be allocated in advance of the display appearing. Gaspelin and Luck (2018) have argued that the signal suppression prevents attention from being captured at the location of the salient item. Furthermore, the suppression indexed by the P_D cannot be implemented simply by an explicit strategy alone. When subjects are told what color the distractor will be, but it varies on a trial-to-trial basis, suppression is *not* observed, and the subjects' attention instead often gets captured at that location (e.g., Cunningham & Egeth, 2016). The suppression instead is dependent on *experience* with the salient distractors and builds up over successive trials (Gaspelin & Luck, 2018). Although there is strong evidence for a suppressive mechanism, as indexed by the P_D, there are important issues outstanding, as reviewed by Luck, Gaspelin, Folk, Remington, and Theeuwes (2021). Among these are how the suppressive mechanism may work outside of the realm of color singletons and whether the suppression is strong enough to overcome the capture of attention by new perceptual objects. (The relation of new perceptual objects to attentional capture is discussed in the next section.) In addition, there has been controversy around the signal suppression hypothesis, with some authors arguing that global enhancement of target properties may explain much of the behavioral evidence, and that suppression is neither necessary nor sufficient to explain the lack of capture to salient items (Oxner, Martinovic, Forschack, Lempe, Gundlach, & Müller, 2023). Other recent research has provided ERP evidence suggesting that the P_D is affected by target feature regularities, independent of distractor type or consistency, suggesting that the P_D is not simply an index of distractor suppression but can also show the upweighting of target features (van Moorselaar, Huang, & Theeuwes, 2023). Thus, the potentially important role of suppression as a way to mediate the conflict between top-down goals and bottom-up capture, and the ability of the P_D component to index this process, remain active areas of debate and research.

6.4   Salience, Emotions, and Social Influences on Attentional Capture

When people talk about something capturing their attention in real life, they are usually referring to something especially big or bright or loud – something that has a physical intensity that is greater than other things in the environment at the time. Even a strong smell or an intense flavor may immediately draw our attention away from other things. The concept of "salience" has been used in research to describe why a stimulus can capture our attention (Koch & Ullman, 1985). Across our sensory modalities, the exact attribute that is responsible for capturing attention varies (e.g., a brighter stimulus in visual attention, a louder stimulus in auditory attention, a colder stimulus in somatosensory attention), but salience refers to how much a stimulus stands out from other stimuli, both within and across modalities. It has been suggested that the brain may have separate saliency maps within each modality plus an overarching saliency map that receives input from all the senses (Itti & Koch, 2001; Koch & Ullman, 1985; Parkhurst, Law, & Niebur, 2002). This conceptualization of a saliency map is reminiscent of the highest-level map in Treisman and Gelade's (1980) feature integration theory, which assumed separate feature maps fed into to an overarching map that was used to guide attention and bind features together. An important aspect of such saliency maps is that the determination of salience is always relative. There are not thought to be specific thresholds of brightness or loudness that are salient versus not salient. Rather, the exact same physical stimulus will be associated with different levels of salience depending on the other stimuli present at that moment, as well as stimuli in close temporal proximity. For example, as described in Chapter 2, some individuals prefer to study while listening to music because nearby conversations are less distracting to them amid the noise of concurrent music compared to those same conversations in an otherwise quiet room. While much of the research on exogenous attention utilizes cue stimuli that are clearly highly salient in their physical properties, other research has revealed that some stimuli capture attention not because of their physical intensity but because of higher-level properties. Indeed, this was the focus of Moray's (1959) study of the cocktail party phenomenon in which a person's own name in a nearby conversation captured their attention away from the speech they were supposed to be shadowing. As detailed in earlier chapters, this finding was influential for the levels of processing debate and the development of theories of attentional selection. More recent research has revealed that there are other special classes of stimuli – besides just our own name – that seem to have the power to grab our attention involuntarily and reflexively. A few of these are detailed in the following subsections.

6.4.1 Faces and Social Gaze

One type of visual stimulus that seems to have a uniquely strong pull on our attention is the face. Behavioral studies have shown that faces capture attention involuntarily. In visual search paradigms, faces are found faster than other items as targets, and irrelevant faces produce interference when searching for other items (e.g., Langton, Law, Burton, & Schweinberger, 2008). The effects of faces capturing attention during visual search can be found in both covert and overt measures of attention (Hershler & Hochstein, 2005; Morrisey, Hofrichter, & Rutherford, 2019; Scheerer, Birmingham, Boucher, & Iarocci, 2021). As noted in Section 6.1.1, exogenous attention produces IOR in cuing paradigms at long cue-to-target intervals. Research has shown that faces also produce IOR at long SOAs, providing more evidence that the orienting to faces is a reflexive mechanism (Theeuwes & Van der Stigchel, 2006). Critically, these studies have controlled for the overall physical intensity of the face versus nonface stimuli, so these behavioral effects suggest that faces are special in capturing attention because of them being faces. Relating back to the N2pc component described earlier, recent work has shown that the task-irrelevant appearance of one's own face in a display triggers an N2pc, suggesting that attentional resources are engaged by this stimulus (Bola, Paź, Doradzińska, & Nowicka, 2021). Finally, it is worth noting that the capture of attention by faces is dependent on the face being in its normal upright position; when faces are inverted, the capture of attention is typically greatly reduced or eliminated (e.g., Langton et al., 2008). Thus, there is something special about the configuration of an upright face that provides it with attentional priority. This is likely related to the enhanced processing of upright faces compared to all other stimuli (including inverted faces) in the human fusiform face area (FFA; Yovel & Kanwisher, 2005). In addition to the initial capture of attention, behavioral studies have found evidence that faces also *hold* attention longer than other stimuli (Parks, Kim, & Hopfinger, 2014). This effect of faces on the involuntary *holding* of attention will be covered in more detail in Chapter 8 on the temporal aspects of attention.

In addition to faces capturing attention to the location of the face, the sudden movement of another person's eyes has also been shown to have a powerful and involuntary influence on attention. As introduced in Chapter 2, this "joint attention" or "social gaze orienting" refers to our involuntary orienting toward what another person looks at. In everyday conversations, this type of "shared attention" facilitates communication by ensuring shared mental states regarding relevant items. In the laboratory, it has been shown that this orienting to the location of another person's gaze is involuntary. Figure 6.6 shows an example of a social gaze-orienting experiment. In that experiment, other factors were also being investigated (i.e., emotion of the face and congruency between the emotion of the face and the target item in the periphery). Those other factors will be explained in the following subsection, but the basic setup of a gaze-cuing experiment can be seen in Figure 6.6. These studies often use a variant of a cuing study, but in these studies there is a face presented at the center of the screen, and the "cue" is the direction that the eyes of that central face look toward. A target stimulus appears in the periphery shortly after the eye gaze movement occurs. A valid (i.e., cued-location) target is one that occurs where the eyes looked, whereas an invalid (i.e., uncued-location) target is one that occurs at the opposite location from where the eyes looked. Early studies using this type of paradigm found that valid targets were responded to significantly faster than invalid targets and, importantly, that this occurred even when the eye gaze was known to be nonpredictive of the target location (Driver, Davis, Ricciardelli, Kidd, Maxwell, & Baron-Cohen, 1999; Friesen & Kingstone, 1998). Whether the eye gaze was predictive, nonpredictive, or even counterpredictive of the upcoming target location, reaction times were consistently faster to gazed-at-location targets (Friesen, Ristic, & Kingstone, 2004). Although this provides evidence that social gaze orienting isn't voluntary, there are also differences between it and typical exogenous orienting. Specifically, gaze orienting doesn't produce IOR at the typical cue-to-target intervals for which exogenous attention does (Friesen & Kingstone, 2003). As compared to typical attention capture to onsets in the periphery, social gaze orienting results in a longer period of facilitation and a delayed onset of inhibitory processes as measured by reaction times (Frischen, Smilek, Eastwood, & Tipper, 2007). Although these behavioral studies provide strong evidence for the special status of gaze orienting in allocating attention, neuroscience studies have investigated whether social gaze orienting can trigger the same mechanisms as typical exogenous attention – specifically, the modulation of early sensory-level processing.

In order to investigate the neural effects of social gaze orienting and to compare these directly to the effects of typical exogenous attention triggered by peripheral flashes and to typical endogenous attention induced by instructive central cues, Chanon and Hopfinger (2011) recorded ERPs while subjects performed each of these types of task. Although social gaze

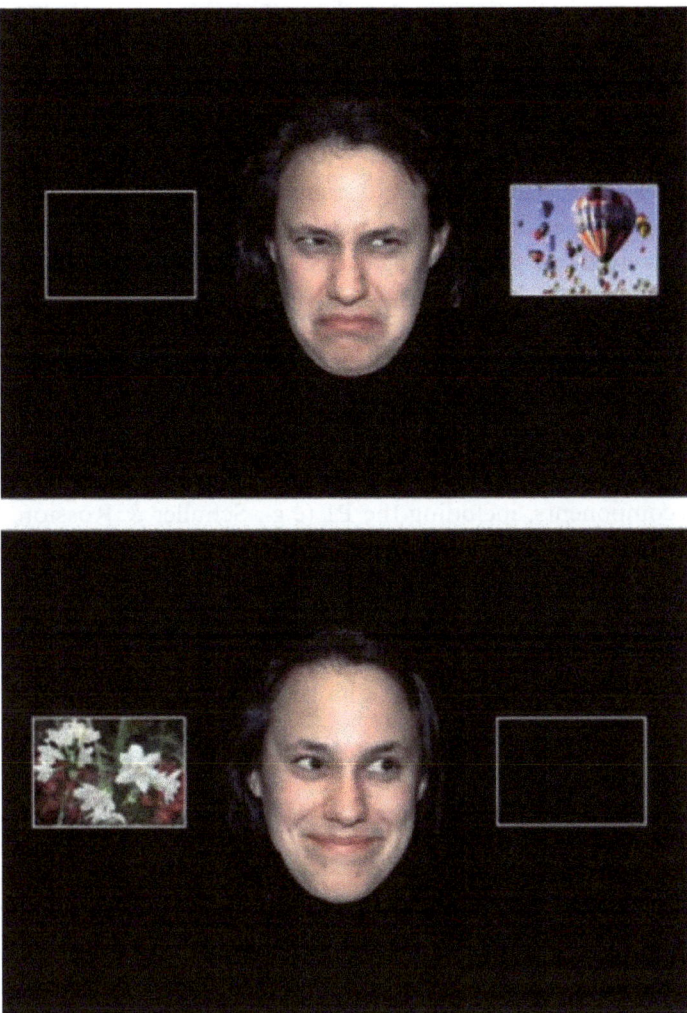

Figure 6.6 Example of a social gaze-orienting paradigm. Each example here shows a target display, and subjects would be responding to the peripheral pictures in these displays. The peripheral target picture would be preceded by a nonpredictive cue that consisted of the face at the center of the screen looking to either the left or the right. Valid trials (shown at top) and invalid trials (shown at bottom) were equally likely. This experiment also manipulated the emotion on the face (disgusted versus happy) and the congruence between the emotion on the face and the emotion associated with the peripheral picture (positive, neutral, or negative). Figure adapted from Bayliss et al. (2010), *Visual Cognition,* with permission from Taylor & Francis Group.

orienting has been more strongly associated with exogenous attention based on it being triggered involuntarily, these ERP results dissociated these two forms of attention. In this experiment, gaze orienting was induced by the eyes of a central face looking unpredictably to the left or right, typical exogenous orienting was triggered by a nonpredictive flash at one of the target locations, and endogenous attention was evoked by an instructive change of color of the central fixation cross indicating to attend to the left or right. Behavioral effects replicated previous studies, showing faster reaction times at valid locations regardless of the type of attention. However, despite their similar behavioral effects, social gaze orienting and reflexive capture produced different neural effects. Whereas the P1 was enhanced by exogenous attention (i.e., flash in the periphery), gaze cuing produced no effect on early sensory processing (P1/N1 components). Furthermore, whereas exogenous attention triggered the typical enhancement of the later, higher-order P3 component at the valid location, gaze orienting had the opposite effect, with the larger P3 being evoked by invalid location targets. This effect of social gaze on the P3 was the same as was seen following voluntary attention, suggesting gaze orienting may tap into some of the same mechanisms associated with endogenous attention effects. Thus, with ERPs able to investigate processing before the final overt behavior, these studies were able to reveal that social gaze orienting, despite being an involuntary process, is distinct from the typical type of exogenous capture of attention to highly salient events in the environment. It should be noted that some ERP studies have found effects of gaze orienting on early sensory components, including the P1 (e.g., Schuller & Rossion, 2001, 2004), but these studies did not directly compare the gaze-orienting effect to exogenous capture effects. Furthermore, other studies have found no evidence of amplitude modulation of early visual processing by social gaze cuing (Fichtenholtz, Hopfinger, Graham, Detwiler, & LaBar, 2007, 2009; Magnée, Kahn, Cahn, & Kemner, 2011). Therefore, although gaze orienting may be able to affect visual processing under some conditions, this early processing effect is not as automatically produced as the typical exogenous capture effect, and the behavioral effect of gaze orienting has been dissociated from effects on the P1.

6.4.2 Interactions of Emotions, Faces, and Gaze

Another class of stimuli that has been proposed to have privileged priority for attentional resources consists of stimuli that are associated with strong emotion. Competing theories on emotion hold that there are a limited set of distinct emotions (e.g., anger, disgust, fear, happiness, sadness, and surprise; Ekman, 1992) versus emotions being a product of our interpretation and cognitive appraisal of our body's physiological responses to stimuli within the current context (e.g., "theory of constructed emotion"; Barret, 2017). According to many theories, two important dimensions of emotions include the valence (very negative to very positive feelings) and arousal (low to high intensity) of the stimuli. Regardless of exactly how emotional states arise in the brain, studies of attention have found that highly emotional stimuli capture attention in an involuntary manner.

Across many different paradigms, highly emotional stimuli have been found to facilitate responses when they are the targets in a task and to distract subjects when they are not targets (reviewed in Carretié, 2014). These studies have found effects on manual response times when measuring **covert attention** and on fixation patterns when measuring overt shifts of attention. A number of studies have also found that threatening and disgusting stimuli have an even stronger power to capture attention than other equally intense emotional stimuli, and it has been suggested that this may bestow survival advantages in terms of ensuring quick responses to potentially harmful or threatening stimuli (e.g., Nummenmaa, Hyönä, & Calvo, 2006).

The capture of attention by emotional stimuli could be at least partially due to the enhanced processing of the emotional stimuli. fMRI studies have found that visual emotional stimuli, relative to nonemotional stimuli, evoke higher levels of blood oxygenation-level dependent (**BOLD**) activity throughout visual processing regions of the brain (e.g., Bradley, Sabatinelli, Lang, Fitzsimmons, King, & Desai, 2003). Padmala and Pessoa (2008) further found that fMRI activity to emotional stimuli in the striate cortex was correlated with the subjects' behavioral performance in response to those stimuli. Magnetoencephalography (MEG) studies have found the magnetic counterpart to the N2pc ERP component (described earlier) to be generated in response to lateralized emotional stimuli in the periphery, even when completely task-irrelevant, providing further evidence that attention is involuntarily captured by emotional stimuli (Fenker, Heipertz, Boehler, Schoenfeld, Noesselt, Heinze, Duezel, & Hopf, 2010).

In regards to how emotional stimuli affect the processing of other concurrent stimuli, electro-encephalography (EEG) activity in visual regions evoked by a central main task was reduced when task-irrelevant, emotionally arousing images were shown in the background, providing further evidence of the ability of emotional stimuli to involuntarily draw attention away from intended goals (Müller, Andersen, & Keil, 2008). Another recent EEG study has provided evidence of the neural mechanisms by which emotional items can distract us from our goals (Arana, Melcón, Kessel, Hoyos, Albert, Carretié, & Capilla, 2022). As described in Chapter 5, voluntary visual attention is associated with reduced alpha band (8–12 Hz) activity recorded by EEG or MEG over visual occipital areas, contralateral to the focus of endogenous attention (e.g., Thut, Nietzel, Brandt, & Pascual-Leone, 2006). This reduction in alpha-band activity is thought to be important for reducing background coordinated phasic activity across the visual cortex, increasing sensitivity to incoming sensory inputs (Kelly, Gomez-Ramirez, & Foxe, 2009; Thut et al., 2006). Arana, Melcón, Kessel, Hoyos, Albert, Carretié, and Capilla (2022) found that emotional distractors also produced this reduction in alpha-band activity over contralateral occipital scalp sites. Furthermore, the amount of alpha-band activity reduction was correlated with behavioral performance: The greater the alpha-band activity reduction to the emotional distractor, the greater the impairment in responding to the (nonemotional) target. These results provide evidence that attentional capture by emotional stimuli affects sensory processing in a similar way to how endogenous attention can affect activity in visual processing regions.

Other research has suggested that the ability to lessen distraction from emotional stimuli may depend on experience with those stimuli. On the basis of a well-controlled eye movement study in which subjects rated all of the stimuli beforehand to account for individual differences in evoked emotions, Ono and Taniguchi (2017) suggested that memory may also play a role in how emotional stimuli affect attention. Compared to previous studies, they found the same level of capture to highly emotional stimuli when those stimuli were targets, but they found *less distraction* than previous studies when the emotional stimuli were nontarget distractors. They proposed that the reduced level of distraction in their experiment was due to their subjects being familiar with the stimuli because they had rated all the items beforehand. This could relate to the signal suppression hypothesis described earlier (Sawaki & Luck, 2010) and to Gaspelin and Luck's (2018) findings that the suppression of potentially distracting stimuli depends on experience with those stimuli. These effects of experience on attentional capture will be discussed further in Sections 6.5 and 6.6.

In addition to highly emotional nonsocial stimuli having a stronger influence on attention than neutral stimuli, research has also shown that social stimuli have a strong effect on attention; specifically, the emotions of faces also affect the capture of attention. Eye movement and manual reaction time studies have found that emotional faces in the periphery capture attention more than neutral faces in the periphery (e.g., Bradley, Mogg, Millar, Bonham-Carter, Fergusson, Jenkins, & Parr, 1997). Furthermore, results have shown that the capture is larger by faces expressing *negative* emotions than faces expressing *positive* emotions (Eastwood, Smilek, & Merikle, 2001; Mogg & Bradley, 1999), providing more evidence that negative emotional stimuli receive higher attentional priority.

Given the findings that emotions affect the allocation of attention, it is a natural next step to wonder if the emotion perceived on a face interacts with social gaze orienting. Is our natural tendency to orient toward where another person is looking enhanced or inhibited if the person is expressing fear versus happiness? Early research provided evidence that, compared to faces expressing no emotion, the eye gaze of a fearful face produced a stronger gaze cuing effect (Mathews, Fox, Yiend, & Calder, 2003). Given this interaction, the next question is whether the congruency between the face's emotion and what it is looking at also affects how our attention is distributed. As shown in Figure 6.6, researchers have investigated these questions with cuing tasks that manipulate the direction of social gaze cuing, the emotion of the face doing the gazing, and the congruency of the emotion on the central face with the emotional content of the target item in the periphery. Bayliss, Schuch, and Tipper (2010), using the paradigm shown in Figure 6.6, replicated the typical effect of gaze orienting, in that gazed-at targets were responded to more quickly than opposite-location targets. Their results further showed that the specific type of emotion on the face (happy versus disgusted) didn't interact with gaze cuing when the peripheral targets were *neutral* in emotion, as the effect was equally strong for both happy and disgusted faces. However, a significant three-way interaction showed that the size of the gaze cuing effect was affected by the congruency between the emotion on the face and the type of emotional target. Specifically, the gaze cuing effect was found to be almost twice as large when the facial expression matched the valence of the target as compared to when those were mismatched. In ERP studies using a similar type of cuing paradigm, Fichtenholtz et al. (2007, 2009) tested the effects of dynamic changes in gaze and expression (changing gradually across three steps from neutral emotion and eyes straight ahead to either fearful or happy and looking to the left or right). The dynamic facial changes were followed by a target stimulus (happy baby or threatening snake) to the right or left of the face, and all combinations were equally likely. Similar to the gaze cuing studies discussed earlier that used neutral faces, Fichtenholtz et al. (2007, 2009) found no effects of *gaze direction* on early visual processing, as indexed by the P1 component to the target stimuli. However, these studies did find that the target P1 was affected by the emotion expressed by the face. The target P1 was enhanced more following happy than fearful faces in the experiment that directly compared these (Fichtenholtz et al., 2007) and was enhanced for fearful versus neutral faces in the second study (Fichtenholtz et al., 2009). These effects of facial expression did not interact with gaze direction or target emotion; the same effect was present for valid (i.e., gazed at) and invalid location (i.e., gazed away from) targets. At the level of the P3 component, there were effects of face emotion (targets preceded by happy faces produced larger P3) and target emotion (positive targets produced the P3 earlier than negative targets). There was also an effect of gaze orienting on the P3, in agreement with previous studies of gaze cuing, but here the effect interacted with facial expression, as the P3 gaze cuing effect (larger to invalid-location targets) was larger following fearful faces (Fichtenholtz et al., 2007). Overall, ERP studies have revealed that emotions and gaze orienting of irrelevant faces can modulate the neural processing of target stimuli at multiple stages, with facial emotional expressions exerting the earliest effects and gaze orienting and target emotions affecting higher-order processing.

6.4.3 Special Status of "New Objects"

Very loud sounds, bright flashes, and uniquely colored objects have all been found to involuntarily capture attention. However, as mentioned in previous sections, top-down attentional control settings may determine under what conditions these stimuli trigger a completely automatic capture of attention. Across a number of studies, however, one particular type of visual stimulus seems to have a special status in automatically triggering a capture of attention to its location: the abrupt appearance of a new object. In some of the earliest research in this area, Yantis and Jonides (1984) found that abruptly appearing new objects within multi-item displays captured attention, even when the new objects did not predict target location in any way. Using a visual search-type paradigm, the authors found that the location of the new object (i.e., the abrupt "onset") was searched first, regardless of the number of no-onset distractors in the display. In studies of this kind, new object distractors appear in an otherwise empty location; the other items, sometimes called "no-onset" items, are revealed by removing parts of the existing object or perceptual mask at that location (Jonides & Yantis, 1988). Some early research had suggested that what looked like a preference for new-onset items may simply be caused by the large amount of visual change at one location. Miller (1989) found that when a substantial number of elements were removed from a stimulus, this also led to a strong draw on exogenous attention. However, other studies that have controlled for the amount of visual change have found that new objects capture attention more readily than the same amount of visual change to an existing object, supporting the claim that it is the "new object" property that captures attention, not simply the amount of visual change at a location (i.e., removing or adding pieces to an existing object does not trigger the same automatic capture of attention as the appearance of a new object). Regarding the contingent capture debate discussed earlier, Chua (2013) found that onset cues (i.e., new objects appearing) captured attention automatically, regardless of whether the expected targets were onsets or offsets; in contrast, offsets (disappearances of objects) only captured attention when subjects were specifically expecting offset targets or when any luminance change defined a target.

Although abruptly appearing new objects have been shown to enjoy some priority in terms of what's likely to capture attention, evidence also shows that highly focused voluntary spatial attention can sometimes prevent such capture. Yantis and Jonides (1990) found that if endogenous attention was directed to the spatial location of an upcoming target with enough time before the targets and distractors appeared, subjects were able to maintain their attention at that location and not show any reduction in performance when a new object appeared simultaneously with the target. This lack of distraction provides evidence that a new object doesn't always capture attention away from where endogenous attention has already been well-focused. It is important to point out, however, that this ability to not be distracted by the new object was only observed when the instruction of where to attend was 100% valid (the target *always* occurred at the indicated location) and when that instruction occurred well before (200 ms in advance of) the target display. Therefore, although not being completely automatic in the sense of being able to always overcome the focus of voluntary attention, the overall pattern across studies suggests that new objects have a uniquely strong pull on exogenous attention (Theeuwes, 1994).

In order to understand how appearing and disappearing objects affect attention, Hopfinger and Maxwell (2005) conducted an ERP study in which they analyzed the neural activity evoked by completely task-irrelevant stimuli that could appear or disappear. Subjects performed a color discrimination task on infrequently occurring circle stimuli while the irrelevant stimuli (bars and dots) appeared and disappeared. The irrelevant stimuli were presented in a paradigm similar to a cuing study, in which the appearance or disappearance of a set of dots at one location (the "cue") was followed by a brief rectangular bar at either the same ("cued") location or the opposite ("uncued") location. The rectangular bar is referred to as a "probe" stimulus here because it allows the researchers to probe the neural response without requiring a behavioral response. In this experiment, the "cues" and "probes" were all completely irrelevant to the task. In regard to new objects potentially being special in their ability to trigger an involuntary capture, the ERPs evoked by the "cue" stimuli showed that the appearance of a new object evoked a positivity over central parietal scalp sites at 150–210 ms that wasn't evoked when the same stimuli disappeared (Figure 6.7). Notably, the typical series of early visual-evoked components (e.g., C1, P1, N1) were evoked by both the transient appearance and the transient disappearance of this task-irrelevant object, and the amplitude of these visual processing components was larger for the onset (appearance) than the offset (disappearance). Later components (e.g., P2) were also evoked by both onset and offset of the stimuli, and the amplitude of the P2 was similar for both. However, the component recorded at 150–210 ms (shown in Figure 6.7B,C) was only evoked by the *appearance* of the stimuli; there wasn't any similar activity to the disappearance of those stimuli. These data provide support for the idea that new objects may be treated as special by the brain. However, it should also be noted that the subsequent probe stimuli showed evidence of capture following both onsets and offsets of the cue – the enhancement of the P1 to targets (described in Section 6.1.2 on the neural effects of exogenous attention) was equally strong here to the probe stimuli following either onset or offset cues. Therefore, this would suggest that the boosting of processing at the level of the P1 may be an automatic consequence of any salient event, and that any special processing of new objects may uniquely affect other higher-order processes. This study, however, couldn't assess any possible effects on the P3 component because no P3 was generated to these irrelevant cue and probe items, confirming that subjects were treating those items as irrelevant to their task.

Subsequent fMRI studies have corroborated the finding that new objects may exert their special influence at higher levels of processing. For instance, Kim and Hopfinger (2010) used fMRI to compare the processing of task-irrelevant luminance changes that either created a new object or simply changed the luminance of an existing object. In that study, subjects performed a continuous discrimination task on a small, periodically rotating stimulus at fixation while an irrelevant luminance change could occur in the periphery at random times. The irrelevant luminance change (a brief brightening of a square-shaped region) either defined a new object

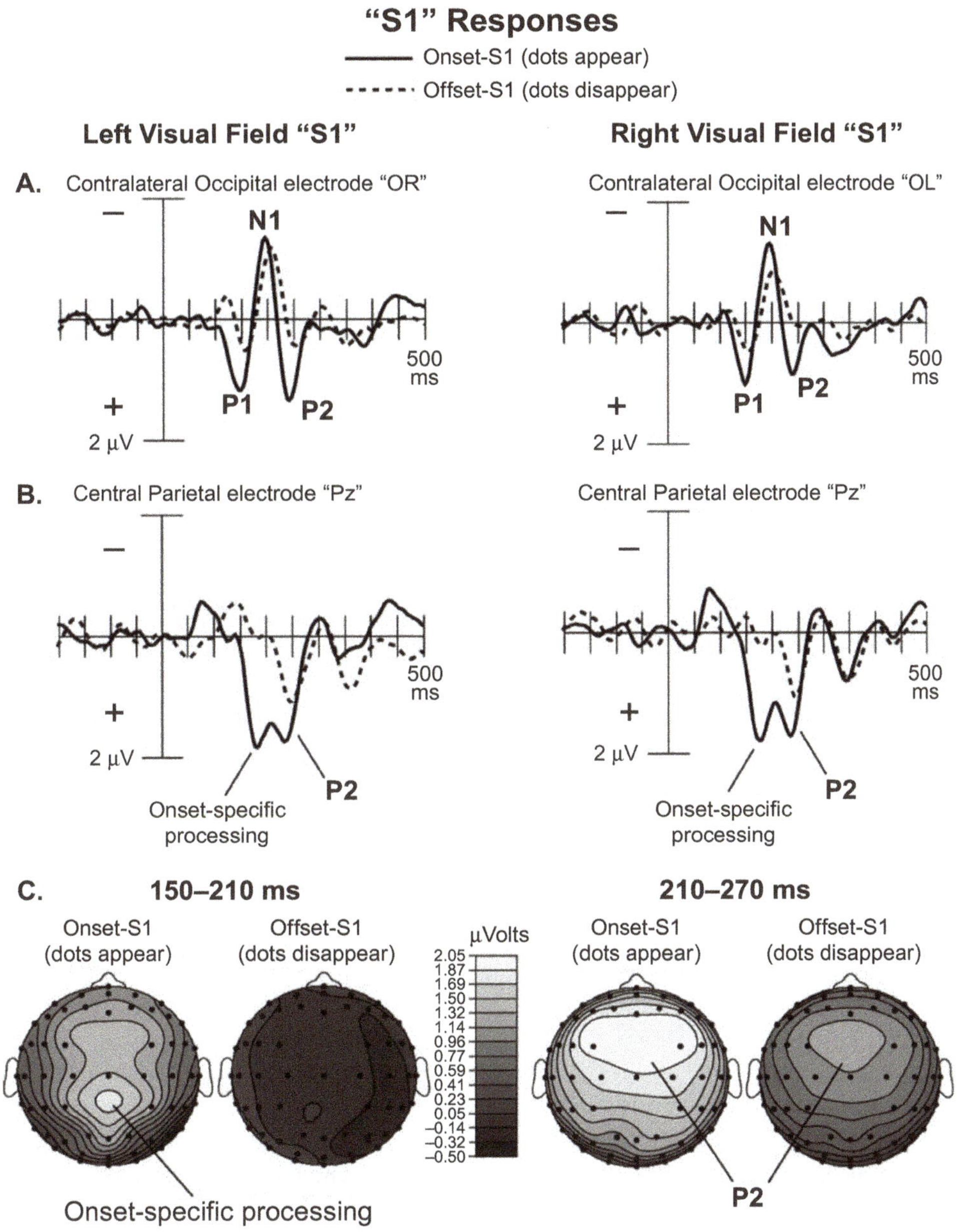

"S1" Responses

— Onset-S1 (dots appear)
----- Offset-S1 (dots disappear)

Left Visual Field "S1" — **Right Visual Field "S1"**

A. Contralateral Occipital electrode "OR" / Contralateral Occipital electrode "OL"

N1, P1, P2, 500 ms, 2 µV

B. Central Parietal electrode "Pz" / Central Parietal electrode "Pz"

P2, 500 ms, 2 µV, Onset-specific processing

C. 150–210 ms / 210–270 ms

Onset-S1 (dots appear) / Offset-S1 (dots disappear) / µVolts / Onset-S1 (dots appear) / Offset-S1 (dots disappear)

2.05 1.87 1.69 1.50 1.32 1.14 0.96 0.77 0.59 0.41 0.23 0.05 −0.14 −0.32 −0.50

Onset-specific processing / P2

Figure 6.7 ERP study showing unique activity evoked by a new object. "S1" refers to the first of the two successive task-irrelevant stimuli that could occur on each trial. In a typical cuing study, S1 would be labeled the "cue." Here, S1 was either the appearance of a new object ("Onset-S1") or the disappearance of that object ("Offset-S1") in the upper left or upper right visual field. Both locations are shown here simply to show that all effects were replicated across both visual fields. (A) Posterior lateral-occipital electrodes show that visual sensory ERP components are evoked by both the onset and offset of this stimulus but are larger for onsets. (B) The central electrode shows a large positive potential (labeled "onset-specific processing") occurring between 150 and 200 ms that is evoked only by the appearance of the stimulus, not by the luminance change of its disappearance. (C) Scalp topographic maps of the top view of the head showing the brain activity at 150–200 ms (left panel) that is unique to appearing objects and the activity at 210–270 ms (right panel; P2 component) that is somewhat larger for onsets but is evoked by both onsets and offsets. Reprinted from Hopfinger and Maxwell (2005), *Cognitive Brain Research*, with permission from Elsevier.

(the newly brightened square appeared in a previously empty region of space) or was a change to an existing object (the brightening occurred within the borders of an outlined square that was present throughout the run). Visual regions representing the location of this distractor showed greater fMRI activity in the new object condition compared to the luminance change condition, providing evidence that visual processing evoked by the luminance change was greater when it created a new object. In addition, regions representing the central target location showed a transient reduction in activity shortly following the appearance of a new object in the periphery, suggesting that attention was being pulled away from the central location of the target when the distractor appeared. Critically, parietal regions showed a marked increase in activity specifically in the new object condition, suggesting greater attentional control was needed to reorient attention back to the central task after a *new object* captured attention. These effects in parietal regions will be discussed more in Chapter 8 on the "holding" of attention, and attentional control regions of the parietal and frontal lobe are described in more detail in the next chapter.

Finally, regarding stimuli that may have a special status in triggering involuntary attention, some authors have argued that *looming* stimuli automatically capture attention. As opposed to motion in other directions (e.g., up, down, left, right, or receding), stimuli that are rapidly approaching (i.e., "looming") have been found to receive special attentional priority (Franconeri & Simons, 2003; Lin, Franconeri, & Enns, 2008). A recent ERP study found that nonpredictive looming stimuli produced greater activity in early visual processing (in the P1–N1 latency range) compared to the same stimuli when static (Fernández-Folgueiras, Hernández-Lorca, Méndez-Bértolo, Álvarez, Giménez-Fernández, & Carretié, 2022). The authors suggested this enhanced sensory processing may relate to the automatic capture of attention, although that study did not find significant behavioral effects. Further research is needed to understand whether this ability of looming stimuli to trigger involuntary attention mechanisms is related to the attentional capture by threatening stimuli described earlier.

6.5   Memory and Attention

The interaction of memory and attention is complex and bidirectional. There is a long history of research investigating how attention during studying affects later memory. These studies often assign subjects to either a "full attention" condition, in which the subjects' only task is to study the items for a later memory test, or to a "divided attention" condition, in which the subjects must do another concurrent task while also trying to encode the items. The results from such studies consistently find much better performance on a variety of subsequent memory tasks (e.g., free or cued recall; item or associative recognition) when the items were studied under full attention conditions compared to divided attention (Craik, Govoni, Naveh-Benjamin, & Anderson, 1996; Kilb & Naveh-Benjamin, 2007; Naveh-Benjamin, Guez, & Marom, 2003). Similar to the role attention plays in binding together the features of currently perceived objects (i.e., feature integration theory; Treisman & Gelade, 1980), recent research has investigated whether attention also affects the binding together of multiple different source elements of a memory event (e.g., the color of the background behind an item and the spatial location of that item, in addition to simple item memory). Greene, Martin, and Naveh-Benjamin (2021) found that attention was critical for binding multiple source dimensions during *encoding*, but that *retrieving* those bound sources didn't require attention. This latter null result is consistent with other studies investigating the effects of attention during the retrieval of memories. Whereas attention plays an important and strong role in the encoding of memory, dividing attention during retrieval often shows no effects on memory performance (Craik, Eftekhari, & Binns, 2018).

The mechanisms responsible for how attention enhances encoding into memory are still being investigated, but the effects of attention described in this and previous chapters might be involved. The enhancement of early sensory-level processes by attention could lead to a more robust memory trace, and the engagement of higher-level processes specifically for attended stimuli could extend the depth and duration of processing for those stimuli, helping create a richer encoding. Although understanding the mechanisms by which attention affects memory is an important and ongoing area of research, the present section will focus on the reverse direction of this relationship: how memory affects attention. We can, of course, explicitly decide to use our memory and knowledge to help us choose where to voluntarily focus our attention. However, that influence of memory on attention is typically thought to be just another instance of voluntary, goal-directed endogenous orienting; memory, in that case, is simply providing the explicit knowledge that we use to allocate our attention. In the rest of this section, we focus instead on the *involuntary* ways that memory can influence attention. In the following subsections, different types of memory are separately addressed, including the active holding of concepts in working memory, the implicit learning of short-term contingencies, and explicit long-term memories. The effects of each of these on the automatic allocation of attention will be discussed.

6.5.1   Working Memory = Working Attention?

Tests of working memory typically involve the extended holding of information in awareness. Subjects are asked to keep some type of information in mind (e.g., a list of numbers or words; a set of spatial locations; a set of colors or shapes) and are tested after a delay period for their accuracy in remembering that information. Variables may include the number of items being held, as well as how long the information has to be held in mind and the mental operations to be performed with the information (i.e., simple maintenance or manipulation). While there are differences between attention and working memory tasks, at least some of the brain mechanisms involved in working memory and voluntary attention tasks likely overlap (Awh & Jonides, 2001; Chun, Golomb, & Turk-Browne, 2011). Some similar processes and neural mechanisms may be triggered by the initial cue directing a subject where or what to attend and the initial presentation of items to be kept in mind during a working memory task. Being instructed to attend to some location or feature in an upcoming display, as is done in attention cuing paradigms, involves at least a brief process of keeping those details in mind until the target display appears. Working memory paradigms would typically require the information to be held for much longer and would likely include more items, but the initial process after being told to keep some specific thing(s) in working memory is likely quite similar to the process of being told to pay attention to those things in an upcoming display. In both cases, there is a need to keep some information in mind during a delay period – a core aspect of working memory. Indeed, Awh and Jonides (2001) suggested that *sustaining* attention on a location or feature can be interpreted as *holding* that information in working memory. In the other direction, some aspects of working memory tasks could be interpreted as processes of attention. For example, a defining aspect of attention is the selection of just one item or feature among multiple competing items and features. In attention tasks, that competition is between items in the physical display, whereas in working memory tasks the potential competition comes from other elements that could rise to awareness in the mind. In both cases, the selection process involves enhancing some information while suppressing other sources. Much of the research has focused on what makes working memory and attention unique. Working memory studies often investigate how the brain *maintains* something in awareness for extended periods of time and how that information can be *manipulated* while being held in memory. Attention studies typically investigate the mechanisms of *shifting* the focus of the mind's eye and of *selecting* one item among competitors for further processing. However, research has confirmed that there is overlap in some of the mechanisms involved in working memory and attention tasks.

Research has revealed that working memory involves selection processes similar to those found in attention tasks. Specifically, working memory research has shown that neural processing in early sensory areas is modulated by the contents of working memory in a similar manner as for voluntary attention. Using fMRI, Harrison and Tong (2009) found that the pattern of activity in early visual areas from V1 through V4 was associated with the maintenance of visual working memory. Serences and colleagues further found that the pattern of fMRI activation across voxels in V1 during working memory maintenance was like the pattern observed when subjects were attending to and discriminating current sensory stimuli (Serences, Ester, Vogel, & Awh, 2009). Furthermore, studies have also found that object-specific processing, as indexed by the face-sensitive N170 ERP component and activity in the FFA, is modulated in a similar manner when those objects are the focus of attention or are being held in working memory. These effects are similar in terms of both *enhancing relevant* information (i.e., boosting face processing when faces are being held in working memory) and *suppressing irrelevant* information (i.e., inhibiting face processing when competing information is being held in working memory; Gazzaley, Cooney, McEvoy, Knight, & D'Esposito, 2005). In a review of the literature on the effects of attention compared to the effects of working memory, Gazzeley and Nobre (2012) concluded that the modulation of sensory processing during working memory is via the mechanisms of selective attention. Furthermore, attention can be directed to the contents of working memory in a similar way as attention can focus on portions of physical displays. EEG studies have found that the effect of selecting relevant information during encoding into working memory enhances early sensory processing (within 200 ms) in a similar way as does attention to a physically present display, and the size of this selection effect is correlated with later working memory accuracy (Gazzaley, 2011; Rutman, Clapp, Chadick, & Gazzaley, 2010).

A recent fMRI study provides evidence that a similar mechanism of selection occurs in early visual areas for attention and working memory (Zhou, Curtis, Sreenivasan, & Fougnie, 2022). This study used machine learning and patterns of activity across voxels to assess the mechanisms of selection in attention and working memory tasks across occipital, parietal, and frontal regions of interest (ROIs). The results revealed that a similar effect in early visual areas occurred when attending to stimuli present in the external world and when selecting among internal working memory representations. In one set of data machine learning classifiers were trained to separate attended versus unattended trials, and in another set of data the classifiers were trained to separate maintained-in-working-memory versus not-maintained trials. The resulting classifiers from the two datasets, based on patterns across all voxels in each ROI, were so similar that those trained on one of the datasets were able to classify the effects in the other dataset equally well. These findings are consistent with a common mechanism of gain control, in which selected information is enhanced relative to other information, within the ROIs covering areas V1–V3. Although this study provides important evidence for the similarity of attention and working memory effects in relatively large areas of visual cortex, fMRI studies cannot measure the activity of individual neurons. However, in a recent monkey electrophysiology study (Panicehllo & Buschman, 2021), single-unit activity was recorded from neurons across four different areas implicated in attention and working memory: lateral prefrontal cortex (LPFC), frontal eye fields, parietal cortex, and V4. In separate blocks of trials, monkeys performed an attention task or a working memory task, both of which involved the selection of one of two different-colored items at different locations. The results showed that control mechanisms in the LPFC preceded the selective processing changes in V4. Furthermore, the control of selection was found to arise from a shared representation of neurons across the LPFC for both working memory and attention. The effects in area V4 showed a similar type of mechanism at work for attention and working memory, as was found by Zhou et al. (2022); however, Panicehllo and Buschman's (2021) single-unit results revealed that the neural representations of the effects of attention in area V4 were not correlated with the population representation of working memory selection in that area. Thus, although controlled through a shared system in the frontal lobe and having a similar type of selective processing mechanism in visual areas, the effects of attention and working memory were instantiated in somewhat different patterns of activity at the neuronal level in V4. The following chapter on attention control will discuss the shared control mechanisms of attention and working memory in more detail.

Finally, given the similar effects of working memory and attention, it is not surprising that research has found interactions between these mental processes. Providing evidence for a sharing of resources, research has shown that holding locations in working memory can disrupt spatial orienting when there is a mismatch between working memory and attention goals (Dell'Acqua et al., 2006). Furthermore, visual search performance suffers when subjects are maintaining other (noncongruent) spatial information in working memory at the same time (Han & Kim, 2004; Oh & Kim, 2004; Woodman & Luck, 2004). Most relevant to the topic of this chapter, it has also been shown that the active maintenance of information in working memory *involuntarily* biases the allocation of attention (Soto, Hodsoll, Rotshtein, & Humphreys, 2008). Holding spatial locations in working memory draws overt and covert attention toward those locations (Awh & Jonides, 2001; Corbetta & Shulman, 2002; Hollingworth, Richard, & Luck, 2008), and holding features in working memory results in attention-like enhancements to items sharing those features (Downing, 2000; Soto, Heinke, Humphreys, & Blanco, 2005). Critically, attention is biased by working memory even when the contents of working memory are completely nonpredictive of the attention target, and even when the contents of working memory may be disruptive to the overall attention task; therefore, the influence of working memory on attention allocation appears to be automatic and involuntary (Soto et al., 2008). Furthermore, research has shown that the effect on attention is due to working memory *maintenance* of that information, not simply implicit priming from being exposed to the stimuli, since the effect is not observed with passive viewing of the items (e.g., Downing, 2000; Soto et al., 2005). In addition, the effect of working memory on attention is thought to be due to this maintenance, not just the initial encoding into working memory, because the effect is seen equally strongly throughout the entire period of working memory maintenance (from 200 to 4,000 ms in Olivers, Meijer, & Theeuwes, 2006). Overall, these studies show a strong overlap between attention and working memory processes in the brain, and they provide evidence that items matching information being held in working memory exert a strong involuntary pull on attention.

6.5.2 Short-Term, Implicit Memory Effects on Attention

In addition to the influence of active working memory on attention, short-term implicit memory processes also affect the allocation of attention. Implicit memory refers to aspects of learning that are often unintentional and that don't rely on conscious awareness. This can include processes of priming, in which responses to a stimulus are altered after the subject has been recently exposed to it. Memory research has shown that reaction times are faster to targets that have been responded to as targets recently (i.e., priming), and that stimuli that have been actively ignored recently may be suppressed below baseline levels when they become targets (i.e., negative priming). For the present purposes, we are interested in the unintentional effects that priming has on the allocation of attention. In a phenomenon termed "priming of pop-out" (Maljkovic & Nakayama, 1994), researchers have found that when a certain feature of a target stimulus repeats over trials in a visual search task, stimuli with that feature attract attention and automatically pop out of the visual display even more so than a typical salient feature. For example, Maljkovic and Nakayama (1994) found that when subjects perform a visual search task for a uniquely colored, diamond-shaped target, search was faster whenever the unique color repeated across successive trials, despite the color itself not being predictable before the search display and despite the unique color changing throughout the experiment. This type of attentional capture does not require that the subject be aware of the specific set or combination of features that repeats, and thus it represents a case in which unconscious memory processes can influence how attention is allocated (Kristjánsson & Campana, 2010). Indeed, research has found that priming of pop-out includes both an enhanced facilitation of target features that repeat and a better suppression of distractor features that repeat (Lamy, Antebi, Aviani, & Carmel, 2008). Since the distractor features are not the focus of attention yet are ignored better with repeated trials, this has provided further evidence for the implicit nature of the effect. In terms of the automatic effects on attention, these studies find both costs and benefits of the priming. Specifically, benefits are seen as faster responses to targets when either target features repeat or when distractor features repeat. Costs are observed as slower responses to targets when a current distractor matches a feature of the previous trial's target or when the current target matches a feature of the previous trial's distractors. The effects of target facilitation and distractor suppression were originally found to be separate independent processes (Lamy et al., 2008), but recent large-scale experiments have found these effects to be correlated (Dent, 2018). Additional research is therefore needed to determine whether there are multiple independent processes involved in priming of pop-out, but research has confirmed that implicit memory can automatically bias the direction and focus of attention. In an ERP study, Eimer, Kiss, and Cheung (2010) found that the N2pc component was delayed on trials in which the target or distractor features were switched compared to trials in which features of the targets and distractors repeated. Since the N2pc has been used as an index of attentional orienting and focus, these data suggest that trial-to-trial priming affects the speed and location of involuntary attentional engagement. Finally, it should be noted that the repetition of target and distractor features has been shown to affect attentional allocation across a number of different paradigms, not only in visual search displays in which one item pops out because of its uniqueness within the display; "intertrial priming" (ITP) has become the generalized term to refer to the effects of such priming on attention (Ramgir & Lamy, 2022).

In addition to the effects of ITP, research has also shown that the automatic spreading of activation throughout semantic memory networks can affect the allocation of attention. Although semantic memory typically is classified as a type of explicit memory, in that the knowledge of the meaning associated with a stimulus is within conscious awareness, the effect on attention referred to here involves unintentional processes. Specifically, language and memory research have provided evidence for semantic networks in the mind. These networks consist of concepts that are related in their meaning, with more closely related concepts typically found to be more strongly activated, as measured by quicker processing times for those stimuli compared to unrelated concepts. The activation of these concepts can be considered another form of priming, but one in which the item itself was not previously presented to the subject; rather, it was activated due to its association with semantically related items. Studies have found that items that are semantically related to previous target items automatically capture eye movements in **overt attention** tasks and produce interference in covert attention tasks to a similar degree as do previous target stimuli (Moores, Laiti, & Chelazzi, 2003). Overall, these studies show that the allocation of attention is biased by short-term memories that are formed implicitly on the basis of recent events.

6.5.3   Long-Term Memory Effects on Attention

In addition to the quick-acting effects of trial-to-trial priming and the rapid automatic activation of semantic memory networks, memory processes that are longer-lasting or that require longer episodes of learning also affect the allocation of attention. These can include explicit item memory or implicit knowledge of sequences of events, scenes, or task structures. For example, in "contextual cueing" experiments, subjects perform demanding search tasks within complex visual displays that contain many distractor elements (Chun & Jiang, 1998). Unbeknownst to subjects, some of the displays repeat multiple times throughout the experiment. The results show significantly better search performance for displays that repeat throughout the experiment relative to displays that were only seen once. Critically, in memory tests given after the search tasks, subjects do not recognize the repeated displays above chance levels. Thus, the enhanced guidance of attention to the targets in these repeated displays is thought to arise at the level of *implicit* memory built up over repeated presentations of the exact configuration of target and distractor elements in the display.

In a contextual cuing study using intracranial ERPs in patients with epilepsy, Olsen, Chun, and Allison (2001) found that repeated displays, compared to new displays, resulted in enhanced processing in early visual areas including the striate cortex (i.e., V1). The enhanced processing of the repeated displays in early visual areas V1 and V2 was not observed until 200 ms of latency, however, suggesting that the effects were due to reentrant processing in these areas. The neural effects were accompanied by the typical behavioral effect of faster search times in the repeated displays, but it is unclear whether the effects in the visual cortex were due to faster capture of attention to the target in the repeated displays. Investigating the neural basis of contextual cuing is complicated because differences between repeated and novel displays may potentially reflect only the memory itself, instead of an effect of the memory on attentional allocation. Johnson, Woodman, Braun, and Luck (2007) used scalp-recorded ERPs and directly tested the N2pc component, an index of attentional focus that is based on differences between hemispheres within a display, not a comparison of overall activity between different display types. The results showed that the N2pc was significantly larger for repeated displays than for new displays, and the authors interpreted this as evidence that the repeated displays were more likely to result in the first shift of attention being to the target location. The size of the N2pc component, when averaged over the many trials in an experiment, is a function of the number of trials in which attention was rapidly focused on the target location – the more trials in which attention was quickly captured, the larger the N2pc in the typical 200–300-ms latency range. This interpretation of the N2pc evidence is in line with evidence from eye-tracking studies that contextual cuing doesn't guarantee capture for every repeated display. Specifically, Peterson and Kramer (2001) found that, during many repeated displays, one of the first eye movements observed was straight to the target, but that during other repeated displays in which eye movements didn't go straight to the target, the search pattern was the same as for new displays. This could be due to the original presentations of those displays not resulting in a strong memory trace or due to the memory not being activated by the current display on some trials. Overall, these results suggest that contextual cuing can facilitate a rapid and automatic orienting to the target location within displays that have been previously encoded well into implicit memory.

In addition to the implicit memory of specific displays that have been previously viewed, semantic categories of scenes can also bias attention (e.g., Torralba, Oliva, Castelhano, & Henderson, 2006). In experiments in which different categories of scenes are associated with targets at different locations, subjects show speeded target search in scenes they've never viewed before if those scenes share global image properties with multiple previous scenes in a category. This relates to both contextual cuing and semantic memory effects on attention, but it also involves the implicit learning of image-wide properties that link scenes at a categorical level. This incidental learning of categories is part of a broader area of research into statistical learning. Studies have found that the brain is sensitive to regularities in the environment, and this may affect how attention is distributed. Research has found that the consistency and predictability of visual stimuli can draw attention toward locations and features exhibiting stability, even when task-irrelevant (Zhao, Al-Aidroos, & Turk-Browne, 2013). MEG and EEG studies of auditory processing have also provided evidence that the brain processes predict auditory streams differently, with repeating patterns of auditory tones eliciting enhanced activity compared to a random presentation of those tones (Barascud, Pearce, Griffiths, Friston, & Chait, 2016; Southwell, Baumann, Gal, Barascud, Friston, & Chait, 2017). This enhanced activity persists throughout much of the duration of the auditory streams (~3 seconds), and the time at which the activity is enhanced for the repeated versus random streams corresponds to how long a repeated sequence takes (i.e., when the repetition occurs after 5 tones the difference in processing is observed sooner than when the repetition occurs after 15 tones; Southwell et al., 2017). However, although measures of neural activity show differences in early sensory processing similar to the effects of attention enhancing the gain of attended inputs, these studies of auditory processing have not found corresponding behavioral measures of attentional capture to the predictable streams (e.g., Southwell et al., 2017). This contrasts with behavioral findings from visual studies that have found evidence that regularities capture attention. Therefore, more research is needed to determine how the brain processes statistical regularities and under what conditions these trigger involuntary shifts of attention. Chapter 9 will discuss the role of prediction in attention and perception in more detail.

In addition to the effects of implicit memories that are built up over many trials, attention may also be affected by explicit memories encoded and associated with a specific learning instance. Early research into the effects of item memory on attention investigated whether old items (i.e., previously studied items) automatically capture attention. The results of these early studies were inconsistent, however, based in part on the types of stimuli and control conditions utilized. In behavioral studies, some research found that new items captured attention (Johnston, Hawley, & Farnham, 1993), whereas other research found that old items captured attention (Christie & Klein, 1995). There were several differences between these studies, however, including how "old" and "new" were operationally defined (in one case, words versus nonwords; in another case, words seen previously in the experiment versus words not presented before in the experiment). Overall, these studies suggested that item memory might influence the orienting of attention, but it wasn't clear whether this was due to the oldness or newness per se of the items or to the uniqueness of the item in terms of it being old or new relative to most other items in the experiment.

Another methodology that has proven useful in studying the allocation of attention is eye tracking. The measurement of overt shifts of attention can be especially useful for tracking not only the initial capture of attention, but also how the distribution of attention over time may be affected by different factors. Early eye-tracking studies investigated whether the consistency of items within a scene would attract attention. Loftus and Mackworth (1978) presented subjects with line drawings of simple scenes, such as a farmyard or underwater environment. Critically, in some scenes, an item was switched in from another scene. For example, an octopus was placed where the tractor was in the farmyard scene, and the tractor was placed where the octopus was in the underwater scene. Subjects had not viewed the original scene, so this was not a test of how their memory for a scene would affect how they looked at it; rather, it was testing whether an item that didn't fit into a well-known *category* of scene would capture attention. Loftus and Mackworth (1978) found that semantically inconsistent items were fixated sooner than the same items when in a semantically consistent scene. While this suggested a form of involuntary orienting, the inconsistent item wasn't usually the *first* item fixated, so the "capture" was not the fast, automatic type of capture triggered by especially salient items such as the abrupt appearance of new objects. In addition, subsequent studies did not replicate the finding of earlier fixations on the inconsistent versus consistent objects (DeGraef, Christiaens, & d'Ydewalle, 1990; Henderson, Weeks, & Hollingworth, 1999). There was, however, consistency across these eye-tracking studies in the finding that inconsistent items were fixated for a *longer* duration, compared to consistent items, when they were fixated.

In order to test how item memory affects the allocation of attention across a *typical* scene, in which all pieces are semantically congruent, Chanon and Hopfinger (2008) tracked eye movements while subjects viewed normal, intact scenes. Before viewing the scenes, subjects studied individual items, extracted from the scenes, for a later memory test. The encoding consisted of asking three "deep" questions about individual extracted items: (1) Is the object heavy or light? (2) Does the object belong inside or outside? (3) Do you own the object? The particular items were counterbalanced across subjects, and the final memory test showed strong explicit memory for each studied item. During scene viewing, different subjects were told either to study the exact layout of the scene for a later test on whole scenes or to try to study all the individual items for a later test on individual extracted items; these instructions made no difference to the effects described in the following. The results, as shown in Figure 6.8, showed that subjects fixated sooner and more often on the items that they had previously studied. These results provide evidence that *item* memory affects the involuntary allocation of attention. It should be noted, however, that these items weren't always the very *first* ones fixated. A similar finding was noted in the Loftus and Mackworth (1978) study described earlier. As opposed to a very highly salient physical event that immediately grabs attention, the ability of item memory to affect attention might not be triggered until fixation brings that item close enough to the parafovea to allow the more detailed processing needed to evoke a memory trace. Most critically, however, the Chanon

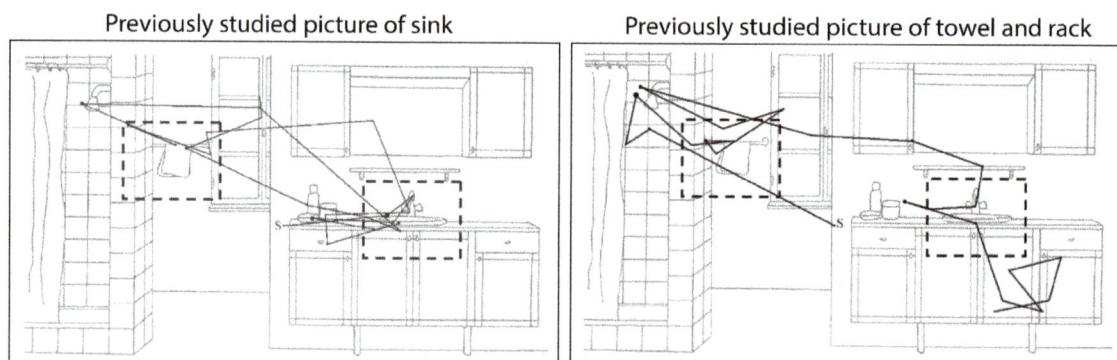

Figure 6.8 Effects of memory on overt attention. Examples of eye movement and fixation patterns (i.e., scan paths) when viewing scenes. The solid black lines represent scan paths, and the size of the dots denotes fixation duration. The letter "S" at the center of the images indicates the starting point of the subjects' scan paths, and the dashed boxes surround the two areas of interest for this experiment. The scan path lines, dashed boxes, and letter "S" were not in the scenes when participants viewed them. The left panel shows a scan path from a participant who had previously studied the picture of the sink; the right panel shows a scan path from a participant who had previously studied a picture of the towel and rack. Figure adapted from Chanon and Hopfinger (2008), *Visual Cognition*, with permission from Taylor & Francis Group.

and Hopfinger (2008) study found that the strongest and most robust effect of item memory on attention was not the *initial orienting* to the old item, but rather the "hold" that the old item kept on attention. Specifically, items were fixated for a *longer duration* when they had been previously studied, and there were more *return fixations* to these old items after the initial fixation. Together, these results suggest that item memory affects how long attention dwells on an item and the continued priority assigned to this item over time. Finally, it should be noted that the scenes in this study were all new to the subjects, and aside from the one studied item from each scene, all the rest of the items in the scene were new to the subjects. Therefore, the results could be due to either the oldness of the item alone or to the memorial uniqueness of the old item among many new items. This issue and the effects of memory on the holding of attention (i.e., "attentional dwell time") will be discussed in more detail in Chapter 8 on the temporal mechanisms of attention.

In another study that looked at the effects of memory on attentional allocation across normal, semantically consistent scenes, Ryan, Althoff, Whitlow, and Cohen (2000) investigated how changes within studied scenes affected the allocation of attention. In their experiments, subjects viewed scenes, some of which would repeat during the experiment. In the scenes that had been viewed previously, some portion of the scene could have changed (e.g., a person walking in a corner of the scene might be removed). The results showed that subjects looked at the manipulated region more often and fixated in that region for longer periods of time compared to when that region remained the same as when originally viewed. Critically, this additional time attending to the region only occurred if subjects were *unaware* of the change. Thus, the effect seems to be driven by implicit memory, not explicit strategies. If the subjects realized what had changed in the scene, they didn't dwell on the manipulated region for as long. Finally, that study addressed the neural mechanisms of this effect by also testing patients with severe amnesia. Unlike healthy control subjects, the patients showed no effect of scene manipulation. Although these patients were capable of showing priming and other forms of basic implicit memories, they exhibited a deficit in binding together the elements of a scene into a holistic memory. The authors interpreted these findings as further evidence for a deficit in *relational memory* processing, even at an implicit level, in patients with amnesia. These studies of scene viewing suggest that memory for individual items and the relation of items within a previously viewed scene both affect the orienting of attention and how long attention dwells there.

6.5.4 Reward-Driven Attentional Biasing

Research has revealed that one particular type of memory has a very powerful effect on attention: specifically, the memory of rewards. Numerous behavioral studies have shown that items or features that have become associated with rewards have a strong influence on attention. These studies have shown that previously rewarded stimuli draw attention involuntarily, speeding responses to targets that share the attributes of the rewarded stimuli and causing distraction when presented as distractors (e.g., Anderson, 2016a; Anderson & Yantis, 2013; Della Libera & Chelazzi, 2006; Failing & Theeuwes, 2017). Critically, the previously rewarded stimuli continue to have this effect on attention well after the subjects are made aware that those stimuli will no longer be rewarded. The effect is thought to be involuntary because the orienting of attention toward these stimuli occurs even when those stimuli do not predict target location in any way and the reward associations disrupt voluntary spatial attention (MacLean, Diaz, & Giesbrecht, 2016). Furthermore, the attentional capturing effects of previously rewarded stimuli transfer across different tasks (e.g., visual search and cuing paradigms; Meyer, Sheridan, & Hopfinger, 2020).

The neural effects of reward-driven attention biasing have also been investigated in fMRI and MEG studies. In Anderson, Laurent, and Yantis (2014), subjects first took part in a training phase in which one of the target stimuli (e.g., a green circle) in a visual search task was consistently associated with a monetary reward, whereas other target stimuli were never associated with monetary rewards. In a second phase, there were no monetary rewards at all, and the targets were defined by shape, with color being completely irrelevant in the task. On some trials, distractor stimuli were the color of the targets from the previous training phase, some of which had been associated with a monetary reward. The behavioral results replicated the standard finding that previously rewarded stimuli produced greater amounts of distraction, significantly slowing down the subjects' ability to find and discriminate the target. The fMRI results showed that early visual areas in the extrastriate cortex showed enhanced activity when the display contained a previously rewarded distractor (Figure 6.9). This enhancement in visual processing occurred despite subjects being aware that the previously rewarded stimuli were irrelevant to the current task, serving only as distractors that would never earn money in the current task. This study also found reward-related enhancements in subcortical regions associated with reward processing (e.g., basal ganglia, caudate nucleus) and in parietal regions associated with attention control. These latter effects will be discussed in more detail in the following chapter on the control of attention. The effect of previously rewarded stimuli enhancing activity in early visual cortical areas has now been replicated across a number of studies (Hopf, Schoenfeld, Buschschulte, Rautzenberg, Krebs, & Boehler, 2015; MacLean &

Distractor RVF

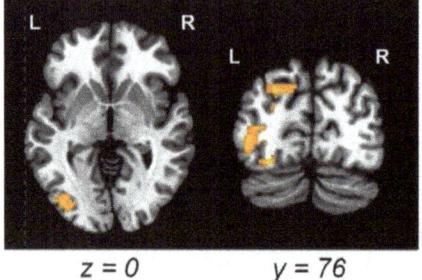

z = 0   y = 76

Distractor LVF

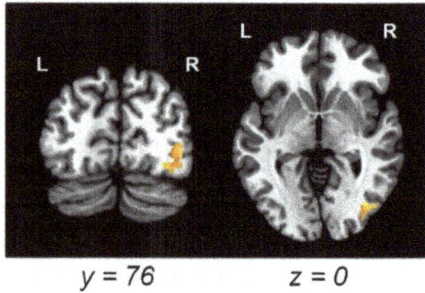

y = 76   z = 0

Figure 6.9 Effects of reward-biased attention in visual processing regions. Shown here are transverse and coronal slices with overlaid fMRI activity. The contrasts here show regions in the extrastriate visual cortex where activity was greater when there was a reward-associated distractor present in the contralateral visual hemifield compared to when there was no previously rewarded distractor in that hemifield. The left panel shows activity when the previously rewarded distractor was present in the right visual field (RVF); the right panel shows activity when the previously rewarded distractor was present in the left visual field (LVF). Figure adapted from Anderson et al. (2014), *Brain Research* with permission from Elsevier.

Giesbrecht, 2015; Serences, 2008; Hickey & Peelen, 2015, 2017). Furthermore, regarding the involuntary nature of reward-biasing effects, when reward contingencies are varied throughout an experiment, activity in the visual cortex tracks reward history, even when subjects aren't aware of the contingencies (Serences, 2008).

Given the poor temporal resolution of fMRI, these findings alone can't determine how early in processing these effects occur and how these relate to the effects of typical spatial attention and feature attention that modulate processing within the first couple of hundred milliseconds. Using MEG, researchers have been able to track the time course of these effects of rewarded stimuli and compare these to other forms of attention. Garcia-Lazaro and colleagues conducted an MEG study in which subjects had to discriminate the curvature (convex versus concave) of the colored half of a bicolored sphere presented to the left visual hemifield while a distractor stimulus (another sphere with two different-colored halves) was presented to the right visual hemifield (Garcia-Lazaro, Bartsch, Boehler, Krebs, Donohue, Harris, Schoenfeld, & Hopf, 2019). Feature-based attention was active because the target was defined by color. The distractor stimuli might share the color of the target, share the color of a rewarded stimulus, share colors with both the target and the rewarded stimulus, or consist of only colors that were neutral (not target or rewarded colors; control condition). The results showed that distractors sharing the rewarded color triggered an enhancement of processing (relative to neutral distractors) beginning at ~200 ms, and these effects were localized to the lateral extrastriate cortex. These results show that reward-driven biases affect visual processing at relatively early levels. Furthermore, the timing and location of the reward effect were the same as those seen for feature-based attention, as measured by the differences in activity between the control condition and the condition in which the distractor matched the color of the target. Although this study showed that reward-based attention affected visual processing at the same early level as feature-based attention, there were differences as well. Specifically, feature-based attention was affected by task difficulty, showing larger selection effects when the task was harder. Reward-based attention, on the other hand, showed the same size effect regardless of task difficulty. It should be noted that although this study found robust neural effects, the behavioral results did not show significant effects of reward-matched or feature-matched distractors relative to control distractors. Further research is needed on these topics, but overall these fMRI and MEG data suggest that reward-based attention triggers an automatic response in early visual areas that is similar to other types of attention. However, feature-based attention may be more dependent on the availability of resources shared with other cognitive processes, as evidenced by the feature-based attention effects being affected by overall task difficulty.

In addition to monetary rewards, other types of feedback can also influence attention. Whereas previous sections in this chapter have focused on positive rewards, research has shown that negative feedback also affects attention, as stimuli associated with aversive outcomes similarly draw attention (Schmidt, Belopolsky, & Theeuwes 2015; Wang, Yu, & Zhou, 2013). Other types of positive reward besides money also affect attention. Certain types of food (e.g., chocolate) being used as rewards can imbue the rewarded stimuli with similar attention-biasing effects as those seen for monetary rewards (Pool, Brosch, Delplanque, & Sander, 2014), and stimuli associated with pleasant sounds have also been shown to draw attention (Miranda & Palmer, 2014). Relating back to Section 6.4 on social stimuli, smiling faces produce reward effects, including an automatic orienting of attention (Anderson, 2016b; Anderson & Kim, 2018). In a related fMRI study, Kim and Anderson (2020) used the same reward training and separate testing sessions with visual search tasks described earlier, but instead of the feedback being monetary, the reward was a picture of a smiling face (with a neutral-emotion face representing the nonrewarded feedback). The results showed that the positive social feedback produced highly similar results to those observed with monetary feedback. Reaction times to targets were slowed when a distractor was present that had previously been associated with the positive reward of a smiling face. Furthermore, distractors that had been associated with social reward evoked enhanced activity in early visual processing areas, as has been seen for monetary reward (Hickey & Peelen, 2015; Serences, 2008). In addition to replicating the finding of enhanced caudate activity to stimuli associated with reward, this study also found that the stimuli associated with social reward elicited additional enhanced activity in the middle frontal gyrus. This frontal activity is not usually seen in studies of monetary rewards, and it could indicate some difference between the systems that control attention in response to social versus monetary rewards. An interesting difference between this study and previous studies of monetary reward is that even stimuli that were only associated with social reward on a small percentage of trials (20%) produced similar behavioral and visual cortex effects to stimuli associated more strongly with reward (80%). In contrast, studies of monetary reward typically find effects only for the stimuli that are strongly associated with reward. Further research is required to understand this difference; it could indicate that we are naturally more attuned to every instance of social reward, or it could be due to the social and monetary rewards used in these experiments not being equivalent (i.e., the reward of a few cents per trial may be less salient than the reward of a happy, smiling face).

Finally, another category of stimuli that are thought to affect attention through reward-related mechanisms are drugs and alcohol. These substances are known to have strong effects within the brain's reward system, and pictures of items associated with these substances (e.g., pictures of cigarettes, bottles of wine or beer, drug paraphernalia) can trigger systems in the brain related to reward and craving. Such pictures have been shown to capture attention in drug-dependent patients in a similar way that monetary and social rewards do across the population (Field & Cox, 2008). A paradigm that has been useful in investigating this type of attention is the "dot-probe paradigm." In such tasks, subjects must detect or discriminate a briefly presented target stimulus in one of two possible target locations. Just before the target stimulus is presented, task-irrelevant stimuli are presented briefly at both possible target locations. In this way, the low-level visual changes that occur when a new object appears happen at both of these locations simultaneously, equating the two locations in terms of any capture effects due simply to the appearance of a new object. These stimuli presented before the target can be different on a *semantic* level, however, allowing the researcher to investigate how the *meaning* of the stimuli affects attentional allocation. Shin and colleagues used this task with pictures of alcoholic drinks (e.g., beer bottles) and nonalcoholic drinks (e.g., water bottles) preceding the target stimulus (Shin, Hopfinger, Lust, Henry, & Bartholow, 2010). The bilateral prestimulus displays always contained an alcoholic drink image at one location and a nonalcoholic drink image at the other. This study specifically investigated whether images of alcoholic drinks would capture attention differentially for subjects who showed low sensitivity to alcohol versus control subjects. Sensitivity to alcohol has been proposed to be a risk factor for possibly developing alcoholism, since low sensitivity to alcohol has been associated with higher rates of drinking (Schuckit, 1994; Schuckit & Smith, 2000). In this study, subjects with low sensitivity to alcohol showed stronger behavioral effects of being captured by the images of alcoholic drinks, whereas control subjects showed no evidence of such capture. In addition, this study found ERP markers of attentional capture. The early P1 component was mildly enhanced for targets occurring at the location where the image of an alcoholic drink was located compared to at the location where the image of a nonalcoholic drink was presented. Importantly, this P1 enhancement was found *only* in the low-sensitivity subjects. The most robust ERP effect, however, was on the IIN component (~200–300 ms), which has been suggested to be involved in the reorienting of attention after being captured by invalid cues (this component was introduced in Section 6.1.2). In this study, low-sensitivity subjects showed an IIN only when the target occurred at the location where an image of a nonalcoholic drink had appeared; there was no IIN when the target appeared where an image of an alcoholic drink appeared. This pattern of results provides evidence that the low-sensitivity subjects' attention had been captured by the image of the alcoholic drink, and that their attention was still there when the target appeared. Therefore, no disengagement or reorienting occurred on those trials for those subjects. In contrast, control subjects showed an IIN to targets at *either* location, providing evidence that they had *not* oriented preferentially to either alcoholic or nonalcoholic drink images, which is the most effective strategy for this task. In further analyses, the effect of *recent alcohol consumption* was also found to be reflected in the IIN effects in some of the control subjects, as subjects who reported the highest amount of drinking in the past month showed the same pattern (i.e., relative absence of an IIN to targets at location of the image of the alcoholic drink) as the group of low-sensitivity subjects.

6.6   Selection History

Many of the influences on attention described in the previous sections could be grouped together in a category of *selection history* effects on attention. The classic dichotomy between top-down attention and bottom-up attention hasn't been sufficient to explain all the different types of attention effects. If bottom-up attention is assumed to include only effects that are driven by the salience of the physical stimuli and if top-down attention is assumed to include only voluntary orienting, then there are indeed many influences on attention and perception left unaccounted for, such as the effects of social gaze, rewards, and all the different effects of memory described earlier. Awh, Belopolsky, and Theeuwes (2012) proposed the term "selection history" as a third category to account for these other influences. Anderson and colleagues reviewed the many studies that could be considered as part of this new category, and they developed criteria and guidelines for what should be considered to represent "selection history" (Anderson, Kim, Kim, Liao, Mrkonja, Clement, & Grégoire, 2021). According to these authors, in order to be considered effects of selection history as opposed to other types of top-down or bottom-up attention effects, three criteria should be satisfied: (1) The effect must depend on *prior experience*; (2) the effect cannot be due to a change in *explicit* strategy because of prior experience; and (3) the effect cannot be accounted for by a specific *type* of stimulus. The third point refers to certain classes of stimuli (e.g., faces, snakes, new perceptual objects) that have higher attentional priority even without any explicit learning experiences; although attention to such stimuli may *also* be affected by experience, selection history effects would not include effects that were likely present *before* the experience. The second criterion basically states that a selection history effect can't be a voluntary attention effect that happens because of a learning experience; if an experience changes how you choose to explicitly allocate attention, then that would still fall under the typical category of voluntary, top-down attention. The first criterion is at the core of this type of influence – that the effect must be related to one's history with the stimulus. This prior experience can be explicit or implicit, and it can induce a long-lasting effect or be quite transient. Future research may dissociate some of these types of experience-dependent effects, but, for now, all of these are grouped under selection history effects.

The Anderson et al. (2021) review also describes a number of different types of selection history effects that relate to Section 6.5 on memory and attention, including ITP (targets and/or distractors repeat over trials), stimulus feature frequency (features that frequently occur as parts of targets or distractors versus novel stimuli), and reward or punishment history (associating a stimulus with a positive or aversive outcome). Other types of selection history include effects that can be created through explicit actions (e.g., consistently searching for the same target feature), as well as implicit learning (e.g., statistical dependencies regarding the sequences and locations of targets and distractors). Taking into account the implicit learning that yields selection history effects can be critical to understanding the mechanisms of selection and how attentional priority is assigned. For instance, in Section 6.2 describing the debate regarding whether attentional capture can be completely salience driven or whether top-down goals determine what captures attention, the influence of mechanisms of suppression was introduced as a way to help settle this debate. Sawaki and Luck (2010) proposed that suppression of known distractors may mediate attentional capture, and the P_D component was used as an index of this suppression process. ERP experiments have found that the P_D component is observed in response to known distractors, and its presence is associated with the successful ignoring of such a distractor. This raised the possibility that bottom-up salience cannot drive attentional orienting when voluntary mechanisms are engaged to suppress the influence of a known distractor. However, Gaspelin and Luck (2018) also found that the P_D component wasn't simply generated due to top-down goals, but rather depended on history with the stimulus. Therefore, the suppression of distractors was at least partially due to recent experience with those stimuli. As is shown in this case, it is critical to account for how memory – even implicit memory – interacts with other top-down and bottom-up influences on selection, because all of these mechanisms contribute to determining where our conscious attention is focused from moment to moment.

The main point of much of the research on memory effects has been to discuss how selection history drives the allocation of attention, but it should also be noted that attention is usually needed to encode the initial learning process that creates those biases in the first place. Specifically, most studies showing history-related effects on attention created the initial memories in previous trials in which those items were either explicitly attended to because they were targets or they were attended to simply because they were the most salient (sometimes only) item in the display. Therefore, attention and selection history likely involve interdependent mechanisms through time, with attention serving to gate what information gets encoded into a memory trace, with those memory traces then affecting the subsequent allocation of attention (Anderson et al., 2021; Gong & Liu, 2018).

Finally, although selection history has an important influence on attention that needs to be understood, recent accounts have pointed out that not all the types of history included under the category of selection history necessarily affect attention in the same way. Ramgir and Lamy (2022) suggested that before accepting a new, single, very broad category that is independent of the classic top-down versus bottom-up attention categories, research findings should be reviewed with regards to the consistency of the different processes included within this new category and their independence from other types of attention. For example, in reviewing findings across studies, Ramgir and Lamy (2022) suggested that ITP may not affect attentional priority independently from top-down attention and working memory processes. An overarching question remains as to how many different types of influences there are on attention and how best to group these into distinct categories. Further research is needed to clarify which influences are interdependent or rely on similar mechanisms versus which represent independent attention processes. This chapter has focused mainly on the *effects* of attention, but in the next chapter the brain regions and neural mechanisms that *control* the allocation of attention will be reviewed, revealing overlapping versus independent brain mechanisms for different forms of attentional control.

CHAPTER SUMMARY

- Endogenous and exogenous attention can similarly affect early levels of sensory processing as indexed by ERP and fMRI measures.
- There are multiple differences between exogenous and endogenous attention, including the time course of how each develops and the effects on higher-order levels of processing.
- Debate continues regarding whether attention capture is contingent on top-down settings or is instead triggered automatically by highly salient stimuli. New theories suggest top-down suppression plays a key role in determining what captures attention.
- Some stimuli (e.g., faces, gazing eyes, highly emotional content, new objects) may have a special status in terms of triggering involuntary shifts of attention.
- Memory, rewards, and selection history have significant effects on the allocation of attention.

REVIEW QUESTIONS

Describe the similarities and differences in the neural effects of endogenous and exogenous attention on visual processing.

Define the N2pc, P_D, and IIN ERP components and describe the mechanism(s) of attention with which each has been associated.

How do faces, emotions, and social gaze affect attention and what are the neural effects of these?

Explain how rewards, working memory, and long-term memory affect attention and result in neural effects that are not simply endogenous or exogenous.

FURTHER READINGS

Anderson, B. A., Kim, H., Kim, A. J., Liao, M. R., Mrkonja, L., Clement, A., & Grégoire, L. (2021). The past, present, and future of selection history. *Neuroscience and Biobehavioral Reviews*, *30*, 326–350.
  – This excellent review paper describes the many areas of research on attention that fall under the category of selection history, and the authors define and explain the different types of selection history results.

Luck, S. J., Gaspelin, N., Folk, C. L., Remington, R. W., & Theeuwes, J. (2021). Progress toward resolving the attentional capture debate. *Visual Cognition*, *29*(1), 1–21.
  – This paper discusses the attentional capture debate and potential solutions from three different perspectives. This paper was also followed by a series of insightful commentaries.

Redden, R. S., MacInnes, W. J., & Klein, R. M. (2021). Inhibition of return: An information processing theory of its natures and significance. *Cortex*, *135*, 30–48.
  – This article reviews behavioral and neuroscience studies of IOR and presents a new model and theory.

7 The Control of Attention

Neural Systems and Mechanisms

Learning Objectives

- Identify the processes of attentional control.
- Compare internally generated shifts of attention to externally triggered shifts of spatial and feature-based attentional control.
- List the brain regions and functions associated with the dorsal versus ventral attention networks.
- Describe the neural regions and connectivity that support the multiple processes of executive control.
- Compare the effectiveness of "brain training" methods, including video games, neurofeedback, and meditation.

7.1 Sources of Attentional Control

The previous two chapters focused on the *effects* of attention on perceptual and higher-order processing and contrasted the *effects* of voluntary versus involuntary attention. In this chapter, we discuss the brain systems and neural mechanisms responsible for the orienting and focusing of attention, referred to as *attentional control*. Aspects of attentional control include the processes involved in moving attention: disengaging from a current object or location, planning the movement, executing the shift, and engaging attention at the new location or object. Higher-order aspects of attentional control, sometimes called "executive control," include the mechanisms of monitoring one's actions, processing feedback, making decisions, updating plans, and inhibiting prepotent responses when no longer appropriate. This chapter will refer to some of the patient populations presented in earlier chapters but will dive deeper into how neuroscience studies of healthy participants are helping us to better understand the neural mechanisms of attention that are disrupted in those populations. Research into therapies and training of attentional control is suggesting that there is some degree of plasticity within the brain's attention systems. Neural stimulation studies are suggesting ways in which attentional control may be manipulated, providing a method to test causal theories of what different brain regions contribute to the control of attention. Finally, this chapter includes a discussion of controversies regarding whether attentional control can be effectively "trained" using online games or neural feedback.

7.2   The Dorsal Attention Network

As described in Chapter 4, some of the earliest evidence for the brain locus of attention mechanisms came from reports of patients suffering unilateral hemineglect (also known as "attentional neglect," or simply "neglect"). These are the patients who seem to ignore an entire side of space, despite having intact sensory processing of those regions. For example, patients with visual hemineglect have an intact early visual processing system and have the ability to see and respond to things in their "bad" hemifield, but they usually ignore things in that affected hemifield and only attend to things in their good hemifield. These are the patients who may eat from only one side of their plate, or shave or apply makeup to only one side of their face. A classic indicator of neglect is a failure in the extinction task, which can be used by a doctor at the patient's bedside. In this task, the doctor asks the patient to look directly at them while the doctor holds up their own index fingers, one in each of the patient's visual hemifields. The doctor then wiggles one or both fingers and asks the patient to detect which finger or fingers move. The patient will respond very well to a wiggling finger in their good field and may even detect some wiggling in their bad field if the finger in their good field isn't moving at the same time. However, they will consistently fail to detect any movement in their bad field when it occurs simultaneously with a movement in their good field. When the patient's attention is focused on a stimulus in their good field, they essentially "extinguish" the stimulus in their bad field, and it doesn't enter their consciousness. A classic paper-and-pencil test for neglect is the line cancelation task. In this task, the patient is given a sheet of paper with several separate horizontal lines drawn across both sides of the page, and they are then asked to draw short, vertical lines bisecting each of the original lines. Patients with neglect typically show a behavioral pattern in which most of the lines in their bad visual field are not touched at all, and the bisection lines drawn by the patients in their good field are drawn as if the half of the line toward their bad field wasn't there. For example, in a patient with neglect of the left hemispace, the lines are bisected as if the left half of each line didn't exist. One of the first indications of hemispheric specialization of attention processes was that left hemineglect (neglect of the left side of space or objects) is much more common than right hemineglect. In addition, the area of damage in most cases of left hemineglect is in the *posterior parietal cortex* in the *right* hemisphere (Vallar, 1998). Studies have suggested that right hemineglect following damage to the left hemisphere is more common than has been previously reported (Kleinman, Newhart, Davis, Heidler-Gary, Gottesman, & Hillis, 2007), but it is generally still agreed that severe cases of unilateral hemineglect are much more common following right hemisphere damage, resulting in left hemineglect. Based on this asymmetry, it has been hypothesized that attentional control may be strongly lateralized to the right hemisphere. It has also been suggested that this lateralization of attention in the human brain may relate to the left hemisphere becoming specialized for language. Another theory for why hemineglect is more common following right hemisphere brain damage is that the parietal regions of the right hemisphere may monitor both visual fields, whereas the left parietal regions monitor only the right visual field. Therefore, if the left hemisphere is damaged, the right visual field can still be monitored by the right hemisphere; however, when the right hemisphere is damaged, the left visual field is no longer represented well in the brain because the left hemisphere is only monitoring the right hemifield. Other regions of the brain, besides the parietal cortex, have also been implicated in unilateral neglect. Studies have reported that lesions to the *superior temporal gyrus* (STG) can lead to neglect, and it has been suggested that lesions here may be the primary cause of neglect symptoms (Karnath, Ferber, & Himmelbach, 2001); however, controversy remains regarding the most critical area(s) underlying hemineglect (Doricchi & Tomaiuolo, 2003; Karnath, Fruhmann Berger, Küker, & Rorden, 2004; Mort, Malhotra, Mannan, Rorden, Pambakian, Kennard, & Husain, 2003). In addition, isolated lesions in *frontal* regions of the right hemisphere have also been associated with neglect (Husain & Kennard, 1996), while yet other studies have reported neglect following lesions to subcortical regions (e.g., the thalamus; Hillis, Newhart, Heidler, Barker, Herskovits, & Degaonkar, 2005). Overall, this research highlights that mechanisms of attention are tied to multiple cortical and subcortical regions. It has remained consistent, however, that lesions to posterior parietal regions of the right hemisphere are strongly associated with unilateral hemineglect (Vallar, 1998), and many of the early neuroimaging studies of attentional control were based on examining the role of this region in healthy subjects performing attention tasks.

7.2.1 Neuroimaging Evidence of Parietal and Frontal Control Regions

With the advent of positron emission tomography (PET) in the early 1990s, researchers were able to study the neural mechanisms of cognitive processes in the intact, healthy brain with spatial precision. Some of the very first PET studies investigated the neural basis of selective attention and control processes. Corbetta and colleagues first tested healthy young adult subjects performing visual attention tasks during PET scanning (Corbetta, Miezin, Shulman, & Petersen, 1993). By comparing PET scans in which subjects attended selectively to stimuli in one visual field (and responded to stimuli in that visual hemifield) to PET scans collected while subjects passively viewed the same stimuli or performed a discrimination task at fixation, they found that posterior parietal and dorsal frontal regions were consistently active when the task required attention to be shifted. In addition, they found a midline region around the supplementary motor area (SMA) to be active during active responding conditions. Another study found that posterior parietal areas were active whenever subjects had to move attention, whether it be to a new spatial location or to a new feature (Petersen, Corbetta, Miezin, & Shulman, 1994). Importantly, this latter study found that the parietal regions were *not* active when subjects maintained attention on one location or feature throughout the scan but were active in all conditions in which attention needed to be shifted. Critically, this activation of the posterior parietal lobes was bilateral and occurred in the same general region whether attention was shifting to a new location, color, shape, or motion. Thus, this posterior parietal area was active across different dimensions of visual attention. Nobre and colleagues further refined the location of the parietal involvement in attentional control to the intraparietal sulcus (IPS), and they found that the IPS and SMA were involved in both *endogenous* and *exogenous* attentional orienting (Nobre, Sebestyen, Gitelman, Mesulam, Frackowiak, & Frith, 1997). With the development of functional magnetic resonance imaging (fMRI) in the mid-1990s, the neural bases of cognitive processes could be studied with even greater precision. Using fMRI, Gitelman and colleagues confirmed that the IPS was involved in attentional control, and they localized dorsal frontal activity to the frontal eye field(s) (FEF; Gitelman, Nobre, Parrish, LaBar, Kim, Meyer, & Mesulam, 1999). The FEF region of the frontal lobe was initially identified by its involvement in generating voluntary eye movements in monkey single-unit research (reviewed in Schall, 1997), and subsequent work showed that it is also involved in shifting attention covertly (i.e., without any eye movements; Kodaka, Mikami, & Kubota, 1997; Thompson, 2005). Neuroimaging studies in humans have confirmed the involvement of this frontal region in moving attention covertly (reviewed in Corbetta & Shulman, 2002), and these findings are important for showing a link between covert attentional control and the brain regions critically involved in overt attention and eye movements. In addition to finding IPS and FEF involvement, Gitelman et al. (1999) replicated the finding of SMA activity during attentional control, and they also identified a superior temporal region as being active during shifts of spatial attention. Reviewing multiple PET and fMRI studies of attentional control, Corbetta and Shulman (2002) identified the regions of the posterior parietal and dorsal frontal cortex consistently activated in experiments involving the shifting of attention between locations, objects, and features. This network of areas is now commonly referred to as the **dorsal attention network (DAN)** (Figure 7.1).

In these neuroimaging studies, the regions of activity associated with attentional control were consistently observed to be bilateral, occurring in each brain hemisphere. As opposed to the findings from unilateral neglect patients that indicate that the right hemisphere is more critically

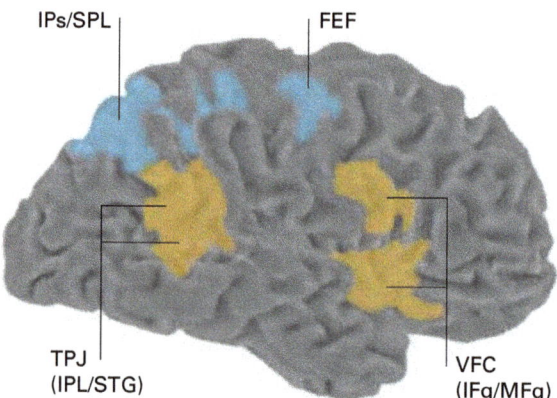

Figure 7.1 Attentional control areas of the cortex. Regions of the DAN are shown in blue and include the FEF, IPS, and superior parietal lobule (SPL). These areas have been consistently found to be active in studies of voluntary orienting of spatial attention. Regions of the ventral attention network (VAN) are shown in orange; the temporoparietal junction (TPJ) area includes the inferior parietal lobule (IPL) and the STG; the ventral frontal cortex (VFC) includes the inferior frontal gyrus (IFg) and the middle frontal gyrus (MFg). Functions of the VAN are described in Section 7.3. Reprinted from Corbetta and Shulman (2002), *Nature Reviews Neuroscience*, with permission from Springer Nature.

involved in attentional control, the neuroimaging results suggest that both hemispheres are strongly involved in attention, regardless of which hemispace is being attended. There are several possible reasons for the discrepant results across methodologies. A critical difference is that neuroimaging evidence is only correlational and can show all areas that are in any way involved in a cognitive process, whereas neuropsychological patient studies indicate what areas are critically *necessary* for a mental operation. Therefore, in terms of attentional control, it could be that both hemispheres are engaged during the shifting of attention, but that an intact and healthy right hemisphere is necessary for normal attention functioning, whereas the left hemisphere is not so critical. Another difference is that neuropsychological patients may experience some neural reorganization following their brain lesion or may develop various compensatory strategies to perform a task. If this is the case, then the differences in hemispheric specialization observed in patients could potentially be due to the relative flexibility/plasticity of the hemispheres as opposed to a preexisting difference in the attentional functions of each hemisphere. Another difference between the findings from these methods is that neuroimaging consistently finds strong FEF involvement, but most studies of unilateral neglect have identified parietal regions as being critical. Since frontal lobe lesions can lead to multiple deficits in motor and executive functions, depending on the size and extent of those lesions, the frontal lobes have not been as uniquely associated with attentional deficits as parietal lesions. However, as mentioned earlier, studies have found attentional deficits in frontal patients as well (Husain & Kennard, 1996), although most cases of unilateral hemineglect involve lesions to the parietal cortex.

As mentioned in Chapter 3, other limitations of neuropsychological patient research include that naturally occurring lesions are variable from subject to subject, the damage is rarely restricted to a single region of interest, and reorganization may occur following each brain infarct. Neuroimaging studies are thus important for assessing brain mechanisms without these confounding effects. With PET, however, activity is summed over a relatively long period of time. Specifically, each PET scan sums activity over at least 60 seconds, and at least 10 minutes is required between experimental conditions to allow for decay of the oxygen-15 radioactive isotope. Therefore, although PET studies consistently found frontal and parietal involvement during attention tasks, it wasn't clear at what stage of processing this activity was produced. Early fMRI studies had similar limitations because those studies used a "block design" (i.e., "epoch design" or "boxcar design") in which the experimental condition was blocked in periods of 15–30 seconds. Within each "block" subjects might be orienting attention to a location, sustaining attention there, and selectively processing stimuli (attended versus unattended). Thus, these early neuroimaging studies could not definitely attribute activity to attentional control processes versus the selective effects of attention on sensory processing that ensue. With the advance of event-related fMRI, however, it became possible to separate the control of attention from the effects of attention. In one of the first studies using this technique, Hopfinger, Buonocore, and Mangun (2000) found that bilateral FEF, IPS, and STG regions were uniquely active in response to cues to shift attention (i.e., attentional control), whereas visual processing regions were active in response to the subsequent peripheral target displays, with the level of activity dependent on where attention had just been oriented (i.e., the effects of attention).

7.2.1.1 The "Free Will" of Attentional Control

As mentioned in Box 7.1, attention is usually experienced as a unitary focus of conscious awareness. This linkage between awareness and attention relates to the distinction between endogenous versus exogenous control. However, most studies of endogenous attention in the laboratory have not been able to investigate purely voluntary acts of attention. This is because of the need to time-lock neural activity to a specific event in order to ensure that brain activity (averaged over many trials) can be associated with the cognitive function of interest. In addition, many experimental paradigms require a certain sequence of events, and allowing subjects to use their free will to decide what to do from trial to trial could compromise the experimental design. Therefore, most studies of endogenous or voluntary attention use an instructive cue stimulus to ensure that the initial engagement of the cognitive process can be locked to a precisely specified point in the task and that the sequence of trials (e.g., attend-right valid trial followed by attend-left invalid trial) can establish the intended experimental design (e.g., a 75% predictive cue). In many early studies, this was done with an arrow cue presented at fixation. Arrows were used in part because they were easy for subjects to understand and didn't require practice or training with the stimuli. However, behavioral studies later revealed that arrows are so well-known, and their meaning so deeply ingrained throughout our daily lives, that they automatically trigger the movement of attention to the area being pointed at, even when subjects are informed that the arrow is uninformative in that task. Therefore, even though they are used to measure *voluntary* orienting, arrow cues likely engage some form of *involuntary* orienting as well.

Box 7.1 Do "split-brain" patients have two spotlights of attention?

One of the most fundamental aspects of attention is that it corresponds to our sense of a single, unified, focus of consciousness. Although we sometimes try to divide our attention across two tasks, this requires a significant cognitive effort, and it feels different than our baseline state of having a single focus of attention. However, studies of callosotomy ("split-brain") patients have suggested that each of our brain hemispheres may support a separate consciousness and separate focus of attention. These rare patients have had surgery to sever their corpus callosum, the large band of fibers connecting the left and right brain hemispheres, in order alleviate intractable epilepsy. The surgery is done to ensure that a seizure in one hemisphere cannot spread to the other hemisphere, but the consequence of a complete callosotomy is that the hemispheres no longer communicate in the usual way. Studies have revealed that these patients seem to have two separate consciousnesses, as each hemisphere is able to respond to stimuli presented to it but is largely unaware of what has been presented to the other hemisphere (Gazzaniga, Bogen, & Sperry, 1962). Although the surgery disconnects the cortical connections between the hemispheres, smaller subcortical pathways still link the hemispheres, and early studies of attention in these patients sought to understand how these connections influence attention. Holtzman (1984) conducted a cuing study that included unilateral cues to either the right or left hemifield or bilateral cues to both hemifields. If attention in each brain hemisphere was completely

Box 7.1 **(cont.)**

independent in these patients, each hemisphere should only "see" the cue in the hemifield it monitors. However, they found that reaction times to the bilaterally cued trials were significantly slower than to unilateral validly cued trials. Since the behavior of these patients was affected by cues to the opposite hemifield even though there were no cortical connections between the hemispheres, Holtzman (1984) suggested that the focus of attention was at least partially coordinated by subcortical systems. However, other work in these patients has provided evidence for two separate spotlights of attention, one in each hemisphere. Luck and colleagues (Luck, Hillyard, Mangun, & Gazzaniga, 1989) found that in visual search tasks these patients were able to search each hemifield independently. When targets are displayed within cluttered fields of stimuli, healthy subjects search through items serially across the entire visual space, but callosotomy patients search each visual hemifield independently, significantly speeding their time to detect targets when spread across the hemifields. Overall, the results from callosotomy patient studies suggest that each brain hemisphere can support an independent focus of attention. In healthy individuals, therefore, the typical experience of having a singular, unified focus of attention likely results from integrated activity across both hemispheres, not simply from one hemisphere being the sole seat of attentional control. Further research is needed to understand the mechanisms by which the coordination across the brain hemispheres results in this singular focus of attention.

Subsequent studies often used other types of informative cues, such as different informative colors or shapes presented at fixation that predicted the location or features of the target. Although this requires more explicit instruction and practice, it avoids the use of cues that may trigger involuntary orienting. Even with such novel cues, however, associations between cues and targets are learned over the many trials in an experiment, and these cues may therefore begin to trigger involuntary processes at some point in the experiment. This could be especially problematic in nonhuman studies in which the animals must be trained over many hundreds of trials to perform the tasks correctly. With long training or experimental sessions, even cues without any preexisting associations with locations or features may begin to trigger involuntary orienting processes. In addition, studies of voluntary attention typically don't allow subjects to decide on their own where to attend. As long as they are being "good" (i.e., compliant) subjects, they follow the instructions on each trial and attend to where they're being told to attend. To assess what, if any, processes may differ when subjects use their own free will to direct their attention, Taylor, Rushworth, and Nobre (2008) modified a typical cuing paradigm. On separate trials, the cue display either instructed subjects where to covertly attend (i.e., a left or a right peripheral location) or instructed them to *choose* which side to covertly attend and then move their attention there. Behavioral measures confirmed that orienting in response to either type of cue (instructed location or self-choice) resulted in improved performance at the single attended location relative to a divided attention condition. Activity in the posterior parietal areas of the DAN was similar for instructed orienting and self-choice orienting, confirming its role in attentional control. In a few medial frontal areas, however, choice-driven orienting resulted in *higher* levels of activity than when subjects were instructed to attend to that location. Specifically, the anterior cingulate cortex (ACC), pre-SMA, presupplementary eye fields (pre-SEF), and lateral FEF were more active following self-choice orienting. These results suggest that orienting attention by choice engages not only the typical attentional regions triggered by instructive cues but may also engage additional frontal regions related to decision-making, working memory, and choice selection of movements. In a combined event-related potential (ERP)–fMRI study using a similar design, Bengson, Kelley, and Mangun (2015) found the FEF and IPS to be activated by both instructive cues and choice cues, and they replicated the finding that the ACC was strongly involved in the process of deciding where to pay attention. In addition, they were able to identify a network including a region in the left middle frontal gyrus and bilateral regions of the anterior insula as being uniquely involved in the processes of "willed attention."

Although Taylor et al. (2008) and Bengson et al. (2015) allowed subjects the freedom to voluntarily choose *where* to attend, the subjects were still presented with a highly salient cue that instructed them exactly *when* to shift their attention. Additionally, the abrupt appearance of a salient stimulus at fixation likely triggered some involuntary processes (e.g., alerting and orienting to the cue stimulus to discriminate what it was instructing them to do). To provide a purer measure of endogenous attentional control, Hopfinger, Camblin, and Parks (2010) asked subjects to orient to one of two locations whenever they wanted. Specifically, subjects were asked to initiate a trial by pressing one button to indicate when they were shifting attention to the left or a second button when they were shifting attention to the right. At a variable time (800–1,500 ms) after the subject indicated their shift of attention, bilateral target displays could be presented, and subjects were to identify (with a button press) the number presented at their attended location (displays consisted of a different number at the left and right visual field locations, each flanked by irrelevant letter stimuli). To observe the attention shift activity without overlap from target-evoked activity, only a third of trials contained a target display; the shift-only trials (with no target) were used to measure internally generated attentional orienting. Before the fMRI session, subjects participated in a practice session to ensure they understood the task and that they would generate an appropriate number of attention shifts (~30–40 shifts per each 5-minute run) at an appropriate pace for an fMRI study (~3–15 seconds between shifts). To compare these self-initiated shifts of attention to the classic cue-instructed shifts of attention, fMRI runs (5 minutes each) alternated between self-initiated shifts and cue-instructed shifts. Furthermore, to ensure that any differences in results would not be due to different timings or sequences, each cue-instructed run used the precise sequence and timing of attention shifts as the preceding self-initiated run for each subject. The results revealed that self-initiated orienting, in the complete absence of any external cue stimulus, was associated with robust activity in the posterior parietal and FEF regions of the DAN (Figure 7.2A,B). These same regions were also active in response to instructive cue stimuli during the separate runs of cued shifts of attention.

Critically, in the Hopfinger et al. (2010) study, the self-initiated shifts revealed a hemispheric *lateralization* that was *not* present in the instructive cue shifts, which is reminiscent of the pattern seen in unilateral neglect. Specifically, when shifts of attention were triggered by instructive cues, neural activity in both the left and right frontal and parietal regions was the same regardless of the direction in which attention was being moved, in line with the results of

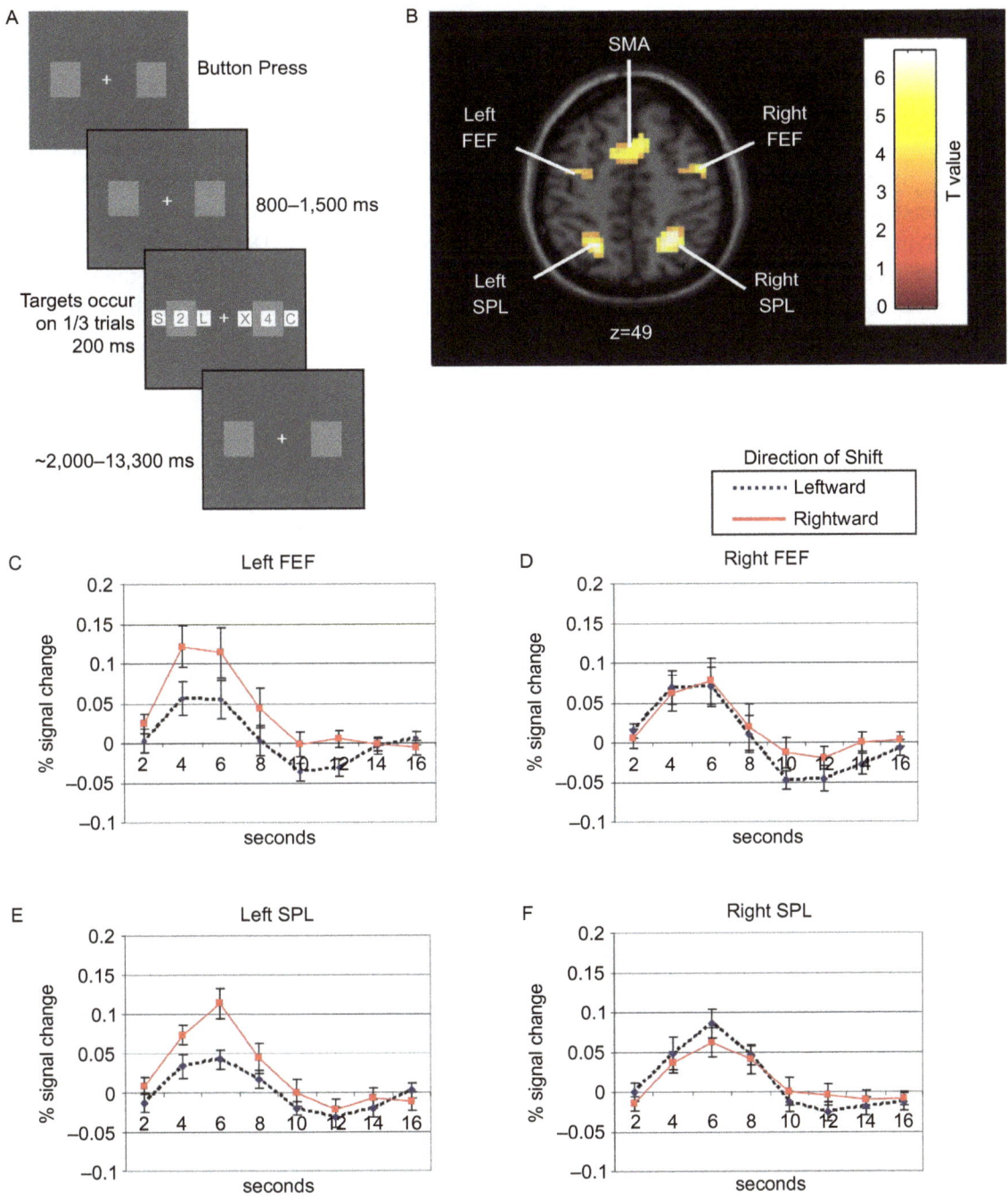

Figure 7.2 Self-initiated shifts of attention. (A) Trial sequence for self-initiated shift runs. (B) Activity within the DAN for self-initiated shifts, thresholded at p < 0.05, corrected for false discovery rate [FDR] over the whole brain). (C–F) Blood oxygenation-level dependent (BOLD) signal time courses for self-initiated shifts of attention. SPL = superior parietal lobule. Reprinted from Hopfinger et al. (2010), *Psychophysiology*, with permission from John Wiley & Sons, Inc.

previous neuroimaging studies. For self-initiated shifts, however, the right hemisphere was equally active when orienting to the left or right visual field, but the level of activity in the left hemisphere was dependent on the visual field to which attention was being directed (Figure 7.2C–F). Activity in the frontal and parietal control regions of the left hemisphere was robust when orienting to the contralateral (right) visual hemifield, but this activity was significantly smaller when orienting to the ipsilateral (left) visual hemifield. This pattern is in line with the explanation of unilateral hemineglect that posits that the right hemisphere monitors both visual hemifields equally, whereas the left hemisphere preferentially monitors the right hemifield. Accordingly, orienting to the left visual hemifield would be more dependent upon intact right hemisphere attentional control regions. This fMRI study showed that when the processes involved in self-initiated orienting of attention can be isolated, the hemispheric specialization seen in unilateral neglect patients can also be observed in the brain activity of healthy young adults.

7.2.1.2 Endogenous versus Exogenous Control within the DAN

As described in earlier chapters, there are several differences in the time courses and behavioral effects of endogenous (i.e., "top-down" or "voluntary") and exogenous (i.e., "bottom-up," "reflexive," or "involuntary") attention. Early PET and fMRI studies found largely overlapping regions of the DAN to be triggered during both endogenous and exogenous control of attention (Corbetta et al., 1993; Kim, Gitelman, Nobre, Parrish, LaBar, & Mesulam, 1999; Kincade, Abrams, Astafiev, Shulman, & Corbetta, 2005; Nobre et al., 1997; Peelen, Heslenfeld, & Theeuwes, 2004). In a subsequent fMRI study, Meyer, Du, Parks, and Hopfinger (2018) tested whether there were differences in the *degree* to which frontal versus parietal regions were engaged for endogenous versus exogenous attention. In this study, exogenous attention was cued by a brief, salient, *nonpredictive* luminance change at one of the two peripheral locations where a subsequent target could occur. Endogenous attention was engaged using a color change of the fixation cross that indicated which visual hemifield the upcoming target was likely to occur in (83% valid cues). The results showed that frontal and parietal areas of the DAN in both hemispheres were significantly active in response to both exogenous and endogenous cues. However, critically, the *balance* of activity across frontal and parietal regions was found to be different depending on the type of attention being engaged. For endogenous attention, the control of orienting was associated with greater activity in the FEF than the IPS, especially at earlier stages of the **blood oxygenation-level dependent (BOLD)** response (shown in Figure 7.3 as the difference between the solid and dashed blue lines). In contrast, when exogenous attention was triggered (shown by the red lines in Figure 7.3), activity in the IPS was greater than the FEF activity, especially during the falling edge (7–8 seconds of latency) of the BOLD response.

In a subsequent analysis of the Meyer et al. (2018) data comparing endogenous and exogenous attentional control across the DAN, Bowling, Friston, and Hopfinger (2020) investigated the directed connectivity patterns across these regions. Most methods of measuring functional connectivity from fMRI data assess only the correlated strength of activity between regions and cannot determine the direction of those connections. **Dynamic causal modeling (DCM)**, however, was designed to disambiguate the direction of connectivity. DCM has the limitation of

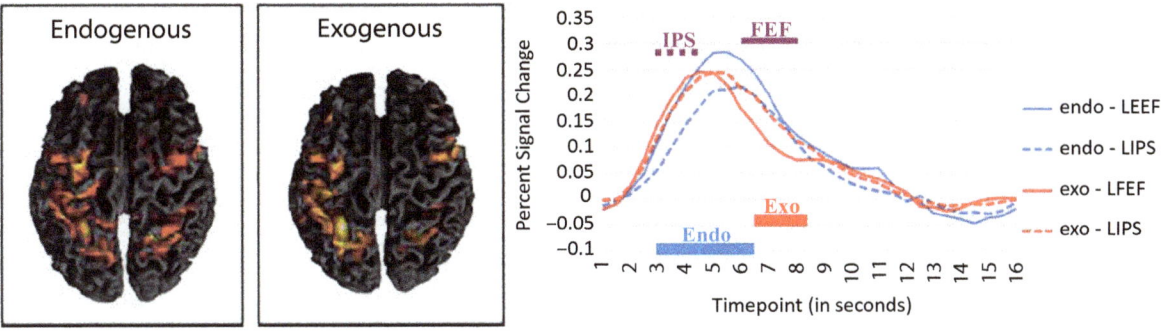

Figure 7.3 Endogenous versus exogenous attention control in the DAN. Cue-evoked fMRI activity for the endogenous (left panel) and exogenous (middle panel) attention conditions, displayed on a 3D rendered cortical surface, showing bilateral FEF and IPS activity for both conditions. Time course plots of the BOLD percentage signal change (right panel) show that the relative activities of frontal versus parietal regions were different for the two types of attention. Blue lines show the BOLD response following endogenous cues ("endo"); red lines show the BOLD response following exogenous cues ("exo"). Solid lines represent FEF activity and dashed lines represent IPS activity. The bars inserted on the graph indicate the time points at which there were significant differences within the area or condition indicated by the label on the bar (e.g., the "IPS" bar indicates when activities in the IPS were different for the endo and exo conditions). Figure adapted from Meyer et al. (2018), *Neuropsychologia*, with permission from Elsevier.

being feasible for only a relatively few areas (due to its high computational demands), whereas functional connectivity methods assessing basic correlated activity may include hundreds of regions. However, when a network can be defined as primarily composed of a few select regions, DCM is a powerful technique for assessing the causal influence of one area upon another. Bowling and colleagues used DCM to test whether the frontoparietal network activated by both endogenous and exogenous attention would show different patterns of directed connectivity depending on which type of attentional control was engaged. The results revealed that the IPS had a baseline inhibitory effect on the FEF, and, most critically, this causal effect was significantly modulated by the type of attention that was engaged. During exogenous orienting, this inhibitory effect was *strengthened*, whereas the causal influence of the IPS on the FEF was significantly *reduced* during endogenous orienting. Furthermore, the effect of endogenous attention was observed across endogenous cue types. In addition to the conditions with nonpredictive peripheral cues or predictive central cues described earlier, this experiment also had a condition in which the peripheral cue was counterpredictive (with 83% of trials in that condition containing targets at the *opposite* location from the cue location) and therefore would engage top-down control following the initial exogenous capture. The DCM results showed that the attenuation of the IPS-to-FEF connectivity by endogenous attention occurred when cued by *either* central symbolic predictive cues or *antipredictive* peripheral cues. This attenuation of the IPS-to-FEF connectivity whenever endogenous control is engaged is the opposite of the strengthening of the influence of the IPS over the FEF observed in the purely exogenous condition described earlier, when those same peripheral cues were known to be noninformative and nonpredictive of target location. These results are in line with theories suggesting that a critical difference between these types of attention is that the parietal cortex has a stronger influence over frontal regions during exogenous attention, whereas endogenous attention is associated with a relatively stronger influence arising from frontal regions.

7.2.1.3 Feature- and Object-Based Attentional Control

As described in earlier chapters, attention can be focused on locations, objects, or features. Most studies of the neural mechanisms of attentional control have focused on the orienting of attention to spatial locations, but several studies have investigated whether attention to non-spatial dimensions recruits the same control regions as does orienting to spatial locations. Some of the first fMRI studies to investigate feature- and object-based attentional control processes revealed activity in the DAN that was highly similar to previous studies of spatial attentional control. Shulman and colleagues (Shulman, Ollinger, Akbudak, Conturo, Snyder, Petersen, & Corbetta, 1999) found that attention to a specific direction of motion evoked activity in the same area of the IPS found in previous studies of spatial attentional control. Weissman and colleagues (Weissman, Woldorff, & Mangun, 2002) studied the control processes when shifting attention between local and global levels of a visual display and found activity in the FEF and IPS that was highly similar to previous studies of spatial attention (e.g., Corbetta, Kincade, Ollinger, McAvoy, & Shulman, 2000; Hopfinger et al., 2000).

In an fMRI study that nicely intermixed spatial and nonspatial attention in a way that was not predictable to subjects, Giesbrecht and colleagues (Giesbrecht, Woldorff, Song, & Mangun, 2003) investigated the neural basis of feature versus spatial attention. Each trial began with the presentation of a letter cue that instructed subjects to attend to either a specific location or a specific color (i.e., "L" for attend left location; "R" for attend right location; "B" for attend blue stimulus; "Y" for attend yellow stimulus). The task was to discriminate the orientation of a subsequent rectangle in the location or color that was cued. On spatial trials, a rectangle occurred in both left and right peripheral locations, and subjects had to discriminate the orientation occurring at only the attended location. In color trials, two overlapping rectangles were displayed centered on the fixation point, and subjects had to discriminate the orientation of the rectangle in the cued color. As shown in Figure 7.4, the results revealed that orienting attention to location or color produced activity in a largely overlapping network of frontal–parietal and occipital–temporal regions. Some of these areas likely relate to aspects of the tasks that involve sensory processing and decoding of the cue stimulus (occipital–temporal activity) and the preparation of manual responses (premotor areas of the frontal lobe). However, regions of the parietal lobe previously shown to be related specifically to the orienting of spatial attention (e.g., bilateral IPS) were active to both types of cues, suggesting overlap in parts of the DAN for spatial and feature attention. In a direct statistical comparison of the two types of cues, color cues produced greater activity in some inferior occipital–temporal regions relating to visual feature processing, but no areas of the DAN showed significantly greater activity in the color versus location condition. In contrast, activity in the FEF and IPS was greater for spatial than color cues. The FEF results accord with theories that the covert orienting of spatial attention is strongly linked to the overt movement of the eyes, as the FEF is known to also be involved in the generation of voluntary saccades. Overall, the results from these neuroimaging studies provide

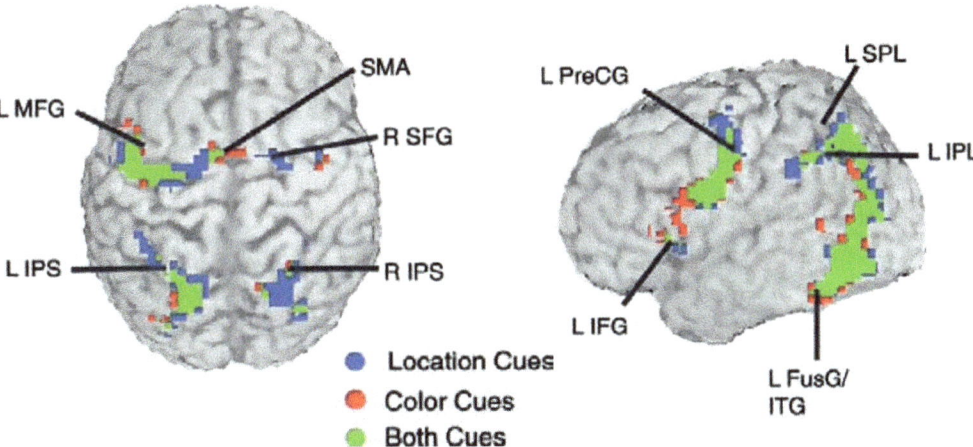

Figure 7.4 Feature-based versus location-based attentional control activity assessed with fMRI. Regions showing significant activity in response to cues that directed attention to either a spatial location (indicated by blue areas) or to a specified color (indicated by red areas) are highlighted. Areas active to both types of cues are shown in green. FusG = fusiform gyrus; IFG = inferior frontal gyrus; IPL = inferior parietal lobule; ITG = inferior temporal gyrus; MFG = midfrontal gyrus; PreCG = precentral gyrus; SFG = superior frontal gyrus. Reprinted from Giesbrecht et al. (2003), *NeuroImage*, with permission from Elsevier.

evidence that some regions implicated in attentional control are similar for spatial and nonspatial attentional control, but they also reveal that spatial attention most strongly engages the DAN in these visual attention tasks.

7.2.2 Neurostimulation Studies of the DAN

Neuroimaging methods are invaluable for identifying networks of regions that are involved in cognitive processes. However, a limitation of these methods is that their results are correlational. They can reveal networks and regions with high spatial precision, but neuroimaging alone cannot determine whether a region is critically necessary for a task or whether activity in that region is simply associated with the task in some way. Neurostimulation methods, on the other hand, are ideally suited to assessing whether a region is critical to performing a cognitive task. However, noninvasive stimulation methods are typically limited to only one or two regions, and so these methods rely on neuroimaging and neuropsychological results to specify the region(s) to be tested. Whereas results from neuropsychology patients are typically interpreted as showing that a particular brain region is critical for a particular process, these studies suffer from limitations, such as there rarely being data available from before the lesion, so it cannot be known for certain how much the lesion affected cognitive performance relative to when the brain was intact. In addition, naturally occurring brain lesions vary greatly from one patient to the next, and many patients may have more than one focal area of damage. Brain plasticity and reorganization following brain damage can also be serious potential confounds in ascribing the current task performance of a patient to the loss of brain activity in a single region.

In contrast, neurostimulation methods disrupt the activity in a brain region for only a brief and transient period, and task performance can be assessed both before and after the stimulation period, allowing excellent assessment of how the stimulation changed the subject's cognitive abilities. In addition, many such studies include a control site of stimulation to ensure that any effects are due to the disruption of that specific region of interest.

Building upon neuropsychological and neuroimaging data suggesting the importance of dorsal frontal and parietal regions in controlling the movements of attention, a number of studies have used transcranial magnetic stimulation (TMS) to investigate the roles of the FEF and IPS in attentional control. Studies that transiently disrupted activity in the FEF have found that the ability to orient attention endogenously is impaired during or shortly after stimulation (Duecker, Formisano, & Sack, 2013; Grosbras & Paus, 2002). These studies have generally found that TMS to the FEF in either hemisphere impairs performance on spatial attention tasks. Other studies have used TMS to disrupt the activity of parietal regions and again found that stimulation of this area impaired attentional orienting (Thut, Nietzel, & Pascual-Leone, 2005) and led to patterns of performance on extinction tasks similar to those of patients with parietal lesions (Battelli, Alvarez, Carlson, & Pascual-Leone, 2008; Hilgetag, Theoret, & Pascual-Leone, 2001).

Although TMS studies are typically used in an attempt to isolate the region of disruption, other regions that are connected to the stimulated region can also be affected. Gallotto and colleagues (Gallotto, Schuhmann, Duecker, Middag-van Spanje, de Graaf, & Sack, 2022) investigated whether attention was affected more severely when either the FEF, IPS, or both regions concurrently were stimulated. In this study, subjects performed a variation of the attention network task (ANT). As described in Chapter 1 (and shown in Figure 1.9), the ANT contains multiple different types of cue and target displays that allow it to assess three components of attention: alerting, orienting, and executive control (inhibition of distractors). Gallotto et al. found that the alerting and executive control components were not significantly affected by **continuous theta burst stimulation (cTBS)** (a type of TMS found to be very effective at impairing cortical activity) to either the FEF, IPS, or both regions, further indicating that the role of these regions in attention is specifically related to the orienting component. Unlike some previous neurostimulation studies, orienting was not significantly modulated if either region was stimulated *alone*, but orienting was significantly affected when *both* the FEF and IPS were stimulated concurrently. There are numerous differences between this and previous stimulation studies that could account for the different results in the single-site stimulation conditions (e.g., most previous studies used simple Posner cuing paradigms that only tested orienting, whereas the ANT task is more difficult, and relatively small differences in the intensity and site of stimulation can affect when the effects of stimulation can be seen). The critical result from the Gallotto et al. study was that the same site and intensity of stimulation that produced significant effects on orienting when *both* areas were stimulated did not produce effects when only one of the areas was stimulated. This suggests that a certain level of disruption to either area can be compensated for, but when both areas are disrupted attentional orienting is significantly impaired. This provides evidence that the orienting of spatial attention is supported by the interplay between the FEF and IPS regions. Of note, this study only stimulated regions of the right hemisphere, and further research is needed to determine whether the left hemisphere attentional control regions would show the same effects.

Transcranial alternating current stimulation (tACS) is a type of neurostimulation that is very well-suited for assessing the *type of activity* that is most relevant to a cognitive process. Since tACS uses electrodes placed on the scalp, it cannot achieve the spatial precision of TMS. However, because it uses continuous electric currents, it can achieve stable and consistent stimulation at a specific *frequency*. Thus, tACS can test not just whether a region of the brain is involved in a cognitive process, but also which frequency of activity is most critical. By entraining the brain region to a particular frequency, this method can essentially prime activity in the stimulated frequency band. In a tACS study, Hopfinger, Parsons, and Fröhlich (2017) investigated the role of the right posterior parietal cortex in attentional orienting by stimulating this region while subjects performed separate exogenous and endogenous cuing tasks. The subjects' task in each condition was to discriminate whether the target (a partial circle outline that could appear briefly at either a left or right visual field location) had a "large" (2/8ths of the circumference) or "small" (1/8th of the circumference) gap in the outline. In the exogenous run, every trial began with a brief luminance change at one of the two possible locations. In the endogenous condition, an arrow instructed subjects where to attend. The endogenous arrow cue was predictive of target location (80% valid trials) and the exogenous cue was nonpredictive of target location (equally likely at each location). The stimulation frequencies of 10 and 40 Hz were chosen based on previous electroencephalography (EEG) studies of attentional cuing that have found that alpha (8–12 Hz) and gamma (>30 Hz) activity is consistently associated with aspects of attentional orienting and target detection (e.g., Doesburg, Roggeveen, Kitajo, & Ward, 2008; Gruber, Müller, Keil, & Elbert, 1999; Worden, Foxe, Wang, & Simpson, 2000). Since gamma is a wide band that encompasses many frequencies, 40 Hz was selected for this experiment based on previous EEG studies that had associated low gamma (~30–60 Hz) activity over parietal sites with the focusing and reorienting of attention (e.g., Landau, Esterman, Robertson, Bentin, & Prinzmetal, 2007; Vidal, Chaumon, O'Regan, & Tallon-Baudry, 2006). In addition to the two active stimulation conditions, Hopfinger et al. (2017) also used a "sham'" stimulation condition in which the stimulation was ramped up for the first 30 seconds, but then no stimulation occurred during the rest of the duration of each run (~225 seconds). The sham frequency was set at 25 Hz so that it was part of a different frequency band (beta) and was equally spaced from each of the active stimulation frequencies. Some subjects in tACS studies report a mild tingling of the scalp just when the currents are first turned on, so the initial ramping up period is used in sham conditions during tACS studies to ensure subjects do not realize which conditions are no-stimulation control conditions. The results from the Hopfinger et al. (2017) study showed that *gamma* stimulation over the right posterior parietal cortex significantly *improved* performance in the endogenous cuing task (Figure 7.5).

Specifically, the Hopfinger et al. (2017) study found that reaction times to the invalid targets were significantly faster during gamma stimulation compared to both the sham stimulation condition and the alpha stimulation condition (Figure 7.5). The latter finding reveals that it is not simply *any* priming of the parietal lobe that will enhance performance, but specifically priming at the *gamma* frequency. In addition, the specificity of the effect was further isolated to invalid trials because there were no differences between any of the conditions for valid trials. These results were interpreted as representing a facilitation of

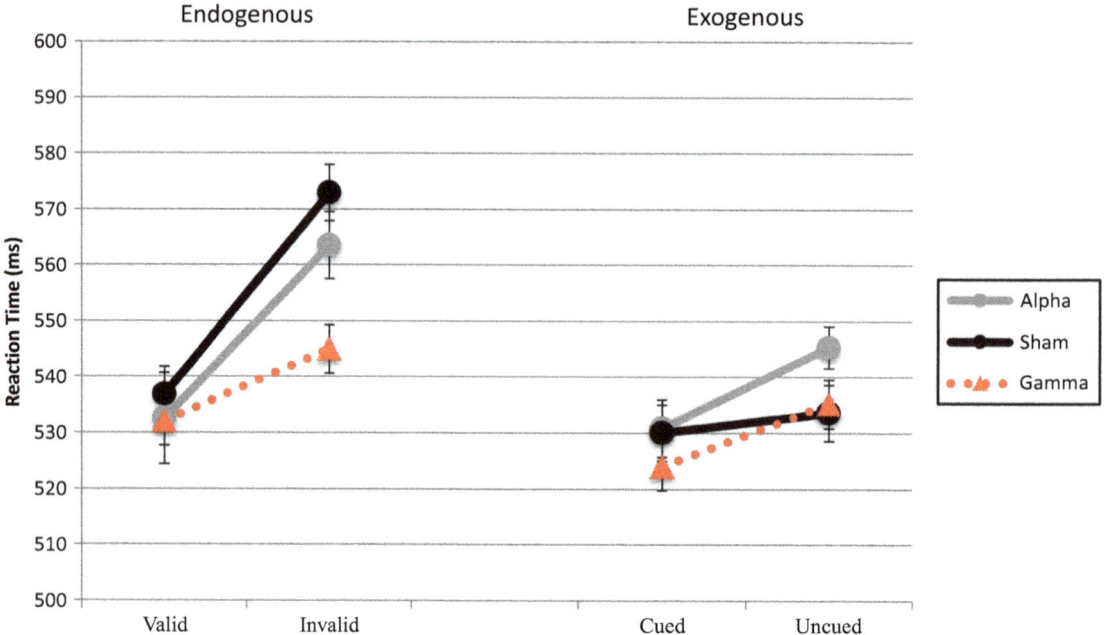

Figure 7.5 Results from a tACS study of attentional disengagement. Data from Hopfinger et al. (2017) replotted and showing the effects of tACS stimulation over right hemisphere posterior parietal regions on endogenous and exogenous attention tasks. Gamma stimulation (red dashed lines) was at 40 Hz; alpha stimulation (gray lines) was 10 Hz; the sham condition (black lines) was used as a baseline with no continuous stimulation. Gamma stimulation significantly sped up reaction times to invalid targets in the endogenous attention task compared to the sham and alpha stimulation conditions.

the "disengagement" process that occurs when attention must be moved from a location where the subject had just voluntarily engaged their attentional focus. This process has been hypothesized to be severely impaired in attentional neglect patients: When those patients orient to a stimulus in their good field, it is very difficult for them to disengage attention from that location in order to move attention to the opposite visual field. In this tACS study, priming the region of cortex that's typically damaged in neglect *boosted* the ability of healthy subjects to *disengage* attention and orient rapidly to the opposite hemifield. This improvement in performance following gamma stimulation was not seen during the exogenous orienting runs, further specifying the effect of gamma simulation to the endogenous condition. Whereas alpha stimulation had no effect in the endogenous runs, it did produce a slightly worse performance in the exogenous attention condition. Specifically, subjects were slightly slower to respond to uncued-location targets during alpha stimulation compared to the sham and the gamma stimulation conditions. This effect of alpha stimulation impairing performance may be related to previous studies that found that alpha activity in occipital regions is usually related to reduced attentional focus and worse behavioral performance (Worden et al., 2000).

7.2.3   ERP Indices of Attentional Control

EEG and magnetoencephalography (MEG) recordings have also been used to assess the brain basis of attentional control. While lacking the spatial precision of fMRI, EEG and ERPs can track the sequence of activities across the brain related to the orienting of attention (e.g., Yamaguchi, Tsuchiya, & Kobayashi, 1994). Harter and colleagues were among the first to use ERPs to track the processes associated with the orienting of attention (Harter, Miller, Price, Lalonde, & Keyes, 1989). By measuring the neural activity time-locked to instructive cues and by comparing the ERPs for attend-right versus attend-left cues, they were able to identify a sequence of components related to the orienting of spatial attention. The first effect they identified in this way was a negativity that occurred ~200–400 ms after the attention-directing cue, and they labeled this the "early directing attention negativity" (EDAN). At a longer latency (~400–700 ms after the cue) they found a positive-going difference and labeled this this "late directing attention positivity" (LDAP). Both components arise from the difference between orienting attention leftward versus rightward, and the polarity refers to the difference between the hemisphere contralateral versus ipsilateral to the shift direction. Subsequent studies replicated the finding of the EDAN over parietal–occipital scalp sites shortly following instructive cue stimuli (as early as ~200 ms; Figure 7.6, top row), and this has been interpreted as indexing the shifting process of spatial attention (Harter et al., 1989; Nobre, Sebestyen, & Miniussi, 2000; Van Velzen & Eimer, 2003). Subsequent research identified another lateralized component occurring between the typical latencies of the EDAN and LDAP. This component usually occurs ~300–500 ms after the instructive cue, and it is referred to as the "anterior directing attention negativity" (ADAN; Figure 7.6, middle row) since it occurs at more anterior scalp sites overlaying the frontal cortex (Eimer, van Velzen, & Driver, 2002; Praamstra, Boutsen, & Humphreys, 2005). This component has been thought to be related to supramodal attentional control in lateral frontal brain areas, but some authors have questioned whether it truly indexes control processes, and they have suggested that multiple brain regions may contribute to its generation (Green, Conder, & McDonald, 2008). The LDAP (Figure 7.6, bottom row) has also been replicated across studies, with its latency often extending until the expected time of the target appearance (e.g., Lasaponara, D'Onofrio, Pinto, Dragone, Menicagli, Bueti, De Lucia, Tomaiuolo, & Doricchi, 2018). The LDAP is thought to index the biasing of sensory processing areas of the occipital–temporal lobes in preparation for the selective processing of the attended stimulus (Harter et al., 1989; Hopf & Mangun, 2000) and possibly also control processes in the parietal cortex related to the maintenance of attention at a specific spatial location (Eimer et al., 2002).

Another component that has been used as a marker that attention has shifted in space is the N2pc (Luck & Hillyard, 1990, 1994a). The N2pc is one of the components that is calculated as a difference wave – in this case, it is observed as a negativity within the N2 latency range that is greater at posterior scalp sites that are contralateral versus ipsilateral to where attention has shifted. The N2pc has been used as an index that attention has been moved to a spatial location, and studies have even found successive (and inverted) N2pc components on the same trial when attention shifts first to one and then to the other visual field location in visual search tasks (Luck & Woodman, 1999). Finally, although the N2pc is typically used to measure a shift of spatial attention, Woodman and colleagues (Woodman, Arita, & Luck, 2009) showed that it requires the presence

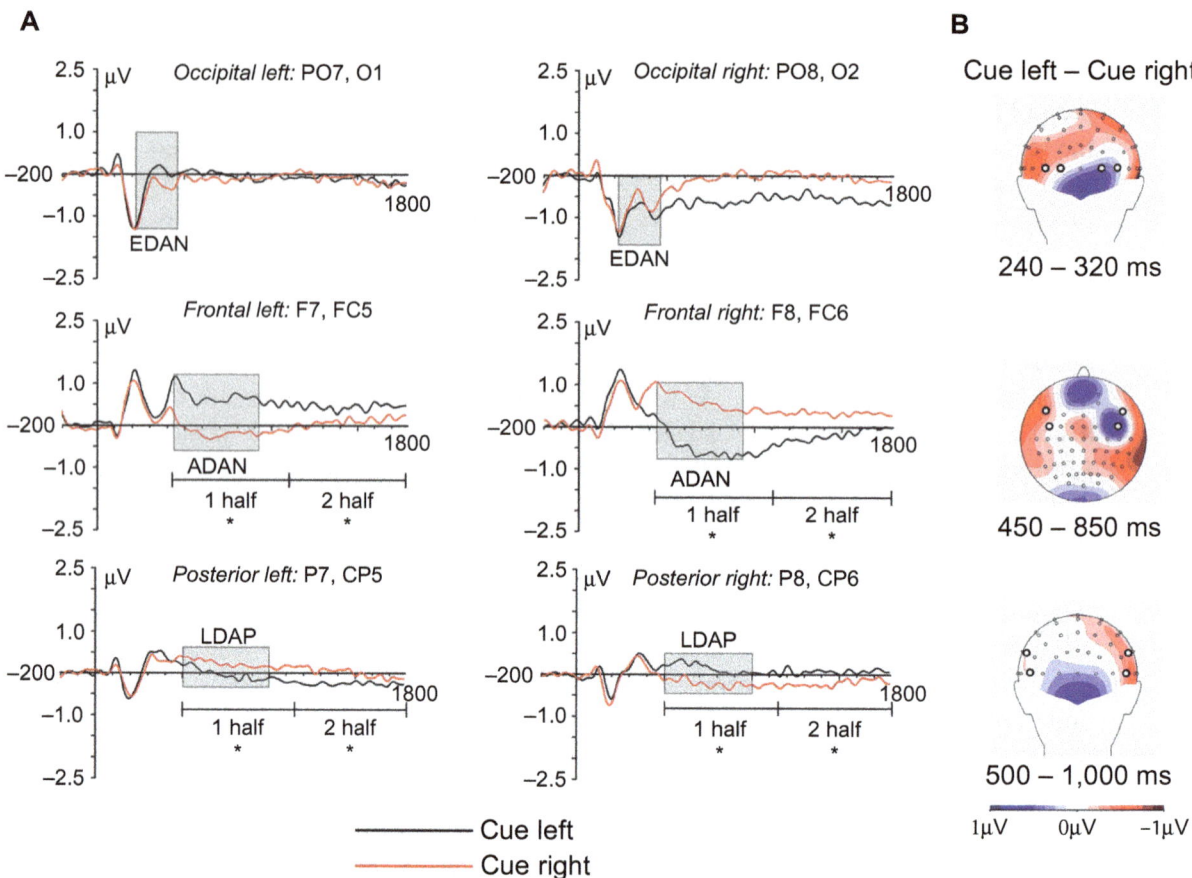

Figure 7.6 ERP indices of attentional control processes. ERP waveforms and scalp topographic maps time-locked to the cue stimuli instructing subjects where to attend. Gray squares on the ERP plots indicate the typical latency ranges for the ERP components labeled on each plot. Scalp topographies (right column) show the distribution of the components, and the black-outlined white circles on each map indicate the scalp locations of the electrodes plotted in the ERP waveforms in that row (and labeled above each ERP plot). The line below the middle and bottom row of ERP plots indicates the first half and second half of the cue period, and asterisks indicate significant differences between ipsilateral and contralateral activity. Reprinted from Lasaponara et al. (2018), *Journal of Neuroscience*. Creative Commons CC BY 4.0.

of an object. When attention had to be shifted to an empty space, no N2pc was found. In addition, the N2pc is sensitive to the presence of distractors in the display, being increased in amplitude when there are more distractors and being absent when no distractors are present (reviewed in Luck, 2012). This has led to the theory that the N2pc may be a measure of the *selection* process of attention.

7.2.3.1 EEG Studies of Internally Generated Attentional Control

Neuroscience methods are also beginning to be used to investigate the neural basis of *internally generated* shifts of attention in the absence of cues instructing subjects where to attend, as introduced in Section 7.2.1.1. Bengson and Mangun have described this type of attention as "willed attention" and have used EEG in a series of studies to investigate its brain basis. Bengson, Kelley, Zhang, Wang, and Mangun (2014) recorded EEG while subjects took part in a cuing paradigm in which they were equally likely to receive cues to attend left, attend right, or to choose which side to attend to. In this way, subjects were unlikely to decide where to shift attention before the cue appeared because, on most trials, an instructive cue would require them to shift in a specified direction. Thereby, all cue types could be used to time-lock to the process of engaging a shift of attention. To validate that shifts of attention were occurring in the willed attention condition, the study authors compared its effects to the effects of instructed cues. In all cases, the prompt to begin a shift of attention was followed a few hundred milliseconds later by a relative reduction in alpha-band activity over occipital sites contralateral (versus ipsilateral) to the direction of the attention shift, in line with previous studies showing that alpha activity is reduced over sensory processing regions coding the attended location (e.g., Worden et al., 2000). Bengson et al. (2014) further found that willed attention shifts produced the same enhancement of subsequent target-evoked processing as did instructive cue shifts (e.g., enhanced target-evoked P1 and N1 components). Therefore, both types of attention were found to produce similar *effects* on processing, and the critical question was whether willed attention involved distinct processes of attentional *control*. Indeed, this paradigm has revealed a couple of components that are unique to willed attention. Bengson et al. (2015) reported a component over the frontal cortex – at a latency of 250–350 ms after the cue – that was present for only the willed attention condition. In addition, another component unique to willed attention was found over central scalp sites at 400–800 ms after the cue. Further research is needed to determine the functions of these activities evoked specifically by the internal choice of where to attend, but these findings reveal that theories of attentional control must account for the differences between the shifts of attention triggered in the typical cuing paradigm and processes involved in generating one's own choice for where to attend. Interestingly, the processes of willed attention have also been linked to brain activity *before* the appearance of the prompt to shift attention. Bengson et al. (2015) found that alpha-band activity within ~1 second before the cue appeared was correlated with the direction in which subjects subsequently chose to shift their attention. This alpha-band activity was observed over parietal–occipital sites and was distinct from the occipital alpha effects observed after the attention shift. This pre-cue effect was interpreted as a natural fluctuation of attentional priority across locations that biases the subjects' choice of where they "want" to attend (Bengson, Liu, Khodayari, & Mangun, 2020; reviewed in Nadra & Mangun, 2023). The idea of attention obeying natural rhythmic oscillations will be discussed in more detail in Chapter 8 on the temporal dynamics of attention.

7.2.4   Single-Unit and Multiunit Studies of Attentional Control

Invasive electrophysiology methods can achieve precision in spatial and temporal dimensions that is not possible using fMRI and EEG methods. By taking single-unit and multiunit recordings from depth electrodes inserted into the brain, these methods can track neural processing in real time, and the location of the electrode can be specified with submillimeter precision. Except in rare cases of patients with severe epilepsy or tumors, these methods are only used in nonhuman animals, and most such studies of attention have used macaque monkeys due to the similarity of their visual system to that of humans. Besides this method being used primarily in only nonhuman species, another limitation is that the recordings are typically from only one or two regions at a time, as opposed to the whole-brain coverage of fMRI and EEG. However, if previous research has identified the regions most strongly involved in a cognitive process, single-unit recordings can provide critical information regarding the timing of neural activity in those regions that support the mental events of interest.

Single-unit studies of the control of attention have revealed how the dorsal frontal and parietal regions instantiate attentional control and modulate subsequent sensory processing in visual areas. Gregoriou and colleagues (Gregoriou, Gotts, Zhou, & Desimone, 2009) trained macaque monkeys to perform the attention task shown in Figure 7.7A while maintaining fixation on a central point. The monkeys would initiate a trial by squeezing a lever, at which point a visual display would appear, consisting of three Gabor patches of different colors. The central fixation point would then change to the color of one of the patches, which was the signal to shift attention to focus on that one patch. The monkey's task was to detect, as quickly and accurately as possible, when the attended patch's luminance was reduced, and then release the lever. By recording from the FEF and from a visual processing area that is known to show robust *effects* of attention (area V4), the investigators were able to assess how these areas may implement attentional control processes. It is important to note that neurons in these areas have **receptive fields (RFs)**, so the stimuli used in the experiment are specifically placed to be either inside or outside the RF of the neuron being recorded. The results showed that neurons in both the FEF and V4 show a burst of activity when attention is directed into their RF, confirming previous results that these areas are involved in the control of visual spatial attention. Critically, the results revealed that the burst of enhanced activity in the FEF (at ~80 ms; Figure 7.7B) preceded the burst of activity in V4 (Figure 7.7C) by about 40 ms. This relative timing difference between the regions provides support for the theory that the FEF initiates the shift of attention that results in the biasing of the sensory processing region. In addition to showing a temporal sequence of FEF followed by V4, this study also showed that coherence between the FEF and V4 also increased when attention was directed to the RF of the FEF neurons, providing more evidence for the functional connectivity between these regions. Furthermore, the rise in coherence between these areas was especially strong in the *gamma* frequency band (peaking at ~50 Hz in this study), in line with research described earlier suggesting that gamma-band activity is especially important for attentional control processes.

Further evidence for the causal role of the FEF in attentional control comes from studies in which the FEF was transiently deactivated. Monkey studies allow for the recording of neural activity in some regions while other areas are chemically deactivated or lesioned. Chemical

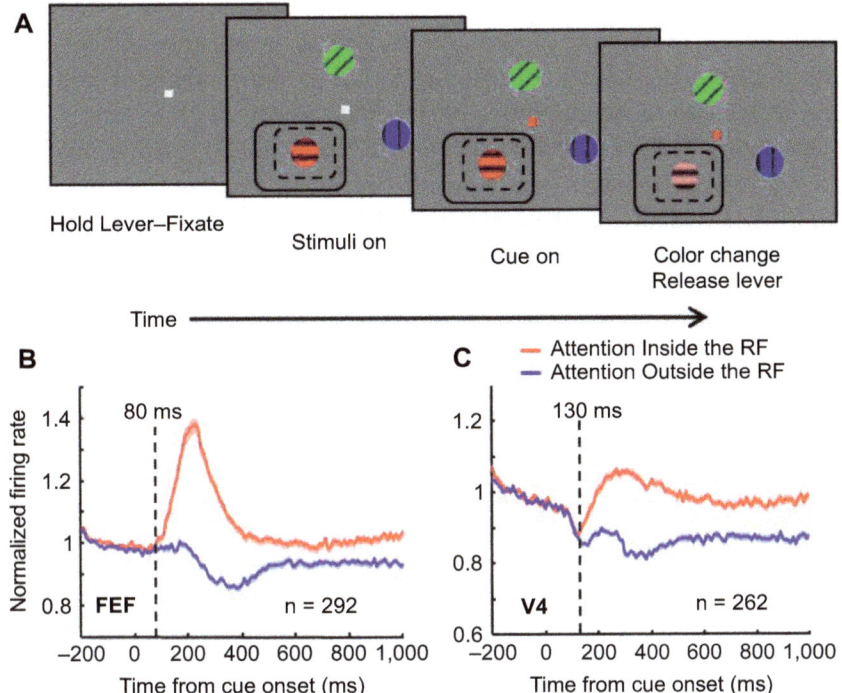

Figure 7.7 Single-unit recordings during spatial attentional orienting. (A) Depiction of task and stimuli. Solid and dashed rectangular regions around the red Gabor patch were not displayed on the screen but are shown here to indicated overlapping RFs of neurons in V4 (dashed-line rectangle) and FEF (solid-line rectangle). (B) Firing rates averaged over trials and over all cells recorded from FEF. Red line presents results from trials in which attention was directed into the RF of the cell; blue line presents results from trials when attention was directed to a location outside of the cell's RF. Vertical dashed line indicates when the cells first began to show a difference between attention within versus outside of the RF. (C) Same as in (B) but for cells in V4. Reprinted from Gregoriou et al. (2009), *Science*, with permission from the American Association for the Advancement of Science.

deactivation studies can be spatially precise, and, since it is a transient effect, the animal can be tested before and after the deactivation to ensure any effects were due specifically to that area being unavailable. When the FEF is deactivated, neural recordings from V4 show that the normal attention effects in this area are no longer produced. This provides converging evidence that the FEF forms a critical part of the attentional control network that can bias processing in early visual regions (Bichot, Xu, Ghadooshahy, Williams, & Desimone, 2019; Rossi, Bichot, Desimone, & Ungerleider, 2007).

As described in Section 7.2.1, the posterior parietal cortex is another important node in the DAN of humans. Single-unit recordings in monkeys have identified the lateral intraparietal area (LIP) – a subregion within the posterior parietal cortex – as being especially involved in the control of attention. Topographically, the LIP in the macaque is located in a similar position to the subregion of the human IPS that has been labeled "IPS2," and these regions represent visual space in a similar topographic organization (Arcaro, Pinsk, Li, & Kastner, 2011; Ben Hamed,

Duhamel, Bremmer, & Graf, 2001). Both of these regions have also been associated with movements of attention and of the eyes (reviewed in Kastner, Chen, Jeong, & Mruczek, 2017). In monkeys, the LIP is consistently found to show enhanced firing during attentional orienting, as well as in oculomotor control (e.g., Bisley & Goldberg, 2003, 2010; Gottlieb, Balan, Oristaglio, & Suzuki, 2009; Mirpour & Bisley, 2021), further linking the processes of overt and covert movements of attention. Chemical deactivation studies in monkeys have provided further evidence for the role of the LIP in attention, as covert orienting is significantly slowed during the period when the LIP has been temporarily deactivated (Davidson & Marrocco, 2000). Neurostimulation studies have provided even stronger support for the causal role of the LIP in guiding attention, as microstimulation of this region has been found to produce shifts of covert attention (Cutrell & Marrocco, 2002). These studies in nonhuman primates have confirmed the involvement of dorsal frontal and parietal areas in attentional control and have provided important insights into the causal mechanisms supporting the orienting of spatial attention.

7.2.4.1 Macaque Studies of Feature versus Spatial Attention

As discussed in previous sections, attention can be allocated based on features as well as spatial locations. Some conceptualizations of attention hold that the control processes of attentional allocation come from a common region or set of regions, regardless of whether attention is being allocated to space, objects, or features. Other theories, however, suggest that the control of feature or object attention may come from separate regions than the control of spatial attention. Yet others suggest that spatial attention must always be involved because features and objects in the outside world must be located in a specific region of space. It's not clear whether there needs to be a hierarchy in which spatial attention is always a part of feature attention or whether those systems can be independent. In a study in which multiunit recordings were collected across several brain regions while macaque monkeys performed a visual attention task, Bichot and colleagues (Bichot, Heard, DeGennaro, & Desimone, 2015) were able to assess the relative time course and influence of multiple frontal regions. On each trial, the monkeys were initially presented with a lone object at fixation that was the target in the upcoming visual search array. The search array consisted of eight complex objects distributed across both visual fields, and the monkeys had to quickly make a saccade to the target item. For each set of cells being recorded, the display was carefully set up to ensure that a cell's RF sometimes included the target location and other times included a distractor item. Thus, this study was able to assess both object and feature attention. The regions recorded from included the FEF, a higher-order visual processing region known to be involved in object perception (the inferior temporal cortex (IT)), and the ventral prearcuate gyrus (VPA), a more ventral region of the prefrontal cortex (PFC) that has been found to be selectively active during object-based attention. The results confirmed previous findings that a burst of activity in the FEF precedes spatial attentional effects in visual processing regions, as increased firing in the FEF was observed in this experiment *before* increased firing in IT. Furthermore, for spatial attention, the burst of activity in the FEF preceded activity in the VPA as well, providing further support that the FEF plays a primary role in the control of *spatial* attention. For feature attention,

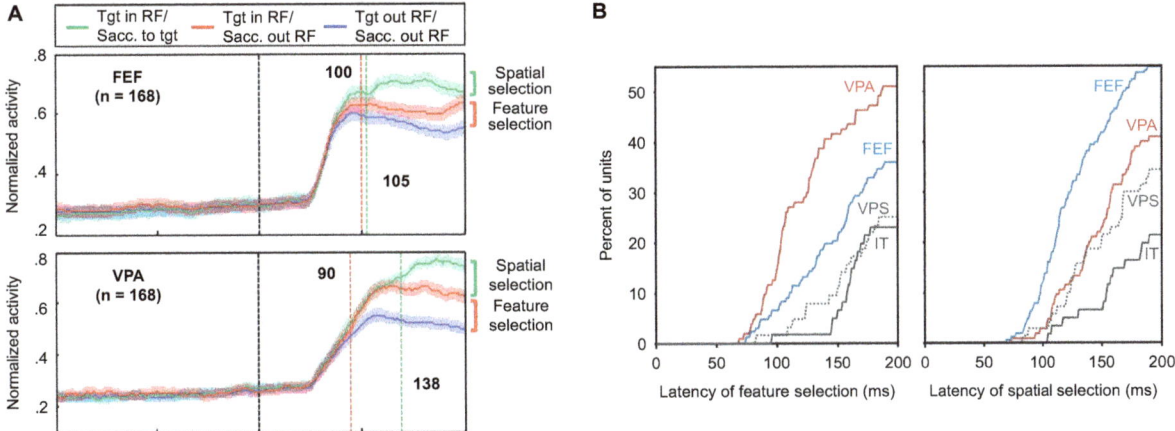

Figure 7.8 Multiunit recordings of space-based versus feature-based attentional control. (A) Normalized activity averaged over all trials and cells in the labeled areas. These plots show a window of 400 ms, with the black dashed line in the middle of each plot indicating the onset of the visual array. Green time course lines show correct trials in which the first saccade was made to the target and that target was in the cell's RF. Red lines show trials in which the target was in the RF but the monkey made the first saccade to a distractor stimulus outside of the cell's RF. Blue lines show trials in which the target was outside the cell's RF (with a distractor in the cell's RF) and the monkey made a saccade to a distractor outside its RF. The vertical red and green dashed lines indicate the onset of feature-based selection (red vertical line) and space-based selection (green vertical line). (B) Plots of the cumulative percentage of cells in each area according to the latency of when effects of feature attention (left plot) or spatial attention (right plot) were observed in those cells. The IT is a visual processing area involved in object recognition. The ventral bank of the principal sulcus (VPS) is another area of the frontal cortex. Figure adapted from Bichot et al. (2015), *Neuron*, with permission from Elsevier.

however, the VPA showed an increase in firing before the FEF and IT (Figure 7.8). Thus, the timing of these effects suggests that the FEF is responsible for spatial attention effects in the VPA and IT, whereas the VPA may be responsible for feature-based attention effects in the FEF and IT.

It should be noted that in this study both the FEF and VPA eventually showed effects of both spatial and object attention, supporting the view that these areas interact and that attention to objects in the real world involves both object- and space-based attention. As shown in Figure 7.8, the effects of spatial attention are observed first in the FEF and only later in the VPA, whereas the opposite order is seen for the control of object attention. Further evidence for the specific roles of the FEF and VPA in spatial and object attention comes from deactivation experiments (Bichot et al., 2015, 2019). During deactivation of the VPA, responses in the FEF no longer showed feature attention effects but still showed spatial attention effects. This suggests that these two frontal regions interact but can perform their primary function alone (i.e., the FEF shows effects of spatial attention even when the VPA has been deactivated). However, the FEF completely loses the ability to code *feature* selection without input from the VPA. The role of the VPA in producing feature-based attention effects in early visual processing regions is also strongly supported by the finding that *feature* attention effects in V4 are *eliminated* when the VPA is deactivated but are not impaired at all by deactivation of the FEF (Bichot et al., 2019). *Spatial* attention effects in V4, however, are dependent upon the FEF and are severely impaired by deactivation of that region. Human MEG studies have provided converging evidence that different frontal attentional control regions are involved in spatial versus feature attention. Baldauf and Desmione (2014) localized MEG activity indexing object attention to a ventral region of the frontal lobe in the inferior frontal junction (IFJ). In a study in which subjects could shift attention between two transparent overlapping objects (a face and a place on each trial), activity in the IFJ preceded the effects of attention in higher-order visual processing regions specialized for the processing of faces (the fusiform face area) and places (the parahippocampal place area). Overall, these data provided further evidence that the regions of the frontal cortex that support object- and feature-based attentional control are separate from the FEF region involved in the control of spatial attention.

7.3   The Ventral Attention Network

In addition to the DAN that is involved in the processes of orienting and engaging attention, more ventral regions of the parietal and frontal lobes have been implicated in other mechanisms of attentional control. Neuroimaging studies have revealed that inferior parietal regions extending into the STG and the temporoparietal junction (TPJ), as well as lateral areas of the ventral prefrontal cortex (VFC) and inferior frontal gyrus (IFG), are active when attention must be rapidly *disengaged* from its present focus. The regions that are thought to be involved in this effortful *reorienting* of attention are referred to as the **ventral attention network (VAN)** (reviewed in Corbetta & Shulman, 2002). This network is sometimes called the "ventral frontoparietal network," and it has also been linked to the "saliency network" of the brain that includes the anterior insula and cingulate cortex. Early work suggested that the VAN, and the TPJ in particular, may be specifically involved when attention had been exogenously captured by a highly salient stimulus at a new location and spatial attention had to be reoriented back to the desired location. However, subsequent work revealed that the TPJ is also active when there are unexpected changes to features or objects, even at a single location where spatial attention is already focused (Marois, Leung, & Gore, 2000). Furthermore, surprising events within any sensory modality trigger the TPJ (Downar, Crawley, Mikulis, & Davis, 2000). Therefore, the TPJ is thought to play a role in the rapid processing of unexpected events of all sorts, disengaging attention from its current focus to process potentially important task-relevant information. Another interesting aspect of research into the functions of the TPJ is that most of the data come from human studies, as it is not clear what the homologous areas might be in nonhuman animals. Human neuroimaging studies have revealed that the TPJ is also active in processes of social cognition, such as in "theory of mind" (ToM) experiments in which the subject must infer the contents of another person's mind (or an animated character on a screen) based on observing their current actions and their history (reviewed in Decety & Lamm, 2007). When making ToM judgments, the subject must distinguish between their own knowledge versus what the person being observed would know in a certain situation. It has been suggested that this switch in perspective, especially when trying to make sense of another person's surprising actions, relates to the role of the TPJ in processing unexpected events and disengaging attention from its current focus in order to attend to other critical information (Corbetta, Patel, & Shulman, 2008).

Although most neuroimaging studies find the VAN to be lateralized more – or sometimes completely – to the right hemisphere, other studies have found the left TPJ to also be involved in the processes of disengaging attention and processing unexpected stimuli (Downar, Crawley, Mikulis, & Davis, 2000; Geng & Mangun, 2011; Vossel, Weidner, Thiel, & Fink, 2009). The exact role of the TPJ may differ between the hemispheres, however. Comparing this region across hemispheres, Doricchi and colleagues found that the right TPJ was more strongly connected to the insula, whereas the left TPJ was more strongly connected to the IFG (Doricchi, Macci, Silvetti, & Macaluso, 2010). This study also found that the right TPJ was more active to invalid trials compared to validly cued trials, in line with the suggestion that the TPJ is involved in processing unexpected events, whereas the left TPJ was equally active to both types of trials.

Connectivity studies have also provided evidence for interactions between the DAN and the VAN. Analyses of white matter tracts have revealed not only separate bands of tracts that connect only dorsal frontoparietal areas and ventral frontoparietal areas, but also a band of tracts connecting the parietal regions of the VAN to the frontal regions of the DAN (de Schotten, Dell'Acqua, Forkel, Simmons, Vergani, Murphy, & Catani, 2011). Although there are separate functions of the DAN and VAN, these networks must also interact to support optimal attention control. One way in which these regions interact is that activity in the TPJ is actively suppressed during highly focused, top-down-guided spatial attention when the DAN is highly active (e.g., Shulman, Astafiev, McAvoy, D'Avossa, & Corbetta, 2007). Using DCM of directed connectivity, DiQuattro and Geng (2011) found that the FEF region of the DAN inhibited activity in the TPJ. When highly salient, potentially task-relevant stimuli appear unexpectedly, however, activity throughout the VAN is triggered, and this activity is thought to interrupt the current focus of attention and subsequently reengage the DAN to orient to the desired focus of attention. Such results have been interpreted as the VAN serving as a "circuit breaker" in processing unexpected events and disengaging attention from its current focus, whereas the DAN is more involved with the top-down control and guidance of attention according to current goals (e.g., Corbetta & Shulman, 2002).

7.3.1 Neuropsychological Evidence for the Roles of VAN Areas

In a landmark study that included over 170 patients with right hemisphere lesions tested within 48 hours of their stroke, Medina and colleagues provided evidence for the critical role of different parietal–temporal areas in subtypes of unilateral neglect (Medina, Kannan, Pawlak, Kleinman, Newhart, Davis, Heidler-Gary, Herskovits, & Hillis, 2009). They found that spatial neglect (also known as "viewer-centered" or "egocentric" neglect) was associated more with lesions to the right supramarginal gyrus (SMG) in inferior parietal regions, whereas object-based neglect (also known as "stimulus-centered" or "allocentric" neglect) was associated more with lesions to the posterior inferior temporal gyrus. For example, when asked to copy a simple line drawing scene consisting of a house, fence, and a couple trees, patients with *spatial* neglect will copy the right side of the scene fine but will not copy anything from the left side of the page. In contrast, patients with *object-based* neglect will copy items from across the entire page but will only copy the left half of each object. In another task, called the gap test, patients are presented with a sheet of paper with both circles with a gap on one side and complete circles distributed evenly across the page. The task is to circle the complete circles and put an "X" in all partial circles that contain a gap. Patients with *spatial* neglect will perform well on the right side of the page but will ignore all the items on the left side of the page. Patients with *object-based* neglect, in contrast, will make a mark on all items across the whole page but will make errors on the circles with gaps on the left side – those will be mistakenly circled by the patient, indicating that they are ignoring the left side of those objects. It should be noted that most such patients exhibit some signs of both space-based and object-based neglect, and of the 171 patients tested in this study, only 32 were identified as having purely space-based neglect (without evidence of object or stimulus-centered neglect), and only 12 were identified with the opposite pattern. Thus, some previous studies of neglect may have contained a mixture of these two forms of neglect, but through a large sample size and strict inclusion criteria it is possible to identify the neural structures that are most critical for these different types of attention.

7.3.2   ERP Studies of Disengaging Attention Following Attentional Capture

Studies using ERPs have also identified components that have been linked to the reorienting processes of attention. Hopfinger and Mangun (2001) identified a component over lateral temporal–parietal sites that was generated only in response to invalid cues in an exogenous attentional capture task. This component was labeled the "ipsilateral invalid negativity" (IIN) because it was a *negativity* observed at electrodes *ipsilateral* to the target location, ~200–250 ms after an *invalidly* cued target. Being ipsilateral to an invalidly cued target means that it is contralateral to the peripheral cue that captured attention to the wrong location. Thus, the IIN has been interpreted as a part of the processes involved with disengaging attention from the wrong location where attention had just been captured in order to reorient attention rapidly to the location of the target. In studies using the contingent capture paradigm developed by Folk and colleagues (Folk, Remington, & Johnston, 1992), in which the noninformative cue display matches or mismatches the target dimension in different runs (e.g., a single onset item or a unique color item among multiple onsetting items), Hopfinger and Ries (2005) found that the IIN was uniquely tied to the *match* between cue and target displays. Specifically, whereas other ERP effects of attentional capture (e.g., enhanced P1 and P3 to cued-location targets) were found to be automatic and linked to the *salience* of the cue, the IIN was linked to *contingent capture*, being observed even following nonsalient cues *if* the cue matched the target type. Shin and colleagues (Shin, Hopfinger, Lust, Henry, & Bartholow, 2010) used a variation of the dot-probe paradigm to investigate whether subjects at higher risk of developing alcohol dependency would show unique EEG indices of differential attentional capture to images of alcoholic drinks relative to images of nonalcoholic drinks. In this experiment, a brief display of two peripheral images (one in each hemifield) preceded the display of a target checkerboard; the visual images preceding the display were completely nonpredictive of the target type or location. The results showed that the most robust difference between high- and low-risk subjects was the presence of an IIN only in the high-risk group, evoked following invalid cues of alcohol-related images. This suggests that the high-risk group was more likely to be captured by the noninformative, task-irrelevant alcohol-related images (relative to the nonalcohol-related images), and that these subjects needed to effortfully disengage attention from those cues to reorient to the target.

7.3.3 Neurostimulation Studies of the VAN

Transcranial neurostimulation techniques have further advanced our knowledge of the timing and functioning of brain regions in the VAN. Some early TMS studies showed that disrupting posterior parietal areas in the right hemisphere led to a rightward bias on attention tasks, similar to the pattern of performance in unilateral neglect patients (e.g., Ellison, Schindler, Pattison, & Milner, 2004; Muggleton, Postma, Moutsopoulou, Nimmo-Smith, Marcel, & Walsh, 2006). Subsequent studies further localized the critical region to the SMG – a portion of the inferior parietal lobe – in the right hemisphere (Oliveri & Vallar, 2009). More recently, Shah-Basak and colleagues (Shah-Basak, Chen, Caulfield, Medina, & Hamilton, 2018) used a form of repetitive transcranial magnetic stimulation (rTMS) to inhibit the activity in the STG of healthy young adults while they performed a version of the line bisection task used with neglect patients. In this version of the task, subjects were shown horizontal lines, one at a time, that had a vertical line drawn through them somewhere along their length. Subjects had to judge whether the vertical line was drawn in the very center of the horizontal line (bisecting it into two even halves) or whether the vertical line was drawn to the left or to the right of the midline. Compared to prestimulation control runs, accuracy on the task was significantly impaired when the STG was suppressed by rTMS. During this condition, subjects made more errors compared to a control condition in which stimulation was delivered to a central location (vertex) on top of the head. Errors were most increased for the lines that had been bisected too far to the right, indicating that suppressing the right STG resulted in the farthest left sides of the lines being ignored. This pattern aligns with the behavior of unilateral neglect patients, who ignore the left side of space, or objects, following damage to the right STG (e.g., Hillis, Newhart, Heidler, Barker, Herskovits, & Degaonkar, 2005; Verdon, Schwartz, Lovblad, Hauert, & Vuilleumier, 2010).

In addition to allowing researchers to test whether a specific region is necessary for a cognitive function, TMS can also be used to investigate exactly *when* an area is involved in a mental process. Most forms of rTMS are designed to essentially create "temporary lesions" that last a short period of time following the extended stimulation period. These methods are useful for ensuring that a region is unable to perform normally for a few minutes, allowing the experimenters to use different tests during the period of time when the brain area is affected to test its effects. However, since individual TMS pulses are created very rapidly and the effect of a single TMS pulse lasts only a few milliseconds, "single-pulse TMS," when combined with a tightly coordinated experimental task, can reveal at exactly what time (or times) a region is needed to perform a task. Chambers, Payne, Stokes, and Mattingley (2004) used this method of TMS to investigate whether and when different regions of the parietal lobe are involved in the reorienting of spatial attention following an invalid cue. In their study, healthy young adults engaged in a typical exogenous attention cuing paradigm in which a single nonpredictive peripheral cue occurred a short time before a target was presented at one of two possible target locations (one in each visual hemifield). The researchers stimulated, one at a time, two different subregions in the inferior parietal lobule: the SMG and the angular gyrus (AG). Each of these regions was stimulated in each hemisphere during separate runs. On each trial, the single-pulse stimulation was delivered at a randomly selected time between 30 and 360 ms after the target appeared, and the main focus of the study was on how the stimulation affected performance on *invalid* trials. This was done to test the theory that these inferior parietal regions were involved in the *disengagement* and reorienting of attention that is needed when a target occurs at the opposite location to where an exogenous cue has just captured attention. Across trials, it was then possible to assess when these areas were critically involved in performing the reorienting process. The results showed that stimulation of the SMG had no effect on performance, regardless of the hemisphere being stimulated. However, stimulation of the AG in the right hemisphere significantly impaired performance. Of note, stimulation of the AG in the *left* hemisphere had no effect on performance. This finding provided converging evidence that this region of the inferior parietal lobe, and specifically in the right hemisphere, is critical for the reorienting of attention. Furthermore, this study revealed that the right AG was critical during *two* separate, noncontiguous time periods following the appearance of the invalid target. As shown in Figure 7.9, stimulation of the right AG impaired performance from 90 to 120 ms after the target, and again at 210–240 ms. At all the other times tested, there was no effect of stimulating the AG. These results provide important temporal specificity for identifying *when* this region is critically involved, and they suggest that visual information is sent to the AG via both a "fast" and a "slow" pathway. It has been suggested that higher processing areas, such as in the parietal lobe, may receive more rudimentary visual information quickly via a fast route through the superior colliculus (SC), whereas the information from the more highly detailed processing through the lateral geniculate nucleus (LGN) and occipital–temporal visual areas arrives in the parietal lobe at a slightly later time (via a "slow pathway"). This study reveals that disrupting processing in the AG at either of these times impairs subjects' ability to quickly reorient attention to the correct target location.

TMS has also been used to investigate inhibition of return (IOR), the phenomenon in which the cued location is inhibited for a short time after the initial facilitation that occurs following an exogenous capture of attention to the cued location. Using the standard rTMS procedure that disrupts activity for a few minutes at a time, Van Koningsbruggen and colleagues (Van Koningsbruggen, Gabay, Sapir, Henik, & Rafal, 2010) found that stimulating the anterior intraparietal cortex diminished IOR. This effect was lateralized, as stimulation of the right parietal cortex reduced IOR in both visual fields, whereas stimulation of that region in the left hemisphere had no effect on IOR (in either visual field). These results support the theory that attention is more strongly lateralized to the right hemisphere, possibly due to the right hemisphere representing shifts of attention to both visual fields. TMS experiments have also identified other brain areas as being critical for the generation of IOR. For example, IOR was found to be disrupted by TMS to the FEF (Ro, Farnè, & Chang, 2003) and to the TPJ (Chica, Bartolomeo, & Valero-Cabré, 2011). Similar to the IPS results described earlier, TMS studies of IOR have provided evidence for a hemispheric specialization of the TPJ, as IOR was

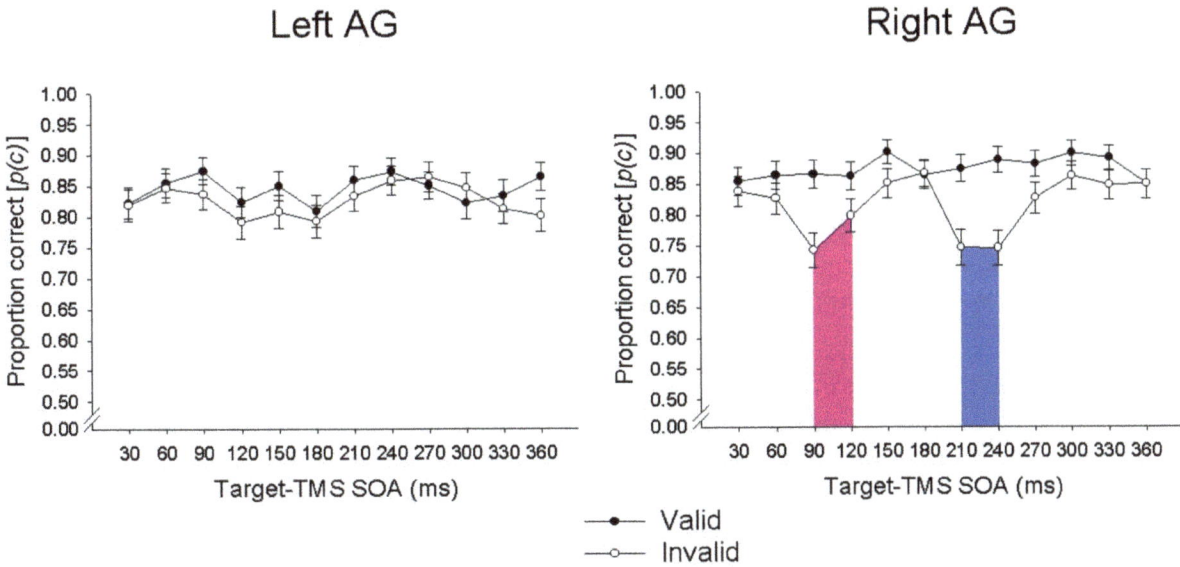

Figure 7.9 TMS study of inferior parietal involvement in attentional control. Results from a study in which single-pulse TMS was delivered to the AG at a random time between 30 and 360 ms following the appearance of a peripheral target stimulus. Results from stimulation of the AG in the left hemisphere are shown in the left plot; stimulation of the right hemisphere AG is presented in the right plot. Accuracy on target discrimination was impaired only by stimulation of the right AG, and only during two brief time windows (highlighted in pink and blue) following targets that had been invalidly cued by the preceding nonpredictive peripheral cue. SOA = stimulus onset asynchrony (here referring to the time between the onset of the target stimulus and the time of the TMS pulse). Reprinted from Chambers et al. (2004), *Nature Neuroscience*, with permission from Springer Nature.

significantly affected by disruption of the TPJ in the right hemisphere, but TMS to the left TPJ had no effect on IOR (Bourgeois, Chica, Valero-Cabré, & Bartolomeo, 2013).

Finally, neurostimulation studies of the VAN have also investigated whether this network is specific to visual attention or whether it represents cross-modal attentional control. In a study investigating whether the inferior parietal lobe is equally involved in tactile and visual attention, Chambers and colleagues (Chambers, Stokes, & Mattingley, 2004) used TMS to disrupt activity in the right SMG during an endogenous cuing task. In the task, visual arrow cues presented at fixation predicted (with a validity of 75%) the location of a subsequent target that was presented either as a visual or vibrotactile stimulus. Stimulating the SMG in the right hemisphere during the presentation of the visual cue impaired orienting to visual targets but not to tactile targets. Such results suggest that the role of the SMG in *endogenous* attentional control may be specific to the visual modality. In a study investigating exogenous cuing, however, evidence was found supporting a supramodal attentional control mechanism in inferior parietal regions (Chambers, Payne, & Mattingley, 2007). In that study, subjects performed a discrimination task on visual or tactile targets, but the cues were now salient, nonpredictive, lateralized stimuli delivered as either tactile or visual cues. Disruption of the SMG or AG area during the cue period significantly reduced the spatial cuing effect for both tactile and visual targets. These results provide support for the view that at least some attentional control mechanisms are cross-modal – in this case, that inferior parietal regions support exogenous attentional orienting across visual and tactile modalities.

7.4   Subcortical Attentional Control Regions

7.4.1   Superior Colliculus

In addition to the cortical areas that support attentional control processes, subcortical regions have also been found to be involved in orienting attention. Two of the subcortical regions most often implicated are the SC and the thalamus. The SC is well-known to play a central role in producing saccadic eye movements. During tasks that require a monkey to move its eyes voluntarily, single-unit recordings consistently find a burst of activity in the SC following the cue to move the eyes, and the activity in the SC remains strong until the eye movement is executed (Kustov & Robinson, 1996). In contrast to recordings from parietal areas, however, which show equally strong bursts for both overt and covert attention, SC neurons do not show as strong an increase in activity when attention is only moving covertly in response to instructive central endogenous cues. However, human patient studies have revealed that the SC plays a critical role in the IOR pattern typically observed following exogenously triggered orienting. Human patients with lesions to the SC are exceedingly rare, due in part to it being located close to other subcortical structures that often cause death if damaged. However, in a few cases in which small lesions selectively damaged the SC, its role in attention has been evaluated. Sapir and colleagues (Sapir, Soroker, Berger, & Henik, 1999) studied a patient who had sustained a small unilateral lesion subsequent to a focal hemorrhage that included the SC in the right hemisphere. Using a standard cuing paradigm with nonpredictive peripheral cues, the authors found that the patient showed the typical pattern of attentional capture (faster manual reaction times to cued-location targets than to uncued-location targets) at short cue-to-target intervals. However, at the longer intervals when IOR is usually found the patient showed no evidence of IOR in their affected left visual hemifield (contralateral to their SC lesion). The patient showed IOR in the right hemifield, which was represented by the intact left SC. These results provide evidence that the SC plays a critical role in the generation of IOR following exogenous cues. Sereno, Briand, Amador, and Szapiel (2006) further reported evidence that overt orienting of exogenous visual spatial attention was also impaired in an individual with an SC lesion. Specifically, exogenously triggered eye movements were impaired, despite voluntary saccades remaining largely intact. This patient also showed no IOR, in confirmation of Sapir et al.'s (1999) conclusion that the SC is necessary for IOR. Of note, another rare patient study reported evidence that IOR can be dissociated from exogenous attentional capture. In a study of a patient with congenital ophthalmoplegia, who could not make eye movements (due to undeveloped eye muscles), Smith and colleagues (Smith, Rorden, & Jackson, 2004) found that the patient was able to covertly orient attention in response to informative cues presented at fixation but showed no signs of attentional capture by peripheral cues. Critically, however, although the uninformative peripheral cues did not cause an initial reflexive orienting to their location, these cues did generate IOR.

7.4.2 Reticular Nucleus of the Thalamus

Another subcortical region that has been proposed to play a critical role in controlling attention is the thalamus. Crick (1984) originally proposed that the *reticular nucleus* of the thalamus was a critical node in the attention processes of selection and the binding of features. The reticular nucleus surrounds the lateral portion of the thalamus, and cortical connections to and from the other regions of the thalamus must pass through this structure. In addition, this nucleus has bidirectional connections with other thalamic nuclei. Thus, Crick suggested it was ideally suited to perform functions of attention that require coordination across dimensions and modalities. In addition, since this nucleus is composed of inhibitory neurons, it could also perform the sort of modulatory process that produces the selective processing of attended versus unattended sensory inputs. McAlonan, Cavanaugh, and Wurtz (2006) provided support for this idea through studies recorded from cells in this region in monkeys. Their results showed that shifts of visual attention were associated with activity in the reticular nucleus and with corresponding inhibition of sensory processing of ignored input. Other studies have further mapped out the connections from prefrontal regions to the reticular nucleus (Zikopoulos & Barbas, 2006), providing additional evidence for its widespread connectivity. Research has also suggested that the attention issues in some clinical groups may be related to the functioning of this region. Due to the very close proximity of this nucleus to other regions of the thalamus, very precise stimulation methods are required to ensure that only this nucleus is being stimulated. As reviewed in Young and Wimmer (2017), new methods of optogenetic stimulation of the reticular nucleus in mice have been found to affect performance on attention tasks in a pattern that could potentially be related to impairments in individuals with schizophrenia.

7.4.3 Pulvinar Nucleus of the Thalamus

The pulvinar nucleus of the thalamus is an interesting subcortical structure, in that it is the largest nucleus within the thalamus of primates, and it shows evidence of the same evolutionary expansion across mammals and primates as the PFC (Baldwin, Balaram, & Kaas, 2017; Smaers, Gómez-Robles, Parks, & Sherwood, 2017). Therefore, much of the most relevant research on the pulvinar nucleus' role in attentional control comes from human and nonhuman primate studies (but see work by Yu, Li, Stitt, Zhou, Sellers, & Frohlich (2018) suggesting that the ferret provides another good model for investigating the role of the pulvinar nucleus in attention). Some early work on the pulvinar nucleus in cats, however, first revealed that this structure is critical for the coordination and modulation of oscillatory activity between cortical areas (Wrobel, Ghazaryan, Bekisz, Bogdan, & Kaminski, 2007). These findings have been replicated and extended in recordings from macaque monkeys, supporting the theory that the pulvinar nucleus is involved in the synchronization of neural activity with shifts of attention (Saalmann, Pinsk, Wang, Li, & Kastner, 2012). Monkey studies have also demonstrated that neural firing in V4 and synchrony, specifically in the gamma band, are reduced across visual areas when the pulvinar nucleus is deactivated (Zhou, Schafer, & Desimone, 2016). Recent single-unit recordings in ferrets have provided additional specificity regarding the functions of the pulvinar nucleus in attentional control. Stitt and colleagues (Stitt, Zhou, Radtke-Schuller, & Fröhlich, 2018) found that the type of synchrony between the pulvinar nucleus and the posterior parietal cortex depended on the level of arousal. During high arousal, the synchrony was strongest in the theta band, whereas enhanced synchrony in the alpha band was observed during states of low arousal. Subsequent work in macaques has shown a similar pattern of coordinated rhythmic activity between the FEF, LIP, and pulvinar nucleus (Fiebelkorn, Pinsk, & Kastner, 2019). Overall, these patterns of activity and synchronization have led to theories that the pulvinar nucleus may work in part to suppress external sensory input during low arousal and to enhance attentional selection during high arousal (e.g., Casanova & Chalupa, 2023).

The importance of the pulvinar nucleus to the functioning of attention is also supported by work with human neuropsychology patients. Karnath and colleagues (Karnath, Himmelbach, & Rorden, 2002) suggested that the pulvinar nucleus would be implicated in unilateral neglect because of its strong connections to the parietal and temporal regions that are most strongly associated with the neglect syndrome. Indeed, in a study of rare patients with lesions restricted to the pulvinar nucleus, Arend, Rafal, and Ward (2008) found that a patient with damage to the anterior pulvinar nucleus exhibited profound deficits in tests of spatial attention. Interestingly, the patient didn't show deficits on temporal attention tasks, in contrast to another patient with a lesion restricted to the posterior pulvinar nucleus, who showed significant deficits in the temporal attention tasks but much less of a deficit in spatial attention tasks. In a subsequent study, Snow and colleagues (Snow, Allen, Rafal, & Humphreys, 2009) reported that these patients were able to perform a target discrimination task when the target was presented alone, but their performance was significantly impaired when salient distractors were present in the visual display. This lends further support to the view that the pulvinar nucleus is especially important for the selective processing of attention needed when irrelevant stimuli must be actively suppressed.

7.5   Executive Control: Processes, Regions, and Networks

The term "attentional control" usually refers to the specific mechanisms involved in the orienting and reorienting of the focus of attention, be it across space or between features or objects. However, as discussed in earlier chapters, attention can also refer to the processes of coordinating higher-order plans and decision-making. These aspects of attention are often referred to as "executive control," and they include things such as planning and executing complex behaviors, monitoring ongoing task performance, correcting mistakes, selecting among possible actions, and using inhibitory processes to overcome habitual responses. According to some theories, these various aspects of control may be coordinated by a *supervisory attention system* (SAS; Norman & Shallice, 1986), whereas other work seeks to understand each of these functions independently. The following subsections will describe a few of the major executive functions and what cognitive neuroscience studies are revealing about the neural systems and processes supporting these high-level components of cognitive control.

7.5.1   "Supervisory Attention" and the Anterior Cingulate Cortex

Norman and Shallice (1986) proposed that a SAS was critically involved in executive functioning, specifically the coordination of thoughts and actions. According to their theory, the response to simple environmental situations that are frequently encountered can be initiated and executed through perception–action schemas – or scripts – that specify the sequence of actions necessary to complete a task. Such schemas allow us to quickly complete well-learned tasks without much cognitive effort. In some situations where multiple schemas may be activated, however, Norman and Shallice's model proposed that "contention scheduling" processes coordinate the order of responses. This contention scheduling is thought to be a lower-level process in which activation of one action schema briefly inhibits other competing schemas to ensure efficient responses without needing to engage in cognitively demanding deliberation and conscious decision-making. However, in other situations, contention scheduling isn't sufficient. Such situations include: when a well-learned response is no longer appropriate and must be actively inhibited; when a novel situation is encountered for which there is no established schema; when preventing errors becomes more critical than quick responding; or any time when a deliberate planning of actions is required. Compared to well-learned schemas, scripts, and the action of contention scheduling, the SAS is a relatively slow but flexible method of control that involves voluntary, conscious effort.

One of the main areas implicated in an SAS is the ACC. The cingulate cortex is the central, medial portion of each hemisphere, directly above the corpus callosum. The cingulate gyrus stretches over the entire length of the corpus callosum, extending around and beneath the most anterior and posterior sections of the corpus callosum as well. The cingulate cortex has been divided into subregions based on location and anatomical connections. As shown in Figure 7.10, the most frontal portions of the cingulate are labeled as the ACC, with a further division between the most anterior portion (the rostral ACC) and a more posterior portion (the caudal ACC). The half of the cingulate cortex posterior to the caudal ACC is labeled as the posterior cingulate cortex (PCC), and this region has also been divided into subregions. A more anterior and dorsal region of the PCC is involved in internally directed cognitive processing (e.g., mental arithmetic, ToM), and a more posterior and ventral region of the PCC is involved in memory-related functions (Rolls, Wirth, Deco, Huang, & Feng, 2023; Vogt, 2019). Recent reviews have suggested that the PCC can be divided into three subregions, with the retrosplenial cortex surrounding the most posterior and ventral region of the corpus callosum being a separate region of the PCC that is more specifically involved in linking episodic memory, perception, and spatial navigation (Foster, Koslov, Aponik-Gremillion, Monko, Hayden, & Heilbronner, 2023). While these areas of the PCC are important for many cognitive functions, the executive processes more directly related to attention and supervisory control have been associated with the regions of the ACC.

The ACC has been found to be active across a wide variety of tasks, and it is densely connected to regions throughout the cortex, making it an ideal candidate for integrating and coordinating control processes (Allman, Hakeem, Erwin, Nimchinsky, Hof, & Hixon, 2001;

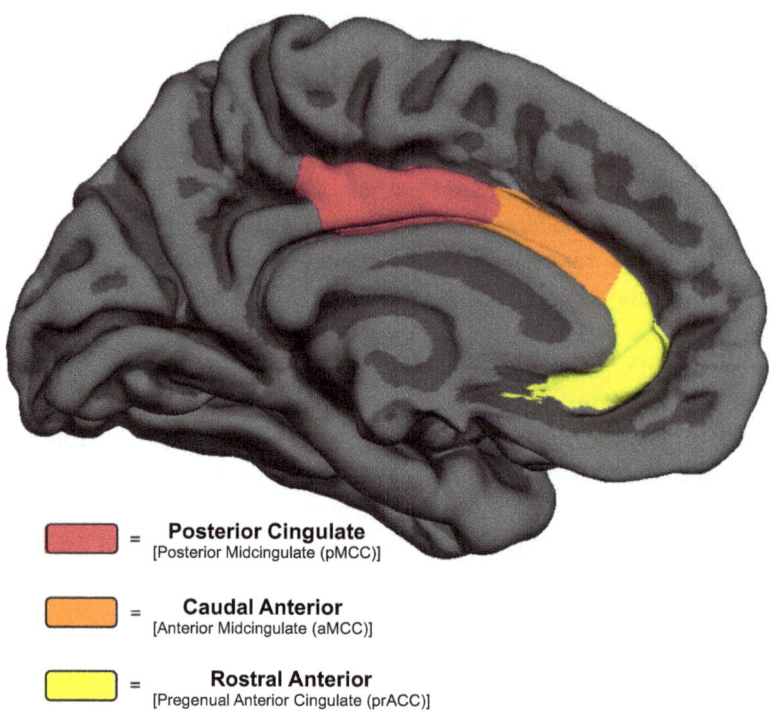

Figure 7.10 Major subdivisions of the cingulate cortex displayed on a 3D rendering of the brain. The ACC is subdivided into two portions: the rostral anterior cingulate (colored yellow) and the caudal anterior cingulate (colored orange). The PCC is colored red. Reprinted from Gefen et al. (2015), *Journal of Neuroscience*, with permission from the Society for Neuroscience.

Bush, Luu, & Posner, 2000). Anatomical connections between the ACC and the FEF and IPS regions support there being a role for the ACC in helping control the allocation of spatial attention. Its connections with the anterior insula, amygdala, and hippocampus may be critical in assessing emotional salience, monitoring feedback during learning, and coordinating where to best allocate resources during problem-solving. PET and fMRI studies have revealed robust activation in the ACC during tasks that require the inhibition of competing influences on attention and action in order complete a task successfully. For example, in the classic Stroop task, subjects must respond to the ink color of a printed word and ignore the meaning of the word (Stroop, 1935). Subjects typically are greatly slowed and/or make errors when the printed word is the name of a color that is different than that of the ink color that must be reported (e.g., the word "BLUE" printed in red ink), relative to when the ink and word color match or when the printed word is not a color word. The Stroop task is used to assess a subject's level of control, and neuroimaging studies find significantly more activity in the ACC during conflict trials (Pardo, Pardo, Janer, & Raichle, 1990; Taylor, 1997).

Multiple sites within the ACC, as well as in more lateral prefrontal regions, are active during the performance of a Stroop task, and further neuroimaging studies have sought to isolate the multiple processes involved in the performance of the Stroop task (Gruber,

Rogowska, Holcomb, Soraci, & Yurgelun-Todd, 2002; Milham, 2003). Posterior regions of the dorsolateral prefrontal cortex (DLPFC) have been linked to the creation and maintenance of the appropriate rule – in this case, counteracting the automatic reading of the word in order to facilitate processing of color perception – while more anterior regions of the DLPFC have been found to be involved in the selection process of filtering, focusing on the color and filtering out the irrelevant but salient processing of lexical access. There also appears to be some hemispheric specialization here, as the left DLPFC is tied more to a subject's expectation of how likely conflict will be to occur (and therefore how likely it will be that control needs to be allocated), while the right DLPFC is associated more with resolving the conflict regardless of expectation. The ACC has been found to be most associated with the processes of resolving conflict, selecting motor responses, evaluating whether the correct response was made, and allocating attentional resources. In a review of neuroimaging studies using the Stroop task, Banich (2019) argued that these multiple levels of executive control proceed in a cascade-like manner, and that the amount of activity during one stage of processing is partially determined by how effective the control processes were at earlier stages; relatively poor control at early stages results in a greater need for control – and greater amounts of control-related neural activity – at later stages. Finally, a recent study linked the processing of perceptual conflict, even when completely irrelevant to the task, to activity in the attentional control regions of the IPS. Specifically, Oren and colleagues (Oren, Abecasis, Inbar, Glik, Steiner, & Shapira-Lichter, 2023) pointed out that in a typical Stroop task the conflict between dimensions is task-relevant, as both the ink color and the color word are related to the subject's task of naming the color. In their task, Oren et al. instead had subjects perform an animacy task ("living" or "nonliving") on pictures that had words superimposed that could be congruent or incongruent (e.g., a picture of a happy face with either "happy" or "sad" typed across it); critically, the conflicting information (e.g., emotion of word versus picture) was unrelated to the animacy judgment task. The results revealed that the IPS, but not medial frontal regions, was significantly more active on incongruent trials (compared to congruent trials), despite the conflict dimension (emotion) being completely irrelevant to the task. These results could be interpreted as supporting there being a role for the IPS in the automatic processing of sensory information, even when task-irrelevant, which helps define the *salience* of external information and the likelihood of attention being captured by it. The medial frontal regions, such as the ACC, work to resolve the conflict *if* it rises to the level of competing for action plans or resources.

7.5.2 ERP Indices of Error Monitoring and Feedback Processing

Research using ERPs has revealed control processes directly related to the self-monitoring of task performance and the detection of errors in one own's responses. For example, in the Eriksen flanker task (Eriksen & Eriksen, 1974), the participant must respond to the direction of a central arrow flanked on the left and the right by arrows that could be pointing in the same or the opposite direction. For example, on a "congruent" trial the flanking arrows point in the same direction as the central target arrow (e.g., < < < < <), whereas on "incongruent" trials the flanking arrows all point in the opposite direction from the central target (e.g., > > < > >). Thus, on congruent trials the flankers and the target elicit the same motor response, whereas opposing motor responses are triggered on incongruent trials. When speed is emphasized on such a task, subjects will typically make errors on some proportion of incongruent trials and will typically be immediately aware of their mistake. In the early 1990s, researchers began using ERPs to investigate the processes taking place during these types of tasks. Gehring, Coles, Meyer, and Donchin (1990) used a version of the flanker task in which a series of "H" or "S" letters appeared (e.g., "HHHHH"; "SSSSS"; "HHSHH"; "SSHSS"), and subjects had to respond to the central letter, with one button for a central H and a different button for a central S. As shown in Figure 7.11, when the EEG is time-locked to the onset of the motor response, as measured by electromyography (EMG) recorded on the responding forearm or hand, the resulting response-locked potential shows a sharp negative-going component shortly after an *incorrect* response. This component has been labeled the **error-related negativity (ERN)**, and it peaks ~150 ms after an incorrect motor response has been initiated. The ERN is found to be maximal over central midline scalp sites, consistent with it being generated in the ACC (Dehaene, Posner, & Tucker, 1994). The ERN component is associated with the process of recognizing that an error has been committed, in line with theories of the ACC being involved in executive control processes such as monitoring ongoing behavior and modifying attention and actions when necessary to correct poor performance. Some studies have found that an ERN may even be produced when the subject is unaware of

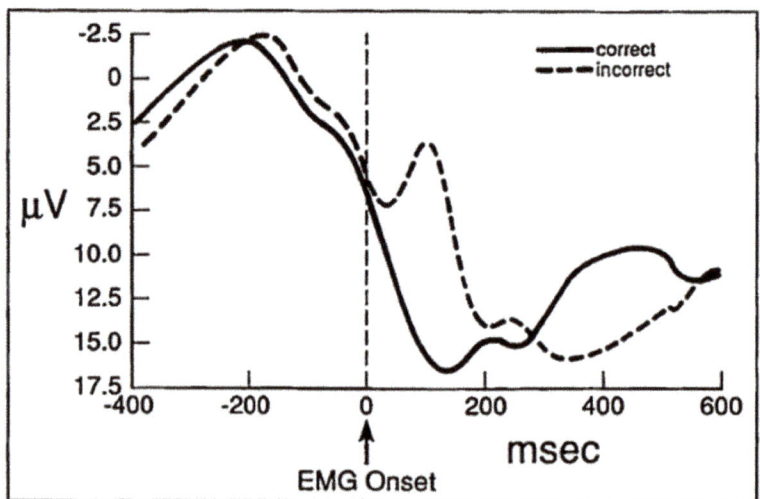

Figure 7.11 Error-related negativity. The ERP waveform recorded at the midline central electrode "Cz" and time-locked to the onset of EMG activity, recording the onset of muscle activity for manual responses. The dashed ERP line shows a prominent negative-going peak about 150 ms after an incorrect manual response was generated – this is referred to as the ERN. Trials that were responded to correctly do not produce this negativity, as shown by the solid ERP line. Reprinted from Gehring et al. (1993), *Psychological Science*, with permission from SAGE.

the error, but other work has found the ERN to be larger and more robust across trials and experiments when the subject is aware of the error (reviewed in Falkenstein, Hoormann, Christ, & Hohnsbein, 2000). While this component is now commonly referred to as the ERN, it should be noted that one of the first studies to report it labeled it the "Ne" ("error negativity"; Falkenstein, Hohnsbein, Hoormann, & Blanke, 1990), and that term is still used in some reports.

The ERN is also produced in "go/no-go" tasks, in which subjects are presented with either a frequent stimulus that requires a set response (e.g., push a button when an "X" appears – the "go" stimulus) or a rare "no-go" stimulus (e.g., a "K" appears – the "no-go" stimulus) that requires the subject to not respond (Scheffers, Coles, Bernstein, Gehring, & Donchin, 1996). Critically, in a go/no-go task, the "go" stimulus occurs much more frequently, and the timing between stimuli is consistent and relatively rapid, so subjects get into a rhythm of responding, and it requires effort and inhibition to stop the execution of the button press when the rare "no-go" stimulus appears. Although the task is different than the flanker task described earlier, both produce a highly similar ERN component in terms of scalp location and timing relative to the error (Wang, Gu, Zhao, & Chen, 2020). Neuroimaging studies have provided further evidence regarding the role of different portions of the ACC in this type of task. Kiehl and colleagues investigated whether the ACC was involved specifically in error detection or whether this region is engaged whenever there is strong response competition and a need for careful monitoring of stimuli and responses (Kiehl, Liddle, & Hopfinger, 2000). That study included one experiment with a standard go/no-go design in which most trials required a response (80% "go" trials, 20% "no-go" trials) and a second experiment that reversed the expectancies, with most trials requiring the inhibition of a response (80% "no-go" trials, 20% "go" trials). The event-related fMRI results revealed that errors of commission (i.e., pressing the button on a "no-go" trial) were associated with a large region of activity that spread across much of the ACC, extending from the more rostral to the more caudal regions of the ACC. The authors noted that these trials could include activity related not only to the recognition of an error being made, but also to the processes of carefully monitoring stimuli and actions under conditions of potential conflict. Indeed, when the brain activity in response to correctly rejected trials (i.e., not pressing a button on "no-go" trials) was analyzed, this revealed activity in the caudal portion of the ACC but no activity in the rostral ACC. Activity in the rostral ACC was found to be significantly greater for errors of commission than for correctly rejected trials in a direct statistical comparison of those two trial types. Finally, analysis of the correctly responded to targets in the second experiment (in which there were rare "go" trials amid frequent "no-go" trials) also showed activity in the caudal ACC but no activity in the rostral ACC. Together, these results suggest that the *caudal* ACC is engaged whenever a very careful monitoring process is required (e.g., under conditions of strong response competition), whereas the *rostral* ACC is engaged more selectively only when errors in response have been made.

ERP studies have also revealed that error-monitoring processes occur at longer latencies after the error as well. While the ERN component can begin as early as 50 ms after an error is produced, a broader, longer-latency positivity can be observed ~250–500 ms after the error has occurred. This error positivity (Pe) has been strongly tied to the subject's awareness of the error, as this component is absent – or at least greatly reduced – when the subject doesn't realize they made a mistake. As opposed to the central midline generator of the ERN, the Pe component is observed over posterior parietal scalp sites, typically encompassing lateral and medial parietal sites. Studies have localized a generator of the Pe in the PCC; however, the Pe is a relatively broad component in terms of duration and topography, and there may be multiple generators contributing to it (Vocat, Pourtois, & Vuilleumier, 2008). In line with the longer latency of this component, compared to the ERN, the Pe has been suggested to index the accumulation of evidence and the setting of criteria for determining that an error has occurred (e.g., Steinhauser & Yeung, 2010; Ullsperger, Harsay, Wessel, & Ridderinkhof, 2010). According to such a model, the ERN may represent a mismatch between a motor response and an initial evaluation of what the correct response should have been, while the Pe indexes a later process in which multiple sources of information (e.g., information from multiple visual pathways; proprioceptive feedback from a motor action; autonomic reactions to feedback) are integrated and evaluated with respect to present expectations and goals in order to conclude whether an important error has been made (Overbeek, Nieuwenhuis, & Ridderinkhof, 2005). It should be noted that the Pe has also been found on trials that don't immediately lead to corrections (Falkenstein, Hoormann, Christ, & Hohnsbein, 2000), and therefore the Pe may not index the specific process of correcting actions, but rather the conscious determination that an error has occurred, which in turn *could* be used to attempt error correction if the subject desires.

In addition to monitoring one's own actions and recognizing our errors by ourselves, we can also learn through feedback from others. Research into how the brain processes external feedback has revealed that medial frontal regions are strongly involved with this process as well. One dimension of feedback that has been hypothesized to play an important role in that type of learning is the distinction between positive and negative feedback. Positive feedback relays confirmation that everything is fine, whereas negative feedback must be attended to and acted upon if performance is to improve. ERP studies have identified a component – the feedback-related negativity (FRN) – that is thought to index the evaluation of feedback, not simply the registration of an error (Holroyd & Coles, 2002). Specifically, the FRN is evoked by feedback and is typically larger following negative feedback (Falkenstein, Hohnsbein, Hoormann, & Blanke, 1991; Miltner, Braun, & Coles, 1997). The FRN peaks ~250 ms after the onset of a feedback event, and its neural generator has been modeled to the ACC region (Figure 7.12; Gehring & Willoughby, 2002). In a study that recorded ERPs during fMRI, the FRN was more specifically localized to the dorsal caudal region of the ACC (Hauser, Iannaccone, Stämpfli, Drechsler, Brandeis, Walitza, & Brem, 2014). Furthermore, this study revealed that the size of the FRN isn't determined simply by the valence of the feedback, but rather by how surprising the feedback is. This suggests that this region of the ACC, and its network of connected brain regions, may be critical in the processes of executive control that determine when significant resources need to be quickly allocated to correct performance.

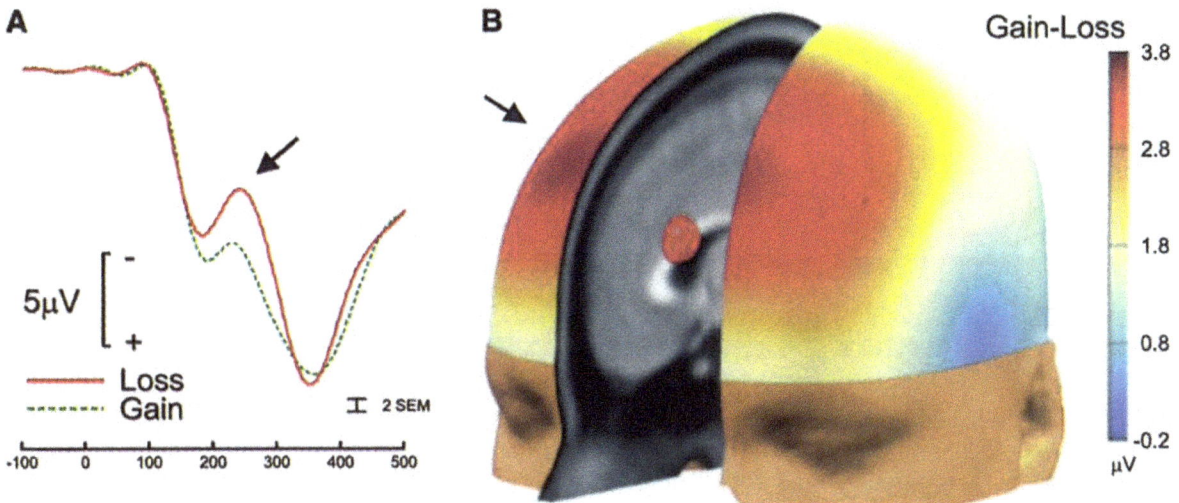

Figure 7.12 Feedback-related negativity. (A) ERP waveform time-locked to the onset of feedback, recorded from the midline frontal electrode "Fz." The arrow points to the FRN component, the negative-going component between 200 and 300 ms, which is larger for feedback indicating a loss than for feedback indicating a gain. (B) Scalp distribution of the peak of the FRN (here calculated as the difference between loss and gain trials) at 265 ms after the onset of feedback. The arrow points toward the location of the electrode "Fz," the waveform of which is shown in (A), which corresponds to the region of maximal voltage at the scalp at this time point. The large red dot on the MRI slice shows the generator of the scalp-recorded FRN estimated by dipole source modeling using the pattern of activity across the scalp. SEM = standard error of the mean (in this case referring to the voltages in the 200–300-ms latency range of the FRN). Figure adapted from Gehring and Willoughby (2002), *Science*, with permission from the American Association for the Advancement of Science.

7.5.3 Cognitive Flexibility, Inhibition, and the Frontal Lobes

In studies of development, aging, and brain injury, lateral regions of the PFC have been found to be critical for other aspects of cognitive control. One classic test of the ability to attend to and select appropriate actions is the **Wisconsin Card Sorting Test (WCST)**. In this task, subjects are presented with cards that display simple geometric shapes, with the color, shape, and number of objects varying on each card (Figure 7.13). Subjects must learn to sort the cards on the basis of color, number, or shape by receiving feedback from the experimenter; the subjects are not given explicit instruction on what dimension by which to sort the cards. At various points throughout the experiment, the rule changes, and subjects must learn the new rule based on feedback from the experimenter. Patients with frontal lobe damage exhibit profound "perseveration" on this task, meaning that they continue to sort the cards based on the initial rule that they learned (Demakis, 2003). These patients have a profound difficulty in inhibiting the old rule and attending to the new dimension that is now critical for the task.

The WCST can be used to assess multiple aspects of attention control, including persever-ation of preceding criterion (PPC; sorting category), perseveration of preceding response

Figure 7.13 Wisconsin Card Sorting Test. On each trial, subjects turn over a new card from a deck of cards (shown in the bottom row) and must choose which pile (on the top row) to sort it into. Cards can be sorted by color, number, or shape. Subjects are not given explicit instructions but must learn the sorting rule through feedback ("correct" or "incorrect"). At unpredictable times, the dimension for sorting changes, and the subject must learn the new rule through feedback. Image from http://pebl.sourceforge.net/battery.html; public domain.

(PPR; repeating last response), and set loss (making an incorrect response after three or more correct responses in a row, suggesting subjects lost track of what the relevant dimension was). Stuss and colleagues compared the performance of patients with different lesion locations (Stuss, Levine, Alexander, Hong, Palumbo, Hamer, Murphy, & Izukawa, 2000). They found that patients with lesions to superior medial frontal areas were most impaired on the perseveration aspects of the task, whereas patients with dorsolateral frontal lesions showed the greatest impairment on set loss. These results are in line with fMRI results in healthy subjects showing that medial frontal structures are critical for selection processes while dorsolateral prefrontal areas are involved in working memory maintenance (Monchi, Petrides, Petre, Worsley, & Dagher, 2001).

Further support for the critical role of the frontal lobes in cognitive flexibility comes from studies of children. Adele Diamond and colleagues have developed simplified versions of this type of task to examine children's abilities across developmental stages (Diamond, 1990; Diamond & Boyer, 1989; Rennie, Bull, & Diamond, 2004). Since connectivity with the frontal lobes is among the last to fully develop, tracking children's performance across ages can help us to understand the functions of these regions. The results show that very young children show a similar pattern of perseveration to patients with frontal lobe damage. For example, when infants (~7–9 months of age) are shown a toy being put into one of two wells in front of them and then both wells are covered with cloths, the infant will quickly remove the cloth from the appropriate well and grab the toy. This reward of getting the toy quickly establishes an association with that well and the toy. On subsequent trials, the child is shown the toy going into a different well before the wells are covered, but the child consistently uncovers the original well, even though they just watched the experimenter put the toy into a different well. It has been suggested that the child quickly formed an association between the original well and the toy and thus must inhibit returning to that well. Their still-developing frontal lobes have trouble with this need to select and inhibit some previously relevant information. In older children (e.g., 2–4 years of age) working memory processes are stronger, and the child is able to easily keep in mind the current location in which the toy was just placed and select the correct well. However, when they need to perform a task that is more similar to the WCST, in which they have to sort cards by either shape or color, they perseverate in responding in the way they initially sorted the cards, even if the experimenter reminds them on each trial how they are supposed to sort on that trial (Diamond, Carlson, & Beck, 2005). These results provide further evidence that our ability to focus attention on the current rule/feature and suppress the originally reinforced action depends on connectivity with the frontal lobes, which develops over the first few years of life. By the time children reach 7–8 years of age, they typically won't make overt errors in sorting cards, but careful recording of their reaction times shows that they respond significantly slower on trials in which a recently relevant dimension must be ignored. Even healthy adults show this milder form of perseveration when tested in versions of the WCST in which the speed of response is strongly emphasized – reaction times are slowed when previously reinforced responses have to be suppressed, and accuracy suffers on switch trials when response windows are severely restricted.

The role of the frontal lobes in enacting selection has also been shown in ERP studies measuring the brain's responses to irrelevant stimuli. Patients with frontal lobe lesions have been found to have deficits in inhibiting irrelevant information. In one study (Knight & Grabowecky, 1995), patients performed a visual task while being presented with task-irrelevant auditory tones. The results showed that the auditory N1 component (~50 ms, generated in the primary auditory cortex and indicating the degree of sensory processing) was significantly *larger* in frontal lobe patients than in healthy controls or a lesion control group (parietal lesions). This provided evidence that the frontal lobes were important for implementing selective processing in early sensory processing regions. In a combined neurostimulation and ERP study, Zanto, Rubens, Thangavel, and Gazzaley (2011) used TMS to temporarily disrupt the PFC during a visual task in which subjects had to attend to color or motion on different trials. The results from the control (no-disruption) condition showed that the visual P1 component to the color stimulus was significantly reduced in amplitude when it was to be ignored (i.e., when subjects were paying attention to the motion stimulus and the color stimulus was irrelevant). However, there was no suppression of the P1 component to the irrelevant color stimulus when the frontal cortex was disrupted by TMS.

7.5.4 Functional Connectivity and the Networks of Executive Control

The previous subsections discussed research that has attempted to isolate specific subcomponent processes of attentional control and to identify brain regions underlying those processes. While much of that work was aimed at localizing functions to a specific brain area, it is also clear that something as complex as executive control involves many areas of the brain working in a coordinated and integrated manner. Indeed, in most cases of deficits in executive functioning,

254   **The Control of Attention**

it is difficult to treat the disorders because there's not simply a single area of the brain that's dysfunctional. Recent work with fMRI has focused on trying to understand how connectivity within and across brain networks relates to attentional control, working memory, and executive functioning (e.g., Cohen & D'Esposito, 2016).

For example, attention deficit hyperactivity disorder (ADHD) is a multifaceted disorder encompassing many aspects of attention, and much remains unknown about its etiology and treatment. However, cognitive neuroscience and genetic studies hold promise for enhancing our understanding of the different subtypes of ADHD, including brain systems and risk factors. These studies are finding differences in white matter tracts, volumes of subcortical nuclei, neurotransmitter levels, and connectivity patterns across cortical structures (reviewed in Yadav, Bhat, Hashem, Nisar, Kamal, Syed, Temanni, Gupta, Kamran, Azeem, Srivastava, Bagga, Chawla, Reddy, Frenneaux, Fakhro, & Haris, 2021). Much work needs to be done to understand the genetic influences on these processes and how the findings at multiple levels of the nervous system can be integrated to understand ADHD. However, resting-state and task-based fMRI studies are already revealing details of the brain connectivity associated with ADHD in children and adults that is advancing our understanding of how connectivity patterns across brain networks may explain these cognitive functions and disorders.

In a meta-review of task-based fMRI studies, Cortese and colleagues found that differences were consistently found in the dorsal attentional control network of individuals with ADHD compared to control subjects (Cortese, Kelly, Chabernaud, Proal, Di Martino, Milham, & Castellanos, 2012). Specifically, children with ADHD were reliably found to have *lower* levels of task-based fMRI activity in the dorsal frontoparietal regions of the attentional control network compared to control groups. Regions of the default mode network, on the other hand, as well as somatosensory–motor regions were found to show significantly *greater* activity in the children with ADHD compared to controls. This review found similar patterns in adults with ADHD, demonstrating lower activity in the frontoparietal control network and higher activity in the default and visual processing networks. Another meta-review dissociated the networks that may be separately associated with hyperactivity versus inattention. Hart and colleagues (Hart, Radua, Nakao, Mataix-Cols, & Rubia, 2013) found that atypical functioning within the ACC and inferior frontal regions was associated with deficits in inhibition, whereas abnormal parietal and dorsal frontal activity was associated with deficits in the focusing of attention.

Newer methods of fMRI connectivity analysis can assess the *directional* influence of one region on another, and incorporating this additional information holds promise for improving our understanding of both the neural bases of attention processes and the patterns of dysfunction that may be responsible for different subtypes of ADHD. A critical requirement for making inferences of directional connectivity is the reliability of the mappings that are revealed. One method that has been shown to have uniquely strong reliability for identifying the presence and direction of connections is the **group iterative multiple model estimation (GIMME)** method (Gates & Molenaar, 2012). Simulations with this method have shown that it can accurately identify directed connectivity in a purely bottom-up manner and differentiate distinct subgroups based on the patterns of connectivity. This method first creates a "group map" that consists of the pattern of directed connections between regions of interest that is consistently present within the majority of individuals in the sample. Then, using the parameters from the group as a starting point, the next step is to identify the strength of these connections at the individual level. These individual-level mappings can then be used to further divide the sample into subgroups that share a specific pattern of directed connections within that network. Gates and colleagues applied this method to analyze resting-state fMRI data in a diverse sample that included both individuals with ADHD and control subjects with no diagnosis of ADHD (Gates, Molenaar, Iyer, Nigg, & Fair, 2014). The sample consisted of 80 children (7–12 years of age), 32 of whom met the diagnostic criteria for ADHD. The regions of interest that were extracted from the resting-state fMRI data included brain areas within the frontoparietal attention networks. Applying the GIMME algorithm to the datasets from the entire sample resulted in five subgroups being established. Interestingly, the data-driven subgrouping of directed connectivity patterns across the regions of interest did not simply divide the group into one ADHD group and one control group. Rather, the bottom-up procedure revealed five distinct patterns of directed connectivity (Figure 7.14). None of the groups consisted of only ADHD or only control subjects, but individuals with ADHD were grouped together significantly more than at chance levels. Two of the subgroups (labeled as subgroups B and D in Figure 7.14) consisted of mainly ADHD individuals, whereas three other subgroups (A, C, and E in Figure 7.14) were composed of mainly individuals without ADHD. Together, these results confirm that there is a good deal of heterogeneity across individuals with ADHD and among control subjects with regards to the patterns of directed connectivity in their frontoparietal attentional control networks. Indeed, diagnosis of ADHD is complex and multidimensional, and the disorder can affect many aspects of attention and executive control. The results from Gates et al. (2014) suggest that there are multiple, distinct patterns of brain connectivity that may be associated with ADHD, possibly with different subtypes. The lack of a single pattern that perfectly dissociates people with ADHD from controls is consistent with the most recent version of the Diagnostic and Statistical Manual of Mental Disorders, 5th edition, Text Revision (DSM-5TR; American Psychiatric Association, 2022), which states that "no biological marker is diagnostic for ADHD" and "no form of neuro-imaging can be used for diagnosis of ADHD" (p. 73). In addition, there are continuing arguments about whether ADHD is overdiagnosed or underdiagnosed in the population, further complicating the comparison of ADHD versus control groups. Gates and colleagues suggested that the patterns of directed connectivity may be informative for understanding whether some patterns of connectivity among attentional control regions make a person more susceptible to developing ADHD versus being more protected from it. Finally, in regard to the use of resting-state fMRI in many connectivity studies, it may be that the directed connections that are most critical for ADHD might not be as strongly engaged during a resting state and might be better assessed via tasks that directly and specifically engage multiple attention processes.

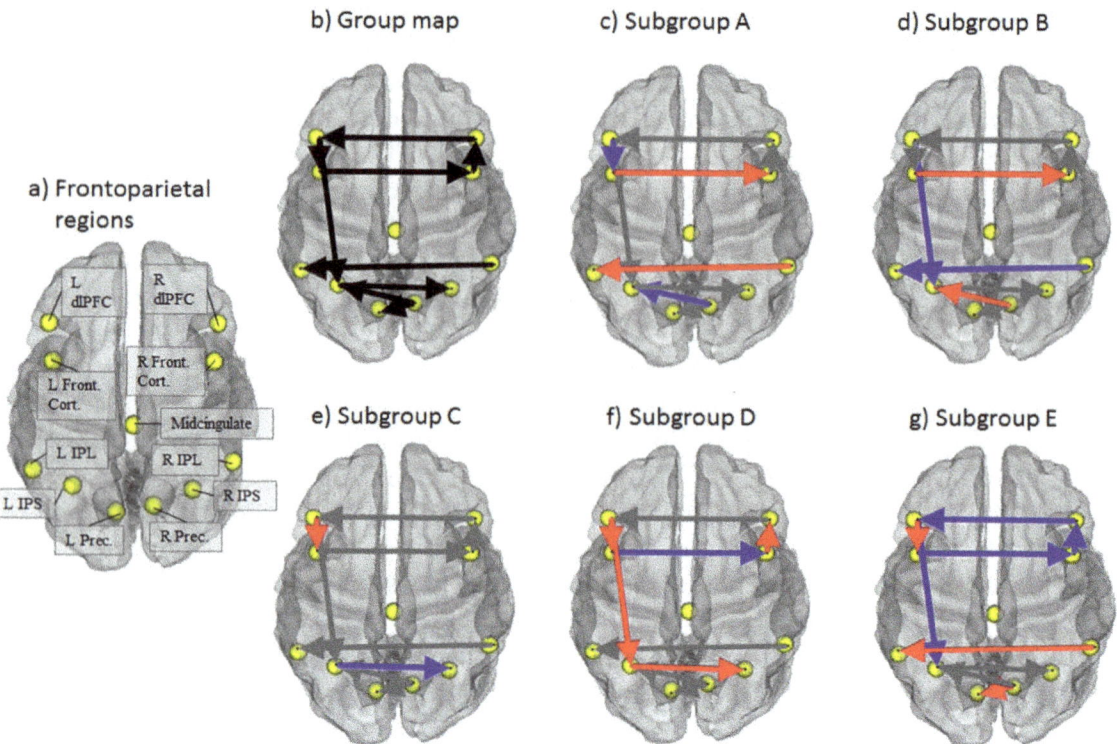

Figure 7.14 Functional connectivity across attentional control regions differentiates subgroups. (a) Yellow spheres indicate the regions of interest that were extracted and analyzed from subjects' resting-state fMRI scans. (b) Overall group results, showing the patterns of directed connectivity between regions that were present for all subgroups. (c–g) The five subgroups that were sorted via data-driven connectivity pattern analysis (using the GIMME algorithm described in the main text). Gray lines indicate the connectivity that was similar across all five subgroups (also shown in the group map); red lines indicate places where connectivity in that subgroup was stronger than the average of the other subgroups; blue lines indicate where that subgroup had lower connectivity than the average of the other subgroups. Subgroups B and D consisted primarily of children who had been diagnosed with ADHD. Subgroups A, C, and E consisted primarily of control subjects. Figure reproduced from Gates et al. (2014), *PLoS One*; Creative Commons CC BY.

7.6   Plasticity and the Training of Attentional Control Networks

The development of attention systems in children and the progressive decline of some attention processes in elderly subjects provides evidence that the brain systems underlying attentional control change over the course of our life. In addition to investigating these long-term changes in attention, researchers have aimed to understand whether attention is malleable over much shorter timescales and which mechanisms of attention may be conducive to therapies or training. This research can be applied to strategies for short-term improvements in specific areas of work or school performance as well as to longer-term enhancements of everyday functioning. These studies also have implications for developing therapies for psychological disorders and the treatment of learning disabilities.

One method for training attentional control is to have participants perform simplified versions of classic laboratory tests of cognitive functioning over multiple days. Rueda and colleagues used this method to train young children over a few sessions on a child version of the ANT, and they then assessed their attention abilities with behavioral tests and EEG measurements (Rueda, Rothbart, McCandliss, Saccomanno, & Posner, 2005). In this study, 4-year-old and 6-year-old children were first assessed with an attention task and a parent-reported child temperament scale. Participants in the training group were then brought back to the lab five times over the next 3 weeks to complete the attention task repeatedly as a means of training. The attention task used here was a variation of the ANT described in Chapter 1 and shown in Figure 1.9. To adapt the test for young children, the target display consisted of cartoon drawings of fish instead of arrowheads. As in the standard task, the target display consisted of a horizontal row of five stimuli, and the child had to push a button to indicate the direction (left or right) in which the middle fish was facing. As in a standard flanker task, there were "congruent trials" in which the flanking fish stimuli were pointed in the same direction as the center fish and "incongruent trials" in which the flanking fish were all pointed in the opposite direction from the center target fish. The differences in reaction time and accuracy between the congruent and incongruent trials provide a measure of executive control, specifically the ability to selectively attend to the center stimulus and suppress the irrelevant flanking stimuli. The children assigned to the control group (no cognitive training) simply watched popular children's videos and pressed a button every 30–60 seconds to continue the video. This control group came to the lab according to the same schedule as the training group. In another experiment within this study, the children in the control group only came to the lab for the initial assessment and the final assessment 2–3 weeks later. The behavioral measure of executive control (better performance for congruent versus incongruent trials) showed the expected effect of development, as 6-year-olds were significantly better than 4-year-olds on the ANT. In addition, however, there was a significant effect of training in both age groups, as the performance on the attention task of the younger children who underwent training improved toward the level of older children. This suggests that even a brief, five-session training period can improve attentional control functioning in young children. The ERP data recorded from these participants further revealed how the training modulated neural activity. Specifically, the frontoparietal N2 component has previously been shown to be associated with attentional control processes in the anterior cingulate during the performance of flanker tasks (van Veen & Carter, 2002). In the Rueda et al. (2005) study, this component showed significant effects of age and of training. Specifically, the standard effect of congruency on the N2 in adults (more negative for incongruent than congruent trials) was observed at central midline sites, being maximal at electrode site Cz at the top of the scalp. In the untrained 6-year-olds this effect was seen at more frontal midline sites (electrode site Fz), and in the untrained 4-year-olds no electrode sites showed the standard congruency effect in the N2 range. Remarkably, the attention training seemed essentially to advance the developmental process. Six-year-old children who had undergone the attention training showed an effect of congruency on the N2 that was similar in timing and scalp distribution to that of adults performing the ANT. Whereas the untrained 6-year-olds produced an N2 effect at more anterior electrode sites, the trained 6-year-olds showed an N2 congruency effect at the same central scalp location as adults. The 4-year-olds in the training group showed a trend toward the standard congruency effect on the N2 with a scalp distribution at the more anterior frontal sites that matched the untrained 6-year-olds, whereas the untrained 4-year-olds didn't show a standard congruency effect on the N2. Thus, the ERPs provide a sensitive measure of brain activity indicating that attention training may rapidly affect the brain's executive control processes, possibly allowing children to exert control in a manner beyond their years. It should be noted, however, that there were no follow-up sessions with these children, so it's not clear if the changes were long-lasting or only short-lived.

Finally, the study by Rueda and colleagues (2005) also included an analysis of genetic influences. Specifically, they collected saliva from a subset of the 6-year-old children in order to investigate whether a specific gene (dopamine transporter type 1, or *DAT1*) that has been linked to executive attention functioning (Fossella, Sommer, Fan, Wu, Swanson, Pfaff, & Posner, 2002) might help explain some individual differences in attention. The results revealed that children carrying the pure long-allele version of the *DAT1* gene showed significantly lower congruency effects than children with a mixed long/short-allele version. Furthermore, analysis of the N2 effect showed a significant difference between these two subgroups, with a robust N2 effect in the expected direction (more negative for incongruent trials) being observed only in the group that showed lower congruency effects (the pure long-allele group). This study only included eight children in the mixed long/short-allele group, so any null effects must be interpreted with caution, but this study highlights that genetic differences play a role in the development of attention mechanisms in the brain. These results suggest that children with the pure long-allele form of the *DAT1* gene express a more mature brain activity pattern in terms of executive control and that children with other variants of this gene may be more in need of attention training.

Of course, the effects of training in the study by Rueda and colleagues (2005) were assessed only narrowly, in terms of using the exact same attention task to both train and test the functioning of attention. Across studies investigating brain plasticity, there is a great deal of variability in how similar the training is to the test(s) used to assess the effectiveness of the training. This has important implications for how we understand the multiple functions of attention and what types of training may be most effective for improving performance on a specific task versus modulating a core process of attention that affects cognition more broadly. At the opposite end of the spectrum from studies that use the same task for training and assessment, some research uses training that involves no cognitive task at all (e.g., meditation, yoga) or attempts to train subjects to implicitly control specific brain circuits. Neurofeedback studies often don't use a cognitive test to train subjects, but rather have subjects attempt to relax and control some specific aspect of their own neural activity through audiovisual feedback displays. The goal of many of these studies is to achieve a certain balance of EEG frequencies or to change the fMRI-measured activity of a selected brain region in a certain direction, with the presumption that the change will result in improvements in everyday functioning or school or work performance. In one of the earliest studies to use real-time fMRI (rt-fMRI) neurofeedback to target attentional control processes, Zilverstand and colleagues tested whether this method of training could improve the brain processes and outcomes of adults with ADHD (Zilverstand, Sorger, Slaats-Willemse, Kan, Goebel, & Buitelaar, 2017). Previous studies had found that activity in the dorsal ACC (dACC) during cognitive control tasks is significantly *reduced* in adults with ADHD compared to control subjects (e.g., Adisetiyo & Gray, 2017; Bush, 2011). Zilverstand et al. utilized rt-fMRI neurofeedback during the performance of a mental calculation task to train subjects to *increase* activity in the dACC. The training consisted of 60-minute fMRI training sessions once per week for 4 weeks. Subjects performed a mental calculation task that varied in its level of difficulty, and those assigned to the active neurofeedback group received continuous feedback on the amount of activity in their dACC. For the mental calculation task, subjects were asked to pick a single-digit number and then, starting at 100, subtract that chosen number repeatedly from the result. Subjects were asked to adjust the difficulty of their task by increasing the tempo/rate of operations or the magnitude of the number being subtracted. In the control group, subjects were simply given visual instructions to increase or decrease the difficulty of the task at randomly selected times. Individuals in the neurofeedback group were shown a scale that continuously displayed the activity level of their dACC, and they were asked to modulate the difficulty of their task until they achieved the level of dACC activity indicated by a marker on the scale and to keep it at that level until instructed to match a different level. Activity in the dACC was found to be increased over the sessions for *both* groups, but the active neurofeedback group showed a greater improvement in task performance and cognitive functioning. In terms of ADHD symptoms and clinical outcomes, however, there were no significant differences between the active feedback group and the control group. Therefore, although the neurofeedback training "succeeded" in boosting the fMRI activity in the dACC, it didn't improve clinical outcomes. There is great interest in the potential of neurofeedback, in part because it is a relatively "easy" type of training for the subject as opposed to therapies that require a high degree of cognitive effort and the completion of challenging tasks. However, there remain critical concerns about the effectiveness of treating disorders with neurofeedback (see Box 7.2).

Box 7.2 Does neurofeedback "work"?

Neurofeedback involves recording the participant's brain activity via EEG or fMRI and then presenting a measure of that activity back to the subject, usually via changes to simple audiovisual displays. For example, in EEG neurofeedback, the visual display may change in degree of clarity as the participant's neural activity gets closer to or farther from the targeted pattern of activity. The targeted EEG measure could be the relative power of different frequency bands (e.g., theta-to-beta ratio) or the level of intensity of an individual band over a certain scalp region (e.g., alpha intensity at occipital scalp sites). In fMRI neurofeedback, the display can be made to change as the activity in a specific targeted region of the brain changes in the desired direction (e.g., increased activity in the ACC). Several studies have reported that neural *activity* can be modified by EEG neurofeedback (reviewed in Pimenta, Brown, Arns, & Enriquez-Geppert, 2021) or fMRI neurofeedback (reviewed in Thibault, MacPherson, Lifshitz, Roth, & Raz, 2018). One could therefore conclude that neurofeedback is "working," because the neural activity has changed in the targeted way with the

Box 7.2 (cont.)

feedback. In this way, neurofeedback could be considered like other types of training, in terms of it changing something about the brain. It differs from many other types of training, however, in that it is generally less effortful and less intentional in terms of exactly what the participant is "trying" to do. There has been criticism that placebo effects may be responsible for at least some of the observed effects (Schönenberg, Weingärtner, Weimer, & Scheeff, 2021), and that the neural changes aren't always associated with lessening of the symptoms of the disorders being treated (Thibault et al., 2018). Another critical issue is what exactly the brain is being trained to do. Neurofeedback for improving attention could be considered to "work" in terms of making a part of the brain more active or modulating a frequency band, but until we completely understand the mechanisms of attention, how can we know exactly how the brain *should* behave? The reason why there are so many methods being used to study the processes and mechanisms of attention is that, in order to fully understand how attention works, we need to measure the brain's activity with high temporal and spatial resolution across many different tasks. EEG neurofeedback trains at a very imprecise spatial level, and studies that use fMRI neurofeedback lack the temporal resolution to account for exactly when a region is optimally active. Therefore, the question can't simply be whether or not neurofeedback "works," but rather: Do we fully understand what the brain should be trained to do? And does the method being used have the temporal and spatial precision to train only the specifically intended cognitive and neural process?

7.6.1 Meditation and Mindfulness Training

A number of methods of self-care for our mental and physical well-being include meditation practices or training. One popular such method that has been used in clinical settings, and which is thought to have beneficial effects on the functioning of attention across ages and individuals, is centered around **mindfulness**-based training. Two mindfulness practices that have been associated with attention are focused attention (FA) training and open monitoring (OM). FA training is often the starting point for beginning mindfulness practices, and it involves maintaining attention over a prolonged period with a very narrow focus. The practitioner seeks to intently sustain attention on a specific aspect of the present moment, focusing on a specific sensation or on an internal process (e.g., breathing). These efforts to sustain attention are thought to help the mind learn to filter out other competing inputs, from external or internal sources, to stay focused on the desired aspect of the present moment. An aspect of FA training can involve self-monitoring of one's own thoughts and exerting control over mind-wandering in order to return attention to the original focus of attention. OM also includes staying in the present moment, but it involves being receptive ("open") to experiencing whatever sensations or thoughts come to mind, without dwelling on any of them beyond the momentary experience. Key to OM is experiencing the events without judgment or prolonged evaluation and to not get stuck dwelling on anything in particular. Paying attention fully to just the current moment is of course more difficult than it sounds, in part because of the involuntary mechanisms of attention that are covered in other sections of this book. Specifically, the automatic capture of attention and the involuntary holding of attention are working against the goal to not dwell on unwanted things. If the internal thoughts holding our attention are negative or stress-inducing, this can produce harmful effects on our physical and emotional well-being. This critical ability to disengage attention and the mechanisms involved with the involuntary holding of attention will be discussed more in the following chapter on the temporal aspects of attention.

It has been proposed that practicing mindfulness trains two primary mechanisms of attention: the ability to sustain a focus of attention amid potential distractions and the ability to engage executive control processes quickly and efficiently, partially through improved meta-awareness of one's moment-to-moment mental states (Bishop, Lau, Shapiro, Carlson, Anderson, Carmody, Segal, Abbey, Speca, Velting, & Devins, 2004). In a review of studies investigating the effects of such training on executive functioning, Gallant (2016) confirmed the beneficial effect of mindfulness on at least one of these mechanisms: the selection process of focusing attention. Specifically, Gallant found that mindfulness training was associated with significantly improved distractor inhibition across labs and attention tasks. In terms of the orienting and reorienting of attention, however, this review indicated only mixed results. In a more recent and larger meta-analysis of studies investigating the behavioral effects of meditation training, Sumantry and Stewart (2021) concluded that mindfulness-based practices were associated with enhancements of multiple different functions of attention. Although the overall effect of mindfulness was small across tasks, looking more closely at the different types of attention being assessed revealed that some functions of attention were significantly enhanced, whereas others showed little or no evidence of being modulated. Measures of alerting, inhibition, reorienting, and executive control were all found to be reliably enhanced across studies. However, these authors did *not* find a reliable effect of meditation on the *orienting* of attention. They suggested that their results are in line with the goals of FA training and OM. Specifically, since FA training involves the maintenance of a singular focus of attention, it helps practitioners learn to modulate their level of alertness and to inhibit the processing of potentially distracting stimuli or intrusive thoughts. OM allows the initial processing of any input without judgment, but it entails not dwelling on any previous thoughts or stimuli in order to be continually aware of the present. Therefore, the disengagement and reorienting of attention away from a stimulus or away from an intrusive thought is continually practiced. Such training does not typically involve practicing shifting attention immediately in response to external instructions, however, which may explain why this review didn't find a robust effect across studies of mindfulness affecting the orienting of attention, which is measured in typical endogenous cuing paradigms.

7.6.1.1 Neuroscience Studies of Meditation

Studies measuring EEG and ERPs have provided additional insights into the mechanisms of attention that may be affected by meditation and mindfulness training. Recent studies have found that even very brief meditation training can significantly affect the perceptual and higher-order processing of stimuli when subjects are experiencing cravings. Previous studies investigating the use of meditation for treating nicotine addiction have found that mindfulness-based training was more effective than many conventional behavioral treatments for smoking cessation (Garrison, Pal, Rojiani, Dallery, O'Malley, & Brewer, 2015; Witkiewitz, Bowen, Douglas, & Hsu, 2013). Andreu and colleagues (Andreu, Cosmelli, Slagter, & Franken, 2018) recorded EEG from chronic cigarette smokers while they performed a version of a go/no-go task that included smoking-related pictures or neutral pictures. In the task, the pictures were irrelevant, as subjects needed only to discriminate the color of the outline frame around each picture, pressing the response button for one of the colors (e.g., a blue frame) as quickly as possible and withholding any response when the other-colored frame (e.g., a yellow frame) appeared. The stimulus that required a speeded response (i.e., the "go" stimulus) was much more frequent (~75%), and therefore the rare "no-go" stimulus provided a measure of the subject's ability to inhibit a prepotent response. Before the task began, cigarette craving was induced in each subject through a cigarette exposure protocol. This involved having each subject progress through a series of steps from opening a pack of cigarettes to bringing a lighter to a cigarette being held in the mouth, but without ever lighting or smoking the cigarette. Immediately afterwards, subjects were randomly assigned to either a brief 15-minute mindfulness session or a control session of equal length. Subjects listened to audio recordings that either guided them through a mindfulness session (e.g., accept feelings and thoughts in a nonjudgmental way and without dwelling on them) or instructed them to use other techniques that they may naturally employ to cope with cravings (e.g., trying to distract oneself from thinking about smoking). Two ERP components were of particular interest in this study: the N2 and P3 components. Both of these components have previously been used to index the amount of effort and inhibitory processing that are exerted in go/no-go tasks. In this study, the amplitude of the N2 component was significantly larger on "no-go" trials than on "go" trials, consistent with previous studies showing that rare "no-go" trials require significant effort in order to overcome the frequent motor response. There was no difference in this N2 effect when comparing the mindfulness group versus the control group, however, suggesting that, at this level, the mindfulness training had no effect. The P3 component was also expected to be significantly larger on "no-go" trials, since previous studies using standard go/no-go tasks have found that the "no-go" trials require more effort to process the rare stimulus and to inhibit the motor response. As shown in Figure 7.15 (left column), this was observed for the control group, as the P3 component was very robust and significantly larger on "no-go" trials than on "go" trials. Critically, mindfulness affected processing at this P3 stage, as the effect across conditions was greatly reduced for subjects in the mindfulness group, and the P3 component on "no-go" trials in the mindfulness group was significantly smaller than the "no-go" P3 component in the control group (compare bottom rows of Figure 7.15). This suggests that even very brief mindfulness training was able to enhance the efficiency and effectiveness of attentional control processes.

In regard to the Andreu et al. (2018) study, it is interesting to note that there were no group differences on behavioral measures from the go/no-go task. In other words, subjects in both the control and mindfulness groups produced more errors on the "no-go" trials, and this error rate didn't differ between groups. The lack of a robust behavioral effect despite the significant effect on the P3 component suggests that ERP indices of neural processing can provide a more sensitive measure of the effects of training than can the overt behavioral

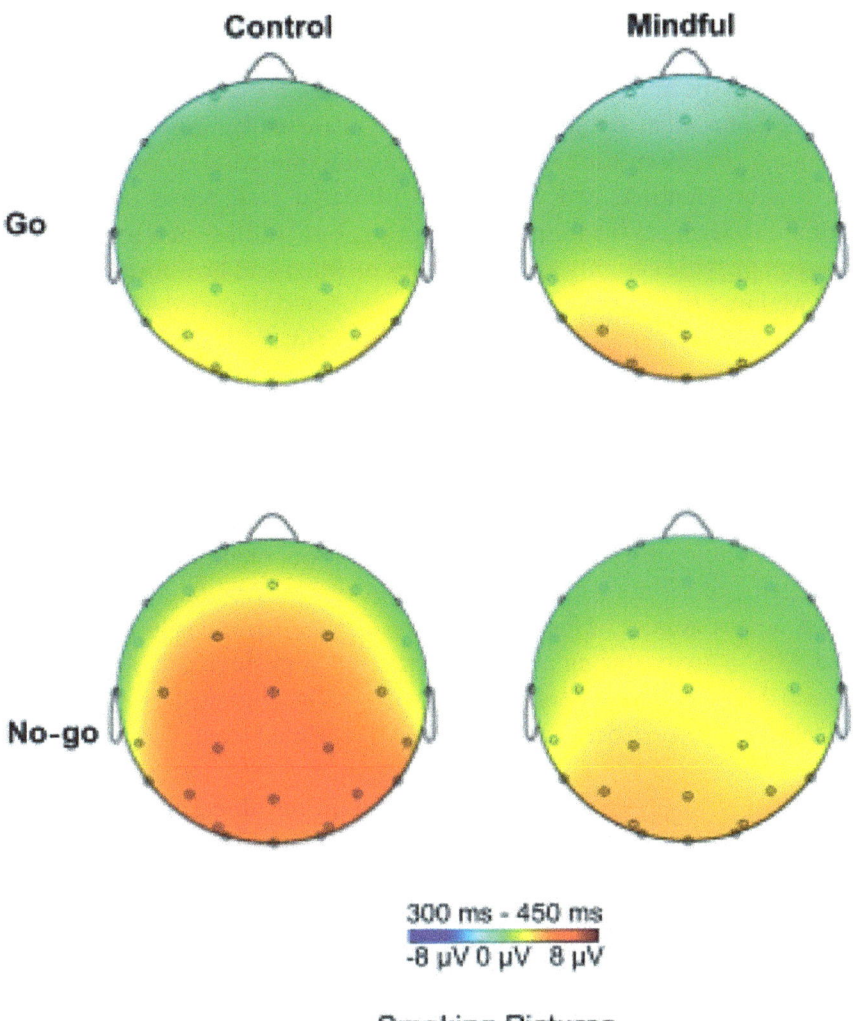

Smoking Pictures

Figure 7.15 ERP effects of mindfulness training on the P3 component. Scalp topographic maps showing the peak of the P3 component (350–400 ms poststimulus latency) during a go/no-go task in a group of individuals who were chronic cigarette smokers. The control group (shown in the left column) showed a prominent P3 component over posterior central sites in response to the "no-go" stimuli (bottom row), which was especially strong for smoking-related pictures. This P3 component is greatly reduced in the group that received a brief period of mindfulness training (shown in the right column) just prior to the go/no-go task, providing evidence that the training facilitated attentional control in terms of making it easier (i.e., less effortful and with less neural activity expended) to inhibit responding to the "no-go" stimuli. Reprinted from Andreu et al. (2018), *PLoS One*; Creative Commons CC BY.

results alone. This may be especially important in studies such as this that use only a single very brief training session. Indeed, in studies that have used much longer mindfulness training sessions (e.g., 8–16 weeks or more), effects have been observed on not only the P3 component (e.g., Moore, Gruber, Derose, & Malinowski, 2012) but also on the earlier N2 component and on overt behavior (e.g., Malinowski, Moore, Mead, & Gruber, 2017). Using a different task, specifically the ANT described in Chapter 1, Norris and colleagues found that the N2 component could be significantly affected by mindfulness training, even with only a single, brief, 10-minute mindfulness session immediately preceding the attention task (Norris, Creem, Hendler, & Kober, 2018). Further research is needed to understand the effects of short-term versus long-term training, but studies such as these confirm that ERP measures can provide a highly sensitive index of when and how attentional control processes can be modulated by meditation practices.

Given that the ACC has been strongly implicated in attentional control, as described in earlier sections, a number of fMRI studies have investigated how this region is affected by meditation practices. In a recent meta-review of neuroimaging results, Zsadanyi and colleagues found that activity in the ACC was usually enhanced during active meditation versus a passive control condition (i.e., rest), and this was consistent across long-term meditators and beginners (Zsadanyi, Kurth, & Luders, 2021). Studies that compared meditation to other *active task* conditions (e.g., color discrimination, word generation, mental arithmetic) were more variable in their patterns of results within the ACC, but this may be due to the very different control tasks used across studies. This meta-review also included longitudinal studies that found increased ACC activity during attention tasks and a positive correlation between amount of training and ACC activity. Overall, these fMRI results align with the view that meditation can affect executive control processes, especially those related to the ability to monitor and control the focus of the mind's eye.

Much of the research on meditation and mindfulness has focused specifically on whether it produces significant and lasting *improvements* in the performance of different cognitive tasks. However, it should be pointed out that making changes to any brain process could have unintended negative consequences as well. There have been reports linking mindfulness to reduced abilities in some other cognitive tasks. For example, Whitmarsh and colleagues (Whitmarsh, Udden, Barendregt, & Petersson, 2013) found that subjects with higher levels of self-reported mindfulness were *worse* at learning artificial grammar regularities compared to those with lower levels of mindfulness. As with all methods of altering the brain (e.g., medications, stimulation treatments, training), we should keep in mind that any change to the brain is likely to affect multiple cognitive and neural processes. Fully assessing both the positive and the negative consequences of any intervention is critical for helping people make informed decisions about whether or when a specific intervention should be initiated. This is not to question whether mindfulness training *can* be beneficial, as there are numerous reports of its positive effects both on cognition and on general health and well-being. Rather, it's simply a reminder that research into both positive and negative effects is important, in that decisions about engaging in any sort of brain-altering activity should include a complete understanding of all the ways that it might change mental processes. This perspective is also critical in the ongoing debate regarding video games: Are they helpful or harmful for young children, and should they be used to help our aging populations?

7.6.2   Video Games and "Brain Training"

As discussed in Chapter 2, there has been an ongoing and active debate about the effects of playing video games on cognitive functions. Studies of video game "experts" (individuals who have played video games for hours per week consistently over years) typically find their scores on attention and perceptual tasks to be higher than those of individuals who don't play video games (reviewed in Sampalo, Lázaro, & Luna, 2023). This comparison could, of course, be due to a "self-selection" confound, in which individuals who are naturally better at visual attention to start with continue to play video games because they're very good at those sorts of things. Therefore, these studies alone cannot address whether there is a *causal* effect of playing video games on the mechanisms of attention. However, randomized controlled studies, in which individuals are randomly assigned to either a video game playing group or a control (no video game playing) group, also find that video games enhance performance on attention tasks. In a meta-analysis of published reports, Bediou and colleagues found that individuals randomly assigned to play video games consistently outperform control subjects (Bediou, Adams, Mayer, Tipton, Green, & Bavelier, 2018). The magnitude of the benefits observed in randomized controlled studies, however, was found to be smaller than what is typically seen when comparing long-term expert gamers to complete novices. Factors contributing to this could include the self-selection bias in "expert" studies, as described earlier, or the selection process for most randomized controlled studies that typically involves the enrollment of only individuals who don't already play video games (thereby sampling from the population that may be less naturally inclined to play video games). Additionally, these results may simply be due to the fact that randomized controlled studies involve only a very brief amount of training compared to the years of play necessary to be included in an "expert" group.

A criticism of some of this research is that the tests of perception and attention often are quite similar to the video games being practiced. Even if the test doesn't share the same task or any specific stimuli with the video games, both typically involve responding to stimuli on a computer screen. This highlights the issue of whether the processes trained by playing video games transfer beyond the task. Some studies suggest that playing video games trains core mechanisms of attention and therefore will translate well beyond the games (Bediou et al., 2018), but other research has found that some attention and perceptual skills *don't* show a benefit from playing video games (e.g., Boot, Kramer, Simons, Fabiani, & Gratton, 2008). Another issue is that there is a great variety of video games, and it would be surprising if all video games had the same effects. Reviews of this area of research generally find that playing action video games provides the largest benefits to attention compared to the other types of video games used in these studies. Action video games include first-person shooter games, in which the player sees the world through their avatar's eyes, and third-person shooter games, in which the player views the back of their avatar moving through the world. What may contribute to the more robust effects on attention seen for both of these types of action video games, compared to other types of video games, is that they are typically more immersive and require more frequent and varied actions, including rapid switching between different objects and tasks. As will be described later, researchers have also been developing new games in an attempt to target specific processes of attentional control in different populations with various attention deficits. Finally, although there is excitement about the potential of video games to help boost attention skills, it must also be pointed out that there is a danger of some individuals becoming addicted to gaming. Indeed, recent updates to the DSM-5TR (American Psychiatric Association, 2022) and the International Classification of Diseases (ICD-11; World Health Organization, 2022) include "gaming disorder." Criteria for this new disorder include many symptoms common to other forms of addiction (e.g., loss of interest in other activities, deceiving family members about frequency of use, needing to spend increasing amounts of time playing to satisfy the urge, withdrawal symptoms). Some of the more severe forms have been linked specifically to *internet* gaming and *gambling*, but this does raise a cautionary point to consider before advising anyone to start playing video games.

To investigate more precisely how playing video games may affect brain processes of attention, a number of studies have used fMRI and ERP measures to track how attentional control systems are affected. In relation to the *effects* of attention on perceptual processing, as covered in the previous chapter, several studies have investigated whether early attention effects, such as enhancement of the P1 and N1 ERP components generated in visual processing areas, would show larger attention effects after video game training. When comparing action video game players to nongamers, these studies found no evidence for a change in attention effects on the P1 or N1 components (Föcker, Mortazavi, Khoe, Hillyard, & Bavelier, 2019; Wu, Cheng, Feng, D'Angelo, Alain, & Spence, 2012), nor for there being any difference in attention effects in fMRI-defined early retinotopic areas (reviewed in Bavelier & Green, 2019). Bavelier and Green (2019) suggested that the effects of playing video games may instead manifest at later levels of attentional selection, such that stimuli are not filtered out at early stages but only after all initial sensory analyses have been completed (e.g., at the level of the P3 component). Some studies using steady-state-evoked potentials (SSVEPs) have found that action video game players are better able to suppress processing of irrelevant stimuli compared to nongamers (Krishnan, Kang, Sperling, & Srinivasan, 2013; Mishra, Zinni, Bavelier, & Hillyard, 2011). SSVEPs measure neural responses, but these data are processed in a way that lacks the high temporal resolution of standard ERP measures and therefore it is difficult to know how early in processing this SSVEP-indexed selection is enacted. Neuroimaging studies have also found that video game players can better inhibit activity to irrelevant moving stimuli in motion-sensitive visual processing areas of the cortex (Bavelier, Achtman, Mani, & Föcker, 2012), but since these fMRI measures have poor temporal resolution, it is again unclear how early in processing these effects occurred. More research is needed to understand how the *effects* of attention at early versus higher levels may be modulated by playing video games.

Neuroscience studies have revealed robust effects of playing video games on *attentional control* processes, however. In terms of the different subcomponent processes of control, behavioral studies have shown that the initial exogenous capture of attention is not affected much by playing video games (Hubert-Wallander, Green, Sugarman, & Bavelier, 2011), but that the ability to quickly disengage and reorient attention is enhanced for video game players (Chisholm & Kingstone, 2012). Structural MRI studies have found that white matter tracts in visual, motor, and prefrontal areas are strengthened in expert video game players (Gong, Ma, Gong, He, Dong, Zhang, Li, Luo, & Yao, 2017; Zhang, Du, Yang, Qin, Li, & Zhang, 2015). In a study of adolescent video game players, Kühn and colleagues found that the cortical thickness of the FEF, which are strongly implicated in attention control, was positively correlated with the hours per week spent playing video games (Kühn, Lorenz, Banaschewski, Barker, Büchel, Conrod, Flor, Garavan, Ittermann, Loth, Mann, Nees, Artiges, Paus, Rietschel, Smolka, Ströhle, Walaszek, Schumann, Heinz, Gallinat, & IMAGEN Consortium et al., 2014). Activity in the frontoparietal attentional control networks has also been found to be affected by playing video games, but in a way that suggests improved efficiency and not a general increase in activity (reviewed in Choi, Shin, Ryu, Jung, Kim, & Park, 2020). Specifically, fMRI studies have reported that expert players show less activity in frontal and parietal regions (e.g., FEF and the superior parietal lobule) compared to nongamers while maintaining the same high level of performance on the attention task, suggesting that the attention control regions in experts simply don't need to work as hard to maintain good performance (Föcker, Cole, Beer, & Bavelier, 2018). The Föcker et al. (2018) study also found that expert gamers had enhanced functional connectivity between parietal and visual areas during the process of orienting of attention. Additionally, expert gamers showed a stronger relation between TPJ activity and behavioral performance on "catch trials" (i.e., trials in which the subject needs to refrain from making any motor response because only nontarget noise stimuli are presented); specifically, the results showed that the greater the TPJ activity, the less likely the subject was to produce a false alarm. Overall, the findings that attentional *control* processes may be modulated by playing video games whereas the *effects* of attention on early perceptual-level processing are not changed with gaming experience suggest that attentional control processes may be more plastic and malleable than are the earliest effects of attention on sensory processing. It is also noteworthy that fMRI studies have found enhanced activity in the striatum and connected frontal regions while participants are actively playing video games, suggesting that reward processing may also play a role in plastic changes to the attentional control system (Lorenz, Gleich, Gallinat, & Kuhn, 2015).

As mentioned in Chapter 2, there has been great interest in understanding whether video games could help people with attention problems. In a meta-review of studies on individuals with ADHD, it was found that some therapies that included video games enhanced cognitive skills in this group (Peñuelas-Calvo, Jiang-Lin, Girela-Serrano, Delgado-Gomez, Navarro-Jimenez, Baca-Garcia, & Porras-Segovia, 2022). Furthermore, this review found that ADHD symptoms outside of the lab were also improved by some therapies that included some video game playing. More research needs to be done to ensure that these benefits can be enjoyed without potential side effects and without danger of addiction, but these studies provide hope that playing video games might be an important addition to treatment options for ADHD.

Other studies have investigated whether video games could be used to help slow the decline in attention functions that occurs with aging. Studies have shown that the ability to perform well in dual-task procedures is especially susceptible to decline in elderly individuals, and that our ability to multitask declines throughout adulthood (Verhaeghen, Steitz, Sliwinski, & Cerella, 2003). Several researchers have investigated whether video games could provide a means to stave off these declines. Since video games are designed to be fun and engaging, the hope is that people may be motivated to play them. Again, however, the critical question is whether such games really help to improve core cognitive processes or just improve performance on a task that is the same or highly similar to the game being played. Studies employing neuroscience methods are important here for measuring these brain processes directly in order to better understand what neural mechanisms are being affected. In a study that assessed both the behavioral and neural effects of multitask training, Anguera and colleagues had older adults (aged 60–85 years) train on a customized driving game and compared their performance and brain activity to younger adults performing the same task (Anguera, Boccanfuso, Rintoul, Al-Hashimi, Faraji, Janowich, Kong, Larraburo, Rolle, Johnston, & Gazzaley, 2013). The older adults were assigned to one of three groups: a no-training (or no-contact) control group, a single-task training group, or a multitask training group. The training here involved playing a custom-designed driving simulation game on a laptop over a period of 4 weeks. Subjects were to play the game for 1 hour a day, three times per week. The game could involve either performing one of two separate tasks or performing both tasks concurrently. The "drive task" involved using a joystick to keep the car in the scene consistently in the center of a winding road. The "sign task" involved discriminating the color and shape of signs presented on the screen and responding as quickly as possible to the predefined target stimulus (e.g., a green circle) and not responding to lures (e.g., green, blue, or red pentagons or squares). The dual-task or multitask condition involved performing both tasks together (the "sign and drive" condition). The single-task training group alternated between conditions in which they performed *either* the "drive-only" task or the "sign-only" task. The multitask training group always performed both tasks concurrently in every session. The older adults were compared to a group of young adults (all 20 years old). After 1 month of training, performance on a test of the multitask version of the game revealed that there was a trend for the single-task training group to perform the task better than the no-training group, but this was not significant. However, the multitask training group showed a significant improvement in task performance compared to their own pretraining baseline, and this group was significantly better after training than both the single-task training and no-contact control groups. This provided evidence that multitask training improved performance to a much greater extent than single-task training. Even more compellingly, the multitask training group showed evidence that some improvement in the task was maintained even 6 months after the training had ended. Although the performance of that group did drop markedly compared to their performance immediately after the training had been completed, it was notable that the multitask training was still improving the performance of subjects even months after the training had ended.

In addition to investigating task performance, this study also measured EEG responses and compared the effects of training for older adults (60–85 years old) to the performance of young adults (20 years old). Before training began, there was a significant difference between younger and older adults, especially in the power (i.e., strength) of theta activity (4–7 Hz) over central frontal regions (Figure 7.16, top row). After 4 weeks of training, the older adults in the multitask training group showed a significant increase in this frontal theta activity, and the scalp distribution of this enhanced activity was highly similar to the pattern seen in younger adults (Figure 7.16, bottom row). This provides further evidence that multitask training can help older adults engage the attentional control processes that younger adults naturally use in the performance of these tasks. The older adults in the single-task training group showed a trend toward this boost in theta activity, but it was not a significant change from the pretraining baseline. This enhanced frontal theta power (i.e., increased activity in the theta frequency band) has been interpreted as indexing the engagement of prefrontal control

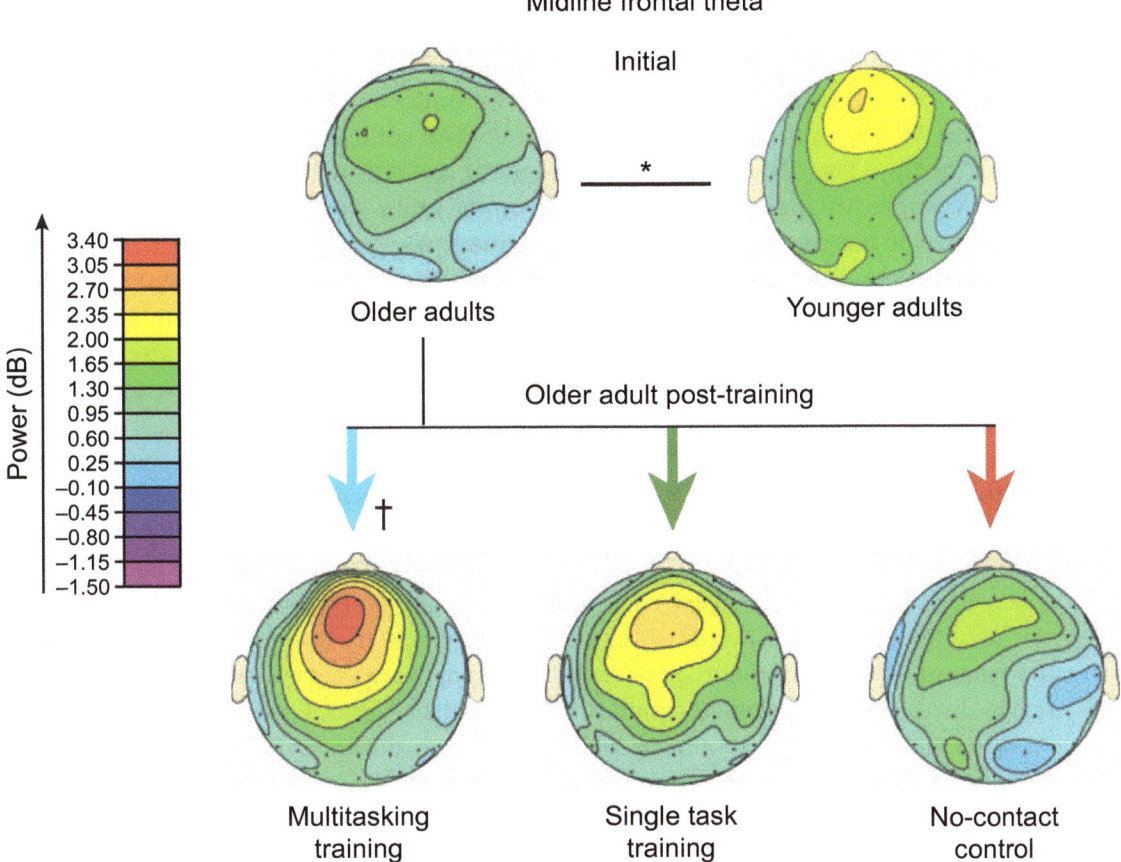

Figure 7.16 EEG effects of video game training. Scalp topographies showing the power (i.e., strength) of neural oscillations in the theta frequency range (4–7 Hz) during an attention task. Top row: Before training, young adults exhibited significantly greater theta power over midline scalp regions compared to older adults. Bottom row: After 4 weeks of training, older adults in the multitask training group showed a significant increase in frontal theta power, matching closely the theta activity of younger adults performing this task. *p < 0.05 between-group comparison; †p < 0.05 post-training > pretraining (within group). Reprinted from Anguera et al. (2013), *Nature*, with permission from Springer Nature.

mechanisms when attentional resources need to be rapidly focused on the appropriate dimension and so selectively process that information for efficient responding. Finally, this study also assessed the coherence of theta activity between frontal and posterior scalp sites. Like the results presented earlier, this measure of connectivity with frontal control regions showed that subjects in the multitask training group but not the single-task or no-contact groups showed a strengthening of this coherence after training. Furthermore, this significant change in the multitask training group after training resulted in a pattern of connectivity that was very similar to that shown in young adults. This modulation of coherence across brain regions suggests that the neuroplastic effects of multitask training may include changing the connectivity across brain networks.

Another recent study has investigated whether training with an augmented reality (AR) program could help treat the impairments of attention in patients with *unilateral spatial neglect* (Stammler, Flammer, Schuster, Lambert, Neumann, Lux, Matuz, & Karnath, 2023). In a randomized controlled study, 20 patients with unilateral spatial neglect were assigned to either the AR experimental condition or to a control condition that involved only standard neglect therapy. The AR program involved having the patients use a tablet to actively explore space. A graphic of a red bird was displayed on the tablet superimposed on a camera view of the section of the room toward which the patient was pointing the tablet. As the patient attempted to move the tablet to center the bird image within a circle in the center of the tablet's screen, the background was updated in real time corresponding to the patient's movements of the tablet. In this way, the AR program could induce the patient to orient attention into and around their neglected hemispace. If the patient was unable to find or track the bird in their neglected field within 30–90 seconds, the trial was aborted to avoid patient frustration, and then a new trial began. The motivation for this intervention was based in part on previous research suggesting that therapies that include overt movements of the head and body to orient attention may be more successful in ameliorating the symptoms of neglect (Wiart, Côme, Debelleix, Petit, Joseph, Mazaux, & Barat, 1997). In having to hold and move a tablet through space, the subject needs to engage overt orienting processes of their whole body, whereas many standard therapies for neglect only instruct the subject to try to attend to the leftward side of drawings or text on a sheet of paper or asking them to move their eyes toward the left side of an image displayed on a computer monitor that is fixed within a small region of space. Both experimental and control groups in this study were evaluated before and during 2 weeks of training with five different attention tasks typically used to assess neglect. During each week of training, the subjects participated in 25-minute therapy sessions five times per week. Whereas both groups showed improvement in neglect symptoms over the course of the 2 weeks of training, the AR group exhibited better performance across four of the five different attention tasks, whereas patients in the control group improved in only one of the five attention tasks. Furthermore, significant improvements were observed in the AR group as soon as the first week of training, and these improvements remained stable when tested a month after training had ended. That 1-month evaluation was the end point of the study, so longer-duration effects were not tested.

Overall, the results of these lines of research provide evidence that repeated playing of video games can have lasting effects on attention functioning and neural processes. Further research is still needed to determine whether those changes generalize widely beyond the games being played, and active debate continues regarding whether these gaming-induced changes in the brain are ultimately beneficial or harmful to the individual. As detailed in the studies presented earlier, there is great potential in customizing games to address specific attentional deficits or to ensure that attentional functions remain strong over the lifespan. What's critical here, however, is the level of understanding of the brain mechanisms of attention. Our knowledge of how attention works dictates how effectively we can customize games to improve attention. Therefore, while the creation of games to help people is an exciting and potentially important application of existing attention research, much remains unknown about the mechanisms of attention, and more research into the core processes of attention is critical for ensuring that therapies and training can accurately target the desired mechanisms in the brain.

This chapter has focused on the multiple neural mechanisms that contribute to the *control* of attention, and much of this research has involved understanding how attention is focused during a specific *moment* in time. However, other research is revealing important *temporal* aspects of attention that are also critical to understanding how attention is allocated. In the next chapter, we will discuss both intrinsically and externally triggered rhythms of brain activity that influence how attention is allocated across time, as well as voluntary and involuntary influences on how long attention dwells on an item or thought.

CHAPTER SUMMARY

- A dorsal attention network, including lateral prefrontal and parietal areas, supports control of the orienting and focusing of attention across space and objects.
- Attentional control, whether triggered internally or externally, biases subsequent perceptual and higher-order processing via multiple preparatory mechanisms.
- The disengagement and reorienting of attention involves ventral regions of the parietal and frontal lobes and neural activity in the gamma frequency band.
- Midline frontal regions, including the anterior cingulate cortex, support multiple processes of executive control.
- The monitoring of actions and the integration of feedback can be indexed with high temporal resolution through distinct ERP components.
- Methods that are being investigated to enhance attention abilities or ameliorate attention deficits include meditation, neurofeedback, and customized video game programs.

REVIEW QUESTIONS

Describe the behavioral evidence from stroke patients and the neuroscience evidence from healthy young adults that a right-hemisphere-lateralized frontoparietal network controls the orienting of attention.

Compare the brain mechanisms of "willed attention" to exogenous orienting and endogenous orienting in response to instructive cues.

Explain how single-unit recordings and deactivation studies in nonhuman primates have advanced our understanding of frontal and parietal control over attention effects in visual processing regions.

List and briefly describe the contributions of ventral frontoparietal and subcortical regions to controlling the allocation of attention.

Describe the "executive control" aspects of attention and the brain regions underlying these processes.

Compare mindfulness, neurofeedback, and video games as ways to "train the brain."

FURTHER READINGS

Chambers, C. D., Payne, J. M., Stokes, M. G., & Mattingley, J. B. (2004). Fast and slow parietal pathways mediate spatial attention. *Nature Neuroscience, 7*(3), 217–218.

– This article illustrates how TMS can be used to test predictions about not only *where*, but also precisely *when* a cortical region is involved in processes of attention.

Corbetta, M., & Shulman, G. L. (2002). Control of goal-directed and stimulus-driven attention in the brain. *Nature Reviews. Neuroscience, 3*(3), 201–215.

– This review article integrates many human neuroimaging studies to provide a model of attention control in the brain that includes networks for top-down and bottom-up attention mechanisms.

Föcker, J., Cole, D., Beer, A. L., & Bavelier, D. (2018). Neural bases of enhanced attentional control: Lessons from action video game players. *Brain and Behavior, 8*(7), e01019.

– This review article summarizes several EEG and fMRI investigations into the effects of playing video games on neural indices of attention.

Martinez-Trujillo, J. (2022). Visual attention in the prefrontal cortex. *Annual Review of Vision Science, 8*, 407–425.

– This article reviews monkey electrophysiology studies that provide evidence for the mechanisms of attentional control enacted in prefrontal regions.

Wu, T., Spagna, A., Mackie, M., & Fan, J. (2023). Resource sharing in cognitive control: Behavioral evidence and neural substrates. *NeuroImage, 273*, 120084.

– This recent fMRI study presents evidence for a network of areas, critically including the ACC, supporting attentional control across tasks through resource sharing and coordination.

8 Temporal Attention
Timing and Rhythms of Brain Mechanisms

Learning Objectives

- Describe how temporal information, at multiple scales, affects attentional allocation.
- Identify the neural processes that are affected when attention "blinks."
- Compare and contrast temporal and spatial attention.
- Describe the various influences on the holding of attention and the mechanisms affecting attentional dwell time.
- Understand the roles of external and internal rhythms in controlling attention.

8.1 Attention and Time

Previous chapters have focused on the distribution of attention across space, features, and objects at specific, isolated *moments* in time. However, the allocation of attention *across* time has important consequences as well. New research is revealing that there are temporal constraints on attention and that there are multiple factors – some involuntary – that contribute to how long attention remains on an object. A comprehensive account of attention is only possible when the temporal aspects of attention are understood as well. Whereas previous chapters discussed the initial orienting of attention, this chapter presents research on the timing of attention, including how long attention dwells on an object and the automaticity of these temporal processes. This dwelling of attention can have a profound effect on our ability to avoid distractions. If we are able to disengage rapidly from a salient stimulus, the initial capture of attention may have only a brief and minor impact on our work; but if involuntary processes cause our attention to dwell upon a stimulus for a prolonged period of time, the impact can end up being much worse. This chapter will revisit processes such as vigilance and sustained attention as they relate to current work on the temporal limits of attention. Finally, new theories suggest that attention is phasic in nature, alternating between periods of high versus low focus and between periods of enhancing perception versus orienting and exploration.

8.2   When Attention "Blinks"

One of the most compelling findings regarding the temporal allocation of attention comes from the **attentional blink (AB)** task. Results from this task reveal that there are temporal limits on our ability to attend, and that without sufficient attentional resources we can be essentially blind to stimuli, even those in plain view. In the classic AB task – a type of **rapid serial visual presentation (RSVP)** paradigm developed by Raymond, Shapiro, and Arnell (1992) – subjects view a sequence of stimuli presented rapidly and sequentially at fixation (Figure 8.1). Each stimulus is presented for only a brief duration, with little time between successive stimuli, and subjects are required to detect certain target stimuli within the stream. In the *single-target* condition, in which subjects are instructed to report on the presence or absence of just a single type of target (e.g., monitor the stream only for the letter "X"), subjects do quite well on the task, rarely missing a target despite the rapid rate of presentation. However, in *dual-target* conditions, in which subjects must also report the presence of another *potential* target (e.g., report the identity of a white letter, if one appears) in addition to reporting the presence or absence of a prespecified target letter (e.g., the "X" in Figure 8.1), subjects perform much worse. Specifically, the second target (T2) is often missed if it occurs shortly after the first target (T1). At longer intervals between the two target stimuli, subjects reliably detect both targets. The term "attentional blink" refers to the finding that attention seems to temporarily close for a brief period while the first target is being processed. Similar to how we cannot see stimuli during a physical blink of the eyes, we're unaware of stimuli presented during a brief period when our attention "blinks." Critically, however, studies have shown that subjects are not physically blinking at the time of the second target in the AB paradigm. Their eyes are open and looking directly at the second target stimulus when it appears, but they're unable to report its presence. Importantly, the finding that the AB only occurs in the dual-target condition shows that the effect isn't simply a form of visual masking or physical interference between the successive stimuli, because the exact same streams of stimuli don't cause the AB when subjects are only looking for a single target. Rather, the deficit in perception only occurs when attention must be allocated to the first target in order to detect and report it. Thus, the AB is linked to limits in attention.

Behavioral studies using this paradigm have refined what is known about the temporal limits of attention, finding that the AB generally occurs most strongly from ~200 to 500 ms after the first target (reviewed in Shapiro, Arnell, & Raymond, 1997). Although the time after the first target is important, research has shown that the temporal position of the subsequent stimuli is also critical. Research has revealed that perception of the second target is often spared if it is the stimulus directly following the first target and it appears within a certain time window. This has been termed "lag 1 sparing" (Potter, Chun, Banks, & Muckenhoupt, 1998). In the AB paradigm, the "lag" of a second target stimulus refers to when it occurs relative to the first target stimulus. "Lag 1" refers to the first item after the initial target, "lag 2" refers to the second item after the initial target, and so on (Figure 8.1). The AB is typically largest at lag 2 and lag 3 (Shapiro et al., 1997). Lag 1 sparing indicates that the "closing" of the attentional window doesn't occur immediately upon the occurrence of the first target. The lag 1 stimulus essentially

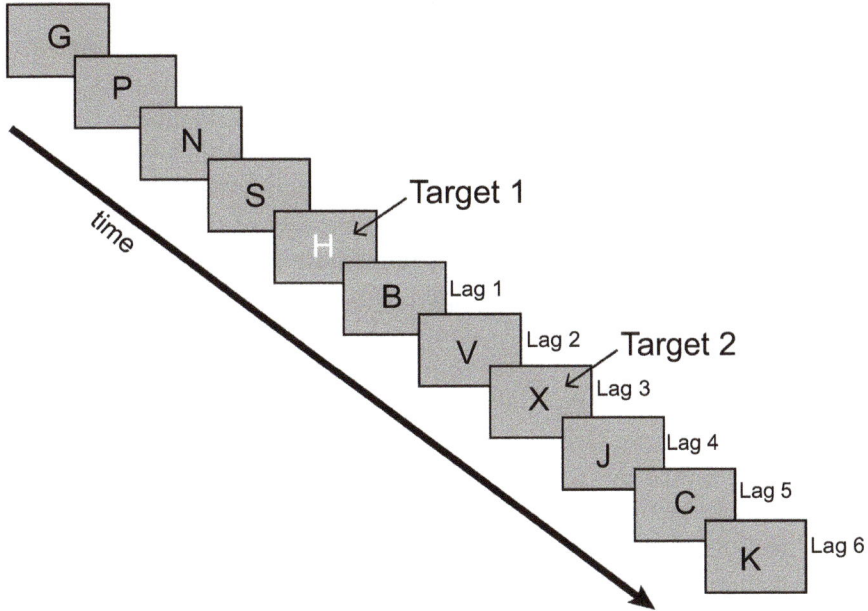

Figure 8.1 Trial sequence for an AB task. In the condition shown here, subjects would have to respond at the end of the trial to identify the letter in white (target 1) and whether or not an "X" occurred anywhere in the stream (here labeled as target 2). The "lag" refers to the ordinal position after target 1.

sneaks in before attention is briefly closed by the AB. This provides further evidence for the timing of attention – while the brain is processing the first target stimulus to the level of understanding whether or not it's a target, subsequent stimuli are still being processed. Only after it's determined that attention must be allocated fully to a stimulus (because it requires encoding and an overt response) does attentional selection become engaged, preventing the full processing of other stimuli for a brief period of time. Critically, the competition for attentional resources appears to be a key element in producing the AB, as research has found that the presence of a lag 1 stimulus shortly after the target is critical for generating the AB. Specifically, if the temporal position of the lag 1 stimulus is replaced with a blank screen, so that there is no immediate competition for attention when processing the first target stimulus, the AB is eliminated or significantly reduced (e.g., Chun & Potter, 1995). In other words, if the first target can be processed and discriminated before a subsequent stimulus is presented to compete with it, the AB is often not produced for a subsequent second target. This aspect of the AB links it to the biased competition theory of attention (Desimone & Duncan, 1995) described in previous chapters. This theory holds that the selection processes of attention are engaged when there is competition. In terms of spatial and object attention, the competition comes from other stimuli present at the same time. The AB shows that the competition also has a temporal element – stimuli appearing within a couple of hundred milliseconds after a target can cause attention to blink. Furthermore, the AB shows that this closing of the access to attentional resources by competition in the temporal domain is a rapid process, as the bottleneck seems to be avoided as long as no competition occurs within a few hundred milliseconds.

Two other phenomena observed in psychology research that appear similar in some ways to the AB are the **psychological refractory period (PRP)** and "repetition blindness." The PRP refers to the finding that when two successive stimuli are targets that must be responded to, the reaction time to the second target is slower the closer it occurs in time to the first target (Telford, 1931; Welford, 1952). A common explanation for the PRP has been that the planning and execution of a response to the first stimulus takes time, and therefore any planning or response to a second stimulus must be pushed back until resources are available (Pashler, 1984). The PRP is dependent on both targets requiring an immediate response, whereas the AB is observed without the need for immediate overt responses. Also, a key difference between the two effects is that the PRP effect on the second target is stronger the closer that target occurred relative to the first target, indicative of a simple refractory period, whereas the AB often shows lag 1 sparing, indicating a slightly delayed critical window when attention is unavailable. The other related effect is repetition blindness, which refers to the finding that when a stimulus is presented a second time in very close temporal proximity to the first, subjects do not report seeing it a second time (e.g., Kanwisher, 1987). However, this effect is specific to the same stimulus appearing twice, and therefore it is thought to be more indicative of how object files are created in the mind and the timing limits of updating those object files. Therefore, although this effect is similar to the AB in terms of subjects being "blind" to a stimulus, the AB is more directly linked to the mechanisms of attention.

8.2.1 Neural Mechanisms of the AB

While behavioral studies have revealed many important aspects of the AB, it is difficult to infer the mechanisms of this process based only on manual response times and accuracy. Using event-related potentials (ERPs) in a modified AB paradigm, Vogel and colleagues (Vogel, Luck, & Shapiro, 1998) were able to directly investigate the stage(s) of neural processing that are affected by the AB. In this study, the typical AB paradigm was modified across a series of experiments in order to evoke robust ERP components that index different stages of processing. As shown in the top parts of each panel in Figure 8.2, these studies found the typical pattern of AB as measured behaviorally, with reduced accuracy of T2 detection for the *dual-target* condition but not for the *single-target* condition, and only at a middle lag between lag 1 sparing and longer lags when the AB isn't found. Critically, the ERP results show that the AB does not completely filter out stimulus processing. Specifically, as shown in Figure 8.2 (left panel), the visual P1 and N1 components were the *same* in amplitude and latency for the stimuli that showed an AB (i.e., the "lag 3" stimuli) as for the stimuli that were consciously perceived (i.e., the "lag 1" and "lag 7" stimuli). As discussed in earlier chapters, the P1 and N1 components are sensory-evoked components generated by visual processing regions of the cortex, and these components are modulated by voluntary and involuntary *spatial* attention. However, these stages of processing are not affected by the *temporal* selection processes of the AB. Processing at these early stages proceeds intact, even for stimuli of which subjects are unaware.

In a different experiment designed to assess later stages of processing, Vogel et al. (1998) tested the P3 component, which is thought to index higher-order stimulus evaluation and

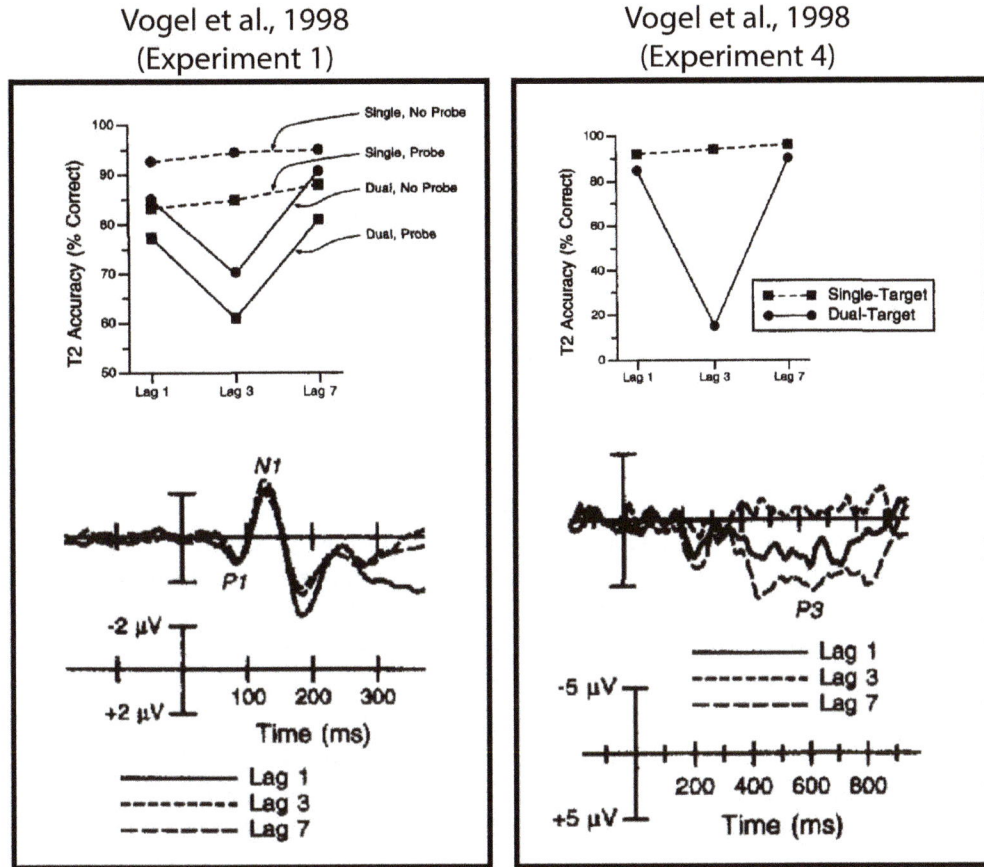

Figure 8.2 ERP studies of the AB. The top of each panel shows the expected AB pattern in behavioral performance, with T2 accuracy at lag 3 being significantly worse than the other lags and significantly worse than the single-target condition when T1 could be ignored. The bottom of each panel shows the ERP components of interest and whether those are affected by the AB (described in the main text). Figure adapted from Vogel et al. (1998), *Journal of Experimental Psychology: Human Perception and Performance*, with permission from the American Psychological Association.

decision-making processes (Chapman & Bragdon, 1964; Sutton, Braren, Zubin, & John, 1965). (Note that the P3 component tested in this study is now often referred to as the P3b, which has been shown to be distinct from a slightly earlier and more frontally located P3a component that is linked to novelty detection [Comerchero & Polich, 1999]; these components are sometimes referred to as P300 components, in reference to their ~300-ms latency.) In Vogel et al.'s (1998) study of the AB, they found that the P3 component was completely absent for the lag 3 stimuli but was robustly generated for both the lag 1 and lag 7 stimuli (Figure 8.2, right panel). The presence of the P3 component for the lag 1 and lag 7 targets provided evidence that the task evoked the level of processing indexed by the P3 component, but that it was blocked during the

AB. In line with theories that the P3 component indexes context updating (Donchin & Coles, 1988), the authors interpreted these results as showing that the encoding of stimuli into working memory and conscious awareness is completely prevented during the AB. However, other research has found that the P3 is not completely eliminated but rather significantly delayed during the AB. Specifically, in studies in which the second target stimulus is masked by the presence of another stimulus in close temporal proximity, the P3 component has been found to be absent during the AB (e.g., Kranczioch, Debener, & Engel, 2003; Vogel et al., 1998); however, in studies in which the second target stimulus is not immediately masked, the P3 component is generated but delayed in time (Brisson & Bourassa, 2014; Sessa, Luria, Verleger, & Dell'Acqua, 2007; Vogel & Luck, 2002). Together, these findings suggest that the encoding of information into working memory is significantly postponed by the AB, and that if the target is masked before this encoding takes place, then those stimuli never get fully encoded.

The effect of the AB on another higher-order stage of processing has also been found to depend on task demands. Specifically, studies have investigated the effect of the AB on the N400, an ERP component that indexes a semantic-level stage of processing and is typically generated to stimuli that violate a semantic context during processing of language stimuli (e.g., Kutas & Federmeier, 2011). In early studies of the AB, it was found that semantically incongruent T2 stimuli produced an N400, regardless of the lag between T1 and T2, suggesting that this level of processing was generated automatically regardless of the resources limited by the AB (Vachon & Jolicoeur, 2011; Vogel et al., 1998). However, subsequent studies have found that the N400 is affected by the AB under certain conditions (e.g., Batterink, Karns, Yamada, & Neville, 2010; Giesbrecht, Sy, & Elliott, 2007). For example, Giesbrecht et al. (2007) replicated the result that the N400 isn't affected by the AB when the task is *easy* (i.e., their "low-load" condition), but when the task was made more difficult (i.e., their "high-load" condition) there was a significant effect on the N400. Specifically, in the high-load condition, the N400 was completely *absent* during AB trials but was still robustly generated outside of the AB window at longer lags. Overall, these studies provide critical information to help us understand how attention in the temporal domain compares to spatial and feature-based attention. Whereas spatial attention has strong effects on modulating early sensory processing (e.g., P1 and N1 components), the limits imposed by the AB do not appear to affect those early stages. Higher-order levels of processing, however, are affected by the AB, and in some cases the AB may completely prevent stimuli from even reaching those higher levels of processing.

In addition to studies using specific ERP components to assess the neural processing associated with the AB, research analyzing electroencephalography (EEG) frequency bands during task performance has also provided important insights. Zauner and colleagues (Zauner, Fellinger, Gross, Hanslmayr, Shapiro, Gruber, Müller, & Klimesch, 2012) suggested that alpha **entrainment** is a critical factor in the production of the AB. Specifically, they noted that most AB paradigms use a temporal pattern in which stimuli appear about every 100 ms, corresponding to the alpha frequency range (8–12 Hz). Since an increase in alpha power over occipital scalp sites has been associated with a relative suppression of visual processing (Foxe & Snyder, 2011; Jensen & Mazaheri, 2010; Klimesch, Sauseng, & Hanslmayr, 2007), Zauner and colleagues (2012) hypothesized that the AB would be associated with an increase in alpha-band activity entrained by the presentation rate of the stimuli. In an EEG study, they found that the AB was associated with the *phase* of the entrained alpha rhythm. Specifically, the AB was more likely to occur when the phase of the alpha rhythm had been tightly entrained and the appearance of the second target occurred during the negative peak (i.e., 180-degree phase) of the alpha rhythm. Importantly, this study also showed that the critical entrainment to the alpha-band phase was "broad band" in the sense that it occurred equally well whether the stimulus presentation was always at precisely 10 Hz or varied from 8.3 to 12.5 Hz. In a subsequent behavioral study, Shapiro and colleagues (Shapiro, Hanslmayr, Enns, & Lleras, 2017) varied the timing of the stimulus presentation in a standard AB paradigm and found that the AB effect was significantly stronger when the stimuli were presented at frequencies within the alpha (10.3 Hz) and beta (16.0 Hz) ranges compared to stimulus presentation rates of both slower (theta: 6.26 Hz) and faster (gamma: 36.0 Hz) frequencies. Together, these studies provide evidence for the importance of timing in attention, and they suggest that the limitations in attentional processes are critically related to the frequency at which the information is being sampled and processed. Findings such as these helped lead researchers to hypothesize a "rhythmic theory of attention," which will be discussed in Section 8.5. Most of the research on the AB has focused on the visual modality, but there is also interest in how this type of attention mechanism may play a role in other sensory modalities as well (see Box 8.1).

Box 8.1 Do the ears "blink"?

Because the original studies of the AB used visual stimuli, the reference to the blinking of the eyes was a useful analogy. However, since the effect is thought to be due to higher-order mechanisms beyond simple sensory processing, it should be possible to observe "blinks" to information coming in through our other senses as well. Indeed, research has shown that auditory and tactile sensations can also be "blinked out" from awareness. Studies have found that responses to streams of auditory stimuli (Arnell & Jolicoeur, 1999; Duncan, Martens, & Ward, 1997) and tactile stimuli (Hillstrom, Shapiro, & Spence, 2002) show a very similar pattern to responses that have been found for visual stimuli, at least when all the stimuli are within a single sensory modality. Consistent with the theory that the blink represents a limitation in higher-order attentional processes, the timing and parameters resulting in an AB are highly similar across visual, auditory, and tactile domains. For example, when brief sounds were presented sequentially in rapid succession, the identification of a first target stimulus resulted in an impaired ability to detect a second target if it followed the first by ~200–500 ms (Tremblay, Vachon, & Jones, 2005). Although the blink has been found in all of these sensory modalities when tested independently, there has been some controversy over whether target stimuli in one sensory modality automatically create a blink across all sensory modalities. Early studies, for example, did not find evidence for a cross-modal auditory-to-visual AB (Duncan et al., 1997; Potter et al., 1998). Subsequent studies, however, have suggested that the existence of a cross-modal blink depends critically on the level of information processing required to identify the target. Soto-Faraco and colleagues (Soto-Faraco, Spence, Fairbank, Kingstone, Hillstrom, & Shapiro, 2002) argued that a cross-

Box 8.1 (cont.)

modal AB occurs when the critical target information is common across modalities (e.g., the location of stimulus), but that such a cross-modal blink would not occur if the critical target information is specific to one modality (e.g., the phonological properties of a sound versus the form of a visual stimulus). This point underscores that the AB is not a singular bottleneck occurring at just one fixed place in the brain, but rather that it can occur at different levels depending on the subjects' task. When critical target properties can be generalized across sensory systems, the AB can be triggered in one modality and affect processing in another. Similarly, Rau and colleagues (Rau, Zheng, Wang, Zhao, & Wang, 2020) found auditory–tactile cross-modal AB effects during a task that was similar for both modalities (i.e., localize the left or right spatial location of stimulus), but they found only within-modality AB during a task requiring within-modality discriminations (i.e., judge the force of a tactile stimulus and the pitch of an auditory stimulus). Ptito and colleagues used ERPs to investigate whether the effects on higher-order processing were the same for cross-modal AB as for single-modality AB (Ptito, Arnell, Jolicœur, & Macleod, 2008). They examined auditory-to-visual and visual-to-auditory AB while subjects detected target digits (regardless of modality) amid streams of nontarget distractors in either modality. The findings revealed that both cross-modal and unimodal AB resulted in the same effect on the second target: specifically, a slower and smaller P3 component for stimuli presented during the AB. Overall, these studies tell us that attention does "blink" for auditory and tactile stimuli, as well as for visual stimuli, and that the blink can suppress processing both within and across sensory modalities.

8.2.2 Neural Sites of Conscious and Unconscious Perception in the AB

Although EEG and ERP studies have provided important information about the timing and stages of processing that are affected by the AB, the neural sources of the AB cannot be localized well with those methods. Neuroimaging, on the other hand, has provided important insights into the neural structures involved in generating the AB and into identifying the sources underlying the conscious perception of stimuli. Marois, Chun, and Gore (2000) used functional magnetic resonance imaging (fMRI) to investigate whether the regions involved in the control of spatial attention are also implicated in controlling the AB. Given the limited temporal resolution of fMRI, their study compared *conditions* that resulted in *relatively* more versus less AB. Their conditions were designed to make use of the finding (described earlier) that the AB is partially dependent on whether there is a distractor item in the lag 1 position (Chun & Potter, 1995). Marois et al. (2000) created two conditions, one in which the AB was very likely to be produced (because a distractor stimulus was presented at the lag 1 position) and another in which the AB was much less likely to be produced (because a blank screen was presented at the lag 1 position). Importantly, each condition had a blank screen at some point in each trial sequence, but only in the "no AB" condition was the blank screen at the critical temporal position that avoided a blink being triggered. As expected, the behavioral results showed a significantly reduced AB in the condition with a blank screen at lag 1. The fMRI results comparing these two conditions revealed that regions of the parietal and frontal lobes that have been associated with spatial attentional orienting were significantly more active in the condition that produced a robust AB in behavior (the condition with a lag 1 distractor stimulus). Specifically, regions of the **dorsal attentional network (DAN)**, including the intraparietal sulcus (IPS) and medial frontal regions, were more active in the condition that showed a larger AB. In another experiment, the authors tested whether these regions showing temporal attentional control would be the same as those involved in spatial attention. Using a similar set of stimuli and task procedure, but with the competition arising in a spatial dimension (comparing trials with near versus far distractors present around the first target), the authors found an effect of competition on the presence of the AB, as well as strong overlap with the brain regions activated in the first experiment in which the competition arose in the temporal dimension. The overlap was especially strong in the right hemisphere, in line with previous research suggesting that the right hemisphere plays a critical role in attention (Mesulam, 1981). The authors suggested that a partially overlapping brain network controls attentional selection processes in the spatial and temporal domains.

In addition to investigating the attentional control regions responsible for engaging the selection processes that lead to the AB, other fMRI studies have helped us to understand the **neural correlates of consciousness (NCC)**. In order to investigate how the AB relates to conscious perception, it is necessary use an *event-related fMRI* approach to be able to sort individual trials according to whether or not the participant detected the stimulus. Marois, Yi, and Chun (2004) used this approach to investigate how object processing regions responded to stimuli that were consciously perceived versus missed in an AB task. In this experiment, pictures of intact faces and intact scenes (e.g., buildings, rooms) were interspersed with scrambled images. The face stimuli served as the T1 and the place stimuli were the T2. The results showed that pictures of scenes activated the **parahippocampal place area (PPA)**, even when the subjects reported *not seeing* any place images in the sequence. Consciously perceiving a place stimulus did result in an even greater amount of activity in the PPA, but the finding of significant activity in the PPA even when the place stimulus was not perceived provides important evidence that this level of processing is engaged in the midst of the AB, even without conscious awareness. In contrast, lateral frontal regions were only active when the place stimuli were consciously perceived. Thus, the AB paradigm has helped us to understand the levels of processing in the brain at which attention gates conscious perception. As discussed in Box 8.2, however, it appears that not all stimuli are equal when it comes to this gating of consciousness.

Whereas fMRI studies of the AB have been able to identify the brain regions associated with conscious processing, the sequence and relative timing of these areas cannot be assessed well with such methods. Magnetoencephalography (MEG), however, can investigate brain processes with both high spatial and temporal resolution. In a study that used an auditory T1 and a visual T2, Marti, Sigman, and Dehaene (2012) found a cross-modal AB, and MEG source localization showed the time course of activity in areas throughout the brain that showed different activity levels depending on whether the T2 stimulus was perceived or not. As shown in Figure 8.3 (top row), early activity (237–277 ms) in occipital–temporal and parietal regions was most enhanced for "seen" T2 stimuli, but these regions still showed some activity during this early time window even for T2 stimuli that had been blinked out of awareness. Later activity,

Box 8.2 To blink or not to blink?

If the AB simply prevents stimuli from being processed to a high level, how can some special stimuli break through? A simple filter mechanism that blocks the processing of all stimuli from reaching consciousness until attentional resources become available again would suggest that all stimuli should be treated the same. However, research has suggested that certain stimuli are relatively immune to the blink. For example, recent studies have shown that the human voice is special in this regard compared to other complex auditory stimuli. Akça and colleagues (Akça, Bishop, Vuoskoski, & Laeng, 2023) presented subjects with rapid streams of everyday environmental sounds (e.g., printers, motorcycles). Within these streams, the T1 was a simple sine wave tone or a bell, and the T2 sounds, when present, were either a human voice, a cello playing, or a dog barking. The cello sound was chosen because it is a complex sound that shares some perceptual similarity to the human voice but is not a biological sound, whereas the dog barking is a biological sound but differs acoustically in many ways from human voices. The results revealed that human voices were unique in being able to escape an AB that was clearly present for the other stimuli. Similar to how human faces seem to have special status in capturing visual spatial attention (as reviewed in Chapter 6), the processing of human *voices* appears to be special in terms of it being immune to the temporal selection mechanisms normally imposed by the AB. Regarding the question posed in the title of this box, whether a blink occurs depends not only on the core mechanisms controlling the temporal window of attention, but also on the nature of the stimuli being processed within that window.

however, especially in the superior frontal regions, was largely absent for stimuli that were not perceived (Figure 8.3, middle and bottom rows). Whereas previous fMRI research had provided evidence linking such frontal activity to awareness, this MEG study was able to further specify that the burst of activity in the superior frontal gyrus starting as early as ~400 ms was associated with conscious awareness of the T2 stimulus. In addition, this study directly compared the effects of the AB to the effects of the PRP (discussed earlier) and found that activity in this same region was present but significantly delayed during PRP trials, consistent with the refractory period associated with the PRP effect. The authors suggested that activity in this frontal region, along with parietal activity at that same time, relates to the conscious access to the T2 stimuli, which is delayed in time during the PRP and prevented entirely during the AB.

Finally, recent work using fMRI-based neural connectivity modeling has identified patterns of connectivity within and across attention networks that are correlated with performance on AB tasks (e.g., Shen, Vuvan, & Alain, 2018; Wu, Liaw, Goh, Chia, Chee, Obana, Rosenberg, Yeo, & Asplund, 2020). Zhang and colleagues (Zhang, Zhang, Zhou, Zhou, & Chang, 2023) calculated connectivity between 218 regions distributed across eight networks (occipital, parietal, prefrontal, temporal, motor, insula, limbic, and subcortical networks). These connectivity indices were related to individual differences in the magnitude of the AB. Furthermore, these indices were correlated with the severity of attention deficit hyperactivity disorder (ADHD) symptoms, extending previous work showing that individuals with ADHD have been found to

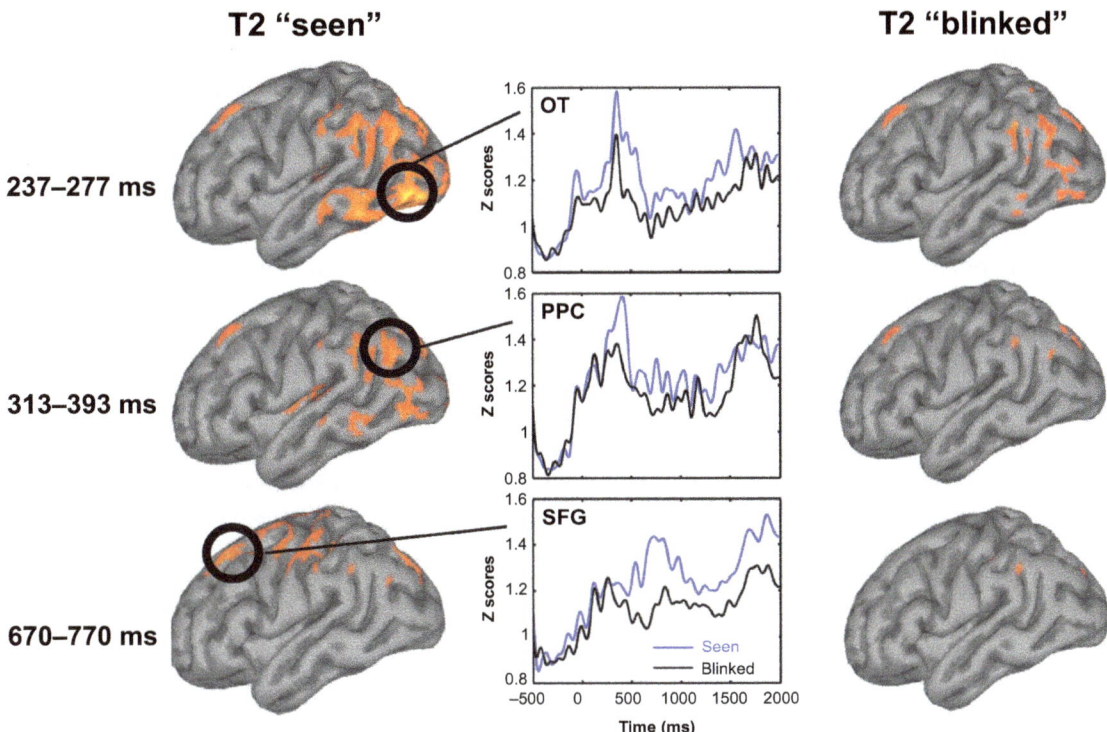

Figure 8.3 MEG study of the AB. The left panels show distributed source models of MEG activity for "seen" T2 stimuli at different time periods overlaid on partially inflated brain models. The middle panels show time courses of activity from those sources. The right panels show activity for T2 stimuli that were not perceived. The top row shows activity in the occipitotemporal cortex (OT), the middle row shows activity in the posterior parietal cortex (PPC); and the bottom row shows activity in the superior frontal gyrus (SFG). Adapted from Marti et al. (2012), *NeuroImage*, with permission from Elsevier.

have larger AB deficits compared to matched controls (Armstrong & Munoz, 2003; Mason, Humphreys, & Kent, 2005). Thus, these findings provide support for the idea that a critical component of ADHD may be the ability to control the temporal aspects of attention. Understanding the control of temporal attention may be an important step in developing new training methods and therapies for individuals with attention deficits, as well as for helping workers whose jobs require high degrees of vigilance. Additionally, as described in the next subsection, neurostimulation research has begun to investigate whether directly stimulating activity in certain nodes of the attention network can cause improvements in temporal attention.

8.2.3 Neurostimulation and Causal Studies of the AB

In order to investigate the causal influences of brain regions involved in the AB, researchers have begun to use neurostimulation techniques. A number of studies have used transcranial magnetic stimulation (TMS) to test the influences of different cortical regions. One of the earliest such studies stimulated the right parietal cortex (Cooper, Humphreys, Hulleman, Praamstra, & Georgeson, 2004), a location implicated in attentional control from fMRI studies of the AB, as discussed earlier. In this study, TMS was delivered in a series of three pulses immediately after the T1 stimulus, during the window when the AB for the T2 is typically largest. The results showed that performance during the peak AB period was significantly *improved* when stimulation was applied over a lateral parietal region in the right hemisphere. In order to ensure that the benefits were not due simply to a generalized arousal triggered by the TMS pulse or associated auditory clicks, the authors showed that stimulation over a control site (midline parietal) produced no effects on AB performance. Also of note, there was no decrement in T1 performance during TMS stimulation, providing evidence that the stimulation didn't simply enhance T2 performance by blocking T1 processing. Although the authors were able to show that the effects of TMS over lateral parietal areas were not simply due to arousal or to disrupting T1 processing, the results could not definitively identify the mechanisms of action. The improvement in T2 identification when the parietal lobe was stimulated after T1 could have been due to: (1) enhanced T1 processing, to the extent that the AB is never triggered because T1 processing completes earlier; (2) an improved disengagement process, allowing attention to move on from T1 more quickly; or (3) enhanced T2 processing. In addition, that study only stimulated the right hemisphere, leaving open the question of whether the critical brain process could be occurring in both hemispheres. Kihara and colleagues (Kihara, Hirose, Mima, Abe, Fukuyama, & Osaka, 2007) replicated and extended these results, including using individual subject MRI scans to more precisely locate the brain regions to target. In that study, the researchers stimulated the IPS in either the right or left hemisphere in separate conditions. The results showed that the AB was significantly reduced (i.e., improved T2 performance) following active stimulation of *either* the right or left IPS, but not during any of the control conditions.

More recently, Kihara and colleagues (Kihara, Ikeda, Matsuyoshi, Hirose, Mima, Fukuyama, & Osaka, 2011) used TMS to test the influences of two different regions of the parietal lobe on the AB. In addition to the IPS, they targeted a region in the inferior parietal lobe corresponding to the temporoparietal junction (TPJ). Previous neuropsychological patient research and fMRI studies of attention have implicated these two regions in different aspects of attention: The IPS has been linked to the control of attention and the creation of an attentional set, whereas the TPJ is thought to be involved in the disengagement and reorienting of attention (as described in Chapter 7). In this study, the researchers found that TMS pulses delivered after the T1 stimulus reduced the magnitude of the AB for T2 stimuli, but only when TMS was applied to the IPS. Stimulation of the TPJ at that same time after T1 did not affect the size of the AB compared to control conditions (i.e., sham stimulation and stimulation at a central top region of the head). These results concur with studies of *spatial* orienting that find that the IPS is a critical region for the control of spatial attention, and they extend the role of this region's involvement in control processes to temporal aspects of attention as well. The authors suggested that the IPS may play a critical role in generating and maintaining an attentional set, and that stimulation of this region may boost the ability to quickly process the T1. The lack of any effect of stimulation of the nearby TPJ region provides further evidence of the specificity of the role of the IPS in temporal attention. TPJ stimulation did have an effect, however, when TMS was delivered later, after T2 appeared. In that condition, T2 detection was impaired by TPJ stimulation, confirming that this region plays some role in visual attention. However, stimulation of the IPS at that later time period produced the same impairment as TPJ stimulation; therefore, this study wasn't able to isolate the role of the TPJ in temporal attention.

Related to the findings discussed earlier in which connectivity patterns across widespread brain networks were linked to the AB, some TMS studies have attempted to stimulate activity throughout a whole attention network. Previous research combining TMS with neuroimaging has shown that stimulation of one region of a distributed network can result in enhanced activity throughout the network (Halko, Farzan, Eldaief, Schmahmann, & Pascual-Leone, 2014; Paus, Jech, Thompson, Comeau, Peters, & Evans, 1997). Building on previous research indicating that midline regions of the cerebellum are linked to the DAN (Brissenden, Levin, Osher, Halko, & Somers, 2016), Esterman and colleagues targeted this region of the cerebellum in order to modulate the DAN (Esterman, Thai, Okabe, DeGutis, Saad, Laganiere, & Halko, 2017). They used individual-subject resting-state scans to identify the midline location of the cerebellum most strongly connected to the DAN and applied intermittent theta burst (iTBS) TMS to this region before subjects performed an AB task. Since iTBS involves repeated trains of stimulation over an extended period (in this case 190 seconds), this method doesn't have the temporal resolution to identify exactly *when* an area is involved in a specific mental process. It can, however, produce modulatory effects that last for some time after stimulation, and therefore it is used in some treatment programs for clinical disorders such as depression. The Esterman et al. (2017) study found *improved* AB performance after iTBS stimulation of the midline cerebellum compared to control conditions, providing evidence for a role of the cerebellum in temporal attention and raising the possibility that this type of stimulation may be helpful in treating some attention problems.

In addition to TMS, other methods such as transcranial direct current stimulation (tDCS) can be used to stimulate the brain in a more subtle manner, increasing or decreasing the excitability of a region without actually triggering neural activity directly. Studies using tDCS have produced potentially important insights into temporal attention and into how individual differences play a role in the effectiveness of neural stimulation. Overall, the results of research using tDCS to study the AB have been somewhat inconsistent. A number of tDCS studies have targeted the dorsolateral prefrontal cortex (DLPFC), since neuroimaging studies have identified this region as being linked to the conscious processing of stimuli that is absent during the AB (e.g., Kranczioch, Debener, Schwarzbach, Goebel, & Engel, 2005). In one of the first studies using tDCS to stimulate the DLPFC, Sdoia and colleagues (Sdoia, Conversi, Pecchinenda, & Ferlazzo, 2019) found significant effects of stimulation relative to a sham stimulation on performance on an AB task. Furthermore, a unique aspect of tDCS is that the underlying neural activity can be biased toward a greater or lesser likelihood of activity depending on the direction of current flow. Specifically, activity under the cathodal electrode (negative terminal) is typically found to be suppressed, whereas activity under the anodal electrode (positive terminal) is typically found to be potentiated. In Sdioa et al. (2019), anodal stimulation over the left DLFPC improved AB performance (greater percentage of correctly reported T2), whereas cathodal stimulation impaired performance. These findings could indicate that the DLPFC plays an important role in gating information into consciousness, and that boosting this process, possibly through more efficient encoding into active working memory, may reduce the competition for resources that typically triggers the AB. A subsequent study (London & Slagter, 2021), however, failed to replicate any main effects of tDCS of the DLPFC on the AB. When that study examined individual differences, however, the researchers found an interesting negative correlation between anodal and cathodal stimulation. Specifically, the larger the increase in AB magnitude during cathodal stimulation, the larger the reduction of AB during anodal stimulation (London & Slagter, 2021). These results raised the intriguing possibility that the relatively subtle stimulation of tDCS could be effective at modulating attention performance *if* individual differences are taken into account. However, in an attempt to extend these findings, Reteig, Newman, Ridderinkhof, and Slagter (2022) failed to replicate the effects of tDCS being contingent on individual differences. Reteig et al. (2022) failed to find either a main effect of DLPFC stimulation or any differential effect of anodal or cathodal stimulation moderated by individual differences. However, there were a few differences between this study and the original London and Slagter (2021) study that could be responsible for this nonreplication, including the use of different lags, different interstimulus intervals, and overall task accuracy. Therefore, although there have been some intriguing findings from tDCS studies, more research is needed to determine the potential links between the DLPFC and temporal attention.

Finally, a recent study found that the AB was affected by a very different means of brain stimulation. Ross and Lopez (2020) used binaural beats (BBs) to entrain neural oscillations. Presenting rhythmic sensory stimulation to a subject has been shown to produce enhanced brain rhythms at specific frequencies. EEG and MEG studies have shown that this effect can be created by rhythmic visual flickers (Herrmann, 2001), somatosensory stimulation (Ross, Jamali, Miyazaki, & Fujioka, 2013), or auditory rhythms (Picton, John, Dimitrijevic, & Purcell, 2003). BBs involve presenting each ear with a sequence of pure tones, with the frequencies being slightly different for each ear. Brain stem regions sensitive to these interaural phase differences then generate signals to the cortex that entrain neural activity as a function of the difference between the frequencies. Ross and Lopez used 40-Hz BBs to stimulate gamma activity, and they used a frequency between alpha and beta (16.3 Hz) as a control frequency. The results showed that BB stimulation at 40 Hz produced a significant improvement in AB performance compared to any other conditions. Notably, the improvement was tested and found in the session *following* the stimulation, not immediately *during* the BB session. Furthermore, there was a full day between sessions. Thus, the authors argued that the ability of this type of stimulation to enhance attention likely depends on automatic learning processes occurring during the consolidation period (i.e., during sleep) between sessions. More research is needed to further determine the possible mechanisms of this type of entrainment, and a recent review of BB studies concluded that the current evidence for a consistent effect of BBs on neural entrainment is inconsistent (Ingendoh, Posny, & Heine, 2023). As such, further research is needed to better determine whether and how this simple, noninvasive way to stimulate the brain could potentially be used to enhance the training of temporal attention.

8.3   Attending to a Moment

Temporal aspects of attention can include both the limits of attention *throughout* time and the ability to attend *to* a specific moment in time. The preceding section on the AB focused on how attention is limited in time in terms of it requiring the sustained allocation of a limited pool of resources over time to fully process a stimulus. Another aspect of temporal attention, however, is the ability to prepare for an upcoming stimulus by ensuring peak attentional focus at the precise moment when the target is expected. Similar to how an informative arrow cue in the Posner cuing paradigm allows one to orient *spatial* attention to the expected location and how the expectation of a certain color or form results in the selective allocation of *feature-based* attention, studies have shown that we can also attend to the specific upcoming time window *when* a target is expected. Behavioral experiments have shown that reaction times and accuracy show significant benefits when targets occur at expected times and relative costs when targets occur at unexpected times (e.g., Coull & Nobre, 1998; Denison, Heeger, & Carrasco, 2017; Griffin, Miniussi, & Nobre, 2002).

ERP studies have investigated attention to time by utilizing a component referred to as the **contingent negative variation (CNV)** to index the brain's anticipation of an upcoming event. Specifically, early studies found a negativity over central scalp sites that steadily increased as the expected time of an upcoming target stimulus neared (Walter, Cooper, Aldridge, McCallum, & Winter, 1964). Furthermore, the CNV does not simply represent the preparing of a motor response, as it is found even when no motor response is required (Frost, Neill, & Fenelon, 1988; Ruchkin, Sutton, Mahaffey, & Glaser, 1986). Thus, the CNV is thought to index a process of preparing to attend to an upcoming stimulus. More recent studies have built upon this to show that when different sets of stimuli become associated with earlier-occurring or later-occurring targets, the CNV tracks *when* attention is being allocated. For example, Cravo, Rohenkohl, Santos, and Nobre (2017) recorded ERPs while subjects performed a task in which different scenes became associated with a target-change event occurring at either a short (800 ms; "early") or long (2,000 ms; "late") interval after the initial onset of the target stimulus. The trial sequence consisted of the presentation of a scene followed by the appearance of the target stimulus and then the target-change event (change of color of the target stimulus) that required a speeded response. The results showed that the onset of the CNV occurred earlier and grew increasingly large as the first ("short") target-change time approached, but only for the scenes that had been previously associated with a target-change event occurring after the short interval. Scenes associated with the long interval didn't show a strong CNV as the early target window approached. Such findings reveal that attention is not simply readying the system in a general way as soon as the scene appears, but instead is allocating attention specifically *when* a target is expected based on previously learned associations. This study also found that behavioral performance was correlated with the size of the CNV, with a larger CNV at target onset being associated with faster reaction times (Cravo et al., 2017). These findings provide evidence that attention can be flexibly allocated in time and that the focusing of this type of attention can be delayed until the expected target time approaches.

Relating back to the AB studies described in the previous section, researchers have found that providing subjects with precise knowledge of when a second target will appear can significantly reduce or eliminate the blink. Martens and Johnson (2005) showed that when subjects were told, before the onset of each sequence, what the precise lag would be between the first and second targets, their accuracy for the second target was much improved. Thus, being able to plan for the precise moment when attention is needed allowed subjects to overcome the AB on a large percentage of trials. Interestingly, they found that this reduction of the AB was observed only if the lag of the second target was indicated on *each* trial. In a separate experiment in which the lag of the second target was constant across all trials but was not indicated on each trial separately, there was no reduction in the AB. Thus, it appears that the ability to use voluntary attention to overcome the AB requires that it be actively engaged anew on each trial. Without reminders on each trial of exactly *when* attention will need to be focused, subjects have difficulty in reengaging this type of attention repeatedly on every trial. Overcoming the AB thus appears to be a highly resource-demanding process. The difficulty in reengaging temporal attention on each trial is reminiscent of early work by Posner and colleagues (Posner, Snyder, & Davidson, 1980) showing that the effects of voluntary *spatial* attention were greater when cued on a trial-to-trial basis. In their study, when the location of the target was simply made consistent across an entire experiment but not cued on each trial, the cuing effects were much reduced or eliminated. This may also relate to studies of vigilance and sustained attention (described in Chapter 2) that have shown the difficulty of maintaining attention consistently for long periods.

8.3.1   Neural Systems Controlling Attention to Time

Whereas previous sections have shown that attention can be allocated to specific points in time, a critical question is whether the control of temporal attention was manifest through the same brain networks responsible for the control of spatial and feature attention. As described in Chapter 7, spatial attention is controlled through coordination across a widespread network of brain regions, including a DAN that is involved in the orienting and focusing of attention and is located primarily in dorsal frontal and parietal regions (e.g., the IPS and frontal eye fields [FEF]). Another coordinated set of regions is referred to as the ventral attention network (VAN), which is thought to be involved in the reorienting of spatial attention and includes the TPJ and lateral ventral frontal regions. Finally, subcortical regions, including the pulvinar and superior colliculus, also play a role in spatial attention, being strongly linked to attentional orienting and selection processes.

In one of the first studies to investigate the neural systems controlling *temporal* attention, Coull and Nobre (1998) used positron emission tomography (PET) *and* fMRI to compare the regions associated with temporal orienting to the areas associated with spatial orienting. They designed a version of the Posner cuing paradigm in which the cue indicated *when* a target was likely to appear and compared results from this task to the standard cuing paradigm in which the cue indicated the likely location of the target. The behavioral results confirmed that task performance was best when subjects had been validly cued to the correct spatial location or onset time of the target. Critically, the neuroimaging results showed overlap in some core regions of the attentional control networks. Specifically, the IPS and FEF, in addition to ventrolateral prefrontal regions and the thalamus, showed activity for both spatial and temporal attention. These results represented the first neuroimaging data to show that temporal orienting relies on some of the same neural architecture as spatial attention. They also showed some differences, however, specifically in the laterality of these control regions. In line with previous research suggesting a right hemisphere lateralization for spatial attention, the researchers found that *spatial* attention was associated more with *right* hemisphere activity in the IPS and lateral intraparietal (LIP) lobule, whereas *temporal* attention was associated more strongly with the IPS in the *left* hemisphere. Of note, this study also found areas of the cerebellum to be involved in both types of attention; this was somewhat expected for temporal attention due to the cerebellum's role in time perception (Ivry & Keele, 1989; Jueptner, Rijntes, Weiller, Faiss, Timmann, Mueller, & Diener, 1995) and showed that spatial attention, at least when measured in a task that includes temporal sequencing, also involves neural processes in the cerebellum. Studies of temporal cuing have also revealed activity in premotor areas, providing evidence that sensorimotor circuits in the frontal lobe play a role in coordinating the allocation of attention across time (Schubotz, 2007). Further support for this proposal comes from similar premotor circuits being active during temporal attention to auditory stimuli as well (Morillon, Schroeder, & Wyart, 2014). Along with previous findings showing that temporal cuing effects are seen across different modalities (Lange & Röder, 2006), these studies provide evidence for cross-modal control of temporal attention. Finally, it has been suggested that the hippocampus may also be involved in temporal aspects of attention, building upon research in rodent studies that has identified cells within the hippocampus that are specifically sensitive to the timing of events and temporal intervals (Eichenbaum, 2014). More research is needed, however, to determine whether and how these hippocampal "time cells" may integrate with other temporal processing regions to affect the allocation of attention across time.

8.3.2 Different Types of "Time" and Their Influences on Attention

The previous subsection described studies that have investigated temporal attention in one specific manner: namely, by providing subjects with informative cues, on a trial-by-trial basis, of exactly when in time a target event is likely to occur. This type of design nicely parallels the types of spatial or feature information provided in versions of the Posner cuing paradigm, and therefore it has been highly useful in comparing the control networks and effects of temporal attention to spatial attention. However, in the real world, the allocation and focusing of our attention are influenced by multiple processes that inform us of how information is likely to unfold over time. These influences include temporal information that can be learned, explicitly or implicitly, over different timescales.

8.3.2.1 Hazard Rate Functions

In addition to pre-cues that explicitly inform the subject about exactly when to expect a target, subjects' temporal attention is also influenced by "hazard rate functions." A *hazard function* essentially refers to the likelihood that something will happen, given that it hasn't happened yet. (Note that the dramatic-sounding term "hazard" in "hazard function" comes from the initial work on this topic in survival analysis, in which the event being modeled was the likelihood of failure of a critical mechanical system or the death of a living organism.) An example from a cognitive study would be an experiment in which most trials include the presentation of a target, with the result being that the longer one waits for the target on a given trial, the higher the expectation grows that the target is likely to occur soon. Given a distribution of target onset times that are variable within some set time window, expectation grows stronger as time proceeds without any target appearing. During the early part of the window, temporal expectation is essentially set to be even across the whole time period because the target could occur at any time within that window. Therefore, attention is unlikely to be at peak focus early on because all points within the window are equally likely. But as more time progresses with no target appearing, peak focus is likely to grow as the possible time window for a target shrinks to whatever time is remaining. Expectation grows until the target appears or after enough times has elapsed such that the subject concludes that the trial has ended without containing any target. A real-world example would be a driver coming to an intersection where the light has just changed to red: When they first pull up to the intersection, they're unlikely to be monitoring the light closely because they would not expect it to change again right away, but as time proceeds, they will become more and more focused on the light and will get themselves ready to step on the accelerator when the "target" (a green light) appears. In the laboratory, studies have shown that behavioral performance follows the expected hazard function, regardless of whether or not subjects are explicitly informed about the target distribution window (Vangkilde, Petersen, & Bundesen, 2013).

Neuroscience experiments have investigated this type of hazard function response in multiple paradigms. The basic CNV component in ERP studies, as described earlier, can be considered as indexing a type of hazard function, although in many CNV studies the timing of the target is defined quite precisely, whereas in most studies of hazard rates the timing is much more variable to allow for a better mapping of the function across time. Single-unit recording studies in nonhuman primates have shown that subjective hazard rate expectancies correlate with changes in the firing rate of neurons in LIP and early visual (V1, V4) areas (Ghose & Maunsell, 2002; Janssen & Shadlen, 2005; Lima, Singer, & Neuenschwander, 2011). Furthermore, temporal expectancies have been found to result in attenuation of neural activity in the alpha (8–12 Hz) and beta frequency bands (15–30 Hz), similar to the type of reduction in those bands observed during spatial attention (van Ede, de Lange, Jensen, & Maris, 2011). In another neuroimaging study, Coull, Cotti, and Vidal (2016) measured activity across attention and time-sensitive areas of the brain and analyzed the correlations with behavioral performance on a temporal attention task in which the target would appear at one of four different time intervals. In their study, five different cue types were used: Four different cues indicated one of the specific time intervals, and a fifth type of cue (the "neutral cue") provided no information on the specific timing of the target. The neutral cue was used to assess how attention might follow a hazard function, as targets could appear at any of the intervals, and if an interval passed without a target, the likelihood of the target appearing would increase for the remaining intervals. The fMRI results replicated the finding that the *left* inferior parietal cortex was strongly activated by orienting attention to specific moments in time. In regards to the neural basis of hazard rate expectations and preparations, the results showed that activity in the lateral inferior parietal cortex of the *left* hemisphere and the lateral inferior frontal area of the *right* hemisphere was significantly correlated with hazard function expectations. More specifically, activity in those brain regions was correlated with the degree to which reaction times were sped up with increasing time intervals in the neutral condition, reflecting subjects' rising expectancies as more time intervals passed without the target appearing (Figure 8.4).

Neurostimulation studies have provided support for the importance of the prefrontal cortex (PFC) in this type of temporal attention, as TMS to this region significantly reduces the typical speeding up of reaction times with increasingly long fore-periods (i.e., the hazard rate benefit; Vallesi, Shallice, & Walsh, 2007). Furthermore, the hemispheric specialization indicated by neuroimaging studies has been supported, in that TMS to the *right* PFC disrupts the expectancy effect, but left hemisphere stimulation does not (Vallesi et al., 2007). However, neuropsychology studies of patients with lateral frontal lesions paint a somewhat different picture. Studies of individuals with naturally occurring lesions to the lateral PFC have found that the hazard rate benefit depends on the lateral PFC being intact and fully functioning (Stuss, Alexander, Shallice, Picton, Binns, Macdonald, Borowiec, & Katz, 2005). However, patient studies have found that lesions in either the left *or* the right hemisphere eliminate the hazard rate benefit at later intervals (Triviño, Arnedo, Lupianez, Chirivella, & Correa, 2011; Triviño, Correa, Arnedo, & Lupiáñez, 2010). Although the issue of hemispheric specialization thus remains to be resolved, the research findings have converged on the conclusion that the inferior parietal cortex and lateral inferior frontal cortex are critical in supporting the type of temporal attention induced by hazard function expectancies.

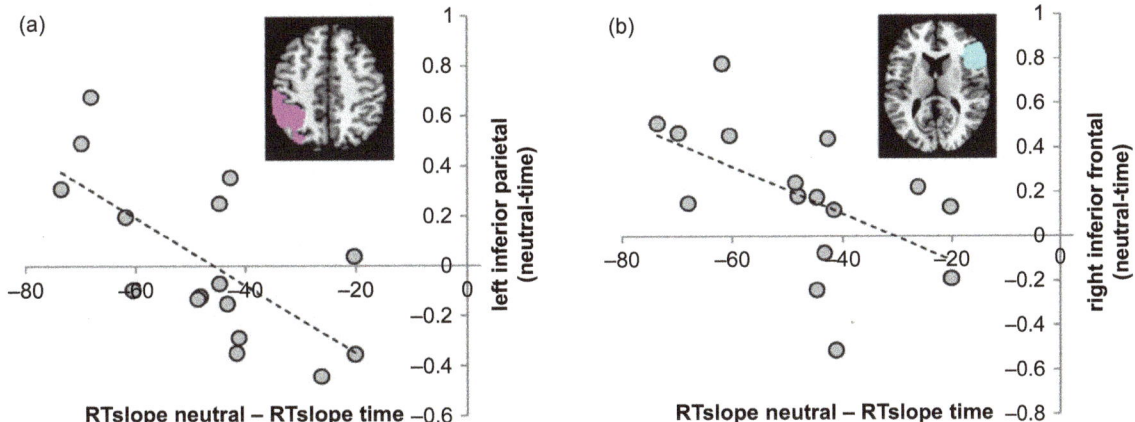

Figure 8.4 Correlating neural activity and performance during hazard function expectancies. Regions of interest drawn in color on structural MRIs, and graphs plotting individual participants' fMRI activities in those regions versus their reaction time (RT) effects. Activity in both (a) the inferior parietal cortex of the left hemisphere and (b) the inferior frontal cortex of the right hemisphere was correlated with perceived hazard rate. The "RTslope" metric refers to how the RTs changed over increasing cue-to-target intervals. Note that a stronger hazard rate is plotted as more negative (left) on these graphs; the "neutral" condition shows the effect of the hazard rate expectancy, and the "time" condition provides a baseline for each subject, controlling for actual time of target presentation. Reprinted from Coull et al. (2016), *NeuroImage*, with permission from Elsevier.

8.3.2.2 Rhythms

In addition to informative cues that provide explicit knowledge of exactly when a target is likely to occur and hazard function distributions that show a buildup of expectancy over time, *rhythms* have also been shown to have a strong effect on how attention is allocated across time. In a pioneering review paper in 1976, Mari Riess Jones argued that time is an integral and essential part of how our minds represent and perceive stimuli, and that theories of attention should include a temporal dimension (Jones, 1976). According to her "rhythmic theory," attention is strongly influenced by the ongoing temporal patterns in the environment, and the effects of attention on perception follow those rhythms. Jones used the example of vowel recognition in speech to illustrate how rapid auditory perception can be when the stream of information follows a natural rhythm. Even though natural speech involves complex rhythms, we quickly attune to the patterns, and vowel recognition is accurate at durations as short as 30 ms in natural speech, whereas it requires durations of 150 ms or more in synthetic speech. In other experiments, Jones and colleagues showed that simple auditory stimuli (e.g., tones of different pitch) were discriminated better when they occurred *precisely* on the beat of an ongoing rhythm (Jones, Johnston, & Puente, 2006). In their study, targets occurring either before or after the beat were not perceived as well, with progressively worsening performance the farther from the beat the target occurred. Jones and colleagues argued that the temporal structure within streams of auditory stimuli entrain neurons in auditory processing and attention regions. Subsequent studies of entrainment have shown that rhythmic presentation of stimuli triggers activity in sensory processing regions at the frequency rate of the stimulus presentation.

Whereas much of the work on rhythms has focused on streams of auditory stimuli, studies have shown that temporal patterns of *visual* stimuli produce a similar effect on attention and perception. Palmer and colleagues showed that visual stimuli presented on the schedule of an ongoing rhythm had enhanced signal-to-noise ratios, as measured by behavioral responses (Palmer, Huk, & Shadlen, 2005). Furthermore, Mathewson and colleagues (Mathewson, Fabiani, Gratton, Beck, & Lleras, 2010) found that entrainment to an ongoing visual temporal pattern (i.e., rhythm) could even overcome visual masking. In their study, subjects were presented with rapid series of visual stimuli with short durations and interstimulus intervals, which resulted in the typical amount of missed targets due to masking from temporally adjacent stimuli. However, on trials in which the target was preceded by a consistent rhythm of presentation and when the target then occurred temporally in phase with that rhythm, there was a dramatic increase in perceptibility compared to conditions in which the target wasn't precisely in phase with the preceding rhythm.

Entrainment to ongoing rhythms has a powerful influence on temporal attention and can even supersede top-down goals. For example, in studies in which targets were more likely to occur "off-beat" (out of phase with regularly flickering visual stimuli), perception was still best when targets occurred in phase with the established rhythm (e.g., Breska & Deouell, 2014). Even though subjects knew to expect the target sometime between beats, they still performed better when the target occurred on the beat. This powerful effect of entrainment has also been shown to occur in the auditory domains (Sanabria, Capizzi, & Correa, 2011). In the Breska and Deouell (2014) study with visual stimuli, ERP results showed that the CNV component (which tracks subjects' expectations, as discussed earlier) was influenced more by the entraining rhythm than by other cues that predicted target onset time. This indicated the strength of this entrainment process, as the rhythmic pattern of stimulation induced unintentional temporal expectancies that influenced attention even when it was detrimental to performance. This study also revealed separate processes that were engaged when rhythmic patterns were predictive. Specifically, when subjects intentionally used predictive rhythms to focus attention, the behavioral effects of the entrainment were even larger, and a different ERP component indexing stimulus evaluation (the P3 component) was modulated only when the rhythm was predictive. This study was important for highlighting that rhythm-induced temporal attention, like spatial attention, can have different effects when engaged automatically versus voluntarily.

8.3.2.3 Implicitly Learned Complex Temporal Sequences

Finally, beyond the entrainment to regular ongoing rhythmic stimulation, studies have shown that the allocation of temporal attention can also be influenced by *learned sequences* of even more complex patterns. When a particular temporal sequence is experienced repeatedly, subjects show a learned sensitivity to that particular sequence. For example, in one version of a **serial reaction time task**, each finger is associated with a different button press that corresponds to a specific stimulus. When a particular sequence repeats over the course of the experiment, subjects become increasingly fast at performing that particular sequence, even though they are often unaware of that specific sequence being repeated across the experiment (Reed & Johnson, 1994). This paradigm is often used in studying implicit memory, but it also reveals how the particular sequence can unintentionally affect attention. When temporal onset of the target is also manipulated, the effects of repeating a precise timing sequence are found to further enhance performance over the effect of only the ordinal sequence repeating alone, and independent effects can be found for each type of repetition (Shin & Ivry, 2002). In an MEG study, Heideman, Ede, and Nobre (2018) had subjects perform a serial reaction time task in which the order of the four stimulus locations/responses repeated in a long 12-item sequence embedded within the ongoing task. The timing of the target onset was also repeated in a 12-item sequence in which the time from manual response to the onset of the next stimulus was one of three different intervals. Overall, the timing interval was not associated with stimulus location/response buttons, but since the repeated 12-item sequences (of target type and timing interval) occurred concurrently, there was a relation between spatial location and timing within just that 12-item sequence. The behavioral data confirmed previous findings that performance was significantly improved for targets within the repeated sequence. MEG results revealed a modulation of power in the beta frequency band, indicative of a preparatory response in premotor regions based on implicitly learned temporal and spatial expectancies. Whereas previous studies had shown a similar modulation of beta frequency power in response to regular isochronous rhythms (Heideman, te Woerd, & Praamstra, 2015; Praamstra, Kourtis, Kwok, & Oostenveld, 2006), the Heideman et al. (2018) study was the first to show that this effect was also produced by implicitly learned temporal patterns that did not follow a regular rhythm. This showed that attention and preparatory processes can be affected by even nonrhythmic patterns, as long as those are repeated enough to establish implicitly learned expectancies.

8.3.3   Interactions between Types of Spatial and Temporal Attention

As noted in previous sections, temporal attention can affect sensory processing at relatively early perceptual levels similarly to spatial attention, and there is some overlap in the brain regions that control these two types of attention. This raises the question of how these systems may interact. In an ERP study, Doherty and colleagues (Doherty, Rao, Mesulam, & Nobre, 2005) investigated the effects of when spatial and temporal attention are engaged concurrently, and they compared these to the effects of when each type of attention is engaged separately. In their study, a red circle moved across the screen toward an occluding vertical bar, and subjects had to detect the presence (or absence) of a black dot that could (at 50% chance) be on the red circle when it reappeared on the other side of the occluding bar. Spatial predictability was manipulated by having the trajectory of the ball be either a straight line (predicting the precise location where the ball would reemerge on the other side of the occluding bar) or be random in terms of having the vertical location at each successive step across the screen be completely unpredictable. Temporal predictability was manipulated by having the duration and time between successive steps across the screen be either entirely consistent (at 500 ms, therefore predicting the precise time when the circle would appear on the other side of the occluding bar) or varied randomly between 200 and 900 ms. The experiment included conditions in which both spatial location and temporal onset were predictable (the "ST" condition), when neither was predictable ("N" for neutral), or when only temporal onset ("T" condition) or only spatial location ("S" condition) was predictable. As shown in Figure 8.5, the results showed the standard effect of spatial attention, with the P1 component being significantly enhanced for targets at the predicted spatial location ("S") compared to when that target was not cued ("N"). When only the time of the target onset was predictable, there was no modulation of this component at all ("T" condition compared to "N" condition). However, despite temporal attention not modulating the P1 component directly, there was a significant interaction between spatial and temporal attention at this

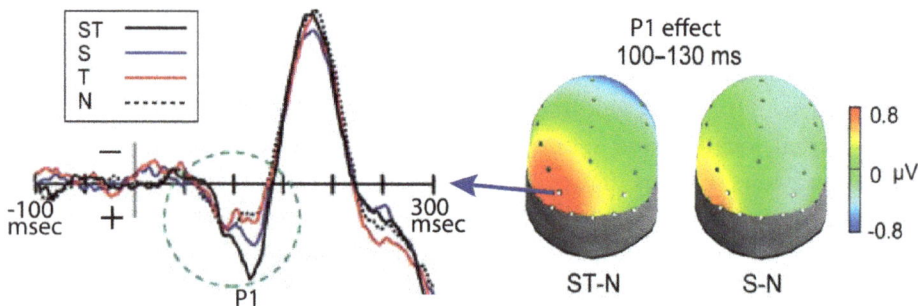

Figure 8.5 ERP study of spatial and temporal attention. ERP waveforms plotted at the left from the electrode indicated on the scalp topographies of the P1 component shown on the right. Spatial attention shows the typical enhancement of the P1 component (S – N), and this effect is further enhanced when temporal attention is also engaged at the same time (ST – N). Temporal attention alone had no effect on this early stage of visual processing. Figure adapted from Doherty et al. (2005), *Journal of Neuroscience*, with permission from the Society for Neuroscience (Copyright, 2005, Society for Neuroscience).

stage of processing, with the P1 component showing an even *greater* enhancement when cuing was accurate in both the temporal *and* spatial dimensions ("ST" condition) compared to when cuing was accurate only in the spatial dimension.

Even just within the realm of temporal attention there is evidence for dissociable and interacting effects of the different varieties of timing expectations. Informative symbolic cues that predict the precise timing of targets can produce robust effects, but the effectiveness of this type of cue is reduced if there are competing task demands. In contrast, the effect of ongoing rhythms is much more automatic and relatively unaffected by task instructions or competing demands (e.g., de la Rosa, Sanabria, Capizzi, & Correa, 2012). Furthermore, the benefits of rhythmic entrainment can be dissociated from the effects of explicit cues. TMS studies have found that external stimulation of the DLPFC modulates the effects of informative symbolic cues but does not modulate rhythmic cuing effects (Correa, Cona, Arbula, Vallesi, & Bisiacchi, 2014). Neuropsychology studies have also found a dissociation between these two types of temporal attention, but following a different pattern. Specifically, deficits in being able to use informative temporal cues have been observed following damage to the PFC in the right hemisphere (Triviño, Arnedo, Lupiáñez, Chirivella, & Correa, 2011), whereas damage to the left PFC affects rhythmic attention but not the ability to use symbolic cues to orient attention to a specific time (Triviño, Correa, Arnedo, & Lupiáñez, 2010). Further research is needed to fully determine the possible hemispheric specialization underlying rhythmic cuing. However, neuropsychology and neurostimulation findings do converge on the conclusion that symbolic cues and rhythms rely on at least somewhat dissociable neural mechanisms. This aligns with behavioral results showing that symbolic cues and rhythms can have additive effects on performance when engaged concurrently (Breska & Deouell, 2014).

8.4   The Holding of Attention

Throughout the history of attention research, the overwhelming majority of studies have investigated the processes of *moving* and *engaging* attention or the selective *effects* of attention on perceptual processing at a specific moment in time. Multitudes of experiments have explored what events trigger an involuntary orienting and the neural underpinnings of our ability to voluntary move our attention across space or objects at will. Yet, the single factor that may have the largest effect on how well we are able to perform tasks and avoid distraction is how long we get stuck attending to irrelevant stimuli or events. The initial "capture" or involuntary orienting to a salient event is a first step in causing distraction, but we're highly capable of briefly noticing things in our environment and quickly continuing on with our task at hand. When a stimulus or event holds our attention, however, we can begin to suffer consequences as we struggle to regain control and disengage from the stimulus in order to reorient back to our task. Indeed, as William James succinctly noted over a century ago, "What holds attention determines action" (James, 1892, p. 7). In this section, we'll discuss recent research that is revealing the different properties of a stimulus or event that involuntarily prolong how long our attention automatically dwells on it.

8.4.1   Memory Holds Attention

Most studies of the effects of memory on attention have focused on the initial orienting of attention toward stimuli that have a long-term memory trace in our minds. Other research, however, has investigated how memory affects both the initial reflexive orienting and the subsequent holding of attention. For example, in the Chanon and Hopfinger (2008) study described in Chapter 6, the researchers used eye tracking to investigate the effects of long-term item memory on the allocation of attention over time. After an initial phase in which the subjects answered a series of deep-encoding-level questions about simple line drawings of isolated objects, they were then presented with scenes that were to be studied for a later memory test. The scenes consisted of mostly new items but could contain "old" items that the subjects had viewed in the initial phase of the experiment. The findings revealed that subjects fixated sooner on an item if it was old. Across subjects, different items were old, so it wasn't simply that particular items were more salient, but rather that whatever item was old for a subject drew their attention more quickly. This effect of memory on the orienting of attention was different, however, from the type of fast and automatic orienting that occurs to an item that is especially salient in terms of physical characteristics. Items that are markedly brighter, bigger, faster, or louder typically capture attention and trigger an immediate orienting. The old items in the Chanon and Hopfinger (2008) study, however, were not usually the first items fixated in the scene, but rather were first fixated about 1.5 seconds after the subject started viewing the scene (i.e., after a few saccades had already been executed). Items did draw a subject's attention sooner if the items were old, but that capture was more subtle than what occurs for a physically salient item. Critically, however, the largest and most robust effect was observed on **dwell time**. Specifically, once a subject had fixated an old object, the eyes remained on that object for a longer period of time (compared to dwell times on that same object for subjects who had not studied it before). Furthermore, the eyes *returned* to the object more often throughout the viewing period when it was old for a subject, suggesting that the item maintained a higher priority for attention and continued to draw the eyes toward it. These results suggest that memory has a subtle influence on the initial orienting of attentional capture, but it has an even stronger effect on how long attention is *held* by an item.

In order to test whether the holding of attention by memory is involuntary and automatic, subsequent research utilized a variation of the AB task. As reviewed earlier, the AB refers to the finding that a second target stimulus (T2) in a stream of stimuli presented at fixation may be missed if it occurs shortly after the first target (T1) in the stream. In the AB task, it is beneficial to disengage attention from the T1 as quickly as possible in order to be able to attend to and perceive the subsequent stimuli. Therefore, dwelling on any item longer than necessary negatively impacts overall performance and should be avoided. In an AB task in which old items could sometimes be the T1, Parks and Hopfinger (2008) found that attention dwelled significantly longer on the T1 if it was unique in terms of subjects' memory of the item. The AB task comprised simple line drawings, some of which had been studied in a previous deep-encoding session. When the sequences of pictures in the AB task were composed of mostly *new* items (i.e., the context of each stream was mostly items that had not been seen previously), the AB effect (deficit in T2 detection) was found to last significantly longer following "old" T1 items compared to the AB following "new" T1 items. Specifically, the typical AB pattern of reduced accuracy for T2 items at lag 2 was found regardless of the memory status of the T1, but "old" T1 items also produced reduced accuracy at lag 3 and lag 4 compared to "new" T1 trials, despite the fact that the T1 had equal probability of being "old" or "new." Interestingly, this pattern reversed in a second experiment in which the sequences were composed of mostly *old* items – in that case, the extended AB was seen after "new" T1 targets, but not after "old" T1 targets. Together, these experiments show that it is the memory *uniqueness* within the current context that is critical, not simply the existence of a memory trace. When an item is notably different from the ongoing context in terms of memory, then that uniqueness has a significant effect on the *holding* of attention. As mentioned in Chapter 6, Ryan, Althoff, Whitlow, and Cohen (2000) found that eye gaze returned more often to areas of a scene that had been changed from a previous viewing of that same scene, regardless of subjects' awareness of the change. Overall, these studies support the view that memory *uniqueness* is a critical factor influencing the involuntary holding of attention.

In addition to memories established from simple previous exposure, stimuli that have a reward history can have an additional degree of hold over attention. As described in Chapter 6, previously *rewarded* stimuli have a strong effect on the capture of attention, involuntarily drawing attention toward them. Other research has shown that rewarded stimuli also affect the *holding* of attention, as experiments have shown that subjects cannot disengage attention as quickly from a previously rewarded stimulus compared to stimuli of similar salience and exposure history (e.g., Muller, Rothermund, & Wentura, 2016; Watson, Pearson, Theeuwes, Most, & Le Pelley, 2020). Koster and colleagues tested whether stimuli associated with threat also show these effects (Koster, Crombez, Van Damme, Verschuere, & De Houwer, 2004). They found that stimuli that had been associated with an aversive white noise captured and held attention longer than previously viewed stimuli without that aversive association. These studies measured the hold of attention by means of testing covert attention and manual reaction times. Eye-tracking studies have found that reward value predicts initial capture of overt attention, but that the amount of time that the eyes dwell on a stimulus is more affected by the uncertainty of the reward. Specifically, overt attention dwells longer on a stimulus that is associated with a 50% likelihood of reward compared to a 100% likelihood of being rewarded (Koenig, Kadel, Uengoer, Schubö, & Lachnit, 2017). These studies highlight that the holding of attention is a complex process that is influenced by multiple aspects of our previous experience with a stimulus.

8.4.2 Special Classes of Stimuli Hold Attention

Besides the effects of memory on attentional dwell time, research has shown that certain classes of stimuli may affect the temporal allocation of attention regardless of a subject's history with those stimuli. Threatening stimuli, in addition to capturing attention quickly, have also been found to hold attention longer than nonthreatening stimuli. For example, snakes, spiders, and angry faces are found more quickly in visual search tasks than are nonthreatening stimuli (e.g., Fox, Lester, Russo, Bowles, Pichler, & Dutton, 2000; Ohman, Flykt, & Esteves, 2001), and such stimuli hold attention longer in cuing paradigms, as measured by slower disengagement following invalid cues (Fox, Russo, & Dutton, 2002; Tipples & Sharma, 2000; Yiend & Mathews, 2001). Their preferential access to emotional and object recognition centers of the brain may explain the ability of these stimuli to rapidly capture attention (e.g., Esteves, Dimberg, & Ohman, 1994), but the mechanisms by which threatening stimuli hold attention are not yet well understood.

As mentioned in previous chapters, one type of stimulus that has an especially strong influence on the capture of attention is the human face. Faces are a special type of stimulus, at least in terms of having a brain region largely dedicated to their processing (the fusiform face area; FFA). The effect of faces has mostly been studied in terms of the initial orienting of attention, with the consistent finding being that subjects orient to face targets more quickly than to other stimuli, even ones that are equally complex. However, in addition to orienting attention automatically, faces have also been found to hold attention longer than other stimuli. In one study designed to investigate the temporal allocation of attention, Parks and colleagues (Parks, Kim, & Hopfinger, 2014) had subjects perform a novel continuous performance task involving discriminating targets in the visual periphery while irrelevant stimuli appeared at random times at fixation (Figure 8.6). The objective of the task was to indicate the orientation of a letter "T" that was always present at one location and that changed orientation about once every second. At random times, a small picture of either a face or a place (e.g., a house) was presented at fixation and would remain on the screen for a variable amount of time, interspersed with periods without any pictures at fixation. The results showed an immediate decrement in performance on the task (discriminating orientation of the "T" in the periphery) at the moment when a distractor first appeared at fixation, and this initial distraction was similar for places and faces. Subjects quickly recovered back to baseline levels of task performance when the distracting picture was a place, with the response to the second target after the distractor's appearance already being back to baseline levels. Critically, however, when the irrelevant distractor was a picture of a face, performance was significantly impaired not only at the initial moment when the face appeared, but also for the second and third targets after it appeared. In contrast to other complex stimuli, attention could not so easily be disengaged from images of human faces, providing evidence that faces may be special in their ability to involuntarily *hold* attention.

The task shown in Figure 8.6 is similar to that used in the Kim and Hopfinger (2010) study (introduced in Chapter 6) that investigated another type of special stimulus: a *new object*. In this study, subjects performed the same discrimination task described earlier of responding to the orientation of a red letter "T" that changed orientation about once per second, but here the target was presented at fixation, and subjects had to ignore a potentially distracting event in the periphery. That event was a luminance change of a small square-shaped region in the periphery. In the new object condition, the *increase* in luminance occurred at a location in space that was previously empty, therefore creating a new object at that location. In the control condition, the same luminance increment occurred within the boundary of a preexisting black-outlined square; the luminance increment in this case simply changed the color (from dark to light gray) of the square that already existed. In additional control conditions, there was a luminance *decrement* of the same magnitude, so that the light-gray square returned back to the dark gray color of the background. In the condition with an ever-present black-outlined square this decrement was only a change in luminance, whereas in the condition without a black

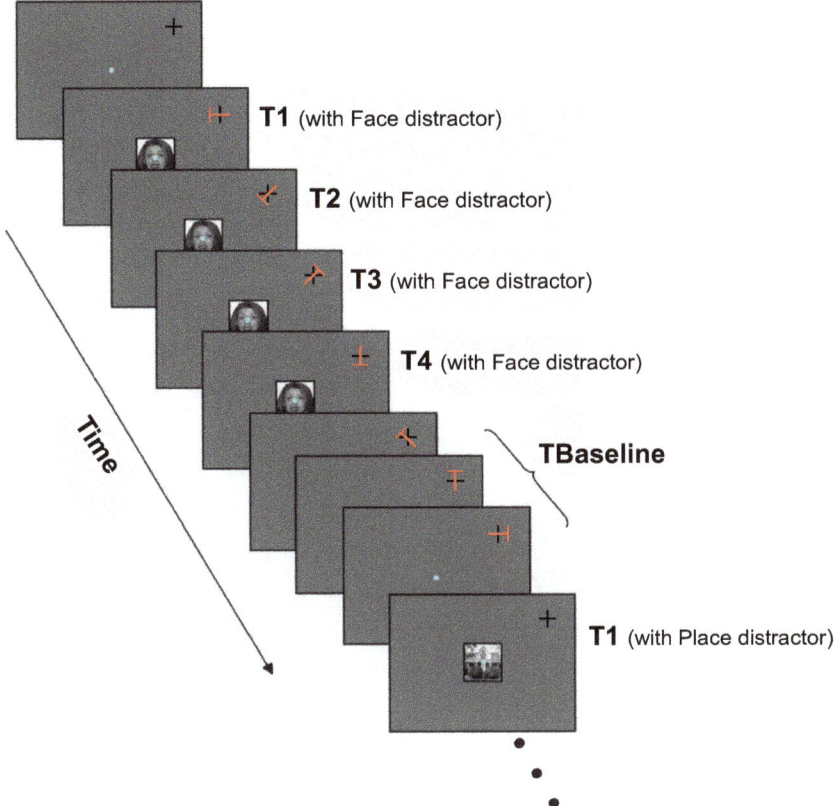

Figure 8.6 Trial sequence of a task designed to measure the holding of attention. Subjects were asked to respond to the change in orientation of a red letter "T" presented in the periphery (laid over a plus sign to help discriminate the orientation) while they kept their eyes fixated on a blue square at the center of the screen. Intermittently, an irrelevant grayscale picture of a face or a place appeared behind the blue fixation square. The picture had no relevance or relation to the task and was to be ignored. "T1" refers to a target event (change in orientation) that occurred at the same time as the onset of a new distractor image; "T2" refers to the second target event while the same image was still present; "T3" refers to the third target event; and so on. Figure adapted from Parks et al. (2014), *Psychonomic Bulletin & Review*, with permission from Springer Nature.

outline the luminance decrement essentially resulted in the disappearance of the object. Distraction, as measured by worse performance on the task, only occurred when the luminance change at the distractor location *created* a *new object* (i.e., only the luminance increment in the condition with no preexisting outlined squares produced significant distraction on discriminating the target orientation). The fMRI results from that study provided evidence for the neural mechanisms of this type of distraction, finding that the *new object* distractor triggered *enhanced* activity in attentional control regions of the parietal lobe and *reduced* activity in visual processing regions coding the location of the target stimulus. In addition, that study revealed potential neural underpinning of individual differences in attention mechanisms when comparing subjects who were relatively more versus less distractible by the new objects. A notable finding here was that *less distractible* subjects were more *consistent* in activating attentional control regions of the parietal lobe compared to more distractible subjects. Specifically, as shown in Figure 8.7, less distractable subjects activated attentional control regions in the IPS and TPJ to a moderate degree *every time* a potentially distracting event occurred in the periphery, whether it defined a new object or not. In contrast, highly distractible subjects showed a very large increase in activity when a new object appeared (labeled "luminance increment" in Figure 8.7) but no increase in activity when the same luminance change did not

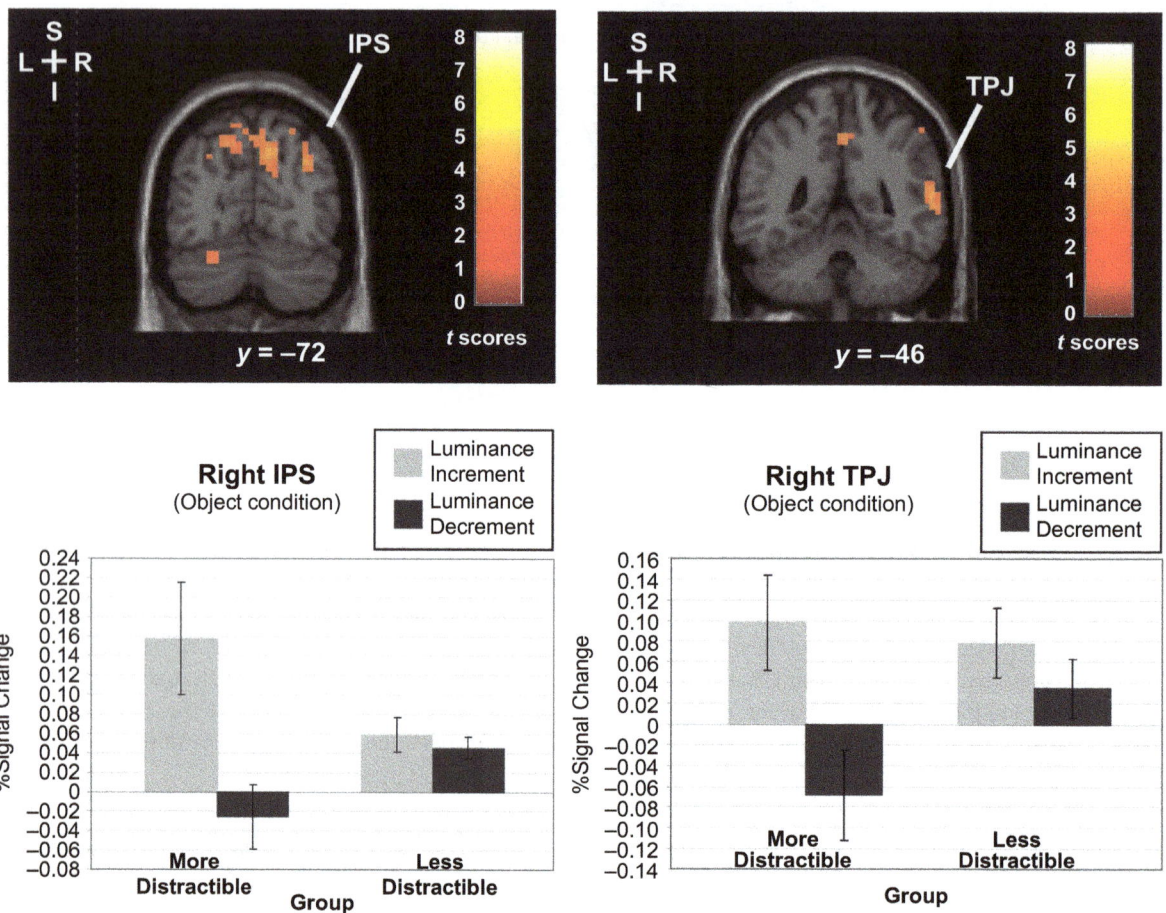

Figure 8.7 Patterns of fMRI activity related to distractibility and disengaging attention. The top panels show areas that were more active for luminance increments than luminance decrements in the object conditions (when increments resulted in the appearance of a new object). The *y*-coordinates refer to positions in the standardized space of the Montreal Neurological Institute (MNI) coordinate system; the *y*-axis refers to the anterior-to-posterior axis. The insets indicate left (L), right (R), superior (S), and inferior (I) directions. The bottom panels show the mean peak BOLD response (6 seconds post-stimulus) in the IPS and TPJ separated by group (more distractible and less distractible) and by luminance increment or decrement. Reprinted from Kim and Hopfinger (2010), *Journal of Cognitive Neuroscience*, with permission from MIT Press.

define a new object (the "luminance decrement" condition). Thus, the greater ability to avoid distraction may be facilitated by more consistently engaging attentional control regions following any change to the display. The more consistent activating of those control regions could enhance the efficiency of the disengagement process, so that when a particularly potent event occurs (such as the appearance of a new object), the brain is well-practiced at reorienting within the context of that task, and distraction can then be quickly overcome. In contrast, for subjects who are not consistently engaging attentional control networks in less urgent situations (i.e., following luminance changes that don't create a new object), the disengagement mechanisms are not well-practiced or primed, and the sudden appearance of a new object then *holds* attention more strongly as they struggle to regain control. Notably, activity levels in early visual processing regions coding the location of the target and the location of the distractor were not different in the two groups. This suggests that the difference in performance was not due to differences in early visual sensory processing, but rather that distractibility was linked to attentional control processes.

Overall, research has provided evidence that certain special stimuli (e.g., threatening stimuli, faces, new perceptual objects) have a disproportionate effect on the *holding* of attention. This research suggests that the appearance of a stimulus triggers a cascade of processes that orient and then hold attention. In the next section, we will discuss theories and data suggesting that these attentional processes are not only time-locked to (and triggered by) the *appearance* of stimuli, but that temporal attention is also influenced by *internal* natural rhythms, oscillating between higher and lower states of focus.

8.5 The Rhythmic Theory of Attention

In addition to top-down control processes that allocate temporal attention at specific moments in time and bottom-up processes that affect how long attention dwells on an item, research has suggested that attention fluctuates periodically according to internal rhythms. Theories regarding the oscillatory nature of brain processes date back at least as early as Hans Berger's (1929) discovery of the alpha wave. It was Berger who coined the term "electroencephalogram" (*Elektrenkephalogramm* in the original German) to describe the brain activity that he recorded in humans. Berger noted a prominent wave at a frequency of ~8–13 Hz and labeled it the "alpha wave." Although there was initial skepticism that the alpha wave reflected cortical activity, further research confirmed Berger's view that the EEG, and the alpha wave specifically, represented a robust rhythm throughout the cortex (e.g., Adrian & Matthews, 1934; Jasper, 1941). Berger had originally labeled all frequencies higher than the 8–13-Hz alpha band as "beta" waves, but he later reported that a particular range of frequencies in those higher frequencies (specifically from 42 to 90 Hz) was increased during intense mental exertion (Berger, 1938). That frequency range is now referred to as "gamma," and it has been consistently found to increase in power during effortful cognitive events, such as those involving attention and working memory (reviewed in Jensen, Kaiser, & Lachaux, 2007). "Beta" is now used to refer to oscillatory neural activity in the narrower 13–30-Hz frequency band just above the alpha band, and gamma is now used to describe all frequencies over 30 Hz (see Table 8.1). In

Table 8.1 Frequency bands of EEG signals.

Band	Range
Delta	1–4 Hz
Theta	4–8 Hz
Alpha	8–13 Hz
Beta	13–30 Hz
Gamma	>30 Hz

the decade after Berger's original report, additional frequency bands *below* alpha were identified and labeled, with frequencies in the 1–4-Hz range labeled as "delta" (Walter, 1936) and activity in the 4–8-Hz range labeled as "theta" (Jung & Kornmüller, 1938).

Typical attention cuing experiments (e.g., the Posner cuing paradigm) that direct attention to a spatial location (or feature) not only indicate what or where subjects are to attend, but also *when* they are to expect a target. Most attention studies use a consistent temporal pattern in which the target occurs at a set time (or at one of just a few possible times) after the start of the trial or after a cue event. Thus, although the main aim of those experiments is to study spatial or feature attention, the trial sequences set up a very strong *temporal* expectation as well. Because of this consistent pattern of stimulus presentation times across trials, any effect of a naturally occurring rhythm on attention processes may not be noticed because the targets occur at only the expected times. More recently, however, studies have begun to investigate whether attention remains engaged at a *constant* level throughout a trial, or if it instead *varies* between periods of higher and lower focus. This research, using a denser sampling of cue-to-target intervals, has revealed that attention appears to fluctuate according to internal rhythms.

Why attention should fluctuate at all could be related to a few different factors. One reason is that attention requires resources: Maintaining attention at peak levels is difficult, and therefore it may be necessary to reduce the allocation of resources at some times to ensure the availability of resources when they are most needed. Relatedly, it's often impossible to know ahead of time exactly when peak attention may be needed, so it may be helpful to have an intrinsic system to periodically reduce focus and preserve resources for those critical moments. Another factor that may be critical in creating fluctuations of attention is that perception and action have contrasting goals. The action of moving the eyes (or head, or whole body) is critical for sampling the environment, but perception is typically suppressed during these movements. For example, the absence of perception during eye movements, called "saccadic suppression," is thought to be necessary to avoid the blurred input that would result as the eyes rapidly jump to a new location. In general, perception is best when the eyes and body are stable, and it is during these stable periods that attention can select the portions of the environment on which to concentrate perceptual resources. Research has also shown that overt exploratory behaviors, including eye movements, often follow a rhythmic pattern of sampling (Buzsaki, 2006). Thus, the alternating periods of *movement* versus *perception* may be a critical foundation upon which attentional rhythms are based. Whether there is a single optimal rhythm for these alternations and how that rhythm may interact with more transient oscillatory activity have been the subjects of much recent research.

Fiebelkorn and Kastner (2019) recently proposed a **rhythmic theory of attention** based on findings from behavioral and neuroscience studies investigating how attention varies over time. Their theory is somewhat different from Jones' "rhythmic theory" of attention discussed earlier (Jones, 1976), in that Jones stressed that *external* rhythms in the current environment were a major factor in determining the fluctuations of attention, whereas Fiebelkorn and Kastner (2019) proposed that *intrinsic* rhythms in the brain represent the primary influence on how attention oscillates. Fiebelkorn and Kastner suggested that an attended location is reassessed every 250 ms or so, resulting in a rhythmic sampling specifically within the *theta* frequency band (3–8 Hz). Results from research supporting – and also questioning – this conclusion are described in the following subsections.

8.5.1   Behavioral Evidence for Rhythmic Attention

Several studies have reported that behavioral measures on attention and perception tasks show alternating periods of better and worse performance, suggesting that attention oscillates between high and low focus. In one of the first studies to investigate whether task performance followed a rhythmic pattern, Landau and Fries (2012) had subjects perform a difficult detection task in which they had to detect a brief contrast decrement to a small region within one of two drifting grating patterns. The moving grating patterns occurred within circumscribed circular regions – one in each visual field – and the target (i.e., contrast decrement) occurred on 90% of trials. Shortly before or after the target, four small dots appeared that briefly surrounded one of the two grating patterns. The "flash" of the four dots at one location lasted 33 ms and was completely random with respect to the target location; the target was equally likely to occur in either the left or right grating pattern, completely independent from where the four dots flashed. The target could occur anywhere from 750 ms before the irrelevant flash of dots to 1,000 ms after it, divided into steps of 16.7 ms, resulting in 105 different intervals. This dense sampling of intervals allowed the researchers to examine whether there were any periodic fluctuations in attention, as measured by accuracy on the task across time. The flash of the dots was used to *reset* ongoing rhythms to ensure that the phase of the rhythm would be consistent across trials. As expected, when the flash occurred *after* the target, it had no effect on target performance. However, when the flash occurred *before* the target, accuracy on the target task was found to oscillate at a theta rhythm (~4 Hz). Furthermore, task performance at the two target locations followed a similar rhythm but in an antiphase relation. In other words, detection accuracy was highest at one location when it was lowest at the other and vice versa. This provided new evidence that attention follows a theta rhythm as it alternates between focusing on different locations that are equally likely to contain a target.

Using a more traditional cuing paradigm, Song, Meng, Lin, Zhou, and Luo (2014) had subjects perform a discrimination task on targets that could occur in one of two locations (to the right or to the left of fixation), and the target was always preceded by a brief irrelevant flash at one location. As in a typical exogenous cuing task, the nonpredictive flash was randomly presented to either the right or left visual field location, and the target was equally likely to occur at either location. In contrast to most previous studies of exogenous cuing that use only two to three cue-to-target intervals, this study used 46 intervals, ranging from 200 to 1,100 ms between the cue and target onset. By utilizing this dense sampling of intervals and by filtering the data to separately investigate oscillations in performance in different frequency bands, the authors found that performance rose and fell in cycles of ~8–15 Hz (in the alpha band) at each location; critically, the results revealed that those bursts of higher and lower alpha activity alternated, in an antiphase manner, between the two possible target locations at a *theta* frequency (3–5 Hz).

In order to test whether oscillatory patterns of attentional focus may relate to how attention shifts *within* versus *across* objects, Fiebelkorn, Saalmann, and Kastner (2013) adapted the paradigm developed by Egly and colleagues (Egly, Driver, & Rafal, 1994) to study how attention involuntarily spreads across an object. As previously discussed in Chapter 5 (and shown in Figure 5.1), the Egly et al. (1994) paradigm is similar to a standard cuing paradigm, but the background of each trial display contains two long rectangles, either to the left and right of fixation or above and below fixation. The cue indicated the likely (75%) target location at one end of one of the two rectangles. Critically, the target could also occur (rarely) at one of two uncued locations that were equally distant from the cued location, with one of those occurring within the same object that was cued and the other occurring in the other (uncued) object. Using this paradigm, but modifying it to include a dense sampling of cue-to-target intervals (randomly selected on each trial from 300 to 1,100 ms), Fiebelkorn et al. (2013) replicated the standard effects of the Egly et al. (1994) paradigm. Specifically, they found that reaction times were fastest at the cued location (i.e., the spatial attention effect), next fastest at the *uncued* location *within* the cued object (i.e., the object-based cuing effect), and slowest to an equally distant uncued location in the *uncued object*. By further analyzing the dense sampling data binned by successive 10-ms cue-to-target intervals, the results revealed that attention periodically sampled locations *within* the cued object at ~8 Hz, and this sampling alternated between objects in a *theta* rhythm (~4 Hz).

However, not all such studies have replicated the finding that attention naturally fluctuates at these frequencies. For example, van der Werf and colleagues (van der Werf, Ten Oever, Schuhmann, & Sack, 2022) used another variation of the Egly et al. (1994) paradigm, in which they added two critical conditions to test how the *predictability* and *behavioral relevance* of the cues affects patterns of attention. In their "moderately informative" condition, they used the standard Egly et al. task procedure in which the target occurred at the cue location on 80% of trials; this is very similar to the Fiebelkorn et al. (2013) study described earlier. The two new conditions in the van der Werf et al. (2022) study included one in which the cue was completely "non-informative" (all possible target locations were equally likely) and another in which the cue was "fully informative" (the target always occurred at the cued location). These three conditions were run in separate blocks. In contrast to some studies, the results showed no rhythmic effects at the *cued* location. As shown in Figure 8.8, performance at the cued location (gray line in Figure 8.8) was maintained at a consistently high level, without any prominent peaks or troughs. The lack of a rhythmic pattern at the cued location was found in all three conditions (noninformative cue, moderately informative cue, and fully predictive cue), and Bayesian analyses supported the conclusion that there was no oscillatory pattern.

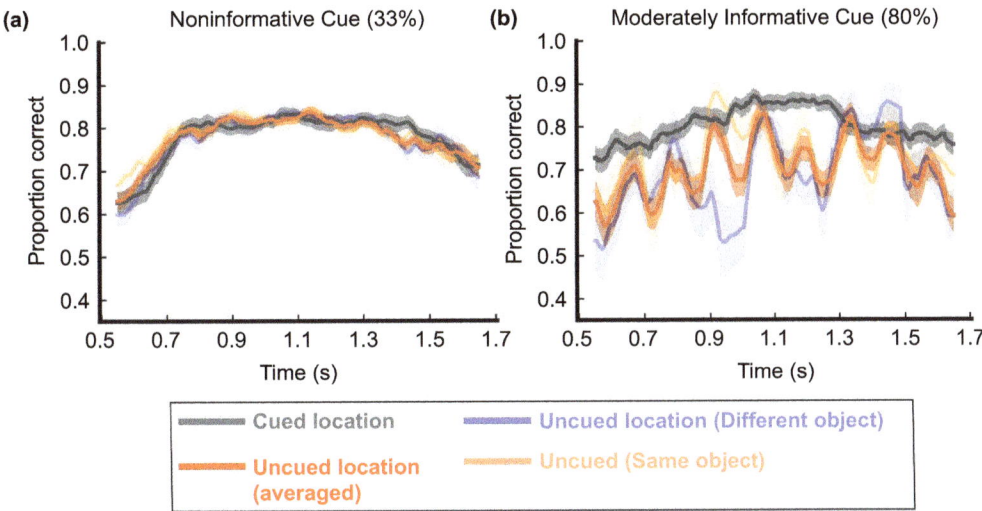

Figure 8.8 Accuracy of target responses across time in a study of spatial and object attention. Panel (a) shows no evidence of rhythmic sampling in the noninformative cue condition (target occurs with equal probability at each of three possible locations). Panel (b) shows the results for the condition in which targets occurred at the cued location 80% of the time and at each of the two uncued locations 10% of the time. In this condition, there was no periodicity for the cued location, but there was a ~7–8-Hz cycle for the invalid locations. Adapted from van der Werf et al. (2022), *The European Journal of Neuroscience*, with permission from John Wiley & Sons, Inc.

Other recent research has also failed to find evidence of an attentional rhythm at the *cued* location (e.g., Peters, Kaiser, Rahm, & Bledowski, 2021). These studies challenge the theory that attention is rhythmic at the *cued* location, and they suggest that it is possible to maintain attention at a consistently high level at the desired location or object. Results at the *uncued* locations have been more consistent across studies. For example, the van der Werf et al. (2022) study did replicate the finding of a robust and significant pattern of rhythmic sampling within the theta band (7–8 Hz) at *uncued* locations. Specifically, in the moderately informative condition (shown as orange lines in Figure 8.8), accuracy at the uncued locations showed clear periods of higher and lower accuracy oscillating at ~133 ms. Notably, this theta rhythm at the uncued locations was only observed in the *moderately informative* condition (Figure 8.8b); it was *not* observed in the *noninformative cue* condition (Figure 8.8a). (Note that the *fully informative* cue condition, not shown in Figure 8.8, had no uncued-location targets, so there were no behavioral responses to uncued targets to measure.) Further confirming the effects of object-based attention, the uncued-location position in the *cued object* showed higher peaks in performance than the uncued-location position in the *uncued object*; however, both positions showed a highly similar oscillatory pattern at a *theta* frequency. Overall, behavioral research provides support for the theory that attention, at least under certain conditions, periodically samples *uncued* locations at a theta frequency, but some findings call into question how automatic this mechanism is and whether the cued location is always sampled in a periodic manner.

Recent studies of inhibition of return (IOR) have investigated whether this mechanism of attention also shows an oscillatory pattern. As discussed in earlier chapters, IOR refers to the finding that shortly after exogenous attention is captured to a location there is a subsequent period during which behavioral responses are *slowed* and early visual ERP components are *reduced* for targets occurring at that location. Attention is thought to be inhibited from returning to the location where it had just been captured. By using a large number of cue-to-target intervals, Michel and Busch (2023) tested whether IOR would show evidence of periodic fluctuations of attention. Their results replicated the basic pattern of an initial period of attention capture (i.e., faster processing at the cued location relative to the uncued location) followed by a prolonged period of IOR. However, despite having a very dense sampling of intervals, there was no evidence of any rhythmicity in the IOR effect. These results support the standard explanation of IOR as being a *sustained* inhibition of the cued location. This finding doesn't contradict the overall theory that *some* aspects of attention may follow a theta frequency, but it does suggest that at least some mechanisms of attention may not vary in an oscillatory manner.

Regarding behavioral studies that have failed to find oscillatory patterns in performance, some authors have argued that this may be due to low-powered experiments or to noise in the data (Fiebelkorn et al., 2013). However, Ten Oever and colleagues (Ten Oever, van def Werf, Schuhmann, & Sack, 2022) pointed out that these null results are highly informative. Specifically, if neural oscillations are important to behavior, then we should try to understand why it has been difficult to consistently replicate behavioral results of rhythmic attention. For example, most studies that find robust behavioral evidence of oscillations typically include a very salient physical event to reset the phase of oscillations. However, single-unit recording studies in nonhuman primates have shown that the "reset" of rhythms by a salient stimulus is not entirely consistent across all trials. The reasons for this are not yet completely understood, but without concurrent neural recordings it cannot simply be assumed that the phase of internal rhythms is being reset consistently on all trials. This may relate to the mechanisms of *involuntary* attention covered in earlier chapters, in that ventral parietal and frontal areas activated by sudden and unexpected events have been proposed to work as "circuit breakers" (Corbetta & Shulman, 2002). In studies of spatial attention, the "break" is thought to involve the quick shift of attention to a highly salient event, often followed by the *reorienting* of attention back to the intended location. In terms of temporal aspects of attention, however, the onset of a salient event may "break" the ongoing *rhythm* and reset it to start again, now time-locked to the appearance of that new salient stimulus. As noted in previous sections, the sudden appearance of a *new object* seems to have a special status in capturing attention, and future work could investigate whether the reason for this special status is that it results from a uniquely effective means of consistently resetting attentional rhythms. Ten Oever and colleagues (2022) suggested that understanding how brain rhythms are manipulated and reset may be important both for understanding the core mechanisms of attention and for developing effective clinical interventions.

Regarding the argument that failures to find evidence of rhythmic sampling are due to insufficient numbers of trials (e.g., Fieblekorn, 2022), this raises questions about how *intrinsic* those rhythms really are. If finding rhythmic effects on behavior is dependent on a having a very large number of trials, it raises the possibility that those effects may actually be related to the "selection history" mechanism covered in Chapter 6; specifically, that the context built up over many trials of an experiment creates predictions (sometimes implicit) that guide attention (e.g., where and when to expect a target). Therefore, as opposed to there being an intrinsic rhythm for attention that simply requires many trials for it to be measured well, it could be that those many trials are what is creating the expectancies, in an externally driven way, about the discrete time windows when targets are likely to occur. In other words, attention may adaptively "learn" to sample at the frequencies most commonly present in a particular experiment. Further work is needed for us to better understand the conditions under which attention is bound to intrinsic rhythms, how these rhythms are reset, and which mechanisms of attention may overcome ongoing rhythms. Neuroscience investigations into rhythmic patterns of brain activity are beginning to shed light on some of these issues.

8.5.2 Human Neuroscience Studies of Attentional Rhythms

The ability to precisely resolve the timing of neural activity is critical for studying brain rhythms, and therefore neuroimaging methods that rely upon the very sluggish hemodynamic response are of limited use in this area. Specifically, the blood flow dynamics measured by PET and fMRI lag behind neural activity by 4–8 *seconds*, so only very slow oscillations can be resolved using those methods. However, EEG and MEG are ideally suited to measure the fast rhythms of the brain, and these methods have begun to be used to study the neural processing underlying oscillatory patterns in perception and attention. Busch and VanRullen (2010) used EEG to study *sustained* attention by employing a difficult target detection task in which each trial consisted of a series of eight successive targets that were preceded by a cue instructing subjects on which visual field to attend. The cue-to-first-target interval and each subsequent target-to-target window were rather long (each interval was randomly sampled from 1,500 to 1,900 ms). This study was thus different from many of the cuing paradigms described earlier in which every target was preceded by a cue and the total trial length was under 2 seconds. In Busch and VanRullen's (2010) study, attention was to be maintained at the cued location over a much longer duration (12–15 seconds). The behavioral results showed the expected *overall* benefit of spatial attention, as perception (measured as the luminance threshold for target detection) was better for attended-location targets than unattended-location targets. Critically, the EEG data provided a means to investigate links between rhythms in brain activity and perception. The results revealed that the *phase* of an intrinsic *theta* oscillation (~7 Hz) recorded over frontocentral sites was strongly linked to detection performance. Furthermore, this relation was observed only for attended-location stimuli; detection of unattended-location targets showed no relation to the phase of this theta rhythm at those scalp sites. This study found that attention and task performance were linked to alpha activity as well. As described in Chapter 5, attention has been linked to alpha activity, in that attended locations show a relative *reduction* of alpha activity compared to unattended locations, and targets are detected best when alpha power is lowest (Ergenoglu, Demiralp, Bayraktaroglu, Ergen, Beydagi, & Uresin, 2004; Thut, Nietzel, Brandt, & Pascual-Leone, 2006; Worden, Foxe, Wang, & Simpson, 2000). In the Busch and VanRullen (2010) study, their dense electrode array allowed them to localize the regions where oscillatory power was most related to behavioral performance. They found that the link to *alpha* power was strongest over *posterior* electrode sites, whereas the relation between performance and *theta* activity was strongest at *frontoparietal* sites. This provides further evidence that the mechanisms of attention involve coordinated activity across multiple different rhythms and brain areas. This study was among the first to directly investigate the effects of oscillatory neural activity in humans on performance during *sustained* attention, and it revealed that when we try to *maintain* our attention at a single location for an extended time, our focus at that location waxes and wanes. Furthermore, this study provided evidence that *intrinsic* rhythms, specifically at a theta frequency, play an important role in the allocation of attention over time. Subsequent EEG studies have confirmed and extended these results (reviewed in Kienitz, Schmid, & Dugué, 2022).

Utilizing a method with the same high temporal resolution as EEG but superior spatial resolution, Landau and colleagues (Landau, Schreyer, van Pelt, & Fries, 2015) used MEG to investigate the relation of these theta oscillations to the effects of attention on the gamma-band brain activity found in previous studies. Previous research has found that bursts of gamma activity recorded over posterior scalp sites are triggered by visual stimuli and are significantly enhanced by attention (reviewed in Fries, 2009). Using MEG, Landau et al. (2015) measured gamma activity from visual processing regions representing the two possible target locations in their study, and they used this as a continuous index of which location was being attended. This allowed them to index where attention was focused without having to rely on overt behavioral responses. Specifically, bursts of gamma activity over the left and right hemispheres were used to calculate a lateralized gamma activity metric that indicated where attention was focused at each moment. Participants performed a similar paradigm to that in the Landau and Fries (2012) study described earlier, in which a brief luminance decrement to a small region within one of two grating patterns on the screen had to be detected. Subjects maintained fixation on a central dot throughout all trials, and the target occurred at unpredictable times. The results replicated previous findings that better performance was associated with higher gamma activity contralateral to the target location, and they provided new evidence that those bursts of higher gamma activity alternated between the brain regions representing the two locations at a theta rhythm (~4 Hz). Furthermore, they localized these effects to early visual processing areas in the calcarine sulcus, lingual gyrus, and precuneus. Another major aim of this study was to investigate whether a salient "reset" event was critical for observing theta oscillations linking brain activity to task performance. Critically, unlike many previous experiments in which there was a consistent temporal trial sequence that could set up an optimal rhythmic sampling for the task, and unlike studies that have used a salient physical event to reset brain oscillations time-locked to that event, this study found that attention followed an *intrinsic* theta rhythm. In this experiment, there was no salient event to trigger a reset, and there wasn't a predictable temporal pattern for when targets should occur. Thus, these results provide support for the view that *intrinsic* theta rhythms in the brain are a critical component of how attention is allocated.

To further investigate the brain regions responsible for the theta rhythms that have been linked to attention sampling, Helfrich and colleagues used **electrocorticography (ECoG)** recorded from subdural electrode arrays (Helfrich, Fiebelkorn, Szczepanski, Lin, Parvizi, Knight, & Kastner, 2018). These recordings were made in 15 pharmacoresistant epilepsy patients who were undergoing presurgical monitoring. Since ECoG records data from an electrode grid placed directly on the cortical surface, it has superior spatial resolution compared to scalp-recorded EEG, with the same high (millisecond) temporal resolution. A limitation of the ECoG method for studying cognitive processes is that, since the electrode arrays are placed specifically to monitor the foci of seizures, the regions sampled are determined by where the seizures are believed to originate, and for any one subject there is only a limited area of the brain being sampled. In the Helfrich et al. (2018) study, occipital sites and the prefrontal pole were not well sampled, but across the 15 participants in their study there was extensive coverage across temporal, parietal, and frontal areas in the left and right hemispheres (758 electrode sites across 8 subjects in one experiment, 614 sites across 7 subjects in the other). In one experiment, subjects performed an attentional cuing task in which they maintained fixation while viewing a dynamic display of appearing and disappearing red dots. At a random time following the beginning of the trial, a cue (arrow at fixation) instructed subjects to attend to one of the two visual hemifields to detect a blue dot target that could appear within the red dots 1,000–2,000 ms after the cue. In this experiment, there were *two* relevant spatial regions – one in each hemifield. In the other experiment, the researchers used a paradigm similar to the Egly et al. (1994) task described earlier, in which there were *three* distinct locations that could contain a target (i.e., cued location, uncued location in the cued object, and uncued location in the uncued object). The analysis of this experiment focused only on the cued location because there were insufficient numbers of trials for both the uncued locations. Across these two experiments, the researchers could assess whether attention sampling would differ when there were two versus three relevant locations to be monitored. In both experiments, the behavioral responses showed a robust oscillatory pattern, cycling between better and worse performance at a theta rhythm (~4 Hz). Critically, this fluctuation in performance at the cued location didn't depend on how many possible locations were relevant. This provides evidence that, at least under some conditions, attention isn't simply maintained at a uniformly high level throughout the entire period of waiting for the target, but rather that it oscillates between higher and lower levels of attentional focus following a theta rhythm. The results also showed that cortical excitability, as indexed by brief bursts of high-frequency-band (70–150-Hz) gamma activity, oscillated at the same theta frequency (~4 Hz) as the behavioral responses. Furthermore, the dense cortical coverage of ECoG allowed the researchers to localize the theta rhythms most strongly linked to perform-ance to frontal and parietal regions previously implicated in attentional control. Specifically, the IPS, superior parietal cortex, and FEF regions were most strongly associated with the phase of theta rhythms associated with higher cortical excitability and better behavioral responses. As noted earlier, the coverage didn't consistently include occipital areas, so the relation between these theta rhythms in attentional control regions and the visual processing regions of the occipital lobe couldn't be directly assessed here. However, this study provided critical new evidence that frontal and parietal regions, which had been previously implicated in attentional control, may enable the coordination of attention through synchronized theta activity.

A few recent studies have used noninvasive *neurostimulation* methods, such as TMS, to test the causal nature of theta rhythms on performance. By stimulating neural activity in focal areas of the brain and measuring its effects on ongoing task performance, the link between neural rhythms and behavior can be directly assessed. Using a visual search experiment, Dugué and colleagues (Dugué, Marque, & VanRullen, 2015) first identified a robust theta oscillation

(~6 Hz) using EEG and localized it to visual processing regions. They then used TMS to stimulate those occipital regions and found that performance was significantly affected by stimulation at a theta frequency (6-Hz stimulation pulses). In another study, Dugué and colleagues (Dugué, Roberts, & Carrasco, 2016) used TMS to test rhythmic attention in a cuing task. The TMS pulses were delivered at a theta rhythm (5 Hz in this experiment) and applied at scalp locations that would target early visual areas (V1/V2) corresponding to the location of the target and distractor stimuli. In an initial phase of the experiment, TMS-induced phosphenes were mapped out to ensure that the TMS pulses delivered in the main experiment were stimulating the portions of early visual areas representing the locations of the targets and distractors. The results showed that task performance at the *invalid* location was significantly affected by neurostimulation of early visual areas at this theta rhythm. These results confirmed that there is a role played by theta rhythms in early visual processing and attention, and more specifically in the *reorienting* of attention that must occur when attention is initially allocated to the wrong (invalid) location and must be quickly disengaged and reoriented to the location of the target. Finally, TMS has also been used to assess the role of theta rhythms in attention control regions of the frontal lobe. Specifically, Dugué, Beck, Marque, & VanRullen (2019) investigated the role of the FEF during a masked visual search task. The TMS pulse was delivered at a random time from 50 to 450 ms following the onset of the visual search display and was positioned either over the right frontal lobe (targeting the right FEF) or at a central scalp location on top of the head (corresponding to the Cz location in EEG coordinates) as a control condition. The results revealed that the FEF stimulation, relative to the control site stimulation, modified performance accuracy on the visual search task, with an oscillatory pattern of interference occurring at ~6 Hz. Combined with the results of the earlier TMS studies, these findings reveal that theta rhythms across frontal and occipital regions play an important role in the control of attention.

8.5.3 Nonhuman Electrophysiology Studies of Attentional Rhythms

Some of the most informative work on rhythmic attention has come from nonhuman electrophysiology studies. By recording from microelectrodes inserted into the brain, single-unit and multiunit recordings provide precise measurement of the location, timing, phase, and amplitude of neural activity. As discussed in Chapter 3, these invasive methods provide the highest spatial and temporal resolution of all the methods typically used in cognitive neuroscience. In addition to providing precise information on the firing of individual neurons, microelectrodes can also be used to measure **local field potentials**, which reflect the summed activity of a population of neurons close to the recording tip consisting of action potentials summed from neurons ~50–350 μm from the tip (Gray, Maldonado, Wilson, & McNaughton, 1995) and ionic changes summed from a wider region ~0.5–3 mm around the tip (Juergens, Guettler, & Eckhorn, 1999).

Recent work using nonhuman primate electrophysiology in studies of attention has further investigated the role of rhythmic activity coordinated across multiple attentional control regions. Fiebelkorn, Pinsk, and Kastner (2018) recorded from the FEF and the LIP region *simultaneously* in order to assess how these regions may work together to allocate attention.

(Note that, as discussed in Chapter 7, the LIP in nonhuman primates has been associated with the attention mechanisms ascribed to the IPS region in human neuroimaging studies.) In this study, monkeys performed a simplified variation of the Egly et al. (1994) task described earlier, in which targets occurred either at the cued location or at an uncued location in the uncued object. Accuracy on the task was better at the cued location than the uncued location, as expected, and performance at the cued location exhibited a theta rhythm, with behavioral accuracy oscillating at ~3–8 Hz. These results are consistent with the behavioral results seen in studies with human subjects, providing evidence that attentional focus in the monkey brain follows a similar periodicity. Along with this pattern of behavioral performance, the recorded activity in the FEF and LIP was found to show a strong theta rhythm (3–8 Hz) that was aligned in phase (peaks and troughs occurring at the same time) across both areas. Figure 8.9 represents this in-phase theta rhythm in the FEF and LIP as a slow oscillation, showing approximately

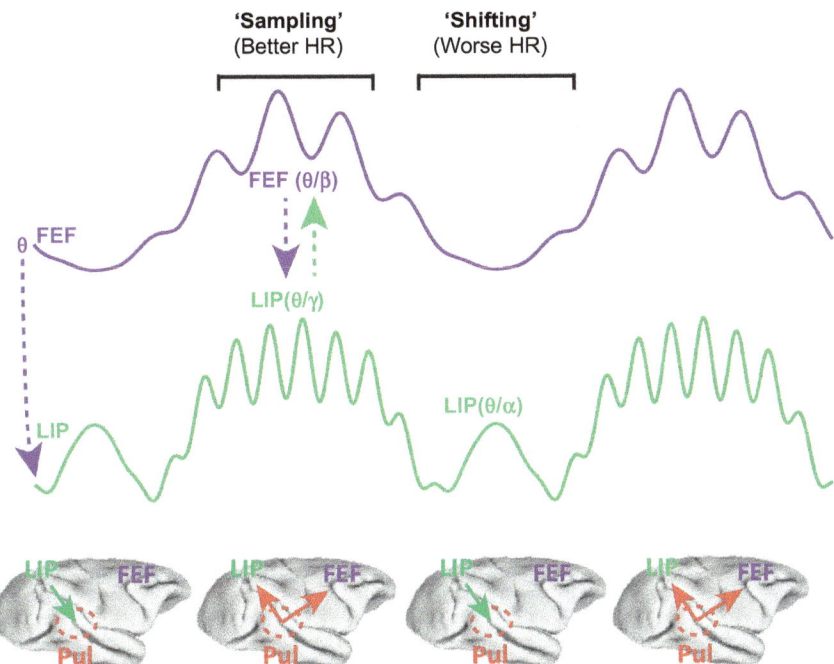

Figure 8.9 Coordinated rhythmic activity across frontal and parietal attention control regions. The model of rhythmic sampling from Fiebelkorn and Kastner's (2019) review, proposing that activity in the FEF and LIP regions, recorded via simultaneous single-unit recording experiments in macaques, exhibits coordinated activity locked to theta-band oscillations. The time course plotted here is illustrative of approximately a third of a second, such that a theta rhythm is shown as the slower oscillation of approximately two complete cycles. During the peak phase of the theta activity shown here, there is an increase in beta activity in the FEF and an increase in gamma activity in the LIP. During this phase, attention is thought to be *less* likely to move and performance on perceptual discrimination to be *enhanced*. During the trough of theta activity in these regions, alpha activity increases in the LIP, task performance is worse, and attention is more likely to move to a new location or object. HR = hit rate (referring to behavioral response); Pul = pulvinar. Adapted from Fiebelkorn and Kastner (2019), *Trends in Cognitive Sciences*, with permission from Elsevier.

two complete cycles in the illustration. Critically, this experiment was able to associate the theta rhythm in the FEF and LIP both with behavior on the task and with other rhythmic neural activity. The phase of the theta rhythm in these regions was associated with two distinct states. One state, as indicated by the relative trough in theta activity in Figure 8.9, was linked to *poorer* performance on the task and to an *increase* in *alpha* activity in the LIP. Previous research has shown that *higher* alpha activity in visual processing regions is related to poor attentional focus and *worse* visual discrimination performance (Foxe & Snyder, 2011; Worden et al., 2000), and this study by Fiebelkorn et al. (2018) directly linked the increase in alpha activity and the decrease in task performance to one particular phase of a theta rhythm occurring in the FEF and LIP. The second state of activity in these two regions was found to be associated with *better* performance on the discrimination task and with an *increase* in two other rhythms in those regions. Specifically, the synchronized theta-phase activity associated with improved performance was also associated with an increase in *gamma* activity (~35–60 Hz) in the LIP and an increase in *beta* activity (~15–35 Hz) in the FEF. Previous research has found increased gamma activity to be linked to better perception and task performance (reviewed in Fries, 2009), whereas increased beta activity has been associated with a suppression of voluntary exploratory movements in humans (Pogosyan, Gaynor, Eusebio, & Brown, 2009) and nonhuman primates (Zhang, Chen, Bressler, & Ding, 2008). By recording from multiple regions simultaneously, the Fiebelkorn et al. (2018) study was able to show that gamma activity in the LIP and beta activity in the FEF were closely linked to the same phase of a theta rhythm. The authors suggested that the different phases of theta activity can be associated with alternating periods of "sampling" (when attentional movements are suppressed and perceptual processing is best) and "shifting" (when perception is worse and attention is more likely to disengage from its current focus and move elsewhere).

If the theory of alternating periods of sampling and shifting is correct, then patterns of overt attention, measured by exploratory saccadic eye movements, should also show a rhythmic pattern in the theta range. Indeed, eye-tracking studies have found that a greater likelihood of making exploratory saccades alternates, following a theta rhythm, with periods of higher perceptual sensitivity (Benedetto, Spinelli, & Morrone, 2016; Hogendoorn, 2016; Tomassini, Spinelli, Jacono, Sandini, & Morrone, 2015). In addition to the large saccadic eye movements of overt attention, recent research has revealed that **microsaccades** can also be informative about the focus of attention. Microsaccades are the tiny movements of the eyes that occur even when we attempt to maintain fixation. The purpose of these small eye movements is debated, but theories suggest it may be to prevent the fading that occurs if visual neurons receive a completely constant input; by shifting the eyes just slightly, different populations of cells in the retina will be stimulated, preventing this fading. Research has shown that even when subjects are attempting to maintain fixation on a central point during a covert attention task, microsaccades around this point are associated with the direction of attention. Specifically, microsaccades toward the cued location are associated with maintaining attention at that location, whereas microsaccades away from that location have been associated with periods of moving attention to a new location (Engbert & Kliegl, 2003; Hafed & Clark, 2002). These microsaccades thus provide converging evidence that the focus of attention oscillates at a theta rhythm, even when we're attempting to sustain attention at a single location.

How activity across the FEF and LIP is coordinated is not yet fully understood, but Fiebelkorn and Kastner (2019) have suggested that the pulvinar nucleus of the thalamus plays a critical role. As illustrated in Figure 8.9 (bottom panel), the pulvinar has been found to coordinate this activity across regions during periods of "sampling," possibly by sending signals simultaneously to the FEF to inhibit movement of attention and to the LIP to ensure attention is highly focused at one location (Fiebelkorn, Pinsk, & Kastner, 2019). During periods of "shifting," in contrast, the LIP is thought to essentially inhibit those actions of the pulvinar, thereby promoting more efficient disengagement and shifting of attention to new locations. This view of there being alternating periods of sampling versus shifting is similar to *predictive coding models* of the mind that propose a process of "active inference" in which the organism alternates between periods of: (1) using current sensory input to update models of the world; and (2) moving the sensory organs (or attention) to change that input and better test those models (e.g., Smith, Badcock, & Friston, 2021). Predictive coding models of perception and attention will be covered in detail in the next chapter.

A final note regarding nonhuman primate studies concerns the finding of attentional rhythms in the *beta* frequency band (15–30 Hz). At first glance, this may seem to conflict with the human ECoG studies described earlier that found robust effects in theta frequencies but *not* in beta frequencies. A reason for this difference may be that ECoG signals come primarily from *superficial* cortical layers, whereas multicontact depth recordings in primates can provide separate measures of activity from each cortical layer. Indeed, in a study of working memory in nonhuman primates, Bastos, Loonis, Kornblith, Lundqvist, and Miller (2018) recorded from **laminar electrodes** and found that beta rhythms originated in deep layers, whereas theta rhythms could be produced in more superficial layers. Since ECoG is recorded from the outer surface of the cortex, it is highly sensitive to neural activity in the superficial layer of cortex but relatively insensitive to activity in deep layers. This further highlights the importance of integrating findings from across different methods. As discussed in previous chapters, there is no perfect method, and it is critical to integrate results from across many different methods in order to build a more complete understanding of the brain basis of attention processes.

CHAPTER SUMMARY

- Attention can be *voluntarily* allocated to specific moments in time when a target is expected, but there are also multiple *involuntary* influences on how attention is allocated over time.
- Neuroscience studies of the attentional blink provide evidence that semantic-level processing and conscious awareness of stimuli depend upon attentional resources that are limited over short timescales.
- How long attention dwells on an object or location is not simply determined by voluntary intention to sustain attention but is affected by stimulus properties that influence the holding of attention and by internal rhythms that may impose alternating periods of higher focus versus exploration.
- Rhythms in the external world and learned history of task structure can involuntarily affect the temporal allocation of attention.

- Electrophysiology and single-unit studies are revealing that internal theta rhythms may play a critical role in coordinating the mechanisms of attention across cortical and subcortical regions.

REVIEW QUESTIONS

Describe the effects of the AB on the multiple ERP and MEG components that have been tested in this paradigm, and explain what these results reveal about attention, perception, and consciousness.

List and describe the different types of information in the environment that can affect the temporal allocation of attention.

Explain how item memory, faces, and new perceptual objects affect attention over time.

Compare the control mechanisms and neural effects of spatial attention, object-based attention, and temporal attention.

Describe how the coordination of different rhythms in the brain may explain the neural and behavioral effects of sustained and transient attention.

FURTHER READINGS

Jones, M. R. (1976). Time, our lost dimension: Toward a new theory of perception, attention, and memory. *Psychological Review*, *83*(5), 323–355.
- This review paper was a seminal and pioneering work in developing the theory that time plays a critical role in attention and cognitive processing.

Nobre, A. C., & Van Ede, F. (2018). Anticipated moments: Temporal structure in attention. *Nature Reviews. Neuroscience*, *19*(1), 34–48.
- This excellent review article covers the neuroscience evidence for the multiple types of temporal patterns that influence how attention is oriented and allocated across time.

Zivony, A., & Lamy, D. (2022). What processes are disrupted during the attentional blink? An integrative review of event-related potential research. *Psychonomic Bulletin & Review*, *29*(2), 394–414.
- This recent article reviews and summarizes the many studies that have used ERPs to investigate the neural mechanisms of the AB.

9   Predictive Coding Models of Attention

Turning Perception on Its Head

Learning Objectives

- Describe the role of top-down processing in perception.
- Contrast the classic model of perception with the predictive coding model.
- Understand why the *balance* of bottom-up and top-down processing is critical.
- Identify how mechanisms of attention might be understood more deeply through the lens of predictive coding models.

This chapter explores "predictive coding" models, which challenge classic theories of perception and brain function. By incorporating details of brain structure from the level of connectivity between brain areas to the level of laminar microcircuitry within cortical regions, these models suggest a radical new way to conceive of perception and cognition. The neural processes revealed in these models will likely lead to revisions of some long-standing theories and should be incorporated into future research on attention and perception.

9.1   Classic Model: Bottom-Up Processing Drives Perception

Most theories of attention are built upon the foundation of models of perception, and much of attention research is aimed at understanding how attention modifies perceptual processing. Attention researchers, therefore, must ensure that models of attention are viable with respect to the theories and advances in the science of perception research. A radical new way of conceiving of perception, and brain function in general, has received increasing support over the past decade and is beginning to be incorporated into theories of attention. This new conception of perception is referred to as "predictive coding," "predictive processing," "Bayesian hierarchical models," or "the free energy principle" (Box 9.1 provides further background and terminology). A comprehensive comparison of these different terms, and what they differentially specify for theories of perception and brain function, is beyond the scope of the current book; however, what they critically have in common is the proposal that "feedback" processing in the brain is much more important than classic models have held. According to classic models of brain function, most of the heavy lifting is

Box 9.1　Bayesian models and predictive coding terminology

"**Predictive coding**," as used in the present chapter, refers to a family of models that utilize concepts initially formalized in Bayesian theory. Bayes' theorem defines the probability of an event as based on prior knowledge combined with current evidence. In Bayesian terms, a "prior" refers to the *initial* belief about the likelihood of some event being true, *before* new evidence has been acquired, whereas a "posterior" refers to the belief *after* accounting for new evidence. Thus, the initial estimate of likelihood is updated with current evidence to provide a new estimate. In mathematical terms, Bayes' theorem is written as

$$P(A|B) = \frac{P(B|A) * P(A)}{P(B)}$$

$P(A|B)$ = **posterior probability** of A given B. This is the probability of hypothesis A being true given the evidence B; also known as the *conditional probability*.

$P(B|A)$ = probability of B given A. This is the probability of observing evidence B if hypothesis A is true; also known as the *likelihood*.

$P(A)$ = **prior probability** of A. The probability of A being true before collecting evidence; also known as *prior knowledge*.

$P(B)$ = prior probability of B. The probability of observing the evidence B; also known as *marginalization*.

It's important to keep in mind that $P(A|B)$ doesn't equal $P(B|A)$. For a real-world example, consider a symptom (B) of having cancer (A).

$$P(cancer|symptoms) = \frac{P(symptoms|cancer) * P(cancer)}{P(symptoms)}$$

Even if we know that most people with cancer have that symptom, and therefore $P(B|A)$ is very high, this doesn't mean that a person with that symptom is highly likely to have cancer. Because the probability is based on the base rate of cancer along with the base rate of that symptom, even if the symptom were to be present in all people with cancer, it could still be unlikely that someone with that symptom has cancer. In other words, the presence of the symptom could have a large false-positive rate. Critically for our purpose, the *priors* (i.e., knowledge of overall base rates) are combined with current *evidence* (e.g., presence of the symptom) to determine the likelihood of the hypothesis being true (that the person has cancer given that they have the symptom).

In predictive coding models, when *posterior probabilities* are calculated, these are used to update the *priors* in a continuous iterative process, so that hypotheses are continually being generated, tested, and revised. Errors of the current hypothesis are propagated forward, causing revision to the current hypothesis, which is then propagated downward, and the new model is then tested against the incoming evidence. This iterative process of continually refining hypotheses about the world continues, with the overall goal being to reduce the prediction error, thereby creating a model that most accurately predicts the external world.

done through bottom-up, "feedforward" processing. In vision, for instance, the classic model posits that perception is driven predominantly by bottom-up processing advancing from the retina, to the lateral geniuculate nucleus of the thalamus (LGN), to V1, to V2, and so on, with more complex processing happening at each successively high stage. The opposite direction of connectivity, going from "higher" to "lower" levels (e.g., V2 to V1), is referred to as "feedback" in the standard model, and this top-down processing is thought to have a relatively minor influence on feedforward processing. According to these models, bottom-up processing predominantly drives perception. Predictive coding models, however, propose a reversal of these roles, with perception being driven by top-down "feedback" processing (e.g., Rao & Ballard, 1999). According to predictive coding, perceptual experience is driven mostly by the top-down model (or "prediction") of the world that is sent "down" the hierarchy. Bottom-up processing, in this account, is mainly just providing information on how the current model of the world needs to be fixed by indicating where that model is in error; bottom-up processing is therefore thought to be transmitting "prediction errors." In this way, the standard model is turned on its head, as the bottom-up processing thought to be driving our perceptual experience is instead relegated to the relatively minor role of "adjustment," while top-down "feedback" processing takes on the role of the primary driver of perception that previously was attributed to bottom-up "feedforward" processing. To put some of this into more concrete terms, we consider a few illusions in the next section, starting with one that, appropriately, involves a turning head.

9.2   Illusions Illustrate Powerful Top-Down Influences on Perception

The hollow mask illusion is one of many illusions that reveal the amazing degree to which top-down processing can overwhelm the bottom-up processing coming from our senses. First presented in 1973 (Gregory & Gombrich, 1973), the phenomenon involves a 3D mask of a human face rotating consistently in one direction (Figure 9.1). The mask can be seen from the front, in which case the face is convex, with the nose coming out toward the viewer, as in an actual human head. From this perspective, the viewer easily perceives the mask as a typical convex face. As the hollow mask is rotated further, so that the mask is now concave (with the nose pointing away from the viewer), the face is *incorrectly* perceived as still being convex, with the nose pointing toward the viewer. The strength of the illusion is driven home when the viewer watches the mask continuously rotate, first seeing the front (convex) view of the mask followed by the concave (hollow; back) view of the mask. When the mask is rotating clockwise and the convex side of the mask is toward the viewer, the participant correctly perceives the clockwise motion without any trouble. As the mask continues to turn clockwise, however, and the participant now is presented with the concave (hollow) side of the mask, the perceived motion *reverses* completely; instead of what they know to be the concave backside of the mask turning clockwise, the participant incorrectly perceives a *convex* face moving in the opposite direction (i.e., counterclockwise in this example). The illusion remains strong despite repeated viewings and despite the viewer having valid knowledge of the actual nature of the mask. This illusion provides evidence of how strongly our top-down knowledge of the world can overcome bottom-up sensory input. In this case, we have strong top-down knowledge that faces in the world are convex – noses point outward, not inward; and lips, eyes,

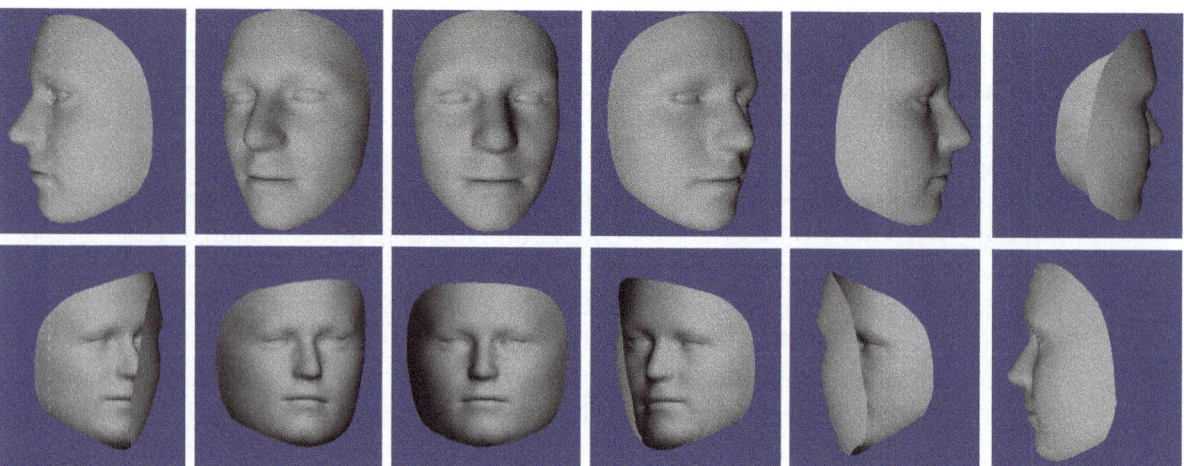

Figure 9.1 The hollow mask illusion. Shown here are still frames from an online illustration of the hollow mask illusion. The stimulus is a mask of a face, in which one side is the outside of a mask, with the features of the face projecting outward (shown in the top row). When viewing that side of the mask rotating, the viewer sees the correct motion (here, the nose is seen at the far left and it rotates toward the viewer until it is straight ahead and then continues in that same direction until the nose is at the far right of the screen). When the mask has rotated around so that we begin to see the inner (hollow) side of the mask (shown here in the bottom row), we don't see the features as hollow and pointing away from us, but rather our mind *flips* the image so that we perceive the face as it "should" be, with the nose, mouth, and forehead pointing toward us. But because we are mistakenly interpreting the image as pointing toward us, we also incorrectly perceive the motion as rotating in the opposite direction. This has been taken as evidence that our top-down models of the world (in this case, that faces must always project outward toward us) dominate our perception, even overriding the sensation of an object rotating continuously in one direction. The screenshots that make up this image were taken from https://commons.wikimedia.org/wiki/File:Hollow-face-illusion.gif; author: Empetrisor, CC BY-SA 4.0 <https://creativecommons.org/licenses/by-sa/4.0>, via Wikimedia Commons.

and foreheads protrude toward you, not away from you. This is fundamental knowledge about the world that confers the important benefit of giving us fast and (normally accurate) perceptions in situations in which poor lighting or occluded stimuli may provide us with impoverished sensory information. But the ability of this knowledge to *overrule* bottom-up processing illustrates the importance of top-down information in our perceptual experience.

Another example is color constancy, in which our world knowledge normally lets us correctly perceive that a color remains constant across different lighting conditions and shadows. In most situations, this allows us to perceive colors of objects quickly and accurately in the real world, without getting confused about why an object is suddenly changing color when a shadow falls across its surface. Although we do consciously perceive that the color is darker within the shadow than outside of it, our attribution of the color is strongly influenced by this top-down knowledge of how shadows affect color, and we can perceive it as maintaining its color across different lighting conditions. However, this ability also means that we are subject to illusions. As shown in Figure 9.2, we perceive square B (within the shadow) as lighter than square A (in the light) even though the two squares are in fact the *identical* color.

Figure 9.2 Color constancy. We perceive the square marked "B" as lighter than the square marked "A," even though they are the exact same color. This has been attributed to the brain's ability to maintain "color constancy" across different lighting conditions. For example, the lighter-colored square directly below "A" has a shadow going across its lower right-hand corner. However, we automatically process the square as being one continuous color, with the actual difference in brightness being due to the shadow not to the square actually being two colors. In this example image, we also perceive all the "light" squares as the same color, which is why it is so hard to see that the "B" square in the shadow is identical in color to the "A" square outside the shadow. This illusion again illustrates the power of top-down processing over bottom-up processing. Image from: https://commons.wikimedia.org/wiki/File:Checker_shadow_illusion.svg, original: Edward H. Adelson, vectorized by Pbroks13, CC BY-SA 4.0 <https://creativecommons.org/licenses/by-sa/4.0>, via Wikimedia Commons.

These illusions provide evidence that top-down knowledge can dominate bottom-up processing. However, they don't directly assess *where* or *how* that higher-level knowledge and top-down processing compete with bottom-up processing. In other words, these examples don't explain whether this knowledge actively transmits information down multiple successive levels of the hierarchy, or whether the competition only occurs at a single, high-level stage after bottom-up processing has propagated all the way up through most of the hierarchy without being adjusted in any way. In the next section, we will go beyond illusions and consider converging evidence from investigations into neural architecture that show the strong influence of top-down models throughout the levels of the sensory processing hierarchy.

9.3   "Feedback" Connections in the Visual System

As noted earlier, the classic model of visual perception is that it is driven by the "bottom-up" flow of information processing through a hierarchy of visual areas, proceeding from the processing of very simple features at lower levels to the processing of complex objects (e.g., faces) at highest levels. The idea of perception being driven in a bottom-up manner is part of the terminology used to describe the nervous system, dating back at least as early as Sherrington's

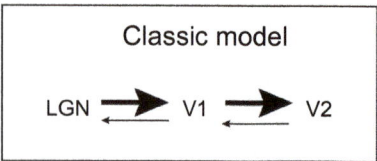 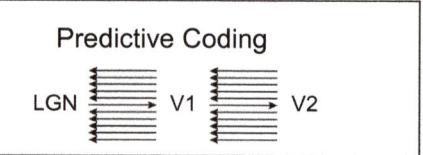

Figure 9.3 Classic model of perception versus predictive coding accounts of early visual processing. The classic model of visual perception typically concentrates on the bottom-up or "feedforward" flow of information from lower to higher brain regions. According to this model, the "feedback" processes from higher to lower areas are relatively minor parts of perception. Newer investigations, however, have shown that there are massive numbers of "feedback" connections, even in very early stages of visual processing, and these can outnumber the feedforward connections. Models such as predictive coding suggest that perception is driven largely through these top-down connections, with the bottom-up connections sending only information on the errors in existing top-down models of the world forward to update those models.

(1906) use of the term "receptive fields," with that term implying that the role of some cells is to passively "receive" input before conveying it upward. Hartline (1938) first extended this into the realm of vision, mapping the receptive fields of ganglion cells. Subsequent research identified neurons in cortical areas that are selectively responsive to increasingly complex objects and combinations of features. The increasing complexity at increasingly high levels of the visual hierarchy has been shown through decades of neuroscience research, starting with the pioneering work of Hubel and Wiesel (1962). However, the idea that our perception of the world is driven *predominantly* by *bottom-up* processing is called into question when considered in light of newer evidence of the neural connectivity between areas. The standard view of perception, with perception being driven by bottom-up processing (with top-down processing serving only a minor, modulatory role), suggests that feedforward connections should be stronger and more prevalent. However, research has now shown that "feedback" connections in some visual processing areas are far greater in *number* than "feedforward" connections (Figure 9.3). In examining the connectivity to area V1, Budd (1998) found that the feedback connections from V2 to V1 outnumbered the feedforward connections from the LGN to V1 by a factor of *10*. Similarly, Erişir, Van Horn, Bickford, and Sherman (1997) and Guillery (1970) provided evidence that there are 10 times as many feedback connections to the LGN (from V1) as compared to the feedforward connections to the LGN from the retina. In addition, studies have shown that the *strength* of feedback connections, as measured as the ability to drive the target area, is often just as strong as that of the feedforward connections in both early visual (Covic & Sherman, 2011) and early auditory processing (De Pasquale & Sherman, 2011) regions. Finally, research has shown that feedback connections are able to drive the target area in the absence of bottom-up input. Mignard and Malpeli (1991) showed that V2 could drive V1 activity when the LGN was completely inactivated. More recent research has further shown that V1 activity can be triggered by areas of the brain even outside of the classical "visual processing" areas; specifically, motor-related signals have been found to be capable of driving V1 activity in the absence of visual input (Keck, Keller, Jacobsen, Eysel, Bonhoeffer, & Hubener, 2013; Keller, Bonhoeffer, & Hubener, 2012; Saleem, Ayaz, Jeffery, Harris, & Carandini, 2013).

Together, these studies paint a picture of perceptual processing in the brain that includes a massive number of feedback connections that can drive the target cells, even in the absence of bottom-up input. The question then is: What are all those feedback connections doing? What are they there for? One potential answer is that the large number of top-down connections simply represent redundancy in the brain. However, compared to the number of feedforward connections, this would mean that the brain maintains a rather massive amount of redundancy for what the classic model holds is the role of feedback connections: namely, just to make minor adjustments to bottom-up processing. That explanation doesn't square well with the idea that the brain needs to be highly efficient. We turn now to the issue of efficiency in the brain, and how large amounts of data might be processed using the least amount of resources.

9.4   Efficiency in Coding and Transmission

It is well established that the brain uses up a tremendous percentage of the body's resources, despite its relatively small size. In spite of this massive resource utilization, however, it runs incredibly efficiently. Compared to the circuitry in even our more high-tech electronic devices, the brain is more efficient in terms of resource utilization (Sengupta & Stemmler, 2014). To appreciate the ways in which information can be efficiently processed, it is useful to consider how computer scientists and data engineers have used various compression techniques to store and transmit information in our modern world. Indeed, methods of efficient coding were being explored by communication engineers as early as the 1950s, and the term "predictive coding" first appeared in an article by Peter Elias in 1955 describing strategies for the transmission of messages (Elias, 1955). These types of efficient coding techniques still impact our daily lives, though often behind the scenes, such as in the many functions of our ever-present cell phones.

Two of the most common uses of our cell phones depend on efficient coding processes: taking high-resolution photos and streaming videos and live events. Regarding our predilection to just keep snapping more and more photos, the memory capacity of even the most expensive phones would be very quickly exceeded if each picture had to be stored at its full resolution; luckily for us, the pictures are stored in a much reduced format. We can continue snapping away with little worry of running out of space, and we can later view our photos in a high-resolution format thanks to the complex coding schemes that store only a fraction of the image without degrading the quality of our pictures when we want to view them. Discussing the full set of these algorithms and data compression techniques is beyond the scope of this book, but the basic idea is that not every pixel in an image needs to be stored. For example, in an image consisting of a beautiful landscape with rocky cliffs, colorful trees, and a clear blue sky, there are portions of the scene that don't require high-resolution storage to be preserved accurately. For example, in Figure 9.4 there are extended sections of the blue sky as well as sections of the cliff in complete darkness (i.e., in shadow) that are the same color across hundreds of successive pixels. As opposed to storing each of those pixels separately in terms of the color at each of those successive coordinates, it is much more efficient to simply code the color as one piece of data, with a second piece of data describing how many pixels (in that line of the image) remain that same color. At the location where the color changes, new pieces of data are required (i.e., the

Figure 9.4 Efficient coding of visual scenes. The picture on the left is at a higher resolution (smaller pixel size) than the one on the right. While some parts of the scene require this higher resolution to be seen clearly (the leaves on the trees; some details in the rock structures), other areas don't need nearly as high a resolution – large areas of the sky and regions in total shadow don't need to be coded with such small pixels because the color stays constant over an extended area. Thus, efficient coding only needs to store (at higher resolution) areas where *change* happens across space. Similarly, in terms of storing and transmitting videos, it is not necessary to encode the whole scene in each frame. Rather, only the parts that are changing across time, such as the waterfall, need to be coded to later recreate the motion. The two pictures here were taken moments apart, and the only changes are in sections showing the waterfall. Thus, a *movie* of this scene could be transmitted much more quickly and efficiently by only sending data for regions that have to be updated from one moment to the next; the receiving device could reconstruct the rest of the image without new data, using those parts of the original image that haven't changed. These concepts of efficient coding relate to predictive coding models of perception, in that it is much more efficient to only code *changes*. For predictive coding models of perception, these changes would be considered *errors* of the existing top-down model of the world, hence the term "prediction errors."

new color must be coded, along with how many successive pixels share that new color). Essentially, the process involves coding *just* where the colors *change*. When we want to view the photo, the coding scheme is reversed, and every single pixel of the high-resolution image can then be painted in the correct color on our screen, but without having to actually store each of the enormous number of pixels that make up its ultimate high-resolution image. Although photo compression involves other techniques as well, this one part of the algorithm illustrates a concept that is fundamental to predictive coding models: Encoding and storing only the *changes* across a scene is much more *efficient* than storing data for each and every pixel separately. (Note that the initial naming formats for files created through these algorithms came from the groups of experts that developed the codes that could be standardized and widely shared: JPEG, the algorithm for compressed visual images, was named after the Joint Photographic Experts Group; MPEG, the format for efficient video coding, was named after the Moving Picture Experts Group.)

Even more relevant to predictive coding models of the brain are the methods allowing for streaming of *videos* over cellular networks. In a movie, or during a live event, most of the scene doesn't change from one instant to the next. If all that unchanging background needed to be transmitted for each and every screen refresh, it simply wouldn't be feasible to stream videos over normal cellular connections. As is the case with photo compression, video compression algorithms allow for highly efficient transmission by encoding only the information that is *changing* from one moment to the next. The entire scene can be quickly created at the receiving end by integrating the relatively few areas of change with the already stored content of the static areas of the scene. This allows full scenes to be recreated without wasting valuable bandwidth on transmitting all of the unchanging information. For example, a movie of the scene in Figure 9.4 might contain long periods in which only the waterfall is changing from moment to moment. All that is needed to reconstruct the scene over multiple frames is to transmit just those relatively small areas of the scene where the water is moving.

Similarly, in predictive coding models of human perception bottom-up processing wouldn't need to include all of the information in the scene at every moment. Rather, bottom-up processing would only send forward, through multiple successive levels, information that is new or changing – information that reveals *errors* in our current top-down model of the world. Furthermore, our current model of the world (also known as our "hypothesis," "prediction," or "priors") is sent "down" from higher levels through the "feedback" connections to multiple earlier levels in the brain. Feedback connections are thought to be important for projecting models of the world ("predictions") *down* through multiple successive levels of the hierarchy. The critical purpose of these top-down connections may not be simply to improve efficiency, however. Section 9.6 discusses further benefits of these top-down connections. Efficiency is most relevant here given that, if top-down connections are already present, possibly to confer other advantages, then bottom-up processing can be made more efficient by processing only the "prediction errors." Bottom-up processing thus contributes to perception in a highly efficient manner by spending processing resources only on highlighting errors in the models, instead of processing all the information across the entire scene at every moment, much of which is completely redundant in the current top-down model of the world.

9.5   Processing "Prediction Errors" in the Brain

This new understanding of perception can also help explain why neurons are found to respond less robustly to repeated stimulation. "Repetition suppression" refers to the finding that the neural response to a stimulus is reduced when that stimulus is repeated a short time later (reviewed in Grill-Spector, Henson, & Martin, 2006). Sokolov (1960) initially coined the term "habituation" to describe the decrease in neural response when a stimulus is presented repeatedly. Critically, however, he noted that it could not be due simply to a low-level sensory adaption because the response rebounded to higher levels if there was any change in the intensity of the stimulus; indeed, even a *decrease* in intensity led to a "dishabituation," or relative increase in processing. According to predictive coding accounts, these results occur because bottom-up processing is strongest when transmitting information that indicates errors to the predicted model (also known as "**prediction errors**"). The very first occurrence of a stimulus is processed more strongly because it is wholly unexpected. After that first occurrence, however, the model of the world is updated to include a higher likelihood of that particular stimulus occurring. Therefore, the next occurrence of that same stimulus is not processed as robustly because it better matches expectations. This makes for more efficient processing, because not much processing need occur for stimuli that already match the current model. Summerfield and Koechlin (2008) provide additional support for this account by showing that the *expectation* of repetition significantly affects the amount of repetition suppression. They found that there was significantly *less* suppression of a repeated stimulus if repetition of that particular stimulus was known to be highly improbable. In other words, top-down knowledge that a repetition of a particular stimulus was *unlikely* to occur counteracted the typical adjustment to the model that a stimulus can be expected to reoccur. The usual suppression of the repeated stimulus wasn't observed in that case because the repetition was made to be *unexpected*. Such results support the theory that top-down information in the mind affects lower-level sensory processing.

Findings from event-related potential (ERP) studies provide further evidence for this explanation. The mismatch negativity (MMN) is an ERP component generated when a stimulus doesn't fit in. For example, if a participant is presented with a sequence of the same "standard" tone being repeatedly presented, then the occurrence of a different tone within that sequence will generate an MMN component (Cammann, 1990; Näätänen, Gaillard, & Mäntysalo, 1978). The MMN has been interpreted as the brain making note of a change, and it appears to be an automatic process, occurring regardless of whether the sequence is being consciously attended to or not. Most relevant to the current topic is that studies have found that the MMN can be triggered by the *absence* of a stimulus (Bullock, Karamürsel, Achimowicz, McClune, & Başar-Eroglu, 1994; Czigler, Weisz, & Winkler, 2006). Specifically, after subjects have learned a particular sequence of stimuli, the omission of one stimulus in that sequence will generate a strong MMN (reviewed in Wacongne, Changeux, & Dehaene, 2012).

Recent functional magnetic resonance imaging (fMRI) studies have found that the neural activity generated by a predicted but absent stimulus can occur at even the earliest cortical levels. Ekman, Kok, and De Lange (2017) presented human subjects with a sequence of four dots appearing sequentially from left to right across a screen. After being exposed to this sequence many times, subjects were then presented, on some trials, with only the first stimulus in the series. Despite the complete absence of stimuli in the other three locations, there was significant activity in the regions of V1 responsive to where those three missing stimuli "should" have been (Figure 9.5). These results provide evidence that top-down expectations ("models" or "hypotheses") can generate activity in even the earliest levels of cortical sensory processing. This study found similar results whether the subjects' task was to attend to a sequence of circles presented in the periphery or to attend to a demanding letter discrimination task at fixation while the circles in the periphery were irrelevant. The activity in the V1 regions representing the

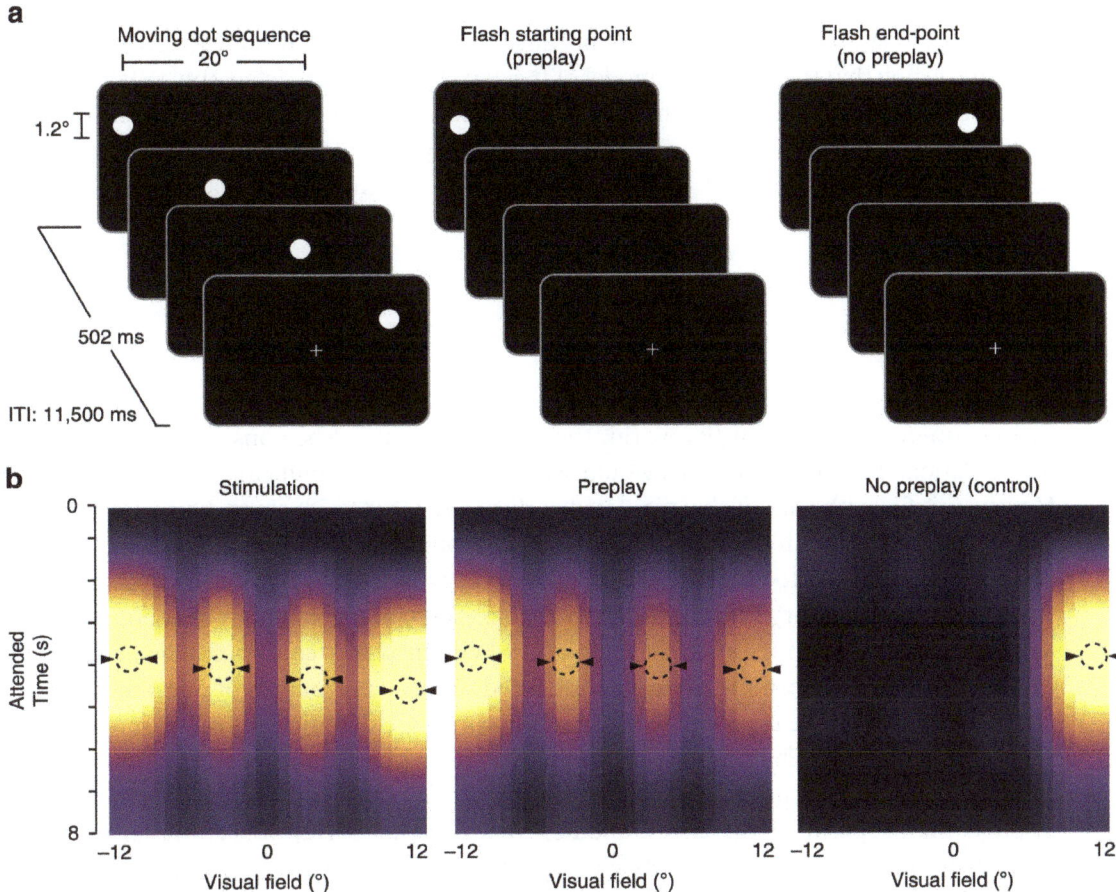

Figure 9.5 fMRI study of expectation-evoked activity in V1. Panel (a) shows the trial sequence. The left-most column shows a "stimulation" trial in which circles appear, one at a time, at each of the four locations in the visual field. During the initial exposure phase, all four circles would occur sequentially in this order, over many trials. During the subsequent phase of the experiment, a single circle could appear on some trials, either only at the location that usually started the sequence ("preplay" condition) or only at the location where the sequence normally ended ("no preplay" control condition). Panel (b) shows the fMRI **blood oxygenation-level dependent (BOLD)** response (greater activity indicated by brighter colors; peak activity indicated by dashed circles and arrowheads) corresponding to the V1 populations coding the visual field regions where the four circles could appear. The y-axis indicates the relative time of the BOLD response (which lags seconds after the neural activity). The stimulation condition shows strong activity in each of the four regions, and the timing of the peak activity shows the successive activation of each region in turn. The preplay condition shows that a similar pattern emerges across the four areas even when no stimulus follows the appearance of the first circle, providing evidence that top-down expectations are driving activity in V1. The lack of activity in the regions of V1 representing the locations of the missing stimuli in the no preplay condition suggests that the significant activity in those regions in the preplay condition is not simply due to the subject trying to remember the sequence on every trial; rather, this is likely due to the presentation of the starting stimulus triggering a specific prediction about what "should" be happening in the world in the upcoming moments. ITI = intertrial interval. Figure reprinted from Ekman et al. (2017), *Nature Communications*, Springer Nature; Creative Commons CC BY 4.0.

missing stimuli (in the "preplay" condition) was greater when attention was being paid to the peripheral circles, but there was still significant activity in the attention-to-fixation condition as well, suggesting that the top-down model of the world was not restricted only to what was being attended to for the task. In both attention conditions there was no activity in the regions representing the missing circles in the "no preplay" condition; thus, the activity seems directly related to the expectation set up by the appearance of the circles that usually started the sequence of four successive circles.

9.6   Benefits of Top-Down Predictions

It has been suggested that if predictive coding models are correct, and our perception is driven by our top-down models of the world, then perception could be considered as a "controlled hallucination" (Clark, 2016). This would seem to be a rather significant cost of such a system. However, it fits with what we've discussed earlier in regards to illusions: Our perception of the world can at times be in contrast to the evidence from our senses. Some aspects of inaccurate perception may relate to algorithms that provide us with a best guess at the state of the world when insufficient sensory information is available. These algorithms can be useful in providing us with at least a working model of the world when our senses can't provide completely clear evidence. Therefore, illusions could be considered simply as rare cases of our minds needing to interpret insufficient sensory data. However, according to predictive coding models, perceptions are always driven by top-down models, even when good sensory information is available. In most situations, the models work well and accurately represent the world, which may have inspired the proposal of the classic models in which our perception is driven primarily by bottom-up sensory input. If predictive coding models are correct, however, and perception is driven primarily through top-down processing, what are the possible advantages of such a system?

One factor that could lead to the evolution of a perceptual system that is dominated by top-down processing is the need for *rapid, real-time perception despite impoverished input*. Since sensory data are often impoverished in some way, waiting until bottom-up information is totally complete and unambiguous would mean that conscious perceptual experience would be far *too slow*. In contrast, by making best approximate guesses in real time and sending these predictions down through the hierarchy, we can be better prepared to respond rapidly, since those levels of processing will have been primed with mostly accurate models of the world (Hohwy, 2013). Even in situations in which the environment provides clear input, the world is always in motion, meaning that we need to be able to respond to where a stimulus *will be* in the near future, not where it *was* when the information first reached our retinas. For example, when returning a shot in a tennis match, we must correctly anticipate where the ball will be, since the limited speed of neural transmission means that higher visual areas are not processing the current state of the world, but rather the world that existed fractions of a second earlier. While this delay may be inconsequential in many circumstances, there are times when working with a model that accurately *predicts* the state of the world has real benefits over responding to a perception of the world that represents just the very recent *past*. For example, in most of the animal world, the ability to accurately predict the location of another animal can be the difference between life and death.

Another potential benefit of robust top-down processing is the ability to have *imagination* (Kirchhoff, 2018). Having a system that creates internal models of the world allows us to plan ahead and play out different scenarios in our minds. Of course, there is then a critical need to be able to recognize when the model is an imagined scenario versus a representation of the external world, and this issue is highlighted by critics of predictive coding models (Orlandi & Lee, 2019). Although the mechanisms that allow us to distinguish real from imagined scenarios are still not well understood (see Jones & Wilkinson, 2020), imagination can confer significant advantages, and having a perceptual system that relies heavily on top-down processing means that much of the neural apparatus for imagination is already in place.

A system that achieves perception through strong top-down processing could also support the ability to have *empathy*. The mechanisms for building models of the world based on top-down information could also allow us to perceive the world from other people's perspectives (Trapp, Schütz-Bosbach, & Bar, 2018). By taking other top-down information into account (e.g., another person's current vantage point; their assumed knowledge of the situation; their background or biases), our imagined model of the world can incorporate these other pieces of knowledge and allow us to better understand how another person may be perceiving the world. Experiencing empathy surely relies on multiple factors and brain mechanisms (e.g., emotions, mirror neurons, etc.; see Bernhardt & Singer, 2012), but having a perceptual system that relies on top-down processing could be an important part of this experience because it provides a neural apparatus for building complex and complete models of the world only from top-down information.

9.7   Balancing Top-Down versus Bottom-Up Influences

The potential advantages, noted earlier, of a system that is built to rely heavily on top-down processing highlights another key feature of predictive coding models – and foreshadows where attention is thought to come into play. Specifically, although perception may be driven primarily by top-down processing, the *balance* between bottom-up and top-down processing is critically important and must be *adjustable*. When evidence for an error in the current top-down model of the world occurs, the brain can: (1) trust the error and update the model; (2) deem the error as unreliable and ignore it; or (3) get better evidence. The first two options are akin to deciding whether the bottom-up information is to be regarded as a true "signal" or rather as "noise." The third option is not mutually exclusive with the first two, but it represents another core aspect of the predictive coding framework, referred to as **active inference** – the idea that we engage in action (e.g., moving our eyes or our bodies) to better test the accuracy of our models of the world (Clark, 2016). Active inference is discussed in more detail in Section 9.8.

The issue of how much to trust bottom-up information relates to how we balance the influences of top-down versus bottom-up processing. Critically, the optimal balance will differ according to the situation. When sensory input is likely to be highly accurate – for example, when coming from a region of space that is well-lit, clear of obstruction, and fixated upon – then bottom-up processing should be allowed to make a strong impact on changing/correcting the current model. In that situation, bottom-up processing is likely to be *highly reliable* (i.e., it has low variance and high *precision*). When sensory input is likely to be noisier, however – for example, for occluded stimuli in the periphery of our visual field viewed at dusk within mottled shadows – then bottom-up processing should have a much more limited influence on revising current models of the world. In this situation, the bottom-up information is likely to be *less precise* and have more variability (i.e., more likely to be noise), and therefore it would be wasteful to immediately change the model of the world in response to such unreliable input.

9.7.1 The Role of Context in Adjusting the Balance

The need to differentially balance the influences of top-down and bottom-up information according to the situation highlights an important concept of predictive coding models: *Context* plays a critical role in modulating this balance. As noted earlier, that context can consist of features of the environment (e.g., good lighting compared to shadows) or regions of the sensorium (e.g., fovea versus periphery; fingertips versus back of the hand) that produce highly precise signals. These particular types of bottom-up processing may be favored because of implicit knowledge that is hard-wired into the brain through evolutionary processes that give the greatest value to sensory information that consistently produces reliable data. This also relates to the idea that especially strong or intense sensory stimulation may be given priority because such stimuli are more likely to provide accurate information with high fidelity. Such hard-wired biases could help explain why illusions (e.g., Figures 9.1 and 9.2) can be so strong and be so resistant to the participant's knowledge of the actual stimulus. However, context in predictive coding models is not limited to only these types of situations that may be hard-wired into the system in most people (and across mammalian visual systems). Instead, the balance of top-down and bottom-up processing is also affected by contexts that are personal and idiosyncratic. These *learned contexts* can include an individual's particular history (e.g., experiences, rewards/punishments, biases) and could also include short-term patterns, such as within a psychology experiment. Indeed, as discussed more in Section 9.8, some standard methods of experimental design may obscure the role of predictive coding in processes of attention and cognition.

9.7.2 Comparing Top-Down and Bottom-Up Processing at *Every* Level

The role of context in adjusting the influence of bottom-up processing highlights the flexibility in predictive coding models. Another important aspect of these models is that the comparison of top-down "predictions" and bottom-up "prediction errors" happens at *every level* of the hierarchy. This means that, at any moment, the model of the world may be accurate at some levels but inaccurate at other levels. At lower levels, prediction errors propagate forward for inaccuracies in any small detail of the model – for example, to update the exact shade of color or the precise shape of an object, or to update the position of the object as it moves. Such details, however, may not require a revision of the model that is active at higher levels, where the model is concerned with the "bigger picture" (i.e., what *type* of object is present and what it *means* for the current situation, regardless of its precise color or shape or precise location). For example, in an encounter with an acquaintance, low-level predictions will have to be updated frequently to account for changing facial expressions and the motion of the mouth while speaking. But higher-level hypotheses would not usually need to be revised, so long as your model of the world has identified the correct person you're speaking to, and so long as the conversation conforms to one's expectation of a casual, friendly encounter. If, however, the low-level information indicates sudden expressions of anger (or, alternatively, unusual levels of smiling and extended eye contact), higher-level models would need to be revised to account for a situation that no longer conforms to a simple, casual conversation with one's acquaintance. As described in Box 9.2, how the balance is set between bottom-up and top-down processing, and consequently when and how often models of the world are updated, could potentially relate to some characteristics of schizophrenia and autism.

Box 9.2 Schizophrenia and autism through the lens of predictive coding

A critical concept of predictive coding models is that the balancing of top-down hypotheses and bottom-up evidence varies across different contexts. This balancing allows the brain to be flexible and to account for situations in which sensory information is highly reliable versus imprecise (e.g., well-lit versus dimly lit environments). Because this balance can be varied, however, we are susceptible to illusions. Even more critically, if the default setting for this balance is set to be more extreme in some individuals, this could lead to atypical perceptual experiences. It has been suggested that some symptoms of schizophrenia and autism could be due to this balancing being set too far in one direction or the other. For example, the hallucinations observed in some types of schizophrenia may occur if top-down hypotheses are given much stronger weighting than bottom-up evidence (Fletcher & Frith, 2009). If bottom-up processing is reduced too much, then these incoming data won't be able to update one's models of the world, and inaccurate top-down hypotheses can dominate perception even in the presence of conflicting real-world evidence. In this way, the setting of the balance too far in the direction of favoring top-down models, and thereby reducing the influence of bottom-up evidence, could potentially lead to the persistence of hallucinations (Horga, Schatz, Abi-Dargham, & Peterson, 2014; Sterzer, Fletcher, Frith, Lawrie, Muckli, Petrovic, Uhlhaas, Voss, & Corlett, 2018).

An imbalance in the other direction, in which bottom-up evidence drives perception with a reduced influence of top-down priors, might account for some characteristics of autism. It has been suggested that the brains of some individuals with autism may put more weight on bottom-up processing (i.e., prediction errors), even at low levels of the visual hierarchy (Pellicano & Burr, 2012; Todorova, Pollick, & Muckli, 2021). As opposed to having some sensory events simply be part of the expected larger model of the world (e.g., a loud laugh in a crowded coffee shop), each low-level "prediction error" triggers the need to establish a new model of the world. The preference of some individuals with autism for established routines and difficulty in dealing with changes to schedules, or the severe disruption triggered by sensory stimulation, could be due to bottom-up processing having too much power to override current models of the world. In this

Box 9.2 (cont.)

way, the individual may appear overly sensitive to sensory stimulation that is innocuous to other people, as it triggers the need to reevaluate and form new hypotheses, at the highest levels, about the current state of the world. It may also be that, for some individuals, high precision is expected throughout the environment at all times, and thus even minor prediction errors signal the need for a new model of the world. Whereas nonautistic individuals process background sounds in the environment as low-level "noise" that can be ignored because this "noise" fits within the overall model of the current situation, those same sounds may be perceived as a significant "signal" for an individual with autism, requiring an overhauling of the entire current model of the world. Thus, without a highly specific and consistent routine, an individual with autism may find situations overwhelming due to this frequent updating and building of new models of the world. This could also produce difficulties in learning generalities across similar situations. If top-down models permit hardly any variance, every low-level "error" triggers the need for a new model, making it more difficult to see how different situations could be grouped together as representing the same general scenario. Therefore, individuals with autism may be surprised much more often by minor "errors" in the model, creating an unsettling world. This overemphasis on bottom-up information could also relate to difficulties in understanding nonliteral speech (e.g., irony, sarcasm). Instead of relying on higher-order models of the situation and what the person using such nonliteral speech probably meant, individuals with autism may sometimes take things literally, resulting in difficulties understanding the social situation. The behavior of humans, in general, can be highly variable and hard to predict in minute detail and therefore could be anxiety-inducing for some individuals. Finally, related to processes of perception, some individuals with autism have been found to show reduced susceptibility to perceiving illusions (Happé, 1996); this could be due to top-down knowledge not exerting the typical amount of influence over sensory input. Specific to processes of attention, research has also revealed that some individuals with autism perform *better* than nonautistic individuals on visual search tasks (Plaisted, O'Riordan, & Baron-Cohen, 1998) and change-blindness tasks (Smith & Milne, 2009). The preference of some individuals with autism to sustain attention on repetitive tasks that others may find tedious or boring could be due to very small details triggering the frequent creation of new models, making those tasks more stimulating. Predictive coding theories are also being applied to post-traumatic stress disorder (Lyndon & Corlett, 2020), drug addiction (Mollick & Kober, 2020), and paranoia (Diaconescu, Wellstein, Kasper, Mathys, & Stephan, 2020). Further research is needed to investigate whether a predictive coding perspective can help advance our understanding and treatment of various special populations, but these models suggest that individual variability in the balance of top-down versus bottom-up processing has important implications for how we perceive and interact with the world (Mirza, Adams, Friston, & Parr, 2019).

One study provides compelling evidence that the comparison of bottom-up information with top-down models occurs at multiple levels (Schwiedrzik & Freiwald, 2017). Using single-unit electrophysiology informed by fMRI localization, these authors recorded from area ML (i.e., middle lateral), a temporal lobe region in macaques known to be selectively responsive to faces. As in humans, the macaque visual system includes a progression of visual areas sensitive to different degrees of abstraction of face stimuli (Hesse & Tsao, 2020). Area ML, a "lower" area in this system, is sensitive to the identity *and* precise orientation of a face, whereas the "higher" area AM (i.e., anterior medial) is selective to identity in a more general "view-independent" sense (i.e., regardless of the orientation of the face). In between these areas, area AL (i.e., anterior lateral) shows "partial view-dependent" selectivity, in that mirror-symmetric views of the face (e.g., right and left profile views) elicit equal activity. The initial part of Schwiedrzik and Freiwald's (2017) study involved repeatedly pairing two different face stimuli (from a set of 18), so that the presentation of the first face reliably predicted the presentation of the paired face as the next stimulus (Figure 9.6). The face images were presented at one of three different orientations: either facing straight ahead or oriented 60 degrees to the right or left. Pairs were always composed of two different identities, and each combination of rotation orientations was used equally often across the pairs. This training/learning phase established a context of what exactly each face predicted. After extensive training (30 days or more), ML neurons were recorded while the macaques were presented with *trained* face pairs and *recombined* face pairs. The recombined pairs could either have only the *orientation* changed (face identity of the pair was exactly as trained), or only the *identity* changed (orientation of the paired face was the same as what had been trained), or both the identity and orientation changed. The results showed that, compared to the trained pair, any violation of the pairing resulted in greater processing in area ML, consistent with a predictive coding account in which bottom-up processing is strongest in response to prediction *errors*. Critically, the different types of violations produced different patterns of enhanced processing. Whereas every type of expectancy violation (orientation or identity) produced an early enhancement of processing (~120–210 ms), only violations of *identity* produced a later extended period of error processing (~300–440 ms). Importantly, these recordings were all done in area ML, and the stimuli in each condition were the same; the experiment manipulated the first member of the pair, so all critical results were comparing responses to the exact same second stimulus but across different conditions. As noted earlier, area ML is sensitive to both identity *and* orientation, so any change (orientation *or* identity *or* both) from what was trained would constitute a violation at this level. This is consistent with the results showing that any change (orientation *or* identity) produced enhanced ML activity. However, when the violation was *only* of the *orientation*, with the *identity* part of the prediction being correct, then the enhanced activity in ML was short-lived, returning to baseline levels after ~250 ms. In contrast, when the violation was of the *identity*, then the enhanced activity in ML remained robust all the way through until ~500 ms.

Regarding this seminal study by Schwiedrzik and Freiwald (2017), an explanation for why ML would continue to process the stimulus as an error for twice as long in the "identity violation" condition compared to the "orientation violation" condition relates to the hypotheses ("models" or "predictions") coming from higher levels. Specifically, when the face is violating expectancy at the level of *identity*, the model is in error at the level represented by both area ML and area AM (two levels "up" from area ML). Therefore, the models remain incompatible with bottom-up input until an updated model is created at the higher level of area AM – indicating this new face identity – and this model has time to propagate further "down" to

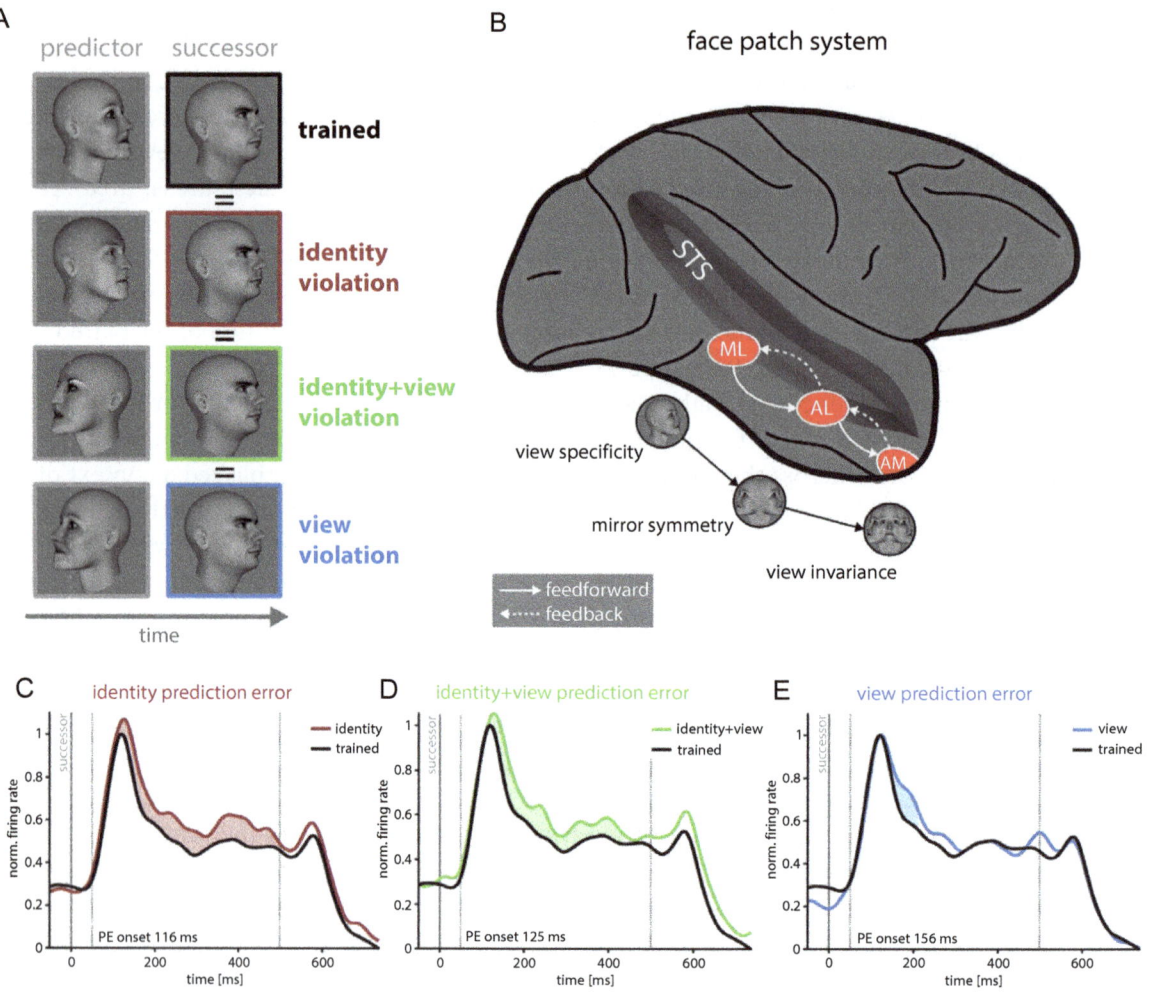

Figure 9.6 Single-unit recordings reveal the influence of multiple higher-level predictions. (A) Stimuli during the testing phase consisted of face pairs that were identical to the pairs seen in the training phase (top) or that differed in the identity, orientation ("view"), or both. Note that the second face is identical to ensure the neural activity being recorded is to the physical stimulus across conditions. (B) Cartoon of a macaque brain showing the relative locations of the areas of interest in this experiment. Recordings were taken from the "lowest"-level area indicated here: area ML (see text for more details on these areas). (C–E) Neural activity in area ML time-locked to the onset of the second of the pair of faces. Shaded regions in each graph indicate significant differences in activity between the trained pair and the type of violation indicted by the title of each graph. PE = prediction error; STS = superior temporal sulcus. Adapted from Schwiedrzik and Freiwald (2017), *Neuron*, with permission from Elsevier.

area ML. However, when the violation is only of *orientation*, with face *identity* matching expectation, then the violation noted in area ML (which produced the early burst of enhanced activity) is quickly quashed by the model being held in area AM because there is no violation at that level (all that matters in area AM is the identity of the face). This time course is important, because it shows that the local violation at area ML will continue to be processed as an error as long as the model at a higher level is still registering an error (in this case, of identify). But that same local violation in area ML can quickly be overcome when a higher-level model signals that "all is well," and an updated model at just the local level (e.g., new orientation) can match the bottom-up input without needing to revise models higher up in the hierarchy. When we consider the impacts of predictive coding models on theories of attention, it is important to keep in mind that these predictions are being tested at many different levels in the brain, and that these different levels may be testing somewhat different hypotheses. This allows for flexibility in the system, in that prediction errors deemed to be highly precise at one level may be propagated up to the next level, even if that error won't ultimately be deemed important enough to update the model being held at that higher level. It is also relevant to point out that although the macaques in the Schwiedrzik and Freiwald (2017) study were trained extensively (for 30 or more days), the learning of context can occur much more quickly. In a behavioral follow-up study with human subjects, the authors showed that training for just 20 minutes produced a similar pattern of behavioral results to that the macaques showed at the end of their experiment, suggesting that the establishment of the learned "context" pairing can be relatively quick.

9.7.3   Where "Top-Down" Predictions Come From

Although theories of predictive coding typically use the standard terminology of hierarchical processing to compare *bottom-up* and *top-down* processing, the models/predictions of the world need not come "down" in a purely vertical, hierarchal manner. Although some "top-down" processing does indeed descend the traditional hierarchy, contributions could also come from regions that are not strictly "higher" in the hierarchy (e.g., cross-modal inputs from different sensory modalities; connections from homologous visual areas across hemispheres; connections between "parallel" visual areas that process different attributes). Areas involved with higher-order cognitive processes could also influence the models/hypotheses that are propagated "down" and tested at multiple levels. Recent studies have shown that *motor* areas of the brain can send "top-down" predictions to visual processing areas of the brain. In a highly innovative study, Jordan and Keller (2020) performed cellular recordings while mice navigated a virtual reality environment (Figure 9.7). The mice were placed on a rollerball and were surrounded by a screen that showed a virtual tunnel, with vertical gratings on the walls. When the mouse ran on the ball, the screen was synced to their rate of movement, providing optical flow that precisely matched their speed. After 5–7 days of training in this environment, the mice began the experimental sessions with an electrode inserted into visual area V1. In the experiment, the mice were again free to run through the virtual maze at their own pace, but now at unpredictable times the visual stimuli (the visual flow synced to their movement) was stopped for 1-second intervals. Thus, every so often the visual flow that should result from the motor movements did not match what was being seen. The results showed a dramatic increase in the activity of some V1 cells throughout the 1-second periods when the screen stopped being synced to the mouse's running. Even though this meant that the visual information on the screen was not changing in any way for a period of time, some V1 cells responded vigorously throughout

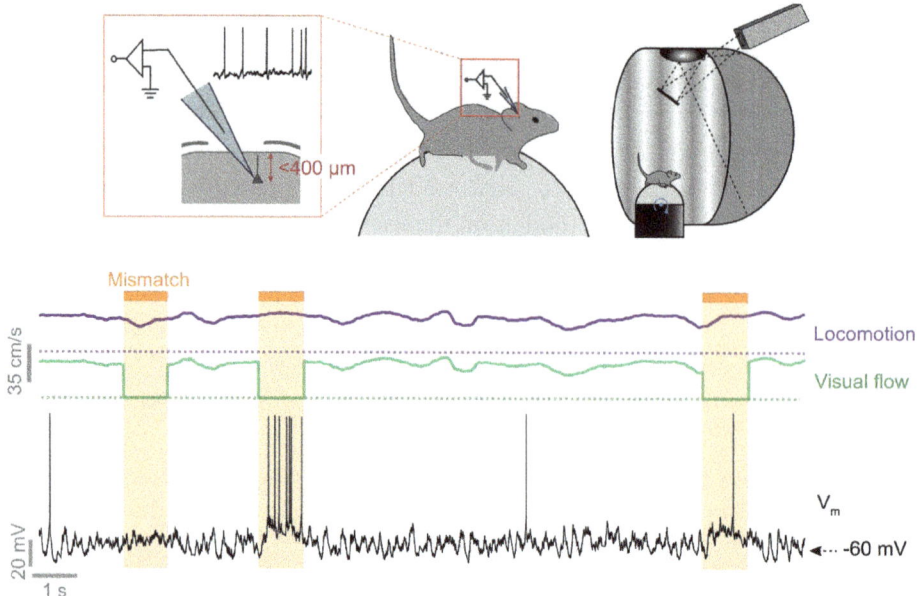

Figure 9.7 Neural activity in area V1 to prediction errors. The top row shows the experimental setup. Single-unit activity was recorded from superficial layers (layers 2 and 3) while the mouse walked on a rollerball that controlled the visual stimuli on the surrounding screens to simulate the visual flow that the mouse would experience of walking through an environment with gratings on the walls. The bottom row shows the activity of a V1 neuron and the middle rows indicate the speed of locomotion (controlled by the mouse) and visual flow (controlled by the experimenter). The orange columns represent the times when the visual flow stopped and therefore created a mismatch between the visual input and what the mouse would expect based on its locomotion. This neuron exhibited a burst of spiking activity on most trials when there was a mismatch between top-down expectations and bottom-up sensory input, specifically when visual flow on the screen was stopped during active locomotion. This is especially noteworthy because the stopping of the visual flow means that the visual field is static, yet a subset of V1 neurons responded vigorously to this lack of motion when it occurred during active locomotion. Reprinted from Jordan and Keller (2020), *Neuron*, with permission from Elsevier.

the time of the *mismatch* between the visual image and the mouse's running speed. Other neurons in this area showed a different type of response, demonstrating a relative *decrease* in activity when the visual gratings weren't moving (thus showing a more bottom-up driving of activity). Together, these results suggest that a subset of neurons in V1 are highly sensitive to the *match* between bottom-up visual stimuli and the *expected* visual flow that should result from their motor activity. These data provide evidence for the existence of a subset of neurons that compare top-down models to incoming sensory input, consistent with theories of predictive coding. The link with motor processes and navigating through the world also introduces another important component of these theories that we discuss in Section 9.8: that we *take action* to test our top-down predictions.

9.7.4 Laminar Microcircuitry and Neural Oscillations

Another reason to consider predictive coding principles when trying to understand the mechanisms of attention is that some recent models have incorporated known aspects of the laminar microcircuitry of the cortex. Since these predictive coding models include detailed neural architecture, they also make predictions about the processes associated with different neural oscillations, which can be highly relevant to interpreting electroencephalography (EEG) and neurostimulation studies of attention. Haeusler and Maass (2007) provided a detailed model of laminar cortical microcircuits based on previous intracellular recordings from somatosensory, motor, and visual areas in rat and cat brains. These models specify cross-layer communications within a cortical column, as well as which layers are most strongly connected to higher versus lower areas of the brain. Bastos, Usrey, Adams, Mangun, Fries, and Friston (2012) built further upon this model to show how predictive coding mechanisms fit into this architecture (Figure 9.8). Specifically, bottom-up "prediction errors" input primarily to layer 4 and are then communicated up to layers 2/3, which also receive top-down "model predictions" from higher cortical areas. Interactive computations within layers 2/3 and 4 compare prediction errors to the current top-down model before sending prediction errors from layers 2/3 "up" to higher brain regions. Those computations between layers 2/3 and 4 are also transferred to layers 5/6, which then send information (potentially updates to the top-down models) "down" to lower brain areas. The Jordan and Keller (2020) study discussed earlier provided new empirical evidence for the laminar microcircuitry that underlies predictive coding processes. Specifically, they found that the *comparison* of top-down information (motor-related expectation) with bottom-up inputs (visual flow) occurred most strongly in layers 2/3. However, there was no evidence in their recordings from deeper layers of any differences in computation occurring in layers 5/6. Although that study was done in mice, exciting advances in fMRI methods are now making human "laminar fMRI" studies possible, in which activity can be attributed to different layers of the cortex. As reviewed in Lawrence, Norris, and de Lange (2019), these methods potentially hold great promise for advancing our understanding of the detailed mechanisms of attention processes.

In addition to modeling which layers of cortex are most associated with top-down predictions versus bottom-up prediction errors, Bastos et al. (2012) also used these sets of equations to estimate the *frequency* of activity that would be generated by these different processes. The results of those simulations suggest that the bottom-up prediction errors from layers 2/3 should primarily be producing activity in the *gamma* frequency range (~40–80 Hz), whereas the top-down information being sent from layers 5/6 should be primarily in the *beta* and *alpha* frequency ranges (~10–30 Hz). These results, suggesting that gamma activity relates to the processing of prediction errors whereas beta/alpha activity relates to transmission of top-down predictions, have potentially important implications for how we interpret electrophysiological and neurostimulation studies of attention.

In line with these predictions, Mayer, Schwiedrzik, Wibral, Singer, & Melloni (2016) found evidence using magnetoencephalography (MEG) to support the idea that slower (alpha/beta) oscillations carry top-down predictions. The researchers presented subjects with sequences of stimuli in which they manipulated the level of visual noise – and the corresponding clarity of the

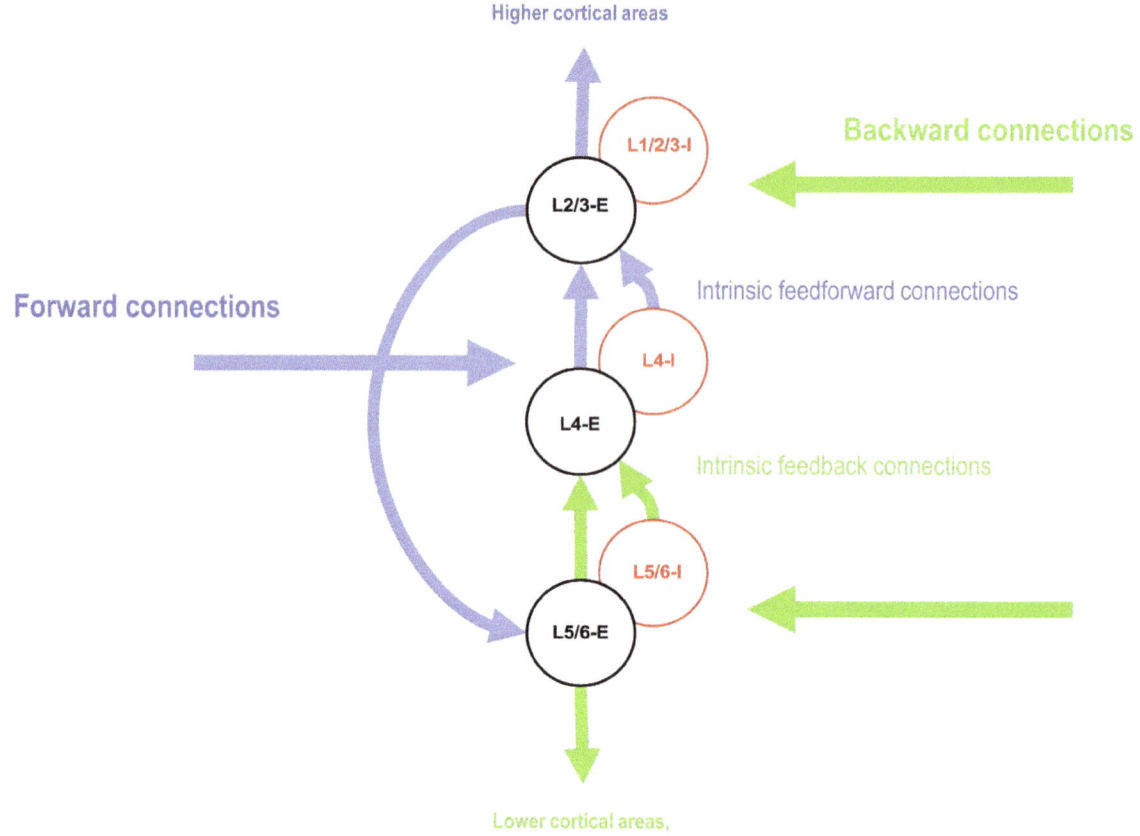

Figure 9.8 Inputs and outputs across cortical layers. Model from Bastos et al. (2012) of the laminar microcircuitry underlying top-down and bottom-up connectivity within and between cortical areas. Within each circle, the "L" term indicates the layer of the cortex, "E" indicates excitatory populations (shown as black circles), and "I" indicates inhibitory populations (shown as red circles). "Forward connections" represent bottom-up prediction errors and "backward connections" indicate top-down model predictions. Reprinted from Bastos et al. (2012), *Neuron*, with permission from Elsevier.

target stimuli – parametrically across six levels. In one condition, the trials started with the lowest signal-to-noise ratio and increased this ratio steadily across trials; in another condition, the trials started with the highest signal-to-noise ratio and decreased this ratio steadily across trials. The behavioral results replicated previous research showing that the ability to perceive a target within noise was greatly enhanced with top-down knowledge (in this case, having seen the target at its clearest at the start of the sequence of trials). The critical advance of this study was that the MEG data showed that *prestimulus* alpha power (8–12 Hz) increased parametrically with increasing signal-to-noise ratio until it plateaued at the signal-to-noise level at which subjects were able to confidently identity the target correctly. In terms of predictive coding, this would be interpreted as the top-down model being strengthened as subsequent trials provide increasingly good information, up to the point at which the model is entirely accurate and no further increase is necessary. These results suggest that the *alpha oscillations* were carrying the *top-down information*, since alpha power increased up until the point at which the model was at maximal information, at which point alpha power remained at this level for all remaining increases in signal-to-noise ratio. In another MEG study, Michalareas, Vezoli, van Pelt, Schoffelen, Kennedy, and Fries (2016) provided further support that lower frequencies are more involved in top-down information flows and higher frequencies are associated with bottom-up transfers. Across seven different visual processing regions within the visual hierarchy, activity in the alpha/beta band was more strongly associated with top-down information flow, whereas gamma activity was associated more strongly with bottom-up information flow.

9.8   Active Inference

According to predictive coding theories, the overarching goal of the brain is to reduce errors in our models of the world, and this is directly related to why bottom-up processing is thought to primarily propagate *prediction errors*. Another consequence of this goal is the motivation to acquire new data to better test these models. In predictive coding theories, this is referred to as "active inference." Note that active inference is a core concept in many theories of predictive coding, and some authors have proposed that it is a critical concept in understanding the functions of the brain in general (Clark, 2016; Friston, Daunizeau, Kilner, & Kiebel, 2010). Here, it is introduced only briefly to help explain how attention relates to predictive coding models. Active inference is typically contrasted with the **perceptual inference** mechanisms described in the previous sections, in which top-down models and bottom-up prediction errors interact to produce our conscious perceptions. Active inference refers to the processes the organism engages in to acquire new data about the world – for example, moving to a new vantage point to gain more information about the stimulus. In the world of visual perception, this can be as simple as moving one's head slightly to get a better view or looking at an object from a different angle. Even more simply, this active inference can involve moving the eyes to fixate on the object of interest.

In the visual system, the quality of data coming from different areas of the visual field varies widely depending on which area of the retina the information lands upon. Information coming from the center of gaze (i.e., the fovea) has better resolution and clarity than information from the periphery (except for the enhanced *sensitivity* to detecting low-luminance stimuli in the periphery). In relation to Section 9.7.1 on *context* and the degree to which prediction errors can be deemed to be *reliable*, information coming from the fovea can be expected to have higher reliability/higher *precision* than that from other visual areas. Thus, when the eyes are preparing to move in order to fixate an item of interest, the system can expect that soon the information from that specific area of space is going to be highly precise. As discussed earlier, when the information is assumed to be highly precise, it is "trusted" more, and the balance between bottom-up and top-down processing is adjusted to allow stronger processing of that particular bottom-up information. In that situation, the bottom-up information is processed more strongly than before, and there is a greater possibility of it causing an adjustment to the top-down model of the world. In other words, in addition to how situational context adjusts the strength of bottom-up processing in general, the knowledge that information is coming from a highly precise area of the visual world (e.g., because it's being processed by the fovea) may also adjust the strength of bottom-up processing in order to change top-down models.

This could potentially relate to attention in the following way: When we decide to move our eyes to a region of space in order to look at an object, we know that this region will very soon be providing information of the highest reliability because we're going to be fixating it. Therefore, it's worth processing this information to a greater extent than other regions of space, even before the eyes have been able to focus on that location, because it is very soon going to be the critical region and is going to be processed by the most reliable section of our eyes. Also, since sensory information is suppressed during the saccade itself ("saccadic suppression"; Bridgeman, Hendry, & Stark, 1975), it could be useful to try to build as accurate a model as possible of that particular region of space, even before the eyes have made that jump. This relates to the numerous findings that attention precedes eyes movements (Deubel & Schneider, 1996; Peterson, Kramer, & Irwin, 2004; Rayner, McConkie, & Ehrlich, 1978). This also relates to research on reading, in which "parafoveal preview" refers to the finding that we begin to process the word at the to-be-fixated location before we actually fixate that word (Rayner, 1998). Together, these areas of research suggest that we preferentially process the information at the region to be fixated, even before our eyes get there. In the normal course of our lives, this sneak preview of the to-be-fixated object is followed up, almost immediately and in almost all cases, by actually fixating that object. This enhanced processing for that location allows for the rapid building of a model that accounts well for this area of interest. Even though that region of space may not be providing more precise information at the moment *preceding* the fixation, the enhancement of processing for that location can accelerate the process of building a better model that will be updated very soon with the highly precise sensory information from the fovea. In this way, covert attention has been likened to a self-fulfilling prophecy, in that creating the *expectation* of better processing *results* in enhanced processing. In most of our daily lives, this sequence of events plays out and works well. In laboratory experiments, however, we break this process in order to provide better experimental control. Specifically, attention experiments often seek to ensure that the physical stimulus is exactly the same across conditions, and that it is entering the sensorium in the identical place across conditions. In that way, any observed differences can be attributed purely to attention, without the "confound" of differences in fidelity across the retina. Much debate has occurred throughout the history of cognitive psychology surrounding the need for precise experimental control versus **ecological validity** – the degree to which laboratory experiments are applicable to the real world. Although well-controlled laboratory experiments have been successful in isolating attention, they may have obscured some mechanisms related to the influence of top-down predictions. It may be particularly hard to appreciate the role of predictive coding models if one only considers experimental settings in which subjects are instructed to *force* themselves to not look where they are attending for *extended* periods. Although we are capable of doing this, it's not what we normally do; we naturally want to shift our gaze to what we're attending. In trying to understand the mechanisms of attention and how they fit with perceptual processes, it is important to consider what they do in the natural course of our daily lives and the evolutionary pressures that would have led to the development of these mechanisms. With that in mind, we turn to research that has attempted to explain attention as part of a predictive coding framework.

9.9   Predictive Coding and Attention

Whereas predictive coding theories have a long history in the field of visual perception research, only relatively recently have theories of attention begun to incorporate these concepts. Rao (2005) and Feldman and Friston (2010) were among the first to describe how the Bayesian principles and series of equations used in predictive coding models of perception could also explain key functions of attention. Most predictive coding models explain attention as adjusting the expected "precision" of bottom-up input. Similar to how context can affect the expected reliability of prediction errors, attention acts to boost or suppress bottom-up input. Rao (2005) first showed that a Bayesian multilevel generative model could reproduce results from electro-physiological recording studies of attention, including the modulation of orientation tuning curves, the overcoming of distractor competition, and the spread of attention to nearby locations. Feldman and Friston (2010) showed that simulations based on equations modeling neural connectivity and response characteristics can reproduce psychophysical and electro-physiological results from previous attention cuing studies. Their simulations were able to reproduce some standard effects of attention, including biased competition, enhanced gain for attended locations, and speed–accuracy trade-offs. Feldman and Friston (2010) described attention in terms of a broader "free-energy system." The interested reader is directed to that seminal paper to better understand how free energy differs from predictive coding; here, we focus on the concepts shared by free energy and predictive coding to discuss in a more general sense how those concepts may explain the functions of attention. Critically, the multilevel hierarchical model used by Feldman and Friston (2010) was based on previous models that had effectively simulated learning (Friston, 2008), auditory sequence categorization (Kiebel, von Kriegstein, Daunizeau, & Friston, 2009), and movement control and action (Friston et al., 2010). This highlights a potentially important reason for considering predictive coding models when trying to understand attention: The effects of attention can arise from within mathematical models that were based on fundamental characteristics of neural activity and connectivity that have previously been shown to account for other aspects of cognition. In this way, predictive coding may be able to explain at least some functions of attention within a more generalized framework of how the brain operates.

Recent studies have extended this work and shown that predictive coding models can simulate multiple attributes of attention. Mirza et al. (2019) found that such a model can reproduce the results from Yarbus' (1967) classic visual search study that showed that the pattern of overt shifts of attention (eye movements) across a complex scene was dependent on the subjects' task, not simply bottom-up saliency. More recently, Holmes, Parr, Griffiths, and Friston (2021) showed that behavioral and ERP results from studies of the "cocktail party phenomenon" (see Chapters 2 and 5) were simulated accurately with mathematical models that instantiate attention as the *modulation of precision*. By treating attention as a state-dependent parameter that *varies over time*, these authors were able to simulate both the effects of preparatory attention and common errors that occur during the cocktail party paradigm.

While simulations and mathematical models show that a predictive coding approach *can* account for many of the effects of attention, empirical studies are needed to directly test predictive coding models of attention. Vossel, Mathys, Daunizeau, Bauer, Driver, Friston, and Stephan (2014) provided some of the first empirical evidence that mechanisms of attention involve the learning of the expected precision of different inputs. In their experiment, subjects performed a modified version of the classic Posner cuing paradigm, and eye movement response times were used to measure attentional orienting. Critically, and unlike standard versions of attention cuing studies, the cue validity (percentage of targets at the cued location versus other locations) varied throughout the experiment. This manipulation was critical for testing whether the modulatory effect of attention involves the learning of the precision (reliability) of the bottom-up input. In standard cuing studies, the validity of the cue is explicitly told to subjects and remains constant throughout the experiment. Vossel and colleagues, in contrast, varied the validity of the cue in order to investigate how dynamic and changing expectations affect the modulation of bottom-up processing. In this work, attention was implemented as a modulatory influence on the postsynaptic gain of prediction error units, adjusting the expected precision of those inputs. By looking at the pattern of eye movement reaction times across time-varying changes in predictability, the authors showed that models using *precision estimation* as the mechanism of attention accounted for the results better than models that were missing that component. This was one of the first studies to provide empirical support for the idea that the mechanism of attentional biasing can be explained as the *setting* of *expected precision* for bottom-up information coming from different locations.

9.9.1 Brain Mechanisms of Attention versus Expectation

In addition to studies measuring overt behavior, a growing number of cognitive neuroscience studies have begun to probe the brain to investigate whether attention may indeed be instantiated in the manner suggested by predictive coding models. Related to the study by Vossel and colleagues (2014), neuroscience studies have investigated the relation between expectation and attention. This relation is of particular interest because predictive coding theories hold that both attention and expectation exert top-down influences, but through different mechanisms. Kok, Rahnev, Jehee, Lau, and de Lange (2012) used fMRI to investigate the possible interaction of these processes in the brain. In a modified cuing task, the investigators instructed subjects to voluntarily orient *attention* on each trial to the spatial location indicated by a central arrow, and to only respond to stimuli at that location. *Expectation* was manipulated by a cue at the start of each block of trials that indicated the likely location (75% predictive) of targets throughout the entire block. The fMRI results from early visual processing regions showed an interaction between attention and expectation, suggesting separate mechanisms were involved. Furthermore, the reduced processing of expected compared to unexpected stimuli at the unattended location was only found in area V1, not in V2 or V3, suggesting that once the "prediction error" had been reconciled at this earliest stage of cortical visual processing, further reconciliation at higher levels was unnecessary.

Jiang, Summerfield, and Egner (2013) similarly used fMRI to investigate the mechanisms of attention versus expectation but at higher levels of the visual hierarchy, in regions showing category specificity to faces (the fusiform face area; FFA) and places (the parahippocampal place area; PPA). In their study, a picture of a face (male or female) or place (outdoor or indoor) was presented on each trial. Attention was manipulated by defining the target type (e.g., female face; male face; outdoor place; indoor place) that required a response for an entire block at the start of each block of trials. Expectation was manipulated at the beginning of each trial by a tone that predicted (75%/25%) what category of stimulus (face or place) would be likely to occur. The results provided evidence that both attention and expectation boost category selectivity, but through different means. Using multivoxel pattern analyses, the authors found that the patterns of voxels associated with enhanced discrimination for attended versus unattended stimuli were different than the voxel patterns associated with enhanced discrimination for expected versus unexpected stimuli. Furthermore, their results provide evidence that attention enhances the ability to discriminate between expected and unexpected stimuli. Predictive coding models might explain this enhanced discriminability as attention *upweighting* the estimated *precision* of bottom-up processing within the category-selective brain region (FFA or PPA) for the attended object type.

Relating to this concept of enhanced processing of prediction errors, Summerfield and Koechlin (2008) provided evidence that different visual processing regions are sensitive to *maintaining* top-down expectations versus detecting *violations* of those expectations. Using fMRI, they found that activity in the inferior temporal gyrus reflected the processing of prediction errors, whereas activity in the fusiform gyrus was associated with top-down expectation. Specifically, fusiform gyrus activity was enhanced in conditions in which there was an expectation for a certain stimulus type, regardless of whether the stimulus ended up being that expected type. In contrast, inferior temporal gyrus activity was enhanced only when the stimulus did not match expectation (i.e., a mismatch, or "prediction error"). Additional analysis of the functional connectivity across frontal and visual processing regions suggested that forward connectivity ("bottom up" from visual to frontal areas) was strongest during *prediction errors*, whereas *expectation* enhanced "top-down" connectivity from higher to lower areas ("top down"). These studies, however, all used fMRI and therefore could not directly assess the relative timing and the sequence of the activities underlying these neural processes.

EEG and ERP results have revealed further insights into the interactions between attention and expectation. In an ERP study of auditory attention, Hsu, Hämäläinen, and Waszak (2014) provided evidence that attention and prediction have interacting influences on perceptual processing as early as 100 ms after the stimulus occurs. In that study, subjects were instructed to attend to one of two interleaved auditory streams, while expectation was manipulated by having the streams follow either a random or predictable sequence. The authors found that the interaction of attention and prediction occurred early in cortical processing, as indexed by the amplitude of the *auditory N1* (~100-ms latency), localized to Heschel's gyrus. Marzecová, Schettino, Widmann, SanMiguel, Kotz, & Schröger (2018) provided converging evidence, from an ERP study of visual processing, for an early locus of the interaction between attention and expectation. In their study, subjects had to discriminate the spatial frequency of a circular Gabor patch on each trial. Attention was manipulated through an instructive cue at the beginning of each block of trials telling subjects to respond to target stimuli only in the indicated visual field throughout the block. Expectation was manipulated by having one of two possible orientations of the Gabor patches (45 or 135 degrees) be more likely (75% versus 25%) across the block. The subjects' task was to respond to one of two possible spatial frequencies (high versus low) of the Gabor patches, which occurred equally often throughout the experiment. The results showed that attention and expectation interacted as early as the anterior *visual N1* component (150–196 ms; thought to arise from parietal areas). This component was enhanced for attended stimuli compared to unattended stimuli, but only for stimuli in the *expected* orientation (Figure 9.9, left); attention did *not* modulate this N1 component for *unexpected* stimuli (Figure 9.9, right). These data suggest that the effects of attention, at least at some stages, may

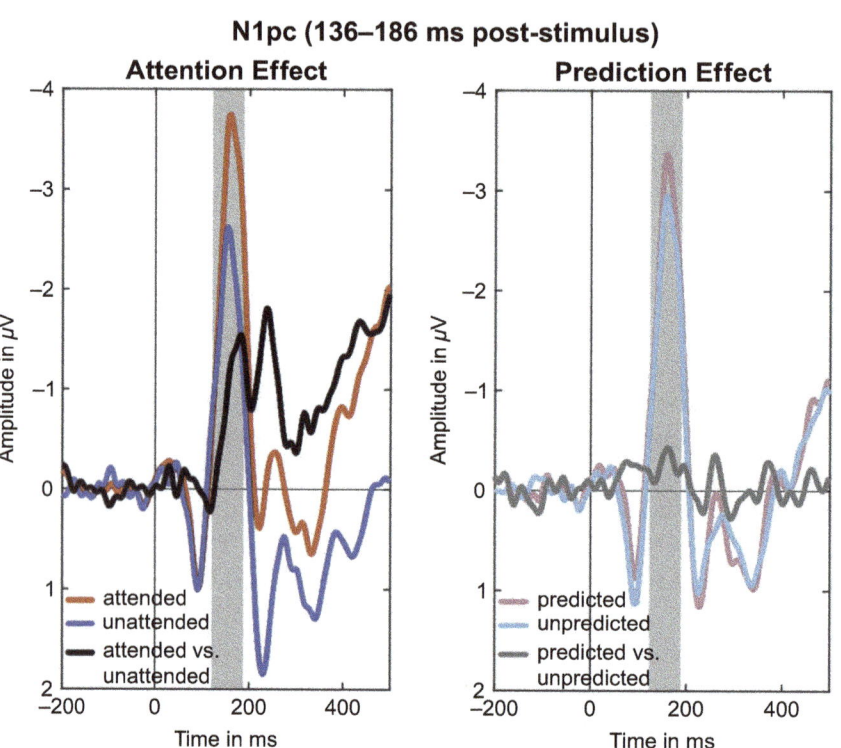

Figure 9.9 ERP study of attention and expectation interactions. Shown here are the ERPs to visual stimuli averaged over a cluster of six electrodes at midline frontal sites. An instructive cue at the beginning of each trial indicated which location to pay attention to (defining the "attended" and "unattended" stimuli), and subjects were informed before each block of trials regarding the likely orientation of the Gabor patches ("predicted" orientation occurred on 75% of trials; "unpredicted" orientation on 25% of trials). The frontal N1 component in these ERP plots is the large negativity from 150 to 196 ms (highlighted in gray in the graphs). This component is significantly enhanced for targets at the attended versus unattended location when the target was of the expected type (left panel), but this stage of processing showed no effect of spatial attention for targets that were of an unexpected type (right panel). Reprinted from Marzecová et al. (2018), *Scientific Reports*, Springer Nature, Creative Commons CC BY 4.0.

be contingent on expectancy. Combined with the findings of Kok et al. (2012), these ERP data suggest that expectation and attention interact at very early levels of cortical sensory processing.

Overall, these neuroscience studies suggest that expectation affects the creation and maintenance of top-down models, or *priors*, whereas attention affects how rapidly and robustly those models are updated by bottom-up prediction errors. Predictive coding accounts suggest that attention achieves this outcome through the setting of *expected precision/reliability* on those bottom-up signals, essentially turning up the gain on those prediction errors. In a recent study examining the multispectral *phase coherence* of EEG data, Gordon, Tsuchiya, Koenig-Robert, and Hohwy (2019) found that expectation was associated with the modulation of descending signals, whereas attention was linked to the modulation of ascending signals. These studies investigating the interactions of expectation and attention have provided support for predictive coding models of attention, although more research is needed. Limitations to some of this research relate to the ways in which expectation versus attention are operationally defined across studies. Furthermore, in some studies, the distinction between attention and expectation may be confounded with level of difficulty, or *cognitive load*, which has also been shown to affect the stages of processing at which attention effects are observed (Lavie, Hirst, de Fockhert, & Viding, 2004).

9.9.2   The Role of Conscious Awareness

Cognitive neuroscience methods have allowed researchers to assess the role of conscious awareness in setting top-down expectations and in processing prediction errors. Meijs, Slagter, de Lange, and van Gaal (2018) used a novel variant of the attentional blink paradigm (see Chapter 8) in which the identity of the first target stimulus (T1) within the stream of letters predicted (75% validity) the identity of the second target stimulus (T2). The behavioral results showed a significant effect of expectation, as T2 identification was significantly enhanced when its identity had been correctly predicted by T1. This provided evidence that expectation (set by T1) increased the access to conscious awareness of a subsequent T2. However, this effect of expectation was found only when the subject was *consciously aware* of the T1 stimulus. When the subject did *not* consciously perceive T1, there was no difference in the identification of T2 stimuli that matched versus mismatched what the T1 stimuli predicted. This result provides evidence that setting the top-down expectation was contingent on the conscious awareness of the stimulus (T1) that defines the expectation; the expectation was not set automatically by the mere occurrence of the T1 stimulus. The propagation of a prediction error when the identity of T2 was incorrectly predicted, however, was *not* dependent on the conscious awareness of T2. In other words, the top-down expectation was only created when T1 was consciously perceived, but once this expectation had been set up, the generation of a bottom-up prediction error (to a nonmatching T2 stimulus) occurred even when the subject was unaware of that T2 stimulus. Thus, these data show that prediction errors can propagate upward, at least to some extent, even *without awareness*. Specifically, the researchers were able to identity a burst of prediction error activity – recorded as a negative-going wave over frontal sites beginning at 175 ms after the presentation of a nonpredicted T2 stimulus – that was equally strong for T2 stimuli that were consciously perceived versus not perceived at all (Figure 9.10). In other analyses, the data replicated previous reports of neural indices of conscious perception, with a separate posterior

Early central negativity

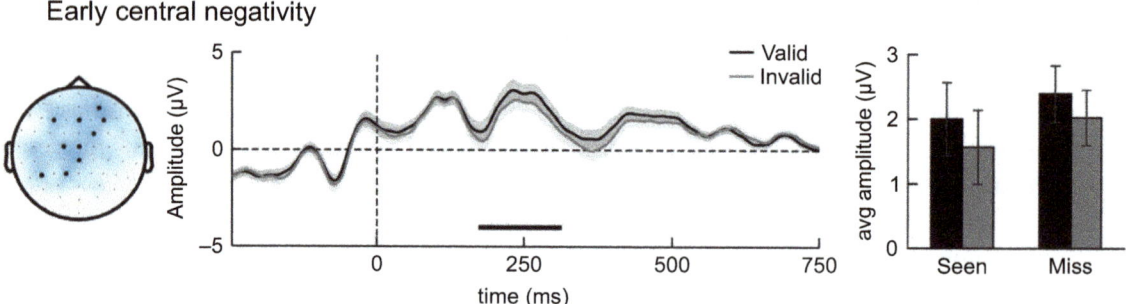

Figure 9.10 ERPs to T2 stimuli in an attentional blink task. The scalp map at far left indicates electrodes and distributions that showed a significant difference between validly predicted and invalidly predicted T2 stimuli. The ERP plot shows activity time-locked to the onset of the T2 stimulus and averaged over the electrodes indicated on the scalp map, indicating that invalidly predicted T2 stimuli produce a negative-going wave (plotted downward here), starting at ~175 ms, compared to validly predicted T2 stimuli. The duration of the significant difference between the waveforms is indicated by the solid black line near the bottom of the graph. Critically, as shown in the graph to the right, this negative shift was observed regardless of whether the T2 stimulus was consciously perceived or not ("seen" and "miss" indicate whether or not subjects were able to identify the T2 stimulus; validly predicted T2 stimuli results are shown in black bars, invalidly predicted T2 stimuli results are shown in gray bars). Error bars indicate standard errors of the mean. Reprinted from Meijs et al. (2018), *Journal of Neuroscience*, with permission from the Society for Neuroscience.

negativity followed by a later enhancement of the P3 component for seen versus missed T2 stimuli. However, the finding of an *early* frontocentral negativity that was sensitive to prediction but not to awareness provided new evidence for the automatic propagation of prediction errors. This experiment highlights the unique and important benefits of cognitive neuroscience techniques that measure the brain's responses to stimuli that remain outside of a subject's awareness.

9.9.3 Saliency Maps and Attention

A prominent component in many theories of attention is that **saliency maps** are instrumental in determining the allocation of attention. These maps are thought to represent the strength of different features, and stimuli that are especially bright, loud, or colorful are assumed to have a strong representation in these maps. According to some theories, an additional "master saliency map" is thought to integrate the strength of different features across lower-level maps and to provide a guide as to the most "important" parts of space where attention should be focused (Triesman & Gelade, 1980). While some work has supported using such maps to model shifts of attention across visual search displays (Itti & Koch, 2000), recent research suggests that these saliency maps do a poor job of predicting search behavior in real-world settings. Jovancevic-Misic and Hayhoe (2009), using a portable eye tracker, measured subjects' gaze while they navigated a novel route that included various objects and other pedestrians. They found that feature-based saliency maps did a poor job predicting shifts of gaze, instead finding that *learned probabilities* of the environment played a very large role in determining when and where gaze was allocated. Rothkopf, Ballard, and Hayhoe (2007) similarly found that feature-based saliency maps made frequent *false* predictions of subjects' gaze in a virtual reality scene, such as allocating gaze to items that were physically intense or unique but irrelevant to the task goals. Furthermore, as reviewed in Tatler, Hayhoe, Land, & Ballard (2011), fixations in the real world are often made to places that we *expect to be relevant* in the very near *future*, even if nothing relevant or salient is currently at that location. For example, when attempting to hit a tennis ball traveling at high speed, we have to look at where we expect the ball to be, not where it is. In more mundane situations, we look slightly ahead of the knife when we cut a sandwich (Hayhoe, Shrivastava, Mruczek, & Pelz, 2003), and experienced drivers look further down the road than do inexperienced drivers while driving (Land & Tatler, 2009). Although some research still assumes saliency maps play a role in determining perception and action, it may be more parsimonious to remove the separate stage/region of processing that calculates a master saliency map and instead to explain the allocation of attention as simply a natural part of multilayer, hierarchical predictive coding models of the brain in which top-down models, incorporating expectation and experience, play a significant role.

In an electrophysiology study in macaques, Lee, Yang, Romero, and Mumford (2002) concluded that "the representation of perceptual saliency of objects in a visual scene is distributed across multiple cortical areas and its computation is interactive in nature." Their experiments investigated the phenomenon of 3D shape–from–shading pop-out, in which shading provides the impression of convex or concave shapes. If presented with an array of such items with one item shaded in the opposite way, the uniquely shaded item pops-out and "captures attention." This type of pop-out has been of particular interest because the determination of the 3D shape is thought to occur at a visual stage of processing beyond V1. Here, the authors found that the earliest stage of processing to show this type of pop-out, before training, involved neurons located in area V2. After training with the stimulus arrays, however, the pop-out was observed in V1 as well. The authors suggested that V1 may become involved in such pop-out over time in order to sharpen the pop-out response and better localize it. The results of this study provide evidence for the flexible nature of the saliency calculation: that it can be partially driven by top-down processes, that it occurs in multiple areas, and that it is affected by experience (i.e., it changes with training). Data such as these suggest that including a master saliency map in theories of attention may be misguided, since a single high-level map that summarizes the bottom-up strength of processing from all lower levels isn't needed. Instead, saliency may be distributed across multiple areas, and top-down predictions appear to affect the saliency computation at multiple levels.

In addition, as described by Clark (2016), if we consider attention to be intimately intertwined with action, then the idea of a saliency map may be better accounted for as a function of top-down knowledge rather than simple bottom-up features. As opposed to the typical explanation of a saliency map as coding locations in a bottom-up manner in terms of the most intense or unique physical features, a predictive coding model would suggest that higher-order models define the most critical objects and features of the current situation. In this view, the driving force for orienting attention is in fact what information can be most informative for evaluating the current top-down model. In some situations this may be a bright, loud, or uniquely colored object, but in other situations a less intense object will attract attention because it is more relevant to testing the current top-down model. Accordingly, schemas may also play a major role in determining what is most highly prioritized in a given context.

9.9.4   Object-Based Attention and Feature Binding

This conceptualization of saliency in a predictive coding framework could also account for other aspects of attention, such as object-based attention and feature integration (binding). Related to the concept of active inference described earlier, the boosting of bottom-up signals need not be limited to only the location of the upcoming saccade. Since the top-down model would include the object to be fixated and not just the single location that will be the next fixation, the benefits of attention (expected high precision) would naturally be conferred on the whole object. In this way, object-based attention may also be explained by predictive coding models, in terms of how the top-down predictive models relate to active inference and the setting of estimated precision. In regards to how features are bound to objects, feature integration theory (Treisman & Gelade, 1980) has been highly influential in suggesting that attention is necessary to bind features together by means of a master saliency map. According to this view, the multiple features of an object (e.g., color, shape, motion) are processed independently by different cortical areas and only bound together onto that one object after attention has been allocated to their location in the master map. However, in predictive coding models, the binding of features is a natural part of the top-down model. The different features that are part of an object are already bound together in the higher-level, top-down model of that situation. Of course, the model may be wrong, and prediction errors will trigger revisions to the model at different levels, but in most cases the models would already be "binding" feature information together due to existing knowledge of those objects. When that object becomes part of the predicted model of the world, the top-down model projects simultaneously down to the different feature levels. In this way, the binding is a top-down process, as opposed to the object and its features being bound as part of the consecutive stages of bottom-up processing. In support of feature integration theory and the critical role of attention in binding features, Triesman and Schmidt (1982) have shown that "illusory conjunctions," in which features are bound to the wrong objects, can occur if attention cannot be allocated to objects fast enough to bind the features correctly. However, these illusory conjunctions only occur in the extreme condition in which multiple stimuli suddenly appear simultaneously for only a few milliseconds. Thus, although attention may be able to play some role in the binding process in laboratory experiments, this may not be an accurate model for how feature binding usually occurs in the typical situations of everyday life.

9.9.5   Automatic versus Contingent Capture of Attention

Predictive coding models may also help explain the controversy regarding the automaticity of attentional capture. While some research has found that highly intense or unique stimuli capture attention automatically (Theeuwes, 1993; Wang & Theeuwes, 2020), other research finds that attentional capture is contingent on the task being performed (Folk, Remington, & Johnston, 1992; Lien, Ruthruff, Goodin, & Remington, 2008). More recently, researchers have suggested that there are additional mechanisms related to the history of recent experiences with certain stimuli (e.g., "selection history"; "reward history") that can also capture attention (Awh, Belopolsky, & Theeuwes, 2012). Under predictive coding theories, these three different views on attentional capture could potentially be integrated. According to predictive coding models, the strength of bottom-up processing depends on multiple simultaneous sources of top-down information. Task set, context, reward history, and selection history could all be viewed as simply subsets of the multiple sources of information that are integrated to determine the current top-down model. In this way, there need not be separate mechanisms to explain how task set determines attentional capture versus how selection history affects attention versus how a uniquely colored item captures attention. Rather, the top-down model in predictive coding theories is always integrating all these pieces of information. In laboratory experiments that attempt to isolate just one such element, this integration may not be observable, but the top-down model would nonetheless always be able to include the various elements. As the top-down model changes according to different goals and tasks, different elements become more important. Similarly, our experience with a given stimulus or environment will change over time, causing updates to the top-down models. Thus, the top-down *priors* regarding a certain stimulus or environment are always changing over time, accounting for how recent history and current task goals affect what captures attention. Although further research is needed to understand how the computations of different sources of information are processed and integrated, predictive coding models can provide more parsimonious accounts by explaining many different experimental results within a single generalized framework.

The predictive coding framework could also potentially account for the special status of new objects. Research has consistently found that the sudden appearance of a new object is unique in its ability to capture attention entirely automatically, regardless of task set (Folk & Remington, 2015; Gaspelin, Ruthruff, & Lien, 2016; Jonides & Yantis, 1988). As discussed earlier, *priors* in predictive coding refer to the predictions of top-down models, and these are continually revised and updated according to context and prediction errors. However, top-down models may also take into account "hyper-priors" – predictions about the world that are stable and overarching, and which may be essentially hard-wired in the brain. Hyper-priors could include such things as information from the fovea being more reliable than visual information from the periphery, that sensory prediction errors should carry more weight when the environment is clear and well-lit, or that faces are always convex objects (leading to the hollow mask illusion described at the beginning of this chapter). The finding that new objects may be unique in automatically capturing attention could be accounted for as a type of *hyper-prior* prioritizing new objects that appear suddenly. There could be strong evolutionary advantages to ensuring that any new stimulus in the environment is very quickly assessed and integrated into the brain's current model of the situation. The sudden occurrence of such a stimulus would thus be a special type of prediction error that requires a rapid revision of current models. The occurrence of gradually changing colors, shapes, or motions is unlikely to be as critical, and therefore those features may only capture attention when they are known to be specifically relevant to the current task. In this way, the allocation of attention is not determined by a single saliency map, but rather is a dynamic and changing process wherein top-down hyper-priors are integrated with current priors that are being continually updated within and across levels. Although further research is needed to better understand these bottom-up and top-down computations, predictive coding theories are opening up new ways of thinking about the neural mechanisms that drive our attention.

CHAPTER SUMMARY

- Illusions help reveal the extent to which our perceptions can be dominated by top-down models of the world, sometimes overruling bottom-up evidence.
- Neuroscience studies have revealed that expectations can drive activity in early sensory processing regions in the absence of external stimulation.
- Top-down predictions and the computation of prediction errors are thought to be assessed at multiple levels within a hierarchical structure in the brain.
- Studies of laminar microcircuitry are providing new insights into the levels of the cortex in which top-down predictions and bottom-up prediction errors interact, and computational models are relating these to different frequencies of neural oscillation.
- Predictive coding models suggest that attention flexibly controls the extent to which top-down models are updated by bottom-up prediction errors by way of setting the expected precision/reliability of bottom-up signals from attended versus unattended inputs.

REVIEW QUESTIONS

Describe the potential benefits versus costs of perception being heavily influenced by top-down processing versus bottom-up processing.

Explain the role of efficiency in predictive coding models and the role of prediction errors.

Contrast active inference with perceptual inference and integrate these types of inference with the models of temporal and spatial attention presented in earlier chapters.

Describe how predictive coding models account for the classic attention concepts of feature binding, saliency maps, and access to consciousness.

FURTHER READINGS

Clark, A. (2016). *Surfing Uncertainty: Prediction, Action, and the Embodied Mind*. Oxford University Press.
- This excellent book integrates philosophy, psychology, and neuroscience to introduce and explain the concepts of predictive coding in perception, action, and thought.

Feldman, H., & Friston, K. J. (2010). Attention, uncertainty, and free-energy. *Frontiers in Human Neuroscience*, *4*, 215.
- This seminal article explains the free-energy principle and its relation to predictive coding and attention.

Holmes, E., Parr, T., Griffiths, T. D., & Friston, K. J. (2021). Active inference, selective attention, and the cocktail party problem. *Neuroscience & Biobehavioral Reviews*, *131*, 1288–1304.
- This article ties recent theories on predictive coding back to the classic phenomenon that spurred early attention research.

Glossary

active inference. Part of a theory of brain function that holds that perception and other cognitive functions are the result of top-down predictive models of the world. The "active" refers to the sampling of the world (through overt movements as well as covert shifts of attention) that is critical in testing and updating the predictive models that give rise to conscious awareness.

allocentric neglect. Sometimes referred to as "object-based neglect," this is a type of unilateral neglect syndrome in which neglect is centered on the object being perceived. A patient with leftward allocentric neglect would neglect the left half of every object, regardless of whether the object is in their left or right visual field. This type of neglect is contrasted with egocentric neglect (also known as "space-based neglect").

attentional blink (AB). This term refers to the phenomenon and the task that elicits it. The classic AB task is a rapid serial visual presentation task in which one letter at a time is presented at fixation, and subjects are required to identify one or two types of targets in each stream. The AB refers to the finding that the second target is often missed if it occurs two to four items after the first target. The explanation given for this is that attention is consumed with the first target for a short period, during which other stimuli are not fully processed.

behaviorism. The dominant school of thought in psychology in the 1930s–1960s, this method focused on observable actions (behaviors) and events. The theory held that learning and most cognitive processes can be explained as functions of rewards, punishments, conditioning, and reinforcement schedules.

binaural beats (BBs). An auditory sensation in which two tones of different frequencies presented separately to each ear create the sensation of a third tone at the frequency of the difference between the two tones. Electroencephalography studies have found that neural activity can be entrained to the frequency of the BB.

BOLD. Stands for "blood oxygenation-level dependent" imaging. This is a type of functional magnetic resonance imaging that is sensitive to the changes in blood oxygenation level that are coupled with neuronal activity. Due to the difference in magnetization between oxygenated hemoglobin (which is diamagnetic) and deoxygenated hemoglobin (which is paramagnetic), the ratio of deoxygenated to oxygenated hemoglobin initially dips as active neurons use up more of the oxygen in the blood, and then the ratio rises significantly as the blood supply subsequently increases to the previously active area.

computerized tomography (CT). A type of biomedical imaging developed initially in the 1970s, this technique uses a series of X-rays taken through a sample at many angles and reconstructed using computer algorithms. The typical human CT scanner consists of an X-ray tube and photon detector positioned to face each other and rotate 360 degrees around the patient. This method was originally referred to as computerized axial tomography (CAT) because of the axial (i.e., transverse, horizontal) orientation of the data collection.

conjunction search. A type of search in which the target is an item that has a combination of features in the visual display – for example, searching for a single blue circle in a display of blue squares and green circles. Conjunction search is typically slower and more difficult than searching for a single unique feature (see "feature search").

contingent capture. Theory that holds that the nonvoluntary orienting of attention (capture) is not determined solely by the salience of physical stimuli but rather depends (is contingent) on top-down settings for the task and the expected target type.

contingent negative variation (CNV). An electroencephalography negativity located over central scalp sites that steadily grows as the expected time of an upcoming target stimulus nears. The CNV does not simply represent the preparing of a motor response, as it is found even when no motor response is required, and therefore it is linked to expectancy and the preparation to attend and perceive the upcoming target stimulus.

continuous theta burst stimulation (cTBS). A method of repetitive transcranial magnetic stimulation in which pulses are delivered in a pulsed pattern at a theta frequency (~5 Hz), and the area of cortex being stimulated is typically suppressed for 45–60 minutes following the stimulation protocol. For example, a series of three pulses at 50 Hz repeated every 200 ms (5 Hz) for 40 seconds has been found to have an inhibitory effect on the area of stimulation that lasts nearly an hour. These protocols produce a longer-lasting suppressive effect than typical transcranial magnetic stimulation (TMS) procedures of the same stimulation time (e.g., nonpatterned TMS for 40 seconds produces effects that last only 10–20 minutes).

covert attention. Type of attention that involves moving the focus of attention without moving the eyes or head; attending to something that is not being fixated.

diffusion-tensor imaging (DTI). A magnetic resonance imaging method that utilizes diffusion-weighted imaging data to estimate the rate and direction of the diffusion of water molecules at each voxel. DTI allows the connections between brain regions to be imaged by providing estimates of the white matter tracts between areas.

diffusion-weighted imaging (DWI). Types of magnetic resonance imaging sequences that are sensitive to the mobility of water and its rate of diffusion at each voxel. In conjunction with diffusion-tensor analysis algorithms, DWI can provide maps of white matter tracts throughout the brain.

dorsal attention network (DAN). This network of brain areas is thought to be critically involved in the control of attention, most specifically the voluntary orienting of spatial attention. Primary regions in this network include the frontal eye fields and the intraparietal sulcus. It has alternatively been referred to as the "dorsal frontoparietal network," the "frontoparietal attention network," or the "task-positive network."

double dissociation. Used in neuropsychology to associate two different cognitive functions with two different areas of the brain. In a double dissociation, there are two different types of patients with nonoverlapping lesions, and they show opposite patterns of performance on two tasks. For example, one patient type would show impaired performance on Task A and normal performance on Task B, while a second patient type would be impaired on Task B without any impairment on Task A.

dwell time. In studies of attention, dwell time refers to the length of time attention remains focused on a particular stimulus before disengaging and moving on to other stimuli or thoughts.

dynamic causal modeling (DCM). An advanced method for assessing the causal direction of functional connectivity between brain regions and how these directed couplings change across time or experimental conditions. Primarily used with functional magnetic resonance imaging (fMRI) data, it has also been used with electro-encephalography and magnetoencephalography. DCM procedures involve using nonlinear state-spaced models to model the interactions of neural populations and a forward model to specify how that neural activity gives rise to the measured response (e.g., the BOLD response in fMRI). Competing models are generated to specify how hidden states, time, and experimental conditions may modulate the effective connectivity within the network being examined. Bayesian model comparison is used to assess the evidence for the competing models and to select which model best accounts for the pattern of observed connectivity.

early selection. Theory that attention acts at a relatively early stage of processing to filter what information proceeds to higher levels of analysis. Channels of information based on basic features (e.g., spatial location, color, pitch) could be selected for further processing, and information from nonselected channels can thus be filtered out. More recent theories suggest that the selection of information at early levels may affect the strength of processing without completely filtering out the nonselected information.

ecological validity. How well a laboratory experiment represents the real-word situation it is meant to be testing.

egocentric neglect. Sometimes referred to as "space-based neglect," this is a type of unilateral neglect syndrome in which neglect is centered on the patient's viewpoint in space. A patient with leftward egocentric neglect would neglect everything on their left side, possibly including all stimuli in their left visual field and the left side of their own body. They are able to attend to both sides of objects within the intact side of their perceptual space. This type of neglect is contrasted with allocentric neglect (also known as "object-based neglect").

electrocorticography (ECoG). An electrophysiological method in which a grid of electrodes is placed directly on the surface of the brain. Similar to electroencephalography (EEG), ECoG reflects primarily postsynaptic potentials in large populations of cortical pyramidal cells. ECoG has the same excellent temporal resolution as EEG but much higher spatial resolution because, whereas EEG is recorded on the scalp, ECoG is recorded on the surface of the cortex and therefore isn't attenuated or spatially distorted by the skull and scalp. However, ECoG placement requires a craniotomy to remove a part of the scalp and thus is a highly invasive procedure that is conducted only when required prior to surgery (typically only for treatment of severe epilepsy). ECoG is also being developed for possible use in brain–computer interfaces in patients.

electroencephalography (EEG). Method of recording brain activity noninvasively with electrodes placed on the scalp. This method has excellent temporal resolution, as it can record activity with millisecond precision, but it has limited spatial resolution due to the distance between the neural activity and the electrodes and the effects of the intervening materials.

endogenous attention. Type of attention that is controlled by the person's internal goals and intentions. Also commonly referred to as "voluntary attention" or "goal-driven attention." May also be referred to as "internal attention" or "willed attention."

entrainment. A term used across a wide range of disciplines to describe the alignment or synchronization of temporal periods and phases between two (or more) systems or processes, even without any direct physical coupling. In neuroscience, it often refers to the activity of a set of neurons becoming aligned with an external stimulus being presented at a regular periodic frequency or rhythm.

error-related negativity (ERN). An event-related potential component that is observed shortly after the subject has made an error. The ERN is a response-locked component that is observed as a sharp negative-polarity spike that typically peaks 40–150 ms after an incorrect response. The ERN is observed over midline frontal scalp locations and is thought to be generated primarily in the anterior cingulate cortex, with a possible contribution from the dorsolateral prefrontal cortex.

event-related magnetic field (ERF). Created by extracting and averaging sections of the raw magnetoencephalography waveforms to measure the magnetic activity related to a specific mental event. ERFs are the magnetic counterpart to the event-related potential.

event-related optical signals (EROS). Neuroimaging technique that uses similar equipment to functional near-infrared spectroscopy to measure the reflectance and absorption of infrared light from diodes and sensors placed on the scalp. The amount of light scattering has been associated with neural activity. EROS is thought to have good spatial and temporal resolution but limited depth.

event-related potential (ERP). Created by extracting and averaging sections of raw electroencephalography waveforms to measure the electrical activity related to specific mental events.

exogenous attention. Type of attention that is controlled by properties of the stimuli and the environment, external to the individual. Classically, this type of attention is thought to be independent of goals and intentions. May also be referred to as "involuntary attention," "reflexive attention," or "automatic attention," although these terms could include different definitions.

extinction task. A classic bedside test for attentional neglect. The doctor holds up both index fingers, one in each of the patient's peripheral visual hemifields, and wiggles one or both fingers while the patient is looking directly at the doctor. Patients with neglect consistently fail to detect any movement of the finger in their bad field when it occurs simultaneously with a movement in their good field. They essentially "extinguish" the stimulus in their bad field when their attention is focused on a stimulus in their good field.

feature-based attention. Type of attention that involves focusing on basic sensory and perceptual properties of stimuli (e.g., color, motion, form, pitch), regardless of spatial location.

feature integration theory (FIT). Theory that a function of attention is to bind together the different basic features (e.g., color, motion) of an item with that object. According to FIT, features from across a scene are processed and represented in parallel in separate areas of the brain, and attention is needed to bind the multiple features of an object together onto that object.

feature search. A type of search in which the target is defined by a simple unique physical property (e.g., color, motion, form, pitch) – for example, searching for a single blue item in an array of many red items. Feature search is typically faster and easier than searching for a target that is defined by a unique combination of features (see "conjunction search").

forward solution. Refers to the calculation of the activity at electroencephalography electrodes or magnetoencephalography sensors that should result when a specific area of the brain is active. Given a known location for activity in the brain and accurate models of head geometry, brain anatomy, and conductivity values, there is a single unique solution for how that activity would be observed at the scalp.

functional magnetic resonance imaging (fMRI). Type of MRI that uses sequences sensitive to blood oxygenation levels (see "BOLD") to measure neural activity indirectly. This noninvasive method provides excellent spatial precision but is limited in terms of its temporal resolution due to the sluggish nature of the hemodynamic response.

functional near-infrared spectroscopy (fNIRS). Noninvasive method of measuring the changes in reflectance of near-infrared light caused by the changes in the ratio of deoxygenated and oxygenated hemoglobin that are associated with neural activity. This method uses light diodes and sensors placed on the scalp.

fusiform face area (FFA). Part of the brain that is especially sensitive to the visual presentation of a face.

This area is located in the ventral visual pathway, along the lateral fusiform gyrus in the inferior temporal cortex, within Brodmann Area 37. It has been argued to be specialized for faces and therefore indicative of domain specificity in the cortex.

group iterative multiple model estimation (GIMME). An advanced algorithm for assessing group commonalities and individual differences in functional connectivity across brain regions. Using a unified structural equation modeling approach, GIMME first identifies the lagged and contemporaneous directed connections that exist for the group as whole. Using this mapping of connections for the majority of the group as a starting point, GIMME then identifies individual differences and allows for bottom-up sorting into subgroups based on these patterns.

Heschl's gyrus. Anatomically defined as the transverse temporal gyrus, it lies in the superior temporal lobe and includes the first auditory processing region of the cortex (i.e., primary auditory cortex, "A1").

information processing model. An instrumental theory in the development of cognitive psychology as a field of study that holds that complex cognitive processes can be broken down into a sequence of simpler functions occurring in series. Through careful experimental design, the timing and computations of different stages of processing can be studied.

inhibition of return (IOR). This term refers to the slowed or suppressed response to a stimulus that occurs at the location where attention has very recently been captured. Attention is thought to be inhibited from returning to a location where it had recently been if there is no target, or desire to maintain attention, at that location. IOR is seen in manual reaction times in covert attention tasks and in eye movements in overt attention tasks.

inverse problem. Refers to the estimation of the neural source of activity recorded at electroencephalography electrodes or magnetoencephalography sensors. As compared to the unique solution of a pattern of activity on the scalp when a neural source is known (i.e., the "forward solution"), estimation of the source of neural activity from sensors located outside the head is always a "problem" because there is never a single unique solution for any pattern of activity recorded at a distance from the source.

laminar electrodes. Microelectrodes that each have a linear array of nearby contacts (e.g., spaced ~100 microns apart) that allow for simultaneous recording of neural activity in different cortical layers. This method is critical for measuring activity across cortical layers and has been especially useful in assessing feedforward and feedback activity in a region. Most often used in nonhuman animal research, laminar electrodes are sometimes used in presurgical procedures in human epilepsy patients.

late-selection theories. Theory that attention acts only relatively late in the stream of information processing to filter out unattended information. According to these theories, information processing can proceed to high levels, including full semantic analysis of the meaning of a stimulus, before being affected by attention.

local field potential (LFP). Electrical signals representing the summed activity of a relatively small population of neurons. Recorded using a microelectrode inserted into the brain, LFPs are a measure of the changes in electrical potential produced by transient changes in extracellular ion concentrations resulting from neural activity. Typically low-pass filtered to remove transient effects of individual neuron outputs, LFPs are thought to index synaptic inputs to a region. Depending on recording and processing methods, LFPs are thought to be able to index potentials generated ~0.5 mm from the electrode tip.

magnetic resonance imaging (MRI). A noninvasive medical imaging technology that uses magnetic fields and radio waves to produce highly detailed anatomical images of structures throughout the body. MRI scans involve no ionizing radiation, but patients must be screened carefully before entering the scanner because of the very strong magnetic fields employed (typically 1.5 or 3.0 Tesla).

magnetic resonance spectroscopy (MRS). A noninvasive method for measuring the chemical composition of biological material. MRS uses the equipment of a standard magnetic resonance imaging machine, but specialized scanning sequences are also employed that are sensitive to the presence of different chemicals and metabolites in the brain (e.g., amino acids, choline, lactate). MRS is often used to test for the presence of a possible tumor.

magnetoencephalography (MEG). A noninvasive functional neuroimaging method that measures the magnetic fields produced by populations of active neurons. MEG has excellent (millisecond) temporal resolution and can achieve good spatial resolution with appropriate head models. Compared to electroencephalography, MEG equipment is much more expensive because it requires a magnetically quiet region around the head and specialized sensors to measure the miniscule magnetic fields produced by neuronal activity.

mental chronometry. Measuring the sequence and timing of mental events. Behavioral responses (e.g., reaction times to pushing a button to indicate the detection of a stimulus) are used to measure the duration and sequence of mental operations involved in cognitive processing.

microsaccades. Tiny eye movements that occur even when the subject is trying to maintain fixation. Sometimes referred to as "fixational saccades," they are thought to be involuntary. Although very small in comparison to regular saccades, the movements can be large enough to cause a stimulus to move into or out of a receptive field of neurons in the lateral geniculate nucleus or in V1.

mindfulness. A type of meditation that involves focusing on the current moment and being aware of sensations and internal states without judgment. It is used in a number of therapeutic interventions and often includes guided imagery and breathing techniques to promote relaxation and reduce stress.

mind-wandering. The cognitive state associated with losing focus on the task being performed and attending to spontaneous internal thoughts instead.

motor theories of attention. Sometimes referred to as the "premotor theory of attention." This theory holds that the shifting of attention is a result of motor plans to overtly orient toward a stimulus of interest. Accordingly, covert attention is thought to be intimately related to overt attention.

neglect. Also referred to as "unilateral neglect," this term refers to a syndrome in which patients, following a brain injury, ignore a large portion of space (egocentric neglect) or one side of objects (allocentric neglect), despite their basic sensory processing being intact.

neural correlates of consciousness (NCC). The minimal set of neuronal activity and mechanisms that are sufficient for experiencing conscious perception. The "minimal" qualifier is necessary in the definition because any mental state will be associated with widespread neural activity, and the search for a neural correlate of consciousness is an attempt to identify precisely the neural activity underlying the conscious experience of a given percept.

neuropsychology. A method of investigating the brain basis of cognitive operations by testing neurological patients (e.g., patients with brain lesions, often due to stroke or disease). By measuring the performance of neurological patients on cognitive tests and comparing this to that of healthy controls, the functions of the affected brain region(s) are inferred.

overt attention. Shifting the focus of attention by moving the sensory organs toward the stimulus of interest (e.g., moving the head and/or eyes to fixate an item; orienting the head to align an ear with the source of a sound; reaching out to touch an item).

parahippocampal place area (PPA). A region of the brain in the medial inferior temporo-occipital cortex – surrounding the hippocampus – that is especially sensitive to the presence of environmental scenes, landscapes, cityscapes, or rooms ("places"). Similarly to how the fusiform face area is thought to be specialized for processing faces, the PPA provides additional evidence for domain specificity in the brain.

perceptual inference. Part of the predictive coding model of brain function that holds that conscious perception involves the inferring of sensory stimuli based on expectations and prior experience.

positron emission tomography (PET). A biomedical imaging method that measures physiological functions such as blood flow and metabolism in the body by injecting a radioactive tracer into the bloodstream. Using ^{15}O, a radioactive isotope of oxygen, PET can measure the increase in blood flow associated with neural activity. By comparing scans in different experimental conditions, PET can provide high-spatial-resolution images of brain regions associated with different cognitive functions.

posterior probability. A part of Bayesian statistical modeling, the posterior probability is the updated probability distribution after evidence has been taken into account.

precision. A critical component of predictive coding models of perception, precision refers to the uncertainty associated with a given input or model.

Inputs associated with low uncertainty (high precision) are afforded more confidence and are weighted more strongly when updating internal models with prediction errors.

prediction errors. The difference between expectation and the current sampling of the environment. According to predictive coding models, bottom-up processing in the brain consists of propagating these errors in order to update the models of the world on which conscious inference is built.

prior probability. A part of Bayesian statistical modeling, the prior probability is the assumed probability distribution before evidence is taken into account. Sometimes this is referred to simply as the "prior."

psychological refractory period (PRP). The finding that the reaction time to manually respond to the second of two successive target stimuli is slower the closer it occurs in time to the first target. The PRP is dependent on both targets requiring an immediate response, and it is thought to occur because the planning of a response to a second target must be delayed until the planning and execution of a response to the first stimulus has been completed.

rapid serial visual presentation (RSVP). A type of cognitive experimental task in which stimuli are presented one at a time (serially) in quick succession. Often the stimuli are presented directly at fixation, so visual acuity should be high, but the rapid presentation taxes higher cognitive resources.

receptive field (RF). The region over which a neuron is responsive. For visual neurons in early sensory areas, the region is typically defined as a region of space. RFs for other senses are typically defined by what those neurons primarily code (e.g., for neurons in the primary somatosensory cortex, the RF is defined as the region of the body that produces activity in the cell; in the primary auditory cortex, the RF may be defined as the range of frequencies to which the neuron is responsive).

rhythmic theory of attention. This theory posits that low-frequency oscillations of neural activity across the brain organize regions into alternating states of sensory acquisition and oculomotor movement. This oscillation is thought to be important for avoiding conflicts between perception and action, and it suggests that attention naturally alternates between high and low degrees of focus.

saliency map. A theoretical map of environmental space in the brain that represents the relative uniqueness and strength of current stimuli in the environment. Such a map has been hypothesized to play an important role in determining the distribution of attention, with items that are especially unique in the environment capturing attention. The saliency map is thought to integrate information from multiple, lower-level maps of basic features (e.g., color, motion, pitch).

serial reaction time task. A task often used in cognitive experiments to study implicit memory. In a typical version of the task, subjects responded to serially presented visual stimuli, and each finger is associated with a different button press that corresponds to one of the stimuli. Sequences that repeat over the course of the experiment are responded to faster, even though subjects are often unaware of that specific sequence being repeated across the experiment.

simultanagnosia. A rare syndrome following bilateral damage to parieto-occipital regions in which patients can perceive only one object at a time. These patients have great difficulty disengaging from an attended object to perceive anything else. It is sometimes referred to as "Bálint's syndrome," although that syndrome includes other deficits in addition to simultanagnosia.

single dissociation. Used in neuropsychology to associate a cognitive function with an area of the brain. When a group of patients with lesions overlapping in a particular brain region all show a deficit in performance on a particular cognitive task compared to healthy controls, this can be used to infer that the brain area is involved in that cognitive process. Stronger conclusions are warranted when a double dissociation can be found.

single-unit recording. A highly precise but invasive method of measuring the activity of neurons in the brain. By inserting a microelectrode directly into the depths of the brain, the electrophysiological responses of a single neuron can be recorded. This method provides the highest spatial and temporal resolution available to cognitive neuroscientists, but because it is highly invasive it is mostly limited to nonhuman studies. In rare cases, human patients with intractable epilepsy may have electrodes inserted to determine the foci of seizures.

spotlight model. An early theory that likens the focus of attention to a spotlight, with only information within the spotlight being fully processed and information outside the spotlight being relatively filtered out.

subtractive method. A method for isolating and estimating the timing associated with a particular cognitive process by comparing the reaction time of responses between two experimental conditions that are designed to differ only in the presence of that cognitive process. This method relies on the assumption that the addition of the process does not interact with the timing or duration of other processes.

sustained attention. The ability to maintain focus on the task being performed and on the relevant stimuli for a prolonged period of time. It is related to the concept of vigilance.

transcranial alternating current stimulation (tACS). This method uses a similar setup to transcranial direct current stimulation, but the electrical current passed between the electrodes on the scalp is not constant, instead varying in a sinusoidal manner. This method entrains the underlying brain areas to the frequency of the electrical stimulation being delivered, and thus it can be used to investigate the influence of different types of oscillatory activity on cognitive functions.

transcranial direct current stimulation (tDCS). A method of neuromodulation that uses small electrical currents passed between electrodes on the scalp to modulate the excitability of underlying neural structures. Typically, the excitability of brain areas close to the anodal electrode is enhanced, whereas the excitability of brain areas close to the cathodal electrode is suppressed.

transcranial electrical stimulation (tES). A term encompassing different methods of noninvasive neural stimulation that use small electrical currents passed between electrodes placed on the scalp, including transcranial direct current stimulation, transcranial alternating current stimulation, and transcranial random noise stimulation.

transcranial magnetic stimulation (TMS). A method of brain stimulation that uses a rapid electrical pulse within a coil placed closed to the head to produce a strong, transient magnetic field that induces an electrical current in the brain that can trigger action potentials in neurons. Combined with magnetic resonance imaging-informed head models, this method can stimulate the brain with high spatial and temporal precision.

ventral attention network (VAN). Network of brain regions implicated in the processes of reorienting attention back to the location/object/task at hand after attention has been involuntarily captured by a salient but task-irrelevant event. Primary nodes of this network include the temporoparietal junction and the ventral frontal cortex. It is sometimes characterized as being a right-hemisphere network, although other work suggests that it is a bilateral network. The VAN is sometimes referred to as the "ventral frontoparietal network," and it has been associated with the "saliency network," which includes the anterior insula, anterior cingulate cortex, and ventral striatum.

vigilance. Related to the concept of sustained attention, or the ability to maintain focus on the task being performed for a prolonged period of time. Vigilance may also refer to the state of arousal in which the person is fully alert and ready to respond.

Wisconsin Card Sorting Test (WCST). A classic test of cognitive flexibility, originally designed to assess neuropsychological patients. Subjects are provided with a deck of cards, each of which displays a simple geometric shape, and subjects must sort the cards based on the number of items, the color of the items, or the shape of the items. The experimenter provides no explicit rule, but they provide feedback ("correct" or "incorrect") with every choice, and the subject must learn the rule through trial and error. The critical trials are when the experimenter changes the rule, again by providing only feedback as to the correctness of the subject's choice. Patients with frontal lobe damage often show a pattern of perseveration, in which they continue to sort by the originally correct rule long after being given feedback that their sorting is no longer correct.

zoom-lens model. An early theory of attention, similar to the spotlight theory of attention, but with the addition that the size of the spotlight is flexible. According to this theory, there is a fixed amount of attentional resources that can be allocated across the attentional window, with the effects of attention becoming weaker as the size of the attentional window is widened.

References

Adamo, S. H., Cox, P. H., Kravitz, D. J., & Mitroff, S. R. (2019). How to correctly put the "subsequent" in subsequent search miss errors. *Attention, Perception & Psychophysics*, *81*(8), 2648–2657.

Adisetiyo, V., & Gray, K. M. (2017). Neuroimaging the neural correlates of increased risk for substance use disorders in attention-deficit/hyperactivity disorder – A systematic review. *The American Journal on Addictions*, *26*(2), 99–111.

Adrian, E. D., & Matthews, B. H. (1934). The Berger rhythm: Potential changes from the occipital lobes in man. *Brain*, *57*(4), 355–385.

Agustí, A. I., Satorres, E., Pitarque, A., & Meléndez, J. C. (2017). An emotional Stroop task with faces and words. A comparison of young and older adults. *Consciousness and Cognition*, *53*, 99–104.

Ahlfors, S. P., & Mody, M. (2019). Overview of MEG. *Organizational Research Methods*, *22*(1), 95–115.

Ahveninen, J., Jaaskelainen, I., Raij, T., Bonmassar, G., Devore, S., Hämäläinen, M., Levanen, S., Lin, F., Sams, M., Shinn-Cunningham, B., Witzel, T., & Belliveau, J. (2006). Task-modulated "what" and "where" pathways in human auditory cortex. *Proceedings of the National Academy of Sciences of the United States of America*, *103*, 14608e14613.

Akça, M., Bishop, L., Vuoskoski, J. K., & Laeng, B. (2023). Human voices escape the auditory attentional blink: Evidence from detections and pupil responses. *Brain and Cognition*, *165*, 105928.

Alimoradi, Z., Lotfi, A., Lin, C. Y., Griffiths, M. D., & Pakpour, A. H. (2022). Estimation of behavioral addiction prevalence during COVID-19 pandemic: A systematic review and meta-analysis. *Current Addiction Reports*, *9*(4), 486–517.

Allman, J. M., Hakeem, A., Erwin, J. M., Nimchinsky, E., Hof, P., & Hixon, F. P. (2001). The anterior cingulate cortex: The evolution of an interface between emotion and cognition. *Annals of the New York Academy of Sciences*, *935*, 107–117.

American Psychiatric Association. (2022). *Diagnostic and Statistical Manual of Mental Disorders*, 5th edition, text revision. American Psychiatric Association.

Amso, D., Haas, S., Tenenbaum, E., Markant, J., & Sheinkopf, S. J. (2014). Bottom-up attention orienting in young children with autism. *Journal of Autism and Developmental Disorders*, *44*(3), 664–673.

Anderson, B. A. (2016). Social reward shapes attentional biases. *Cognitive Neuroscience*, *7*(14), 30–36.

Anderson, B. A., & Druker, M. (2013). Attention improves perceptual quality. *Psychonomic Bulletin & Review*, *20*(1), 120–127.

Anderson, B. A., & Folk, C. L. (2010). Variations in the magnitude of attentional capture: Testing a two-process model. *Attention, Perception & Psychophysics*, *72*(2), 342–352.

Anderson, B. A., & Kim, H. (2018). Relating attentional biases for stimuli associated with social reward and punishment to autistic traits. *Collabra: Psychology*, *4*(1), 10.

Anderson, B. A., & Yantis, S. (2013). Persistence of value-driven attentional capture. *Journal* of *Experimental Psychology: Human Perception and Performance, 39*(1), 6–9.

Anderson, B. A., Kim, H., Kim, A. J., Liao, M. R., Mrkonja, L., Clement, A., & Grégoire, L. (2021). The past, present, and future of selection history. *Neuroscience and Biobehavioral Reviews, 30*, 326–350.

Anderson, B.A., Laurent, P.A., & Yantis, S. (2014). Value-driven attentional priority signals in human basal ganglia and visual cortex. *Brain Research, 1587*, 88–96.

Andreu, C. I., Cosmelli, D., Slagter, H. A., & Franken, I. H. A. (2018). Effects of a brief mindfulness-meditation intervention on neural measures of response inhibition in cigarette smokers. *PLoS One, 13*(1), e0191661.

Anguera, J. A., Boccanfuso, J., Rintoul, J. L., Al-Hashimi, O., Faraji, F., Janowich, J., Kong, E., Larraburo, Y., Rolle, C., Johnston, E., & Gazzaley, A. (2013). Video game training enhances cognitive control in older adults. *Nature, 501*(7465), 97–101.

Anllo-Vento, L., Luck, S. J., & Hillyard, S. A. (1998). Spatio-temporal dynamics of attention to color: Evidence from human electrophysiology. *Human Brain Mapping, 6*(4), 216–238.

Arana, L., Melcón, M., Kessel, D., Hoyos, S., Albert, J., Carretié, L., & Capilla, A. (2022). Suppression of alpha-band power underlies exogenous attention to emotional distractors. *Psychophysiology, 59*(9), e14051.

Arasanz, C. P., Staines, W. R., & Schweizer, T. A. (2012). Isolating a cerebellar contribution to rapid visual attention using transcranial magnetic stimulation. *Frontiers in Behavioral Neuroscience, 6*, 55.

Arcaro, M.J., Pinsk, M.A., Li, X., & Kastner, S. (2011). Visuotopic organization of macaque posterior parietal cortex: A functional magnetic resonance imaging study. *Journal of Neuroscience, 31*, 2064–2078.

Arend, I., Rafal, R., & Ward, R. (2008). Spatial and temporal deficits are regionally dissociable in patients with pulvinar lesions. *Brain, 131*(8), 2140–2152.

Ariew, R., & Garber, D. (1989). *G.W. Leibniz: Philosophical Essays.* Hackett Publishing Company.

Armstrong, I. T., & Munoz, D. P. (2003) Attentional blink in adults with attention-deficit hyperactivity disorder. *Experimental Brain Research, 152*, 243–250.

Arnell, K. M., & Jolicœur, P. (1999). The attentional blink across stimulus modalities: Evidence for central processing limitations. *Journal of Experimental Psychology: Human Perception & Performance, 25*, 630–648.

Arnett, A. B., Pennington, B. F., Willcutt, E. G., DeFries, J. C., & Olson, R. K. (2015). Sex differences in ADHD symptom severity. *Journal of Child Psychology and Psychiatry, 56*(6), 632–639.

Arnsten, A. F. (2006). Fundamentals of attention-deficit/hyperactivity disorder: Circuits and pathways. *Journal of Clinical Psychiatry, 67*(Suppl. 8), 7–12.

Arora, S., Lawrence, M. A., & Klein, R. M. (2020). The attention network test database: ADHD and cross-cultural applications. *Frontiers in Psychology, 11*, 388.

Awh, E., & Jonides, J. (2001). Overlapping mechanisms of attention and spatial working memory. *Trends in Cognitive Sciences, 5*, 119–126.

Awh, E., & Pashler, H. (2000). Evidence for split attentional foci. *Journal of Experimental Psychology: Human Perception and Performance, 26*(2), 834–846.

Awh, E., Belopolsky, A. V., & Theeuwes, J. (2012). Top-down versus bottom-up attentional control: A failed theoretical dichotomy. *Trends in Cognitive Sciences, 16* (8), 437–443.

Bain, A. (1888). *The Emotions and the Will*, 3rd edition. Longmans Green and Co.

Baldassi, S., & Burr, D. C. (2000). Feature-based integration of orientation signals in visual search. *Vision Research, 40*(10–12), 1293–1300.

Baldauf, D., & Desimone, R. (2014). Neural mechanisms of object-based attention. *Science, 344*(6182), 424–427.

Baldwin, M. K. L., Balaram, P., & Kaas, J. H. (2017). The evolution and functions of nuclei of the visual pulvinar in primates. *Journal of Comparative Neurology, 525*(15), 3207–3226.

Balestrieri, E., Ronconi, L., Melcher, D., & Foxe, J. (2022). Shared resources between visual attention and visual working memory are allocated through rhythmic sampling. *The European Journal of Neuroscience, 55* (11–12), 3040–3053.

Bálint, R. (1909). Seelenlähmung des 'Schauens', optische Ataxie, räumliche Störung der Aufmerksamkeit. *European Neurology, 25,* 51–66.

Bálint, R., & Harvey, M. (1995). Psychic paralysis of gaze, optic ataxia, and spatial disorder of attention. *Cognitive Neuropsychology, 12*(3), 265–281.

Bangasser, D. A., & Cuarenta, A. (2021). Sex differences in anxiety and depression: Circuits and mechanisms. *Nature Reviews: Neuroscience, 22*(11), 674–684.

Banich, M. T. (2019). The Stroop effect occurs at multiple points along a cascade of control: Evidence from cognitive neuroscience approaches. *Frontiers in Psychology, 10,* 2164.

Banks, M., & Salapatek, P. (1983). Infant visual perception. In: Haith, M. M., & Campos, J. J. (eds.), *Infancy and Developmental Psychobiology. Handbook of Child Psychology, Volume 2* (pp. 435–572). Wiley.

Bantin, T., Stevens, S., Gerlach, A. L., & Hermann, C. (2016). What does the facial dot-probe task tell us about attentional processes in social anxiety? A systematic review. *Journal of Behavior Therapy and Experimental Psychiatry, 50,* 40–51.

Bar-Haim, Y., Lamy, D., Pergamin, L., Bakermans-Kranenburg, M. J., & van IJzendoorn, M. H. (2007). Threat-related attentional bias in anxious and nonanxious individuals: A meta-analytic study. *Psychological Bulletin, 133*(1), 1–24.

Barascud, N., Pearce, M. T., Griffiths, T. D., Friston, K. J., & Chait, M. (2016). Brain responses in humans reveal ideal observer-like sensitivity to complex acoustic patterns. *Proceedings of the National Academy of Sciences of the United States of America, 113,* E616–E625.

Barkley, R. A. (1997). *Attention-Deficit Hyperactivity Disorder and the Nature of Self-Control.* Guilford Press.

Barkley, R. A., & Peters, H. (2012). The earliest reference to ADHD in the medical literature? Melchior Adam Weikard's description in 1775 of "Attention deficit" (mangel der aufmerksamkeit, attentio volubilis). *Journal of Attention Disorders, 16*(8), 623–630.

Barnhart, A. S., Martinez-Conde, S., Macknik, S. L., Costela, F. M., & Goldinger, S. D. (2019). Microsaccades reflect the dynamics of misdirected attention in magic. *Journal of Eye Movement Research, 12*(6), 1–14.

Barrett, L. F. (2017). *How Emotions Are Made.* Houghton Mifflin Harcourt.

Bartsch, M. V., Loewe, K., Merkel, C., Heinze, H., Schoenfeld, M. A., Tsotsos, J. K., & Hopf, J. (2017). Attention to color sharpens neural population tuning via feedback processing in the human visual cortex hierarchy. *The Journal of Neuroscience, 37*(43), 10346–10357.

Baruch, O., & Goldfarb, L. (2020). Mexican hat modulation of visual acuity following an exogenous cue. *Frontiers in Psychology, 11,* 854.

Bastos, A. M., Loonis, R., Kornblith, S., Lundqvist, M., & Miller, E. K. (2018). Laminar recordings in frontal cortex suggest distinct layers for maintenance and control of working memory. *Proceedings of the National Academy of Sciences of the United States of America, 115,* 1117–1122.

Bastos, A. M., Usrey, W. M., Adams, R. A., Mangun, G. R., Fries, P., & Friston, K. J. (2012). Canonical microcircuits for predictive coding. *Neuron, 76,* 695–711.

Battelli, L., Alvarez, G., Carlson, T., & Pascual-Leone, A. (2008). The role of MT and the parietal lobe in visual tracking studied with transcranial magnetic stimulation. *Journal of Vision*, *6*, 822.

Batterink, L., Karns, C. M., Yamada, Y., & Neville, H. (2010). The role of awareness in semantic and syntactic processing: An ERP attentional blink study. *Journal of Cognitive Neuroscience*, *22*(11), 2514–2529.

Baumgartner, H. M., Graulty, C. J., Hillyard, S. A., & Pitts, M. A. (2017). Does spatial attention modulate the earliest component of the visual evoked potential? *Cognitive Neuroscience*, *9*(1–2), 4–19.

Bavelier, D., & Green, C. S. (2019). Enhancing attentional control: Lessons from action video games. *Neuron*, *104*(1), 147–163.

Bavelier, D., Achtman, R. L., Mani, M., & Föcker, J. (2012). Neural bases of selective attention in action video game players. *Vision Research*, *61*, 132–143.

Bayliss, A. P., Schuch, S., & Tipper, S. P. (2010). Gaze cueing elicited by emotional faces is influenced by affective context. *Visual Cognition*, *18*(8), 1214–1232.

Becker, E. S., Rinck, M., Margraf, J., & Roth, W. T. (2001). The emotional Stroop effect in anxiety disorders: General emotional or disorder specificity? *Journal of Anxiety Disorders*, *15*(3), 147–159.

Becker, S. I., Atalla, M., & Folk, C. L. (2020). Conjunction search: Can we simultaneously bias attention to features and relations? *Attention, Perception & Psychophysics*, *82*(1), 246–268.

Bediou, B., Adams, D. M., Mayer, R. E., Tipton, E., Green, C. S., & Bavelier, D. (2018). Meta-analysis of action video game impact on perceptual, attentional, and cognitive skills. *Psychological Bulletin*, *144* (1), 77–110.

Ben Hamed, S., Duhamel, J.R., Bremmer, F., & Graf, W. (2001). Representation of the visual field in the lateral intraparietal area of macaque monkeys: a quantitative receptive field analysis. *Experimental Brain Research*, *140*, 127–144.

Benedetto, A., Spinelli, D., & Morrone, M. C. (2016). Rhythmic modulation of visual contrast discrimination triggered by action. *Proceedings of the Royal Society B: Biological Sciences*, 283(1831), 20160692.

Bengson, J. J., Kelley, T. A., & Mangun, G. R. (2015). The neural correlates of volitional attention: A combined fMRI and ERP study. *Human Brain Mapping*, *36*(7), 2443–2454.

Bengson, J. J., Kelley, T. A., Zhang, X., Wang, J. L., & Mangun, G. R. (2014). Spontaneous neural fluctuations predict decisions to attend. *Journal of Cognitive Neuroscience*, *26*(11), 2578–2584.

Bengson, J. J., Liu, Y., Khodayari, N., & Mangun, G. R. (2020). Gating by inhibition during top-down control of willed attention. *Cognitive Neuroscience*, *11*(1–2), 60–70.

Berger, A., Henik, A., & Rafal, R. (2005). Competition between endogenous and exogenous orienting of visual attention. *Journal of Experimental Psychology: General*, *134*, 207–221.

Berger, H. (1929). Über das Elektroenkephalogramm des Menschen. *Archiv für Psychiatrie und Nervenkrankheiten*, *87*(1), 527–570.

Berger, H. (1938). Über das Elektrenkephalogramm des Menschen. XIV Mitteilung. *Archiv für Psychiatrie und Nervenkrankheiten*, *108*, 407–431.

Bernhardt, B. C., & Singer, T. (2012). The neural basis of empathy. *Annual Review of Neuroscience*, *35*(1), 1–23.

Bichot, N., Heard, M., DeGennaro, E., & Desimone, R. (2015). A source for feature-based attention in the prefrontal cortex. *Neuron*, *88*(4), 832–844.

Bichot, N. P., Rossi, A. F., & Desimone, R. (2005). Parallel and serial neural mechanisms for visual search in macaque area V4. *Science*, *308*, 529–534.

Bichot, N. P., Xu, R., Ghadooshahy, A., Williams, M. L., & Desimone, R. (2019). The role of prefrontal cortex in the control of feature attention in area V4. *Nature Communications*, *10*(1), 5727–5712.

Binet, A. (1896). Psychology of prestidigitation. In *Annual Report of the Board of Regents of the Smithsonian Institution Showing the*

Operations, Expenditures, and Conditions of the Institution to July 1894 (pp. 555–571). US Government Printing Office.

Bishop, S. R., Lau, M., Shapiro, S., Carlson, L., Anderson, N. D., Carmody, J., Segal, Z. V., Abbey, S., Speca, M., Velting, D., & Devins, G. (2004). Mindfulness: A proposed operational definition. *Clinical Psychology (New York, N.Y.)*, *11*(3), 230–241.

Bisley, J. W., & Goldberg, M. E. (2003). Neuronal activity in the lateral intraparietal area and spatial attention. *Science*, *299*, 81–86.

Bisley, J. W., & Goldberg, M. E. (2010). Attention, intention, and priority in the parietal lobe. *Annual Review of Neuroscience*, *33*, 1–21.

Blakemore, S.J. (2008). The social brain in adolescence. *Nature Reviews: Neuroscience*, *9*, 267–277.

Bola, M., Paź, M., Doradzińska, Ł., & Nowicka, A. (2021). The self-face captures attention without consciousness: Evidence from the N2pc ERP component analysis. *Psychophysiology*, *58*(4), e13759.

Bolger, D., Coull, J. T., & Schön, D. (2014). Metrical rhythm implicitly orients attention in time as indexed by improved target detection and left inferior parietal activation. *Journal of Cognitive Neuroscience*, *26*(3), 593–605.

Boot, W. R., Kramer, A. F., Simons, D. J., Fabiani, M., & Gratton, G. (2008). The effects of video game playing on attention, memory, and executive control. *Acta Psychologica*, *129*(3), 387–398.

Boshra, R., & Kastner, S. (2022). Attention control in the primate brain. *Current Opinion in Neurobiology*, *76*, 102605.

Bosman, C. A., Schoffelen, J.-M., Brunet, N., Oostenveld, R., Bastos, A. M., Womelsdorf, T., Rubehn, B., Stieglitz, T., De Weerd, P., & Fries, P. (2012). Attentional stimulus selection through selective synchronization between monkey visual areas. *Neuron*, *75*, 875–888.

Bourgeois, A., Chica, A. B., Valero-Cabré, A., & Bartolomeo, P. (2013). Cortical control of inhibition of return: Exploring the causal contributions of the left parietal cortex. *Cortex*, *49*, 2927–2934.

Bowling, J., Friston, K. J., & Hopfinger, J. B. (2020). Top-down versus bottom-up attention differentially modulate frontal-parietal connectivity. *Human Brain Mapping*, *41*(4), 928–942.

Bradley, B. P., Mogg, K., Millar, N., Bonham-Carter, B., Fergusson, E., Jenkins, J., & Parr, M. (1997). Attentional biases for emotional faces. *Cognition & Emotion*, *11*, 25–42.

Bradley, M. M., Sabatinelli, D., Lang, P. J., Fitzsimmons, J. R., King, W. & Desai, P. (2003). Activation of the visual cortex in motivated attention. *Behavioral Neuroscience*, *117*(2), 369–380.

Brefczynski, J. A., & DeYoe, E. A. (1999). A physiological correlate of the 'spotlight' of visual attention. *Nature Neuroscience*, *2*, 370–374.

Breland, K., & Breland, M. (1961). The misbehavior of organisms. *The American Psychologist*, *16*(11), 681–684.

Breska, A., & Deouell, L. Y. (2014). Automatic bias of temporal expectations following temporally regular input independently of high-level temporal expectation. *Journal of Cognitive Neuroscience*, *26*(7), 1555–1571.

Briand, K. A. (1998). Feature integration and spatial attention: More evidence of a dissociation between endogenous and exogenous orienting. *Journal of Experimental Psychology: Human Perception and Performance*, *24*, 1243–1256.

Briand, K. A., & Klein, R. M. (1987). Is Posner's "beam" the same as Treisman's "glue"? On the relation between visual orienting and feature integration theory. *Journal of Experimental Psychology: Human Perception and Performance*, *13*, 228–241.

Bridgeman, G., Hendry, D., & Stark, L. (1975). Failure to detect displacement of visual world during saccadic eye movements. *Vision Research*, *15*, 719–722.

Brissenden, J. A., Levin, E. J., Osher, D. E., Halko, M. A., & Somers, D. C. (2016).

Functional evidence for a cerebellar node of the dorsal attention network. *Journal of Neuroscience*, *36*(22), 6083–6096.

Brisson, B., & Bourassa, M. È. (2014). Masking of a first target in the attentional blink attenuates the P 3 to the first target and delays the P 3 to the second target. *Psychophysiology*, *51*(7), 611–619.

Broadbent, D. E. (1952). Listening to one of two synchronous messages. *Journal of Experimental Psychology*, 44, 51–55.

Broadbent, D. E. (1958). *Perception and Communication*. Pergamon Press.

Broadbent, D. E. (1982). Task-combination and selective intake of information. *Acta Psychologica*, *50*, 253–290.

Broca, P. (1861). Remarques sur le siège de la faculté du langage articulé, suivies d'une observation d'aphémie (perte de la parole). *Bulletin de la Société d'Anthropologie*, *6*, 330–357.

Broca, P. (1865). Sur le siège de la faculté du langage articulé. *Bulletin de la Société d'Anthropologie*, *6*, 337–393.

Bronson, G. W. (1994). Infants' transitions toward adult-like scanning. *Child Development*, *65*(5), 1243–1261.

Brown, J. A., & Bidelman, G. M. (2022). Familiarity of background music modulates the cortical tracking of target speech at the "cocktail party". *Brain Sciences*, *12*(10), 1320.

Bruya, B., & Tang, Y. (2018). Is attention really effort? Revisiting Daniel Kahneman's influential 1973 book *Attention and Effort*. *Frontiers in Psychology*, *9*, 1133.

Budd, J. M. L. (1998). Extrastriate feedback to primary visual cortex in primates: A quantitative analysis of connectivity. *Proceedings of the Royal Society B: Biological Sciences*, *265*(1400), 1037–1044.

Bullock, T. H., Karamürsel, S., Achimowicz, J. Z., McClune, M. C., & Başar-Eroglu, C. (1994). Dynamic properties of human visual evoked and omitted stimulus potentials. *Electroencephalography and Clinical Neurophysiology*, *91*(1), 42–53.

Busch, N. A., & VanRullen, R. (2010). Spontaneous EEG oscillations reveal periodic sampling of visual attention. *Proceedings of the National Academy of Sciences of the United States of America*, *107*(37), 16048–16053.

Bush, G. (2011). Cingulate, frontal, and parietal cortical dysfunction in attention-deficit/hyperactivity disorder. *Biological Psychiatry (1969)*, *69*(12), 1160–1167.

Bush, G., Luu, P., & Posner, M. I. (2000). Cognitive and emotional influences in anterior cingulate cortex. *Trends in Cognitive Sciences*, *4*(6), 215–222.

Buzsaki, G. (2006). *Rhythms of the Brain*. Oxford University Press.

Cahill, L. (2006). Why sex matters for neuroscience. *Nature Reviews: Neuroscience*, *7*(6), 477–484.

Caird, J. K., Willness, C. R., Steel, P., & Scialfa, C. (2008). A meta-analysis of the effects of cell phones on driver performance. *Accident Analysis & Prevention*, *40*(4), 1282–1293.

Cammann, R. (1990). Is there a mismatch negativity (MMN) in visual modality? *The Behavioral and Brain Sciences*, *13*(2), 234–235.

Capozzi, F., & Ristic, J. (2018). How attention gates social interactions. *Annals of the New York Academy of Sciences*, *1426*, 179–198.

Carper, R. A., & Courchesne, E. (2005). Localized enlargement of the frontal cortex in early autism. *Biological Psychiatry*, *57*, 126–133.

Carr, H. (1952). *Free Precession Techniques in Nuclear Magnetic Resonance*. PhD thesis. Harvard University.

Carrasco, M. (2011). Visual attention: The past 25 years. *Vision Research*, *51*(13), 1484–1525.

Carrasco, M. (2018). How visual spatial attention alters perception. *Cognitive Processes*, *19* (Suppl. 1), 77–88.

Carretié, L. (2014). Exogenous (automatic) attention to emotional stimuli: A review. *Cognitive, Affective, & Behavioral Neuroscience*, *14*(4), 1228–1258.

Carter, C. S., Krener, P., Chaderjian, M., Northcutt, C., & Wolfe, V. (1995). Abnormal processing of irrelevant information in attention deficit hyperactivity disorder. *Psychiatry Research*, *56*(1), 59–70.

Casanova, C., & Chalupa, L. M. (2023). The dorsal lateral geniculate nucleus and the pulvinar as essential partners for visual cortical functions. *Frontiers in Neuroscience*, *17*, 1258393.

Castiello, U., & Umiltà, C. (1992). Splitting focal attention. *Journal of Experimental Psychology: Human Perception and Performance*, *18*(3), 837–848.

Chambers, C. D., Payne, J. M., & Mattingley, J. B. (2007). Parietal disruption impairs reflexive spatial attention within and between sensory modalities. *Neuropsychologia*, *45*, 1715–1724.

Chambers, C. D., Payne, J. M., Stokes, M. G., & Mattingley, J. B. (2004). Fast and slow parietal pathways mediate spatial attention. *Nature Neuroscience*, *7*(3), 217–218.

Chambers, C. D., Stokes, M. G., & Mattingley, J. B. (2004). Modality-specific control of strategic spatial attention in parietal cortex. *Neuron*, *44*, 925–930.

Chan, M. M., & Han, Y. M. (2020). Differential mirror neuron system (MNS) activation during action observation with and without social-emotional components in autism: A meta-analysis of neuroimaging studies. *Molecular Autism*, *11*, 1–18.

Chanon, V. M., & Hopfinger, J. B. (2008). Memory's grip on attention: The influence of item memory on the allocation of attention. *Visual Cognition*, *16*, 325–340.

Chanon, V. W., & Hopfinger, J. B. (2011). ERPs reveal similar effects of social gaze orienting and voluntary attention, and distinguish each from reflexive attention. *Attention, Perception & Psychophysics*, *73*(8), 2502–2513.

Chapman, R. M., & Bragdon, H. R. (1964). Evoked responses to numerical and non-numerical visual stimuli while problem solving. *Nature*, *203*(4950), 1155–1157.

Chawla, D., Rees, G., & Friston, K. J. (1999). The physiological basis of attentional modulation in extrastriate visual areas. *Nature Neuroscience*, *2*, 671–676.

Cheal, M. L., & Lyon, D. R. (1991). Central and peripheral precuing of forced-choice discrimination. *Quarterly Journal of Experimental Psychology*, *43A*, 859–880.

Chen, P., & Goedert, K. M. (2012). Clock drawing in spatial neglect: A comprehensive analysis of clock perimeter, placement, and accuracy. *Journal of Neuropsychology*, *6*, 270–289.

Chen, Y. P., Ehlers, A., Clark, D. M., & Mansell, W. (2002). Patients with generalized social phobia direct their attention away from faces. *Behaviour Research and Therapy*, *40*, 677–687.

Cherry, E. C. (1953). Some experiments on the recognition of speech, with one and with two ears. *Journal of the Acoustical Society of America*, *25*, 975–979.

Chica, A. B., & Lupiáñez, J. (2009). Effects of endogenous and exogenous attention on visual processing: An inhibition of return study. *Brain Research*, *1278*, 75–85.

Chica, A. B., Bartolomeo, P., & Valero-Cabré, A. (2011). Dorsal and ventral parietal contributions to spatial orienting in the human brain. *Journal of Neuroscience*, *31*, 8143–8149.

Chica, A. B., Taylor, T. L., Lupiáñez, J., & Klein, R. M. (2010). Two mechanisms underlying inhibition of return. *Experimental Brain Research*, *201*(1), 25–35.

Chien, H. Y., Lin, H. Y., Lai, M. C., Gau, S. S. F., & Tseng, W. Y. I. (2015). Hyperconnectivity of the right posterior temporo-parietal junction predicts social difficulties in boys with autism spectrum disorder. *Autism Research*, *8*, 427–441.

Chisholm, J. D., & Kingstone, A. (2012). Improved top-down control reduces oculomotor capture: the case of action video game players. *Attention, Perception & Psychophysics*, *74*, 257–262.

Choi, E., Shin, S. H., Ryu, J. K., Jung, K. I., Kim, S. Y., & Park, M. H. (2020).

Commercial video games and cognitive functions: Video game genres and modulating factors of cognitive enhancement. *Behavioral and Brain Functions, 16*(1), 2.

Chomsky, N. (1959). Review of Skinner's verbal behavior. *Language, 35,* 26–58.

Christie, J., & Klein, R. (1995). Familiarity and attention: Does what we know affect what we notice? *Memory and Cognition, 2,* 547550.

Chua, F. K. (2013). Attentional capture by onsets and offsets. *Visual Cognition, 21*(5), 569–598.

Chun, M. M., & Jiang, Y. (1998). Contextual cueing: Implicit learning and memory of visual context guides spatial attention. *Cognitive Psychology, 36,* 28–71.

Chun, M. M., & Potter, M. C. (1995). A two-stage model for multiple target detection in rapid serial visual presentation. *Journal of Experimental Psychology: Human Perception and Performance, 21,* 109–127.

Chun, M. M., Golomb, J. D., & Turk-Browne, N. B. (2011). A taxonomy of external and internal attention. *Annual Review of Psychology, 62*(1), 73–101.

Clark, A. (2016). *Surfing Uncertainty: Prediction, Action, and the Embodied Mind.* Oxford University Press.

Clark, V. P., Fan, S., & Hillyard, S. A. (1994). Identification of early visual evoked potential generators by retinotopic and topographic analyses. *Human Brain Mapping, 2*(3), 170–187.

Cohen, E. H., & Tong, F. (2015). Neural mechanisms of object-based attention. *Cerebral Cortex (New York, N.Y. 1991), 25* (4), 1080–1092.

Cohen, J. R., & D'Esposito, M. (2016). The segregation and integration of distinct brain networks and their relationship to cognition. *The Journal of Neuroscience, 36*(48), 12083–12094.

Comerchero, M. D., & Polich, J. (1999). P3a and P3b from typical auditory and visual stimuli. *Clinical Neurophysiology, 110*(1), 24–30.

Conway, A. R. A., Cowan, N., & Bunting, M. F. (2001). The cocktail party phenomenon revisited: The importance of working memory capacity. *Psychonomic Bulletin & Review, 8*(2), 331–335.

Cooper, A. C. G., Humphreys, G. W., Hulleman, J., Praamstra, P., & Georgeson, M. (2004). Transcranial magnetic stimulation to right parietal cortex modifies the attentional blink. *Experimental Brain Research, 155,* 24–29.

Corbetta, M. (1998). Frontoparietal cortical networks for directing attention and the eye to visual locations: Identical, independent, or overlapping neural systems? *Proceedings of the National Academy of Sciences of the United States of America, 95*(3), 831–838.

Corbetta, M., & Shulman, G. L. (2002). Control of goal-directed and stimulus-driven attention in the brain. *Nature Reviews: Neuroscience, 3*(3), 201–215.

Corbetta, M., Kincade, J. M., Ollinger, J. M., McAvoy, M. P., & Shulman, G. L. (2000). Voluntary orienting is dissociated from target detection in human posterior parietal cortex. *Nature Neuroscience, 3*(3), 292–297.

Corbetta, M., Miezin, F. M., Dobmeyer, S., Shulman, G. L., & Petersen, S. E. (1990). Attentional modulation of neural processing of shape, color, and velocity in humans. *Science, 248*(4962), 1556–1559.

Corbetta, M., Miezin, F., Dobmeyer, S., Shulman, G., & Petersen, S. (1991). Selective and divided attention during visual discriminations of shape, color, and speed: Functional anatomy by positron emission tomography. *The Journal of Neuroscience, 11* (8), 2383–2402.

Corbetta, M., Miezin, F., Shulman, G., & Petersen, S. (1993). A PET study of visuospatial attention. *Journal of Neuroscience, 13,* 1202–1226.

Corbetta, M., Patel, G., & Shulman, G. L. (2008). The reorienting system of the human brain: From environment to theory of mind. *Neuron, 58*(3), 306–324.

Correa, Á., Cona, G., Arbula, S., Vallesi, A., & Bisiacchi, P. (2014). Neural dissociation of automatic and controlled temporal preparation by transcranial magnetic stimulation. *Neuropsychologia, 65,* 131–136.

Cortese, S., Kelly, C., Chabernaud, C., Proal, E., Di Martino, A., Milham, M. P., & Castellanos, F. X. (2012). Toward systems neuroscience of ADHD: A meta-analysis of 55 fMRI studies. *The American Journal of Psychiatry, 169*(10), 1038–1055.

Corthout, E., Uttl, B., Ziemann, U., Cowey, A., & Hallett, M. (1999). Two periods of processing in the (circum)striate visual cortex as revealed by transcranial magnetic stimulation. *Neuropsychologia, 37*(2), 137–145.

Coull, J. T., & Nobre, A. C. (1998). Where and when to pay attention: The neural systems for directing attention to spatial locations and to time intervals as revealed by both PET and fMRI. *Journal of Neuroscience, 18,* 7426–7435.

Coull, J. T., Cotti, J., & Vidal, F. (2016). Differential roles for parietal and frontal cortices in fixed versus evolving temporal expectations: Dissociating prior from posterior temporal probabilities with fMRI. *Neuroimage, 141,* 40–51.

Covic, E. N., & Sherman, S. M. (2011). Synaptic properties of connections between the primary and secondary auditory cortices in mice. *Cerebral Cortex (New York, N. Y. 1991), 21*(11), 2425–2441.

Craik, F. I. M., Eftekhari, E., & Binns, M. A. (2018). Effects of divided attention at encoding and retrieval: Further data. *Memory & Cognition, 46*(8), 1263–1277.

Craik, F. I. M., Govoni, R., Naveh-Benjamin, M., & Anderson, N. D. (1996). The effects of divided attention on encoding and retrieval processes in human memory. *Journal of Experimental Psychology: General, 125*(2), 159–180.

Cravo, A. M., Rohenkohl, G., Santos, K. M., & Nobre, A. C. (2017). Temporal anticipation based on memory. *Journal of Cognitive Neuroscience, 29*(12), 2081–2089.

Crichton, A (1798). An inquiry into the nature and origin of mental derangement: comprehending a concise system of the physiology and pathology of the human mind and a history of the passions and their effects. Printed for T. Cadell, Junior, and W. Davies. [Reprint: Crichton, A. (2008). An inquiry into the nature and origin of mental derangement. On attention and its diseases. *Journal of Attention Disorders, 12,* 200–204.]

Crick, F. (1984). Function of the thalamic reticular complex: The searchlight hypothesis. *Proceedings of the National Academy of Sciences of the United States of America, 81,* 4586–4590.

Cunningham, C. A., & Egeth, H. E. (2016). Taming the white bear: Initial costs and eventual benefits of distractor inhibition. *Psychological Science, 27*(4), 476–485.

Cutrell, E. B., & Marrocco, R. T. (2002). Electrical microstimulation of primate posterior parietal cortex initiates orienting and alerting components of covert attention. *Experimental Brain Research, 144,* 103–113.

Czigler, I., Weisz, J., & Winkler, I. (2006). ERPs and deviance detection: Visual mismatch negativity to repeated visual stimuli. *Neuroscience Letters, 401*(1), 178–182.

Dalebout, S. D., Nelson, N. W., Hletko, P. J., & Frentheway, B. (1991). Selective auditory attention and children with attention-deficit hyperactivity disorder: Effects of repeated measurement with and without methylphenidate. *Language, Speech & Hearing Services in Schools, 22*(4), 219–227.

David, S. P., Naudet, F., Laude, J., Radua, J., Fusar-Poli, P., Chu, I., Steganick, M. L., & Ioannidis, J. P. A. (2018). Potential reporting bias in neuroimaging studies of sex differences. *Scientific Reports, 8*(1), 1–8.

Davidson, M. C., & Marrocco, R. T. (2000). Local infusion of scopolamine into intraparietal cortex slows covert orienting in rhesus monkeys. *Journal of Neurophysiology, 83,* 1536–1549.

Davis, C. J., & Coltheart, M. (2002). Paying attention to reading errors in acquired dyslexia. *Trends in Cognitive Sciences, 6*(9), 359–361.

Dawson, G., Meltzoff, A. N., Osterling, J., Rinaldi, J., & Brown, E. (1998). Children with autism fail to orient to naturally occurring social stimuli. *Journal of Autism and Developmental Disorders, 28*(6), 479–485.

De Fockert, J. W., Rees, G., Frith, C. D., & Lavie, N. (2001). The role of working memory in visual selective attention. *Science, 291*, 1803–1806.

de la Rosa, M. D., Sanabria, D., Capizzi, M., & Correa, A. (2012). Temporal preparation driven by rhythms is resistant to working memory interference. *Frontiers in Psychology, 3*, 308–317.

De Pasquale, R., & Sherman, S. M. (2011). Synaptic properties of corticocortical connections between the primary and secondary visual cortical areas in the mouse. *The Journal of Neuroscience, 31*(46), 16494–16506.

de Schotten, M. T., Dell'Acqua, F., Forkel, S. J., Simmons, A., Vergani, F., Murphy, D. G. M., & Catani, M. (2011). A lateralized brain network for visuospatial attention. *Nature Neuroscience, 14*(10), 1245–1246.

Decety, J., & Lamm, C. (2007). The role of the right temporoparietal junction in social interaction: How low-level computational processes contribute to meta-cognition. *The Neuroscientist, 13*(6), 580–593.

DeGraef, P., Christiaens, D., & d'Ydewalle, G. (1990). Perceptual effects of scene context on object identification. *Psychological Research, 52*, 317329.

Dehaene, S., Posner, M. I., & Tucker, D. M. (1994). Localization of a neural system for error detection and compensation. *Psychological Science, 5*(5), 303–305.

Dell'Acqua, R., Sessa, P., Jolicœur, P., & Robitaille, N. (2006). Spatial attention freezes during the attention blink. *Psychophysiology, 43*(4), 394–400.

Dell'Acqua, R., Sessa, P., Toffanin, P., Luria, R., & Jolicœur, P. (2010). Orienting attention to objects in visual short-term memory. *Neuropsychologia, 48*(2), 419–428.

Della Libera, C., & Chelazzi, L. (2006). Visual selective attention and the effects of monetary rewards. *Psychological Science, 17*(3), 222–227.

Demakis, G. J. (2003). A meta-analytic review of the sensitivity of the Wisconsin card sorting test to frontal and lateralized frontal brain damage. *Neuropsychology, 17*(2), 255–264.

Deng, Z., Lisanby, S. H., & Peterchev, A. V. (2013). Electric field depth–focality tradeoff in transcranial magnetic stimulation: Simulation comparison of 50 coil designs. *Brain Stimulation, 6*(1), 1–13.

Denison, R. N., Heeger, D. J., & Carrasco, M. (2017). Attention flexibly trades off across points in time. *Psychonomic Bulletin & Review, 24*, 1142–1151.

Dent, K. (2018). Priming of pop-out does not provide reliable measures of target activation and distractor inhibition in selective attention: Evidence from a large-scale online study. *Vision Research, 149*, 124–130.

Desimone, R., & Duncan, J. (1995). Neural mechanisms of selective visual attention. *Annual Review of Neuroscience, 18*, 193–222.

Deubel, H., & Schneider, W. X. (1996). Saccade target selection and object recognition: Evidence for a common attentional mechanism. *Visual Research, 36*, 1827–1837.

Deutsch, J. A., & Deutsch, D. (1963). Attention: Some theoretical considerations. *Psychological Review, 70*, 80–90.

Di Russo, F., Martínez, A., & Hillyard, S. A. (2003). Source analysis of event-related cortical activity during visuo-spatial attention. *Cerebral Cortex, 13*(5), 486–499.

Diaconescu, A. O., Wellstein, K. V., Kasper, L., Mathys, C., & Stephan, K. E. (2020). Hierarchical Bayesian models of social inference for probing persecutory. *Journal of Abnormal Psychology, 129*, 556–569.

Diamond, A. (1990). Developmental time course in human infants and infant monkeys, and the neural bases, of inhibitory control in reaching. *Annals of the New York Academy of Sciences, 608*, 637–676.

Diamond, A., & Boyer, K. (1989). A version of the Wisconsin Card Sort Test for use with preschool children, and an exploration of their sources of error. *Journal of Clinical and Experimental Neuropsychology, 11*, 83.

Diamond, A., Carlson, S. M., & Beck, D. M. (2005). Preschool children's performance in task switching on the dimensional change card sort task: Separating the dimensions aids the ability to switch. *Developmental Neuropsychology, 28*, 689–729.

DiQuattro, N. E., & Geng, J. J. (2011). Contextual knowledge configures attentional control networks. *Journal of Neuroscience, 31*(49), 18026–18035.

Doesburg, S. M., Roggeveen, A. B., Kitajo, K., & Ward, L. M. (2008). Large-scale gamma-band phase synchronization and selective attention. *Cerebral Cortex, 18*(2), 386–396.

Doherty, J. R., Rao, A., Mesulam, M. M., & Nobre, A. C. (2005). Synergistic effect of combined temporal and spatial expectations on visual attention. *The Journal of Neuroscience, 25*(36), 8259–8266.

Donchin, E. (1981). Presidential Address, 1980: Surprise! . . . Surprise? *Psychophysiology, 18*(5), 493–513.

Donchin, E., & Coles, M. G. H. (1988). Is the P300 component a manifestation of context updating? *Behavioral and Brain Sciences, 11*, 357–374.

Donders, F. C. (1868/1969). On the speed of mental processes, *Acta Psychologica, 30*, 412–431.

Donohue, S. E., Schoenfeld, M. A., & Hopf, J. M. (2020). Parallel fast and slow recurrent cortical processing mediates target and distractor selection in visual search. *Communications Biology, 3*(1), 689.

Doricchi, F., & Tomaiuolo, F. (2003). The anatomy of neglect without hemianopia: A key role for parietal-frontal disconnection? *Neuroreport 14*, 2239–2243.

Doricchi, F., Macci, E., Silvetti, M., & Macaluso, E. (2010). Neural correlates of the spatial and expectancy components of endogenous and stimulus-driven orienting of attention in the Posner task. *Cerebral Cortex, 20*(7), 1574–1585.

Downar, J., Crawley, A. P., Mikulis, D. J., & Davis, K. D. (2000). A multimodal cortical network for the detection of changes in the sensory environment. *Nature Neuroscience, 3*, 277–283.

Downing, P. E. (2000). Interactions between visual working memory and selective attention. *Psychological Science, 11*, 467–473.

Driver, J., Davis, G., Ricciardelli, P., Kidd, P., Maxwell, E., & Baron-Cohen, S. (1999). Gaze perception triggers reflexive visuospatial orienting. *Visual Cognition, 6*, 509–540.

Dubois, J., Hamker, F. H., & VanRullen, R. (2009). Attentional selection of noncontiguous locations: The spotlight is only transiently "split". *Journal of Vision, 9*(5), 1–11.

Duecker, F., Formisano, E., & Sack, A. T. (2013). Hemispheric differences in the voluntary control of spatial attention: Direct evidence for a right-hemispheric dominance within frontal cortex. *Journal of Cognitive Neuroscience, 25*, 1332–1342.

Dugué, L., Beck, A. A., Marque, P., & VanRullen, R. (2019). Contribution of FEF to attentional periodicity during visual search: A TMS study. *ENeuro, 6*(3), 1–10.

Dugué, L., Marque, P., & VanRullen, R. (2015). Theta oscillations modulate attentional search performance periodically. *Journal of Cognitive Neuroscience, 27*(5), 945–958.

Dugué, L., Roberts, M., & Carrasco, M. (2016). Attention reorients periodically. *Current Biology, 26*(12), 1595–1601.

Duncan, J., Bundesen, C., Olson, A., Humphreys, G., Ward, R., Kyllingsbæk, S., van Raamsdonk, M., Rorden, C., & Chavda, S. (2003). Attentional functions in dorsal and ventral simultanagnosia. *Cognitive Neuropsychology, 20*(8), 675–701.

Duncan, J., Martens, S., & Ward, R. (1997). Restricted attentional capacity within but not between sensory modalities. *Nature, 387*, 808–810.

Duque, J., Petitjean, C., & Swinnen, S. P. (2016). Effect of aging on motor inhibition during action preparation under sensory conflict. *Frontiers in Aging Neuroscience, 8*, 322.

Dye, M. W. G., Green, C. S., & Bavelier, D. (2009). Increasing speed of processing with action video games. *Current Directions in Psychological Science: A Journal of the American Psychological Society, 18*(6), 321–326.

Eason, R. G., Harter, M., & White, C. (1969). Effects of attention and arousal on visually evoked cortical potentials. *Physiology and Behavior, 4*, 283–289.

Eastwood, J. D., Smilek, D., & Merikle, P. M. (2001). Differential attentional guidance by unattended faces expressing positive and negative emotion. *Perception & Psychophysics, 63*, 1004–1013.

Egeland, J., Johansen, S. N., & Ueland, T. (2009). Differentiating between ADHD sub-types on CCPT measures of sustained attention and vigilance. *Scandinavian Journal of Psychology, 50*(4), 347–354.

Egly, R., Driver, J., & Rafal, R. D. (1994). Shifting visual attention between objects and locations: Evidence from normal and parietal lesion subjects. *Journal of Experimental Psychology: General, 123*(2), 161–177.

Eichenbaum, H. (2014). Time cells in the hippocampus: a new dimension for mapping memories. *Nature Reviews: Neuroscience, 15*, 732–744.

Eimer, M. (1995). Event-related potential correlates of transient attention shifts to color and location. *Biological Psychology, 41*(2), 167–182.

Eimer, M. (1996). The N2pc as an indicator of attentional selectivity. *Electroencephalography and Clinical Neurophysiology, 99*(3), 225–234.

Eimer, M. (1999). Attending to quadrants and ring-shaped regions: ERP effects of visual attention in different spatial selection tasks. *Psychophysiology, 36*(4), 491–503.

Eimer, M. (2000). An ERP study of sustained spatial attention to stimulus eccentricity. *Biological Psychology, 52*(3), 205–220.

Eimer, M., & Kiss, M. (2008). Involuntary attentional capture is determined by task set: Evidence from event-related brain potentials. *Journal of Cognitive Neuroscience, 20*(8), 1423–1433.

Eimer, M., Kiss, M., & Cheung, T. (2010). Priming of pop-out modulates attentional target selection in visual search: Behavioural and electrophysiological evidence. *Vision Research, 50*(14), 1353–1361.

Eimer, M., van Velzen, J., & Driver, J. (2002). Cross-modal interactions between audition, touch, and vision in endogenous spatial attention: ERP evidence on preparatory states and sensory modulations. *Journal of Cognitive Neuroscience, 14*, 254–271.

Ekman, M., Kok, P., & De Lange, F. P. (2017). Time-compressed preplay of anticipated events in human primary visual cortex. *Nature Communications, 8*(1), 15276.

Ekman, P. (1992). An argument for basic emotions. *Cognition & Emotion, 6*(3), 169–200.

Elahipanah, A., Christensen, B. K., & Reingold, E. M. (2010). Visual search performance among persons with schizophrenia as a function of target eccentricity. *Neuropsychology, 24*(2), 192–198.

Elias, P. (1955). Predictive coding – I. *IEEE Transactions on Information Theory, 1*(1), 16–24.

Ellison, A., Schindler, I., Pattison, L. L., & Milner, A. D. (2004). An exploration of the role of the superior temporal gyrus in visual search and spatial perception using TMS. *Brain, 127*(10), 2307–2315.

Engbert, R., & Kliegl, R. (2003). Microsaccades uncover the orientation of covert attention. *Vision Research, 43*, 1035–1045

Engel, S. A., Glover, G. H., & Wandell, B. A. (1997). Retinotopic organization in human visual cortex and the spatial precision of functional MRI. *Cerebral Cortex, 7*(2), 181–192.

Enriquez-Geppert, S., Konrad, C., Pantev, C., & Huster, R. J. (2010). Conflict and inhibition differentially affect the N200/P300 complex in a combined go/nogo and stop-signal task. *Neuroimage, 51*(2), 877–887.

Ergenoglu, T., Demiralp, T., Bayraktaroglu, Z., Ergen, M., Beydagi, H., & Uresin, Y. (2004). Alpha rhythm of the EEG modulates visual detection performance in humans. *Brain Research. Cognitive Brain Research, 20*(3), 376–383.

Eriksen, B. A., & Eriksen, C. W. (1974). Effects of noise letters upon identification of a target letter in a non-search task. *Perception and Psychophysics, 16*, 143–149.

Eriksen, C. W., & St. James, J. D. (1986). Visual attention within and around the field of focal attention: A zoom lens model. *Perception & Psychophysics, 40*, 225–240.

Eriksen, C. W., & Yeh, Y. Y. (1985). Allocation of attention in the visual field. *Journal of Experimental Psychology: Human Perception and Performance, 11*, 583–597.

Erişir, A., Van Horn, S. C., Bickford, M. E., & Sherman, S. M. (1997). Immunocytochemistry and distribution of parabrachial terminals in the lateral geniculate nucleus of the cat: A comparison with corticogeniculate terminals. *Journal of Comparative Neurology (1911), 377*(4), 535.

Esterman, M., Thai, M., Okabe, H., DeGutis, J., Saad, E., Laganiere, S. E., & Halko, M. A. (2017). Network-targeted cerebellar transcranial magnetic stimulation improves attentional control. *NeuroImage, 156*, 190–198.

Esteves, F., Dimberg, U., & Ohman, A. (1994). Automatically elicited fear: Conditioned skin conductance responses to masked facial expressions. *Cognition & Emotion, 8*, 393–413.

Failing, M., & Theeuwes, J. (2017). Don't let it distract you: how information about the availability of reward affects attentional selection. *Attention Perception & Psychophysics, 79*(8), 2275–2298.

Falkenstein, M., Hohnsbein, J., Hoormann, J., & Blanke, L. (1990). Effects of errors in choice reaction tasks on the ERP under focused and divided attention. In:

Brunia, C. H. M., Gaillard, A. W. K., & Kok, A. (eds.), *Psychological Brain Research* (pp. 192–195). Tilburg University Press.

Falkenstein, M., Hohnsbein, J., Hoormann, J., & Blanke, L. (1991). Effects of crossmodal divided attention on late ERP components. II. Error processing in choice reaction tasks. *Electroencephalography & Clinical Neurophysiology, 78*, 447–455.

Falkenstein, M., Hoormann, J., Christ, S., & Hohnsbein, J. (2000). ERP components on reaction errors and their functional significance: A tutorial. *Biological Psychology, 51*(2), 87–107.

Fan, J., McCandliss, B. D., Sommer, T., Raz, A., & Posner, M. I. (2002). Testing the efficiency and independence of attentional networks. *Journal of Cognitive Neuroscience, 14*(3), 340–347.

Farahbod, H., Saberi, K., & Hickok, G. (2020). The rhythm of attention: Perceptual modulation via rhythmic entrainment is lowpass and attention mediated. *Attention, Perception & Psychophysics, 82*(7), 3558–3570.

Farrant, K., & Uddin, L. Q. (2015). Asymmetric development of dorsal and ventral attention networks in the human brain. *Developmental Cognitive Neuroscience, 12*, 165–174.

Feigenbaum, E. A., & Feldman, J. (1963). *Computers and Thought*. McGraw-Hill.

Feldman, H., & Friston, K.J. (2010). Attention, uncertainty, and free-energy. *Frontiers in Human Neuroscience, 4*, 215.

Feldmann-Wüstefeld, T., Busch, N. A., & Schubö, A. (2020). Failed suppression of salient stimuli precedes behavioral errors. *Journal of Cognitive Neuroscience, 32*(2), 367–377.

Felleman, D. J., & Van Essen, D. C. (1991). Distributed hierarchical processing in the primate cerebral cortex. *Cerebral Cortex, 1*(1), 1–47.

Fenker, D. B., Heipertz, D., Boehler, C. N., Schoenfeld, M. A., Noesselt, T., Heinze, H., Duezel, E., & Hopf, J. (2010). Mandatory processing of irrelevant fearful face features in visual search. *Journal of Cognitive Neuroscience, 22*(12), 2926–2938.

Fernández, A., Okun, S., & Carrasco, M. (2022). Differential effects of endogenous and exogenous attention on sensory tuning. *The Journal of Neuroscience*, *42*(7), 1316–1327.

Fernández-Folgueiras, U., Hernández-Lorca, M., Méndez-Bértolo, C., Álvarez, F., Giménez-Fernández, T., & Carretié, L. (2022). Exogenous attention to emotional stimuli presenting realistic (3D) looming motion. *Brain Topography*, *35*(5–6), 599–612.

Fichtenholtz, H. M., Hopfinger, J. B., Graham, R., Detwiler, J. M., & LaBar, K. S. (2007). Happy and fearful emotion in cues and targets modulates event-related potential indices of gaze-directed attentional orienting. *Social Cognitive and Affective Neuroscience*, *2*, 323–333.

Fichtenholtz, H. M., Hopfinger, J. B., Graham, R., Detwiler, J. M., & LaBar, K. S. (2009). Event-related potentials reveal temporal staging of dynamic emotional expression and gaze shift effects on attentional orienting. *Social Neuroscience*, *4*(4), 317–331.

Fiebelkorn, I. C. (2022a). Detecting attention-related rhythms: When is behavior not enough? (commentary on van der Werf et al. 2021). *The European Journal of Neuroscience*, *55*(11–12), 3117–3120.

Fiebelkorn, I. C. (2022b). There is more evidence of rhythmic attention than can be found in behavioral studies: Perspective on Brookshire, 2022. *Journal of Cognitive Neuroscience*, *35*(1), 128–134.

Fiebelkorn, I. C., & Kastner, S. (2019). A rhythmic theory of attention. *Trends in Cognitive Sciences*, *23*, 87–101.

Fiebelkorn, I. C., & Kastner, S. (2020). Functional specialization in the attention network. *Annual Review of Psychology*, *71*(1), 221–249.

Fiebelkorn, I. C., Pinsk, M. A., & Kastner, S. (2018). A dynamic interplay within the frontoparietal network underlies rhythmic spatial attention. *Neuron (Cambridge, Mass.)*, *99*(4), 842–853.e8.

Fiebelkorn, I. C., Pinsk, M. A., & Kastner, S. (2019). The mediodorsal pulvinar coordinates the macaque fronto-parietal network during rhythmic spatial attention. *Nature Communications*, *10*(1), 215.

Fiebelkorn, I., Saalmann, Y., & Kastner, S. (2013). Rhythmic sampling within and between objects despite sustained attention at a cued location. *Current Biology*, *23*(24), 2553–2558.

Field, M., & Cox, W. M. (2008). Attentional bias in addictive behaviors: A review of its development, causes, and consequences. *Drug and Alcohol Dependence*, *97*(1), 1–20.

Fischer, B., & Weber, H. (1993). Express saccades and visual attention. *Behavioral and Brain Sciences*, *16*(3), 553–567.

Fitch, A., Lieberman, A. M., Luyster, R. J., & Arunachalam, S. (2020). Toddlers' word learning through overhearing: Others' attention matters. *Journal of Experimental Child Psychology*, *193*, 104793.

Fitzgerald, J., Johnson, K., Kehoe, E., Bokde, A. L., Garavan, H., Gallagher, L., & McGrath, J. (2015). Disrupted functional connectivity in dorsal and ventral attention networks during attention orienting in autism spectrum disorders. *Autism Research*, *8*, 136–152.

Fletcher, P. C., & Frith, C. D. (2009). Perceiving is believing: A Bayesian approach to explaining the positive symptoms of schizophrenia. *Nature Reviews: Neuroscience*, *10*, 48–58.

Föcker, J., Cole, D., Beer, A. L., & Bavelier, D. (2018). Neural bases of enhanced attentional control: Lessons from action video game players. *Brain and Behavior*, *8*(7), e01019.

Föcker, J., Mortazavi, M., Khoe, W., Hillyard, S. A., & Bavelier, D. (2019). Neural correlates of enhanced visual attentional control in action video game players: An event-related potential study. *Journal of Cognitive Neuroscience*, *31*, 377–389.

Foerde, K., Knowlton, B. J., & Poldrack, R. A. (2006). Modulation of competing memory systems by distraction. *Proceedings of the*

National Academy of Sciences of the United States of America, *103*(31), 11778–11783.

Folk, C. L., & Remington, R. W. (1999). Can new objects override attentional control settings? *Perception & Psychophysics*, *61*(4), 727–739.

Folk, C. L., & Remington, R. W. (2015). Unexpected abrupt onsets can override a top-down set for color. *Journal of Experimental Psychology: Human Perception and Performance*, *41*(4), 1153–1165.

Folk, C. L., Remington, R. W., & Johnston, J. C. (1992). Involuntary covert orienting is contingent on attentional control settings. *Journal of Experimental Psychology: Human Perception and Performance*, *18*(4), 1030–1044.

Folk, C. L., Remington, R. W., & Wright, J. H. (1994). The structure of attentional control: Contingent attentional capture by apparent motion, abrupt onset, and color. *Journal of Experimental Psychology: Human Perception and Performance*, *20*, 317–329.

Forster, B., Sambo, C. F., & Pavone, E. F. (2009). ERP correlates of tactile spatial attention differ under intra- and intermodal conditions. *Biological Psychology*, *82*(3), 227–233.

Forte, A. E., Octave, E., & Reichenbach, T. (2017). The human auditory brainstem response to running speech reveals a subcortical mechanism for selective attention. *ELife*, *6*, e27203.

Fortenbaugh, F. C., DeGutis, J., Germine, L., Wilmer, J., Grosso, M., Russo, K., & Esterman, M. (2015). Sustained attention across the life span in a sample of 10,000: Dissociating ability and strategy. *Psychological Science*, *26*(9), 1497–1510.

Fossella, J., Sommer, T., Fan, J., Wu, Y., Swanson, J. M., Pfaff, D. W., & Posner, M. I. (2002). Assessing the molecular genetics of attention networks. *BMC Neuroscience*, *3*(1), 14.

Foster, B. L., Koslov, S. R., Aponik-Gremillion, L., Monko, M. E., Hayden, B. Y., & Heilbronner, S. R. (2023). A tripartite view of the posterior cingulate cortex. *Nature Reviews: Neuroscience*, *24*(3), 173–189.

Foster, J. J., & Ling, S. (2022). Feature-based attention multiplicatively scales the fMRI-BOLD contrast-response function. *The Journal of Neuroscience*, *42*(36), 6894–6906.

Fox, E., Lester, V., Russo, R., Bowles, R. J., Pichler, A., & Dutton, K. (2000). Facial expressions of emotion: Are angry faces detected more efficiently? *Cognition & Emotion*, *14*, 61–92.

Fox, E., Russo, R., & Dutton, K. (2002). Attentional bias for threat: Evidence for delayed disengagement from emotional faces. *Cognition & Emotion*, *16*, 355–379.

Foxe, J. J., & Snyder, A. C. (2011). The role of alpha-band brain oscillations as a sensory suppression mechanism during selective attention. *Frontiers in Psychology*, *2*, 154.

Franconeri, S. L., & Simons, D. J. (2003). Moving and looming stimuli capture attention. *Perception & Psychophysics*, *65*, 999–1010.

Friedman-Hill, S. R., Robertson, L. C., & Treisman, A. (1995). Parietal contributions to visual feature binding: Evidence from a patient with bilateral lesions. *Science*, *269*(5225), 853–855.

Fries, P. (2009). Neuronal gamma-band synchronization as a fundamental process in cortical computation. *Annual Review of Neuroscience*, *32*, 209–224.

Fries, P., Womelsdorf, T., Oostenveld, R., & Desimone, R. (2008). The effects of visual stimulation and selective visual attention on rhythmic neuronal synchronization in macaque area V4. *The Journal of Neuroscience*, *28*(18), 4823–4835.

Friesen, C. K., & Kingstone, A. (1998). The eyes have it! Reflexive orienting is triggered by nonpredictive gaze. *Psychonomic Bulletin & Review*, *5*(3), 490–495.

Friesen, C. K., & Kingstone, A. (2003). Abrupt onsets and gaze direction cues trigger independent reflexive attentional effects. *Cognition*, *87*, 1–10.

Friesen, C. K., Ristic, J., & Kingstone, A. (2004). Attentional effects of counterpredictive gaze and arrow cues. *Journal of Experimental Psychology: Human Perception and Performance*, *30*, 319–329.

Frischen, A., Smilek, D., Eastwood, J. D., & Tipper, S. P. (2007). Inhibition of return in response to gaze cues: The roles of time course and fixation cue. *Visual Cognition*, *15*(8), 881–895.

Friston, K. (2008). Hierarchical models in the brain. *PLoS Computational Biology*, *4*, e1000211.

Friston, K. (2009). The free-energy principle: A rough guide to the brain? *Trends in Cognitive Science*, *13*, 293–301.

Friston, K., & Kiebel, S. (2009). Predictive coding under the free-energy principle. *Philosophical Transactions of the Royal Society of London B: Biological Sciences*, *364*, 1211–1221.

Friston, K. J., Daunizeau, J., Kilner, J., & Kiebel, S. J. (2010). Action and behavior: A free-energy formulation. *Biological Cybernetics*, *102*, 227–260.

Frost, B. G., Neill, R. A., & Fenelon, B. (1988). The determinants of the non-motoric CNV in a complex, variable foreperiod, information processing paradigm. *Biological Psychology*, *27*(1), 1–21.

Fu, S., Fan, S., Chen, L., & Zhuo, Y. (2001). The attentional effects of peripheral cueing as revealed by two event-related potential studies. *Clinical Neurophysiology*, *112*(1), 172–185.

Fu, S., Greenwood, P. M., & Parasuraman, R. (2005). Brain mechanisms of involuntary visuospatial attention: An event-related potential study. *Human Brain Mapping*, *25*(4), 378–390.

Fuller, R. L., Luck, S. J., Braun, E. L., Robinson, B. M., McMahon, R. P. & Gold, J.M. (2006). Impaired control of visual attention in schizophrenia. *Journal of Abnormal Psychology*, *115*(2), 266–275.

Gallant, S. (2016). Mindfulness meditation practice and executive functioning: Breaking down the benefit. *Consciousness and Cognition*, *40*, 116–130.

Gallotto, S., Schuhmann, T., Duecker, F., Middag-van Spanje, M., de Graaf, T. A., & Sack, A. T. (2022). Concurrent frontal and parietal network TMS for modulating attention. *iScience*, *25*(3), 103962.

Gamble, K. R., Howard, J. H., & Howard, D. V. (2014). Not just scenery: Viewing nature pictures improves executive attention in older adults. *Experimental Aging Research*, *40*(5), 513–530.

Ganguly, J., Murgai, A., Sharma, S., Aur, D., & Jog, M. (2020). Non-invasive transcranial electrical stimulation in movement disorders. *Frontiers in Neuroscience*, *14*, 522.

Gao, Y. X., Wang, J. Y., & Dong, G. H. (2022). The prevalence and possible risk factors of internet gaming disorder among adolescents and young adults: Systematic reviews and meta-analyses. *Journal of Psychiatric Research*, *154*, 35–43.

Garcia-Lazaro, H. G., Bartsch, M. V., Boehler, C. N., Krebs, R. M., Donohue, S. E., Harris, J. A., Schoenfeld, M. A., & Hopf, J. (2019). Dissociating reward- and attention-driven biasing of global feature-based selection in human visual cortex. *Journal of Cognitive Neuroscience*, *31*(4), 469–481.

Garrison, K. A., Pal, P., Rojiani, R., Dallery, J., O'Malley, S. S., & Brewer, J. A. (2015). A randomized controlled trial of smart-phone-based mindfulness training for smoking cessation: a study protocol. *BMC Psychiatry*, *15*, 83.

Gaspar, J. M., & McDonald, J. J. (2014). Suppression of salient objects prevents distraction in visual search. *Journal of Neuroscience*, *34*, 5658–5666.

Gaspar, J. M., Christie, G. J., Prime, D. J., Jolicœur, P., & McDonald, J. J. (2016). Inability to suppress salient distractors predicts low visual working memory capacity. *Proceedings of the National Academy of Sciences of the United States of America*, *113*, 3693–3698.

Gaspelin, N., & Luck, S. J. (2018). Combined electrophysiological and behavioral evidence for the suppression of salient distractors. *Journal of Cognitive Neuroscience, 30*(9), 1265–1280.

Gaspelin, N., Ruthruff, E., & Lien, M. (2016). The problem of latent attentional capture: Easy visual search conceals capture by task-irrelevant abrupt onsets. *Journal of Experimental Psychology: Human Perception and Performance, 42*(8), 1104–1120.

Gates, K. M., & Molenaar, P. C. M. (2012). Group search algorithm recovers effective connectivity maps for individuals in homogeneous and heterogeneous samples. *NeuroImage, 63*, 310–319.

Gates, K. M., Molenaar, P. C. M., Iyer, S. P., Nigg, J. T., & Fair, D. A. (2014). Organizing heterogeneous samples using community detection of GIMME-derived resting state functional networks. *PLoS One, 9*(3), e91322.

Gazzaley, A. (2011). Influence of early attentional modulation on working memory. *Neuropsychologia, 49*, 1410–1424.

Gazzaley, A., & Nobre, A. C. (2012). Top-down modulation: Bridging selective attention and working memory. *Trends in Cognitive Sciences, 16*(2), 129–135.

Gazzaley, A., Cooney, J. W., McEvoy, K., Knight, R. T., & D'Esposito, M. (2005). Top-down enhancement and suppression of the magnitude and speed of neural activity. *Journal of Cognitive Neuroscience, 17*(3), 507–517.

Gazzaniga, M. S., Bogen, J. E., & Sperry, R. W. (1962). Some functional effects of sectioning the cerebral commissures in man. *Proceedings of the National Academy of Sciences of the United States of America, 48* (10), 1765–1769.

Gefen, T., Peterson, M., Papastefan, S. T., Martersteck, A., Whitney, K., Rademaker, A., Bigio, E. H., Weintraub, S., Rogalski, E., Mesulam, M. M., & Geula, C. (2015). Morphometric and histologic substrates of cingulate integrity in elders with exceptional memory capacity. *Journal of Neuroscience, 35*(4), 1781–1791.

Gehring, W. J., & Willoughby, A. R. (2002). The medial frontal cortex and the rapid processing of monetary gains and losses. *Science (American Association for the Advancement of Science), 295*(5563), 2279–2282.

Gehring, W. J., Coles, M., Meyer, D., & Donchin, E. (1990). The error-related negativity: An event-related brain potential accompanying errors. *Psychophysiology, 27*, S34.

Gehring, W. J., Goss, B., Coles, M. G. H., Meyer, D. E., & Donchin, E. (1993). A neural system for error detection and compensation. *Psychological Science, 4*, 385–390.

Geng, J. J., & Mangun, G. R. (2011). Right temporoparietal junction activation by a salient contextual cue facilitates target discrimination. *NeuroImage 54*(1), 594–601.

Gennari, F. (1782). *De peculiari structura cerebri nonnullisque ejus morbis.* Ex regio typographeo.

Gershon, J. (2002). Gender differences in ADHD. *The ADHD Report, 10*(4), 8–16.

Geva, R., Zivan, M., Warsha, A., & Olchik, D. (2013). Alerting, orienting or executive attention networks: Differential patters of pupil dilations. *Frontiers in Behavioral Neuroscience, 7*, 145.

Geyer, T., Müller, H. J., & Krummenacher, J. (2008). Expectancies modulate attentional capture by salient color singletons. *Vision Research, 48*(11), 1315–1326.

Ghose, G. M., & Maunsell, J. H. R. (2002). Attentional modulation in visual cortex depends on task timing. *Nature, 419*, 616–620.

Gibb, B. E., McGeary, J. E., & Beevers, C. G. (2016). Attentional biases to emotional stimuli: Key components of the RDoC constructs of sustained threat and loss. *American Journal of Medical Genetics Part B: Neuropsychiatric Genetics, 171*(1), 65–80.

Gibb, B. E., Pollak, S. D., Hajcak, G., & Owens, M. (2016). Attentional biases in children of depressed mothers: An event-related potential (ERP) study. *Journal of Abnormal Psychology, 125*(8), 1166–1178.

Giesbrecht, B., Sy, J. L., & Elliott, J. C. (2007). Electrophysiological evidence for both perceptual and postperceptual selection during the attentional blink. *Journal of Cognitive Neuroscience, 19*(12), 2005–2018.

Giesbrecht, B., Woldorff, M. G., Song, A. W., & Mangun, G. R. (2003). Neural mechanisms of top-down control during spatial and feature attention. *NeuroImage (Orlando, Fla.), 19*(3), 496–512.

Gitelman, D. R., Nobre, A. C., Parrish, T. B., LaBar, K. S., Kim, Y., Meyer, J. R., & Mesulam, M. (1999). A large-scale distributed network for covert spatial attention. further anatomical delineation based on stringent behavioural and cognitive controls. *Brain, 122*(6), 1093–1106.

Gledhill, D., Grimsen, C., Fahle, M., & Wegener, D. (2015). Human feature-based attention consists of two distinct spatiotemporal processes. *Journal of Vision, 15*(8), 1–17.

Goldsmith, M. (1998). What's in a location? Comparing object-based and space-based models of feature integration in visual search. *Journal of Experimental Psychology: General, 127*(2), 189–219.

Gong, D., Ma, W., Gong, J., He, H., Dong, L., Zhang, D., Li, J., Luo, C., & Yao, D. (2017). Action video game experience related to altered large-scale white matter networks. *Neural Plasticity, 2017*, 7543686.

Gong, M., & Liu, T. (2018). Reward differentially interacts with physical salience in feature-based attention. *Journal of Vision, 18*(11), 1–12.

Gonon, F. (2009). The dopaminergic hypothesis of attention-deficit/hyperactivity disorder needs re-examining. *Trends in Neuroscience, 32*, 2–8.

Gordon, N., Tsuchiya, N., Koenig-Robert, R., & Hohwy, J. (2019). Expectation and attention increase the integration of top-down and bottom-up signals in perception through different pathways. *PLoS Biology, 17*(4), e3000233.

Gottlieb, J., Balan, P., Oristaglio, J., & Suzuki, M. (2009). Parietal control of attentional guidance: The significance of sensory, motivational and motor factors. *Neurobiology of Learning and Memory, 91*, 121–128.

Grady, C. L., Van Meter, J. W., Maisog, J. M., Pietrini, P., Krasuski, J., & Rauschecker, J. P. (1997). Attention-related modulation of activity in primary and secondary auditory cortex. *Neuroreport, 8*, 2511–2516.

Grafton, B., MacLeod, C., Rudaizky, D., Holmes, E. A., Salemink, E., Fox, E., & Notebaert, L. (2017). Confusing procedures with process when appraising the impact of cognitive bias modification on emotional vulnerability. *British Journal of Psychiatry, 211*(5), 266–271.

Gratton, G. (1997). Attention and probability effects in the human occipital cortex: An optical imaging study. *Neuroreport, 8*(7), 1749–1753.

Gratton, G., Chiarelli, A. M., & Fabiani, M. (2017). From brain to blood vessels and back: A noninvasive optical imaging approach. *Neurophotonics, 4*(3), 031208.

Gray, C. M., Maldonado, P. E., Wilson, M., & McNaughton, B. (1995). Tetrodes markedly improve the reliability and yield of multiple single-unit isolation from multi-unit recordings in cat striate cortex. *Journal of Neuroscience Methods, 63*(1–2), 43–54.

Green, C. D. (2003). Where did the ventricular localization of mental faculties come from? *Journal of the History of the Behavioral Sciences, 39*(2), 131–142.

Green, J. J., Conder, J. A., & McDonald, J. J. (2008). Lateralized frontal activity elicited by attention-directing visual and auditory cues. *Psychophysiology, 45*, 579–587.

Green, M. (2006). Is the moth effect real? *Accident Reconstruction Journal, 16*(3), 18–19.

Greene, D. J., Colich, N., Iacoboni, M., Zaidel, E., Bookheimer, S. Y., & Dapretto, M. (2011). Atypical neural networks for social orienting in autism spectrum disorders. *Neuroimage, 56,* 354–362.

Greene, N. R., Martin, B. A., & Naveh-Benjamin, M. (2021). The effects of divided attention at encoding and at retrieval on multidimensional source memory. *Journal of Experimental Psychology: Learning, Memory, and Cognition, 47*(11), 1870–1887.

Gregoriou, G. G., Gotts, S. J., Zhou, H., & Desimone, R. (2009). High-frequency, long-range coupling between prefrontal and visual cortex during attention. *Science (American Association for the Advancement of Science), 324*(5931), 1207–1210.

Gregory, R. L., & Gombrich, E. H. (1973). *Illusion in Nature and Art.* Duckworth.

Griffin, I. C., Miniussi, C., & Nobre, A. C. (2002). Multiple mechanisms of selective attention: differential modulation of stimulus processing by attention to space or time. *Neuropsychologia, 40,* 2325–2340.

Grill-Spector, K., Henson, R., & Martin, A. (2006). Repetition and the brain: Neural models of stimulus-specific effects. *Trends in Cognitive Sciences, 10*(1), 14–23.

Grosbras, M.-H., & Paus, T. (2002). Transcranial magnetic stimulation of the human frontal eye field: Effects on visual perception and attention. *Journal of Cognitive Neuroscience, 14,* 1109–1120.

Gruber, S. A., Rogowska, J., Holcomb, P., Soraci, S., & Yurgelun-Todd, D. (2002). Stroop performance in normal control subjects: An fMRI study. *NeuroImage, 16*(2), 349–360.

Gruber, T., Müller, M. M., Keil, A., & Elbert, T. (1999). Selective visual-spatial attention alters induced gamma band responses in the human EEG. *Clinical Neurophysiology, 110*(12), 2074–2085.

Grüner, M., & Ansorge, U. (2017). Mobile eye tracking during real-world night driving: A selective review of findings and recommendations for future research. *Journal of Eye Movement Research, 10*(2), 1–18.

Guarino, A., Favieri, F., Boncompagni, I., Agostini, F., Cantone, M., & Casagrande, M. (2019). Executive functions in Alzheimer disease: A systematic review. *Frontiers in Aging Neuroscience, 10,* 437.

Guillery, R. W. (1970). The laminar distribution of retinal fibers in the dorsal lateral geniculate nucleus of the cat: A new interpretation. *Journal of Comparative Neurology (1911), 138*(3), 339–367.

Haeusler, S., & Maass, W. (2007). A statistical analysis of information-processing properties of lamina-specific cortical microcircuit models. *Cerebral Cortex, 17,* 149–162.

Hafed, Z. M., & Clark, J. J. (2002). Microsaccades as an overt measure of covert attention shifts. *Vision Research, 42,* 2533–2545.

Hahn, B., Robinson, B. M., Kiat, J. E., Geng, J., Bansal, S., Luck, S. J., & Gold, J. M. (2022). Impaired filtering and hyperfocusing: Neural evidence for distinct selective attention abnormalities in people with schizophrenia. *Cerebral Cortex, 32*(9), 1950–1964.

Hahn, E. L. (1950). Nuclear induction due to free Larmor precession. *Physical Review, 77* (2), 297–298.

Haider, M., Lindsley, D. B., & Spong, P. (1965). Selective attentiveness and cortical evoked responses to visual and auditory stimuli (selective short term attentiveness and cortical evoked responses in human beings to alternate flashing light and auditory clicks). *Science, 148,* 395–397.

Halko, M. A., Farzan, F., Eldaief, M. C., Schmahmann, J. D., & Pascual-Leone, A. (2014). Intermittent theta-burst stimulation of the lateral cerebellum increases functional connectivity of the default network. *Journal of Neuroscience 34*(36), 12049–12056.

Hall, D. A., Haggard, M. P., Akeroyd, M. A., Summerfield, A. Q., Palmer, A. R., Elliott, M. R., & Bowtell, R. W. (2000). Modulation and task effects in auditory processing measured using fMRI. *Human Brain Mapping, 10,* 107–119.

Han, S., & Kim, M. (2004). Visual search does not remain efficient when executive working memory is working. *Psychological Science*, *15*, 623–628.

Handy, T. C., & Mangun, G. R. (2000). Attention and spatial selection: Electrophysiological evidence for modulation by perceptual load. *Perception & Psychophysics*, *62*(1), 175–186.

Happé, F. G. (1996). Studying weak central coherence at low levels: children with autism do not succumb to visual illusions. A research note. *Journal of Child Psychology and Psychiatry*, *37*, 873–877.

Harrison, S. A., & Tong, F. (2009). Decoding reveals the contents of visual working memory in early visual areas. *Nature*, *458*, 632–635.

Hart, H., Radua, J., Nakao, T., Mataix-Cols, D., & Rubia, K. (2013). Meta-analysis of functional magnetic resonance imaging studies of inhibition and attention in attention-deficit/hyperactivity disorder: Exploring task-specific, stimulant medication, and age effects. *JAMA Psychiatry*, *70*(2), 185–198.

Harter, M. R., & Guido, W. (1980). Attention to pattern orientation: Negative cortical potentials, reaction time, and the selection process. *Electroencephalography and Clinical Neurophysiology*, *49*(5), 461–475.

Harter, M. R., Miller, S. L., Price, N. J., Lalonde, M. E., & Keyes, A. L. (1989). Neural processes involved in directing attention. *Journal of Cognitive Neuroscience*, *1*, 223–237.

Hartline, H. K. (1938). The response of single optic nerve fibers of the vertebrate eye to illumination of the retina. *American Journal of Physiology*, *121*, 400–415

Hauser, T. U., Iannaccone, R., Stämpfli, P., Drechsler, R., Brandeis, D., Walitza, S., & Brem, S. (2014). The feedback-related negativity (FRN) revisited: New insights into the localization, meaning and network organization. *NeuroImage (Orlando, Fla.)*, *84*, 159–168.

Haut, K. M., & Barch, D. M. (2006). Sex influences on material sensitive functional lateralization in working and episodic memory: Men and women are not all that different. *NeuroImage*, *32*(1), 411–422.

Hayden, B. Y., & Platt, M. L. (2010). Neurons in anterior cingulate cortex multiplex information about reward and action. *Journal of Neuroscience*, *30*, 3339–3346.

Hayhoe, M. M., Shrivastava, A., Mruczek, R., & Pelz, J. B. (2003). Visual memory and motor planning in a natural task. *Journal of Vision (Charlottesville, Va.)*, *3*(1), 49–63.

Heideman, S. G., Ede, F., & Nobre, A. C. (2018). Temporal alignment of anticipatory motor cortical beta lateralisation in hidden visual-motor sequences. *The European Journal of Neuroscience*, *48*(8), 2684–2695.

Heideman, S. G., te Woerd, E. S., & Praamstra, P. (2015). Rhythmic entrainment of slow brain activity preceding leg movements. *Clinical Neurophysiology*, *126*, 348–355.

Heinze, H. J., Luck, S. J., Mangun, G. R., & Hillyard, S. A. (1990). Visual event-related potentials index focused attention within bilateral stimulus arrays: I. Evidence for early selection. *Electroencephalography and Clinical Neurophysiology*, *75*, 511–527.

Heinze, H. J., Luck, S. J., Munte, T. F., Gös, A., Mangun, G. R., & Hillyard, S. A. (1994). Attention to adjacent and separate positions in space: An electrophysiological analysis. *Perception & Psychophysics*, *56*(1), 42–52.

Heinze, H. J., Mangun, G. R., Burchert, W., Hinrichs, H., Scholz, M., Münte, T. F., Gös, A., Scherg, M., Johannes, S., Hundeshagen, H., Gazzaniga, M. S., & Hillyard, S. A. (1994). Combined spatial and temporal imaging of brain activity during visual selective attention in humans. *Nature*, *372*(6506), 543–546.

Helfrich, R. F., Fiebelkorn, I. C., Szczepanski, S. M., Lin, J. J., Parvizi, J., Knight, R. T., & Kastner, S. (2018). Neural mechanisms of sustained attention are rhythmic. *Neuron*, *99*(4), 854–865.e5.

Helmholtz, H. von (1894). Über den Ursprung der richtigen Deutung unserer Sinneseindrücke. *Zeitschrift für Psychologie der Sinnesorgane, 7,* 81–96. [Translation by: Warren, R. M., & Warren, R. P. (1968). The origin of the correct interpretation of our sensory impressions. In: *Helmholtz on Perception: Its Physiology and Development* (pp. 249–260). John Wiley.]

Helmholtz, H. von (1896). *Handbuch der Physiologischen Optik, Dritter Abschnitt.* Voss.

Henderson, C. M., & McClelland, J. L. (2020). Intrusions into the shadow of attention: A new take on illusory conjunctions. *Attention, Perception & Psychophysics, 82*(2), 564–584.

Henderson, J. M., Weeks, P. A., & Hollingworth, A. (1999). The effects of semantic consistency on eye movements during complex scene viewing. *Journal of Experimental Psychology: Human Perception and Performance, 25,* 210228.

Herrmann, C. S. (2001). Human EEG responses to 1-100 Hz flicker: Resonance phenomena in visual cortex and their potential correlation to cognitive phenomena. *Experimental Brain Research, 137,* 346–353.

Hershler, O., & Hochstein, S. (2005). At first sight: A high-level pop out effect for faces. *Vision Research, 45,* 1707–1724.

Hesse, J. K., & Tsao, D. Y. (2020). The macaque face patch system: A turtle's underbelly for the brain. *Nature Reviews: Neuroscience, 21,* 695–716.

Hickey, C., & Peelen, M. (2015). Neural mechanisms of incentive salience in naturalistic human vision. *Neuron (Cambridge, Mass.), 85*(3), 512–518.

Hickey, C., & Peelen, V. (2017). Reward selectively modulates the lingering neural representation of recently attended objects in natural scenes. *Journal of Neuroscience, 37*(31), 7297–7304.

Hickey, C., Di Lollo, V., & McDonald, J. J. (2009). Electrophysiological indices of target and distractor processing in visual search. *Journal of Cognitive Neuroscience, 21*(4), 760–775.

Hilgetag, C. C., Theoret, H., & Pascual-Leone, A. (2001). Enhanced visual spatial attention ipsilateral to rTMS-induced "virtual lesions" of human parietal cortex. *Nature Neuroscience, 4,* 953–957.

Hillis, A. E., Newhart, M., Heidler, J., Barker, P. B., Herskovits, E. H., & Degaonkar, M. (2005). Anatomy of spatial attention: insights from perfusion imaging and hemispatial neglect in acute stroke. *Journal of Neuroscience, 25,* 3161–3167.

Hillstrom, A. P., Shapiro, K. L., & Spence, C. (2002). Attentional limitations in processing sequentially presented vibrotactile targets. *Perception & Psychophysics, 64,* 1068–1082.

Hillyard, S. A., & Anllo-Vento, L. (1998). Event-related brain potentials in the study of visual selective attention. *Proceedings of the National Academy of Sciences of the United States of America, 95*(3), 781–787.

Hillyard, S. A., & Münte, T. F. (1984). Selective attention to color and location: An analysis with event-related brain potentials. *Perception & Psychophysics, 36*(2), 185–198.

Hillyard, S. A., Hink, R. F., Schwent, V. L., & Picton, T. W. (1973). Electrical signs of selective attention in the human brain. *Science (American Association for the Advancement of Science), 182*(4108), 177–180.

Hogendoorn, H. (2016). Voluntary saccadic eye movements ride the attentional rhythm. *Journal of Cognitive Neuroscience, 28*(10), 1625–1635.

Hohwy, J. (2013). *The Predictive Mind.* Oxford University Press.

Hollingworth, A., Richard, A. M., & Luck, S. J. (2008). Understanding the function of visual short-term memory: Transsaccadic memory, object correspondence, and gaze correction. *Journal of Experimental Psychology: General, 137,* 163–181.

Holmes, E., Parr, T., Griffiths, T. D., & Friston, K. J. (2021). Active inference, selective attention, and the cocktail party problem. *Neuroscience & Biobehavioral Reviews, 131*, 1288–1304.

Holroyd, C. B., & Coles, M. G. (2002). The neural basis of human error processing: reinforcement learning, dopamine, and the error-related negativity. *Psychological Review, 109*(4), 679–709.

Holtzman, J. D. (1984). Interactions between cortical and subcortical visual areas: Evidence from human commissurotomy patients. *Vision Research (Oxford), 24*(8), 801–813.

Hong, X., Yang, F., Wang, J., Li, C., Ding, M., & Sheng, J. (2021). Conflict processing in schizophrenia: Dissociable neural mechanisms revealed by the N2 and frontal midline theta. *Neuropsychologia, 155*, 107791.

Hood, B. M., Murray, L., King, F., Hooper, R., Atkinson, J., & Braddick, O. (1996). Habituation changes in early infancy: Longitudinal measures from birth to 6 months. *Journal of Reproductive and Infant Psychology, 14*(3), 177–185.

Hooks, K., Milich, R., & Pugzles Lorch, E. (1994). Sustained and selective attention in boys with attention deficit hyperactivity disorder. *Journal of Clinical Child Psychology, 23*(1), 69–77.

Hopf, J. M., & Mangun, G. R. (2000). Shifting visual attention in space: An electrophysiological analysis using high spatial resolution mapping. *Clinical Neurophysiology, 111*, 1241–1257.

Hopf, J. M., Boehler, C. N., Luck, S. J., Tsotsos, J. K., Heinze, H. J., & Schoenfeld, M. A. (2006). Direct neurophysiological evidence for spatial suppression surrounding the focus of attention in vision. *Proceedings of the National Academy of Sciences of the United States of America, 103*, 1053–1058.

Hopf, J. M., Boelmans, K., Schoenfeld, M. A., Luck, S. J., & Heinze, H. J. (2004). Attention to features precedes attention to locations in visual search: Evidence from electromagnetic brain responses in humans. *Journal of Neuroscience, 24*, 1822–1832.

Hopf, J. M., Schoenfeld, M. A., Buschschulte, A., Rautzenberg, A., Krebs, R. M., & Boehler, C. N. (2015). The modulatory impact of reward and attention on global feature selection in human visual cortex. *Visual Cognition, 23*(1–2), 229–248.

Hopfinger, J. B., & Mangun, G. R. (1998). Reflexive attention modulates processing of visual stimuli in human extrastriate cortex. *Psychological Science, 9*, 441–447.

Hopfinger, J. B., & Mangun, G. R. (2001). Tracking the influence of reflexive attention on sensory and cognitive processing. *Cognitive, Affective, & Behavioral Neuroscience, 1*, 56–65.

Hopfinger, J. B., & Maxwell, J. (2005). Appearing and disappearing stimuli trigger a reflexive modulation of visual cortical activity. *Cognitive Brain Research, 25*, 48–56.

Hopfinger, J. B., & Parks, E. L. (2012). Involuntary attention. In: Mangun, G. R. (ed.), *Neuroscience of Attention: Attentional Control and Selection* (pp. 30–53). Oxford University Press.

Hopfinger, J. B., & Ries, A. J. (2005). Automatic versus contingent mechanisms of sensory-driven neural biasing and reflexive attention. *Journal of Cognitive Neuroscience, 17*(8), 1341–1352.

Hopfinger, J. B., & West, V. M. (2006). Interactions between endogenous and exogenous attention on cortical visual processing. *NeuroImage, 31*, 774–789.

Hopfinger, J. B., Buonocore, M. H., & Mangun, G. R. (2000). The neural mechanisms of top-down attentional control. *Nature Neuroscience, 3*(3), 284–291.

Hopfinger, J. B., Camblin, C. C., & Parks, E. L. (2010). Isolating the internal in endogenous attention. *Psychophysiology, 47*(4), 739–747.

Hopfinger, J. B., Parsons, J., & Fröhlich, F. (2017). Differential effects of 10-Hz and 40-Hz transcranial alternating current stimulation (tACS) on endogenous versus exogenous attention. *Cognitive Neuroscience, 8*(2), 102.

Horga, G., Schatz, K. C., Abi-Dargham, A., & Peterson, B. S. (2014). Deficits in predictive coding underlie hallucinations in schizophrenia. *Journal of Neuroscience, 34* (24), 8072-8082.

Hsu, Y., Hämäläinen, J. A., & Waszak, F. (2014). Both attention and prediction are necessary for adaptive neuronal tuning in sensory processing. *Frontiers in Human Neuroscience, 8*(152), 152.

Hubel, D. H., & Wiesel, T. N. (1959). Receptive fields of single neurones in the cat's striate cortex. *The Journal of Physiology, 148*(3), 574–591.

Hubel, D. H., & Wiesel, T. N. (1962). Receptive fields, binocular interaction and functional architecture in the cat's visual cortex. *The Journal of Physiology, 160* (45), 106–154.

Hubert-Wallander, B., Green, C. S., Sugarman, M., & Bavelier, D. (2011). Changes in search rate but not in the dynamics of exogenous attention in action videogame players. *Attention, Perception & Psychophysics, 73*, 2399–2412.

Husain, M., & Kennard, C. (1996). Visual neglect associated with frontal lobe infarction. *Journal of Neurology 243*, 652–657.

Husain, M., & Rorden, C. (2003). Non-spatially lateralized mechanisms in hemispatial neglect. *Nature Reviews: Neuroscience, 4*(1), 26–36.

Ingendoh, R. M., Posny, E. S., & Heine, A. (2023). Binaural beats to entrain the brain? A systematic review of the effects of binaural beat stimulation on brain oscillatory activity, and the implications for psychological research and intervention. *PLoS One 18*(5), e0286023.

Itti, L., & Koch, C. (2000). A saliency-based search mechanism for overt and covert shifts of visual attention. *Vision Research, 40*(10–12), 1489–1506.

Itti, L., & Koch, C. (2001). Computational modeling of visual attention. *Nature Reviews: Neuroscience, 2*(3), 194–203.

Ivry, R. B., & Keele, S. W. (1989) Timing functions of the cerebellum. *Journal of Cognitive Neuroscience, 1*, 136–152.

James, W. (1890). *The Principles of Psychology*. Harvard University Press.

James, W. (1892). *Psychology: The Briefer Course*. Holt.

Jäncke, L., Gaab, N., Wüstenberg, T., Scheich, H., & Heinze, H. (2001). Short-term functional plasticity in the human auditory cortex: An fMRI study. *Brain Research. Cognitive Brain Research, 12*(3), 479–485.

Jäncke, L., Mirzazade, S., & Shah, N. D. (1999). Attention modulates activity in the primary and the secondary auditory cortex: a functional magnetic resonance imaging study in human subjects. *Neuroscience Letters, 266*(2), 125–128.

Jans, B., Peters, J. C., & de Weerd, P. (2010). Visual spatial attention to multiple locations at once: The jury is still out. *Psychological Review, 117*(2), 637–684.

Janssen, P., & Shadlen, M. N. (2005). A representation of the hazard rate of elapsed time in macaque area LIP. *Nature Neuroscience, 8*, 234–242.

Jasper, H. H. (1941). Electrical activity of the brain. *Annual Review of Physiology, 3*(1), 377–398.

Jensen, O., & Mazaheri, A. (2010). Shaping functional architecture by oscillatory alpha activity: Gating by inhibition. *Frontiers in Human Neuroscience. 4*, 12.

Jensen, O., Kaiser, J., & Lachaux, J. (2007). Human gamma-frequency oscillations associated with attention and memory. *Trends in Neurosciences, 30*(7), 317–324.

Jiang, J., Summerfield, C., & Egner, T. (2013). Attention sharpens the distinction between expected and unexpected percepts in the visual brain. *The Journal of Neuroscience, 33*(47), 18438–18447.

Joel, D. (2011). Male or female? Brains are intersex. *Frontiers in Integrative Neuroscience, 5*, 57.

Johnson, J. S., Woodman, G. F., Braun, E., & Luck, S. J. (2007). Implicit memory influences the allocation of attention in visual cortex. *Psychonomic Bulletin & Review, 14* (5), 834–839.

Johnson, K. A., Robertson, I. H., Barry, E., Mulligan, A., Dáibhis, A., Daly, M., Watchorn, A., Gill, M., & Bellgrove, M. A. (2008). Impaired conflict resolution and alerting in children with ADHD: Evidence from the attention network task (ANT). *Journal of Child Psychology and Psychiatry*, *49*(12), 1339–1347.

Johnson, M. H. (1990). Cortical maturation and the development of visual attention in early infancy. *Journal of Cognitive Neuroscience, 2*(2), 81–95.

Johnson, S. P. (2019). Development of visual-spatial attention. In: Hodgson, T. (ed.), *Current Topics in Behavioral Neurosciences, Volume 41* (pp. 37–58). Springer International Publishing AG.

Johnston, S. J., Linden, D. E. J., & Shapiro, K. L. (2012). Functional imaging reveals working memory and attention interact to produce the attentional blink. *Journal of Cognitive Neuroscience, 24*(1), 28–38.

Johnston, W. A., Hawley, K. J., & Farnham, J. M. (1993). Novel popout: Empirical boundaries and tentative theory. *Journal of Experimental Psychology: Human Perception and Performance, 19*, 140153.

Jones, M., & Wilkinson, S. (2020). From prediction to imagination. In: Abraham, A. (ed.), *The Cambridge Handbook of the Imagination* (Cambridge Handbooks in Psychology, pp. 94–110). Cambridge University Press.

Jones, M. R. (1976). Time, our lost dimension: Toward a new theory of perception, attention, and memory. *Psychological Review, 83*(5), 323–355.

Jones, M. R., Johnston, H. M., & Puente, J. (2006). Effects of auditory pattern structure on anticipatory and reactive attending. *Cognitive Psychology, 53*, 59–96.

Jonides, J. (1981). Voluntary versus automatic control over the mind's eye movement. In Long, J. B., & Baddeley, A. D. (eds.), *Attention and Performance* (vol. IX, pp. 187–203). Erlbaum Associates.

Jonides, J., & Yantis, S. (1988). Uniqueness of abrupt visual onset as an attention-capturing property. *Perception & Psychophysics, 43*, 346–354.

Jordan, R., & Keller, G. B. (2020). Opposing influence of top-down and bottom-up input on excitatory layer 2/3 neurons in mouse primary visual cortex. *Neuron, 108*, 1194–1206.

Joseph, R. M., Keehn, B., Connolly, C., Wolfe, J. M., & Horowitz, T. S. (2009). Why is visual search superior in autism spectrum disorder? *Developmental Science, 12*(6), 1083–1096.

Jovancevic-Misic, J., & Hayhoe, M. (2009). Adaptive gaze control in natural environments. *Journal of Neuroscience, 29*(19), 6234–6238.

Jueptner, I. H., Rijntes, M., Weiller, C., Faiss, J. H., Timmann, D., Mueller, S. P., & Diener, H. C. (1995). Localization of a cerebellar timing process using PET. *Neurology 45*, 1540–1545.

Juergens, E., Guettler, A., & Eckhorn, R. (1999). Visual stimulation elicits locked and induced gamma oscillations in monkey intracortical- and EEG-potentials, but not in human EEG. *Experimental Brain Research, 129*(2), 247–259.

Jung, R., & Kornmüller, A. E. (1938). Eine Methodik der Ableitung Iokalisierter Potentialschwankungen aus subcorticalen Hirngebieten. *Archiv für Psychiatrie und Nervenkrankheiten, 109*, 1–30.

Kahneman, D. (1973). *Attention and Effort.* Prentice-Hall.

Kang, O. E., Huffer, K. E., & Wheatley, T. P. (2014). Pupil dilation dynamics track attention to high-level information. *PLoS One, 9*(8), e102463.

Kanwisher, N. G. (1987). Repetition blindness: Type recognition without token individuation. *Cognition, 27*, 117–143.

Karnath, H., & Rorden, C. (2012). The anatomy of spatial neglect. *Neuropsychologia, 50*(6), 1010–1017.

Karnath, H., Ferber, S., & Himmelbach, M. (2001). Spatial awareness is a function of the temporal not the posterior parietal lobe. *Nature, 411*, 950–953.

Karnath, H., Fruhmann Berger, M., Küker, W., & Rorden, C. (2004). The anatomy of spatial neglect based on voxelwise statistical analysis: A study of 140 patients. *Cerebral Cortex*, *14*(10), 1164–1172.

Karnath, H., Himmelbach, M., & Rorden, C. (2002). The subcortical anatomy of human spatial neglect: Putamen, caudate nucleus and pulvinar. *Brain*, *125*(2), 350–360.

Karns, C. M., & Knight, R. T. (2009). Intermodal auditory, visual, and tactile attention modulates early stages of neural processing. *Journal of Cognitive Neuroscience*, *21*(4), 669–683.

Kastner, S., Chen, Q., Jeong, S. K., & Mruczek, R. E. B. (2017). A brief comparative review of primate posterior parietal cortex: A novel hypothesis on the human toolmaker. *Neuropsychologia*, *105*, 123–134.

Keck, T., Keller, G.B ., Jacobsen, R. I., Eysel, U. T., Bonhoeffer, T., & Hubener, M. (2013). Synaptic scaling and homeostatic plasticity in the mouse visual cortex in vivo. *Neuron*, *80*, 327–334.

Keehn, B., Lincoln, A. J., Müller, R., & Townsend, J. (2010). Attentional networks in children and adolescents with autism spectrum disorder. *Journal of Child Psychology and Psychiatry*, *51*(11), 1251–1259.

Keehn, B., Müller, R.-A., & Townsend, J. (2013). Atypical attentional networks and the emergence of autism. *Neuroscience and Biobehavioral Reviews*, *37*, 164–183.

Keeler, R. (2002) Antique ophthalmic instruments and books: The Royal College Museum. *British Journal of Ophthalmology*, *86*, 602–603.

Keller, G. B., Bonhoeffer, T., & Hubener, M. (2012). Sensorimotor mismatch signals in primary visual cortex of the behaving mouse. *Neuron*, *74*, 809–815.

Kelley, T. A., & Lavie, N. (2011). Working memory load modulates distractor competition in primary visual cortex. *Cerebral Cortex*, *21*, 659–665.

Kelley, T. A., Rees, G., & Lavie, N. (2013). The impact of distractor congruency on stimulus processing in retinotopic visual cortex. *NeuroImage*, *81*, 158–163.

Kelly, S. P., Gomez-Ramirez, M., & Foxe, J. J. (2008). Spatial attention modulates initial afferent activity in human primary visual cortex. *Cerebral Cortex*, *18*, 2629–2636.

Kelly, S. P., Gomez-Ramirez, M., & Foxe, J. J. (2009). The strength of anticipatory spatial biasing predicts target discrimination at attended locations: A high-density EEG study. *European Journal of Neuroscience*, *30*(11), 2224–2234.

Kenemans, J. L., Kok, A., & Smulders, F. T. Y. (1993). Event-related potentials to conjunctions of spatial frequency and orientation as a function of stimulus parameters and response requirements. *Electroencephalography and Clinical Neurophysiology*, *88*(1), 51–63.

Kenklies, K. (2012). Educational theory as topological rhetoric: The concepts of pedagogy of Johann Friedrich Herbart and Friedrich Schleiermacher. *Studies in Philosophy & Education*, *31*, 265–273.

Kessler, K., Schmitz, F., Gross, J., Hommel, B., Shapiro, K., & Schnitzler, A. (2005). Target consolidation under high temporal processing demands as revealed by MEG. *NeuroImage*, *26*(4), 1030–1041.

Kiebel, S. J., von Kriegstein, K., Daunizeau, J., & Friston, K. J. (2009). Recognizing sequences of sequences. *PLoS Computational Biology*, *5*(8), e1000464.

Kiehl, K. A., Liddle, P. F., & Hopfinger, J. B. (2000). Error processing and the rostral anterior cingulate: An event-related fMRI study. *Psychophysiology*, *37*(2), 216–223.

Kienitz, R., Schmid, M. C., & Dugué, L. (2022). Rhythmic sampling revisited: Experimental paradigms and neural mechanisms. *The European Journal of Neuroscience*, *55*(11–12), 3010–3024.

Kihara, K., Hirose, N., Mima, T., Abe, M., Fukuyama, H., & Osaka, N. (2007). The role of left and right intraparietal sulcus in the attentional blink: A transcranial magnetic stimulation study. *Experimental Brain Research*, *178*, 135–140.

Kihara, K., Ikeda, T., Matsuyoshi, D., Hirose, N., Mima, T., Fukuyama, H., & Osaka, N. (2011). Differential contributions of the intraparietal sulcus and the inferior parietal lobe to attentional blink: Evidence from transcranial magnetic stimulation. *Journal of Cognitive Neuroscience, 23*(1), 247–256.

Kilb, A., & Naveh-Benjamin, M. (2007). Paying attention to binding: Further studies assessing the role of reduced attentional resources in the associative deficit of older adults. *Memory & Cognition, 35*(5), 1162–1174.

Kim, A. J., & Anderson, B. A. (2020). Neural correlates of attentional capture by stimuli previously associated with social reward. *Cognitive Neuroscience, 11*, 5–15.

Kim, S., & Hopfinger, J. B. (2010). Neural basis of distraction. *Journal of Cognitive Neuroscience, 22*(8), 1794–1807.

Kim, Y. H., Gitelman, D. R., Nobre, A. C., Parrish, T. B., LaBar, K. S., & Mesulam, M. M. (1999). The large-scale neural network for spatial attention displays multifunctional overlap but differential asymmetry. *NeuroImage 9*, 269–277.

Kincade, J. M., Abrams, R. A., Astafiev, S. V., Shulman, G. L., & Corbetta, M. (2005). An event-related functional magnetic resonance imaging study of voluntary and stimulus-driven orienting of attention. *Journal of Neuroscience, 25*, 4593–4604.

Kirchhoff, M. (2018). Predictive processing, perceiving and imagining: Is to perceive to imagine, or something close to it? *Philosophical Studies, 175*, 751–767.

Kiss, M., Van Velzen, J., & Eimer, M. (2008). The N2pc component and its links to attention shifts and spatially selective visual processing. *Psychophysiology, 45*(2), 240–249.

Kitamura, F., & Matsunaga, K. (1994). Influence of looking at hazard lights on car-driving performance. *Perceptual & Motor Skills, 78*, 1059–1065.

Klawohn, J., Santopetro, N. J., Meyer, A., & Hajcak, G. (2020). Reduced P300 in depression: Evidence from a flanker task and impact on ERN, CRN, and PE. *Psychophysiology, 57*(4), e13520.

Klein, R. M. (2000). Inhibition of return. *Trends in Cognitive Sciences, 4*(4), 138–147.

Klein, R. M., & MacInnes, W. J. (1999) Inhibition of return is a foraging facilitator in visual search. *Psychological Science, 10*, 346–352.

Kleinman, J. T., Newhart, M., Davis, C., Heidler-Gary, J., Gottesman, R. F., & Hillis, A. E. (2007). Right hemispatial neglect: Frequency and characterization following acute left hemisphere stroke. *Brain and Cognition, 64*(1), 50–59.

Kliemann, D., Dziobek, I., Hatri, A., Baudewig, J., & Heekeren, H. R. (2012). The role of the amygdala in atypical gaze on emotional faces in autism spectrum disorders. *The Journal of Neuroscience, 32*(28), 9469–9476.

Klimesch, W., Sauseng, P., & Hanslmayr, S. (2007). EEG alpha oscillations: The inhibition-timing hypothesis. *Brain Research Reviews, 53*, 63–88.

Klin, A., Jones, W., Schultz, R., Volkmar, F., & Cohen, D. (2002). Visual fixation patterns during viewing of naturalistic social situations as predictors of social competence in individuals with autism. *Archives of General Psychiatry, 59*(9), 809–816.

Knight, R. T., & Grabowecky, M. F. (1995). Escape from linear time: Prefrontal cortex and conscious experience. In: Gazzaniga, M. S. (ed.), *The Cognitive Neurosciences* (pp. 1357–1371). MIT Press.

Knudsen, E. I. (2011). Control from below: The role of a midbrain network in spatial attention. *European Journal of Neuroscience, 33*, 1961–1972.

Koch, C., & Ullman, S. (1985). Shifts in selective visual attention: Towards the underlying neural circuitry. *Human Neurobiology, 4*, 219–227.

Kodaka, Y., Mikami, A., & Kubota, K. (1997). Neuronal activity in the frontal eye field of the monkey is modulated while attention is focused on to a stimulus in the peripheral visual field, irrespective of eye movement. *Neuroscience Research, 28*(4), 291–298.

Koenig, S., Kadel, H., Uengoer, M., Schubö, A., & Lachnit, H. (2017). Reward draws the eye, uncertainty holds the eye: Associative learning modulates distractor interference in visual search. *Frontiers in Behavioral Neuroscience, 11*, 128.

Kofler, M. J., Rapport, M. D., Sarver, D. E., Raiker, J. S., Orban, S. A., Friedman, L. M., & Kolomeyer, E. G. (2013). Reaction time variability in ADHD: A meta-analytic review of 319 studies. *Clinical Psychology Review, 33* (6), 795–811.

Kok, P., Rahnev, D., Jehee, J. F., Lau, H. C., & de Lange, F. P. (2012). Attention reverses the effect of prediction in silencing sensory signals. *Cerebral Cortex, 22*, 2197–2206.

Konrad, K., Neufang, S., Hanisch, C., Fink, G. R., & Herpertz-Dahlmann, B. (2006). Dysfunctional attentional networks in children with attention deficit/ hyperactivity disorder: Evidence from an event-related functional magnetic resonance imaging study. *Biological Psychiatry (1969), 59*(7), 643–651.

Koster, E. H. W., Crombez, G., Van Damme, S., Verschuere, B., & De Houwer, J. (2004). Does imminent threat capture and hold attention? *Emotion 4*, 312–317.

Krafnick, A. J., & Evans, T. M. (2019). Neurobiological sex differences in developmental dyslexia. *Frontiers in Psychology*, 9, 2669.

Kramer, A. F., & Hahn, S. (1995). Splitting the beam: Distribution of attention over noncontiguous regions of the visual field. *Psychological Science, 6*(6), 381–386.

Kranczioch, C., Debener, S., & Engel, A. K. (2003). Event-related potential correlates of the attentional blink phenomenon. *Cognitive Brain Research, 17*(1), 177–187.

Kranczioch, C., Debener, S., Schwarzbach, J., Goebel, R., & Engel, A. K. (2005). Neural correlates of conscious perception in the attentional blink. *NeuroImage, 24*, 704–714.

Kreither, J., Lopez-Calderon, J., Leonard, C. J., Robinson, B. M., Ruffle, A., Hahn, B.,

Gold, J. M., & Luck, S. J. (2017). Electrophysiological evidence for hyperfocusing of spatial attention in schizophrenia. *Journal of Neuroscience, 37* (14), 3813–3823.

Krishnan, L., Kang, A., Sperling, G., & Srinivasan, R. (2013). Neural strategies for selective attention distinguish fast-action video game players. *Brain Topography, 26*, 83–97.

Kristjánsson, Á., & Campana, G. (2010). Where perception meets memory: A review of repetition priming in visual search tasks. *Attention, Perception & Psychophysics, 72*(1), 5–18.

Kropotov, J. D., Ponomarev, V. A., Hollup, S., & Mueller, A. (2011). Dissociating action inhibition, conflict monitoring and sensory mismatch into independent components of event related potentials in GO/NOGO task. *NeuroImage, 57*(2), 565–575.

Kropotov, J. D., Pronina, M. V., Ponomarev, V. A., Poliakov, Y. I., Plotnikova, I. V., & Mueller, A. (2019). Latent ERP components of cognitive dysfunctions in ADHD and schizophrenia. *Clinical Neurophysiology, 130*(4), 445–453.

Kuhn, G. (2015). Attention and misdirection: How to use conjuring experience to study attentional processes. In: Fawcett, J. M., Risko, E. F., & Kingstone, A. (eds.), *The Handbook of Attention* (pp. 503–525). MIT Press.

Kühn, S., Lorenz, R., Banaschewski, T., Barker, G. J., Büchel, C., Conrod, P. J., Flor, H., Garavan, H., Ittermann, B., Loth, E., Mann, K., Nees, F., Artiges, E., Paus, T., Rietschel, M., Smolka, M. N., Ströhle, A., Walaszek, B., Schumann, G., Heinz, A., Gallinat, J., & IMAGEN Consortium (2014). Positive association of video game playing with left frontal cortical thickness in adolescents. *PLoS One, 9*, e91506.

Kurki, I., Hyvärinen, A., & Henriksson, L. (2022). Dynamics of retinotopic spatial attention revealed by multifocal MEG. *NeuroImage, 263*, 119643.

Kustov, A., & Robinson, D. (1996). Shared neural control of attentional shifts and eye movements. *Nature, 384*(6604), 74–77.

Kutas, M., & Federmeier, K. D. (2011). Thirty years and counting: Finding meaning in the N400 component of the event-related brain potential (ERP). *Annual Review of Psychology, 62*, 621–647.

Kwak, Y., Hanning, N. M., & Carrasco, M. (2023). Presaccadic attention sharpens visual acuity. *Scientific Reports, 13*(1), 2981.

Lakatos, P., Karmos, G., Mehta, A. D., Ulbert, I., & Schroeder, C. E. (2008). Entrainment of neuronal oscillations as a mechanism of attentional selection. *Science (American Association for the Advancement of Science), 320*(5872), 110–113.

Lamy, D., Antebi, C., Aviani, N., & Carmel, T. (2008). Priming of pop-out provides reliable measures of target activation and distractor inhibition in selective attention. *Vision Research, 48*(1), 30–41.

Land, M. F., & Tatler, B. W. (2009). *Looking and Acting: Vision and Eye Movements in Natural Behavior*. Oxford University Press.

Landau, A., & Fries, P. (2012). Attention samples stimuli rhythmically. *Current Biology, 22*(11), 1000–1004.

Landau, A. N., Esterman, M., Robertson, L. C., Bentin, S., & Prinzmetal, W. (2007). Different effects of voluntary and involuntary attention on EEG activity in the gamma band. *The Journal of Neuroscience, 27*(44), 11986–11990.

Landau, A. N., Schreyer, H., van Pelt, S., & Fries, P. (2015). Distributed attention is implemented through theta-rhythmic gamma modulation. *Current Biology, 25*(17), 2332–2337.

Lange, K., & Röder, B. (2006). Orienting attention to points in time improves stimulus processing both within and across modalities. *Journal of Cognitive Neuroscience, 18*, 715–729.

Lange, K. W., Reichl, S., Lange, K. M., Tucha, L., & Tucha, O. (2010). The history of attention deficit hyperactivity disorder.

Attention Deficit and Hyperactivity Disorders, 2(4), 241–255.

Langton, S. R. H., Law, A. S., Burton, A. M., & Schweinberger, S. R. (2008). Attention capture by faces. *Cognition, 107*(1), 330–342.

Lansbergen, M. M., Kenemans, J. L., & van Engeland, H. (2007). Stroop interference and attention-deficit/hyperactivity disorder: A review and meta-analysis. *Neuropsychology, 21*(2), 251–262.

Lasaponara, S., D'Onofrio, M., Pinto, M., Dragone, A., Menicagli, D., Bueti, D., De Lucia, M., Tomaiuolo, F., & Doricchi, F. (2018). EEG correlates of preparatory orienting, contextual updating, and inhibition of sensory processing in left spatial neglect. *The Journal of Neuroscience, 38*(15), 3792–3808.

Laukkonen, R. E., & Slagter, H. A. (2021). From many to (n)one: Meditation and the plasticity of the predictive mind. *Neuroscience and Biobehavioral Reviews, 128*, 199–217.

Lauterbur, P. C. (1974). Magnetic resonance zeugmatography. *Pure and Applied Chemistry, 40*(1–2), 149–157.

Lavie, N. (1995). Perceptual load as a necessary condition for selective attention. *Journal of Experimental Psychology: Human Perception & Performance, 21*, 451–468.

Lavie, N., & Tsal, Y. (1994). Perceptual load as a major determinant of the locus of selection in visual attention. *Perception & Psychophysics, 56*(2), 183–197.

Lavie, N., Hirst, A, de Fockhert, J., & Viding, E (2004). Load theory of selective attention and cognitive control. *Journal of Experimental Psychology: General, 133*(3), 339–354.

Lawrence, S. J., Norris, D. G., & de Lange, F. P. (2019). Dissociable laminar profiles of concurrent bottom-up and top-down modulation in the human visual cortex. *Elife, 8*, e44422.

Lee, T. S., Yang, C. F., Romero, R. D., & Mumford, D. (2002). Neural activity in early visual cortex reflects behavioral experience and higher-order perceptual saliency. *Nature Neuroscience, 5*(6), 589–597.

Leibniz, G. W. (1714 /1989). *Philosophical Essays*. Edited and translated by R. Ariew & D. Garber. Hackett Publishing Company.

Lepsien, J., Griffin, I. C., Devlin, J. T., & Nobre, A. C. (2005). Directing spatial attention in mental representations: Interactions between attentional orienting and working-memory load. *NeuroImage, 26*(3), 733–743.

Li, X., Strasser, B., Jafari-Khouzani, K., Thapa, B., Small, J., Cahill, D. P., Dietrich, J., Batchelor, T. T., & Andronesi, O. C. (2020). Super-resolution whole-brain 3D MR spectroscopic imaging for mapping D-2-hydroxyglutarate and tumor metabolism in isocitrate dehydrogenase 1-mutated human gliomas. *Radiology, 294*(3), 589–597.

Lien, M.-C., Ruthruff, E., Goodin, Z., & Remington, R. W. (2008). Contingent attentional capture by top-down control settings: Converging evidence from event-related potentials. *Journal of Experimental Psychology: Human Perception and Performance, 34*(3), 509–530.

Liesefeld, H. R., & Müller, H. J. (2020). A theoretical attempt to revive the serial/parallel-search dichotomy. *Attention, Perception & Psychophysics, 82*(1), 228–245.

Lima, B., Singer, W., & Neuenschwander, S. (2011). Gamma responses correlate with temporal expectation in monkey primary visual cortex. *Journal of Neuroscience, 31*, 15919–15931.

Lin, J. Y., Franconeri, S. L., & Enns, J. T. (2008). Objects on a collision path with the observer demand attention. *Psychological Science, 19*, 686–692.

Lin, W. M., Oetringer, D. A., Bakker-Marshall, I., Emmerzaal, J., Wilsch, A., ElShafei, H. A., Rassi, E., & Haegens, S. (2022). No behavioural evidence for rhythmic facilitation of perceptual discrimination. *The European Journal of Neuroscience, 55*(11–12), 3352–3364.

Liss, M., Saulnier, C., Fein, D., & Kinsbourne, M. (2006). Sensory and attention abnormalities in autistic spectrum disorders. *Autism, 10*, 155–172.

Liu, T., & Mance, I. (2011). Constant spread of feature-based attention across the visual field. *Vision Research, 51*(1), 26–33.

Liu, T., Fang, M. W. H., & Saba-Sadiya, S. (2023). Adaptive visual selection in feature space. *Psychonomic Bulletin & Review, 30*(3), 994–1003.

Liu, T., Slotnick, S. D., Serences, J. T., & Yantis, S. (2003). Cortical mechanisms of feature-based attentional control. *Cerebral Cortex, 13*(12), 1334–1343.

Locke, J. (1689). *The Works, Vol. 1: An Essay Concerning Human Understanding Part 1*. Rivington.

Loftus, G. R., & Mackworth, N. H. (1978). Cognitive determinants of fixation location during picture viewing. *Journal of Experimental Psychology: Human Perception and Performance, 4*, 565572.

London, R. E., & Slagter, H. A. (2021). No effect of transcranial direct current stimulation over left dorsolateral prefrontal cortex on temporal attention. *Journal of Cognitive Neuroscience, 33*, 756–768.

Lorenz, R. C., Gleich, T., Gallinat, J., & Kuhn, S. (2015). Video game training and the reward system. *Frontiers in Human Neuroscience, 9*, 40.

Lorusso, M. L., Facoetti, A., Paganoni, P., Pezzani, M., & Molteni, M. (2006). Effects of visual hemisphere-specific stimulation versus reading-focused training in dyslexic children. *Neuropsychological Rehabilitation, 16*(2), 194–212.

Lu, H., Fung, A. W. T., Chan, S. S. M., & Lam, L. C. W. (2016). Disturbance of attention network functions in Chinese healthy older adults: An intra-individual perspective. *International Psychogeriatrics, 28*(2), 291–301.

Luck, S. J. (2012). Electrophysiological correlates of the focusing of attention within complex visual scenes: N2pc and related ERP components. In: Luck, S. J. & Kappenman, E. S. (eds.), *The Oxford Handbook of Event-Related Potential*

Components (pp. 329–360). Oxford University Press.

Luck, S. J., & Ford, M. A. (1998). On the role of selective attention in visual perception. *Proceedings of the National Academy of Sciences of the United States of America, 95* (3), 825–830.

Luck, S. J., & Hillyard, S. A. (1990). Electrophysiological evidence for parallel and serial processing during visual search. *Perception & Psychophysics, 48*(6), 603–617.

Luck, S. J., & Hillyard, S. A. (1994a). Electrophysiological correlates of feature analysis during visual search. *Psychophysiology 31*, 291–308.

Luck, S. J., & Hillyard, S. A. (1994b). Spatial filtering during visual search: Evidence from human electrophysiology. *Journal of Experimental Psychology: Human Perception and Performance, 20*(5), 1000–1014.

Luck, S. J., & Hillyard, S. A. (1995). The role of attention in feature detection and conjunction discrimination: An electrophysiological analysis. *International Journal of Neuroscience, 80*(1–4), 281–297.

Luck, S. J., & Woodman, G. F. (1999). Electrophysiological measurement of rapid shifts of attention during visual search. *Nature, 400*(6747), 867–869.

Luck, S. J., Chelazzi, L., Hillyard, S. A., & Desimone, R. (1997). Neural mechanisms of spatial selective attention in areas V1, V2, and V4 of macaque visual cortex. *Journal of Neurophysiology, 77*(1), 24–42.

Luck, S. J., Gaspelin, N., Folk, C. L., Remington, R. W., & Theeuwes, J. (2021). Progress toward resolving the attentional capture debate. *Visual Cognition, 29*(1), 1–21.

Luck, S. J., Heinze, H. J., Mangun, G. R., & Hillyard, S. A. (1990). Visual event-related potentials index focused attention within bilateral stimulus arrays. II. Functional dissociation of P1 and N1 components. *Electroencephalography and Clinical Neurophysiology, 75*(6), 528–542.

Luck, S. J., Hillyard, S. A., Mangun, G. R., & Gazzaniga, M. S. (1989). Independent hemispheric attentional systems mediate visual search in split-brain patients. *Nature, 342*(6249), 543–545.

Luck, S. J., Hillyard, S. A., Mouloua, M., Woldorff, M. G., Clark, V. P., & Hawkins, H. L. (1994). Effects of spatial cuing on luminance detectability: Psychophysical and electrophysiological evidence for early selection. *Journal of Experimental Psychology: Human Perception and Performance, 20*(4), 887–904.

Luck, S. J., Vogel, E. K., & Shapiro, K. L. (1996). Word meanings can be accessed but not reported during the attentional blink. *Nature, 382*, 616–618.

Luman, M., Oosterlaan, J., & Sergeant, J. A. (2005). The impact of reinforcement contingencies on AD/HD: A review and theoretical appraisal. *Clinical Psychology Review, 25*(2), 183–213.

Lunven, M., & Bartolomeo, P. (2017). Attention and spatial cognition: Neural and anatomical substrates of visual neglect. *Annals of Physical and Rehabilitation Medicine, 60*(3), 124–129.

Luria, A. R. (1959). Disorders of "simultaneous perception" in a case of bilateral occipito-parietal brain injury. *Brain, 82*, 437–449.

Lutz, A., Slagter, H. A., Rawlings, N. B., Francis, A. D., Greischar, L. L., & Davidson, R. J. (2009). Mental training enhances attentional stability: Neural and behavioral evidence. *The Journal of Neuroscience, 29*(42), 13418–13427.

Lyndon, S., & Corlett, P. R. (2020). Hallucinations in posttraumatic stress disorder: Insights from predictive coding. *Journal of Abnormal Psychology, 129*, 534–543.

Mack, A., & Rock, I. (1998). *Inattentional Blindness*. MIT Press.

Macknik, S. L., King, M., Randi, J., Robbins, A., Teller, Thompson, J., & Martinez-Conde, S. (2008). Attention and awareness in stage magic: Turning tricks into research. *Nature Reviews: Neuroscience, 9* (11), 871–879.

Mackworth, N. H. (1948). The breakdown of vigilance during prolonged visual search.

Quarterly Journal of Experimental Psychology, *1*, 6–21.

MacLean, M. H., & Giesbrecht, B. (2015). Neural evidence reveals the rapid effects of reward history on selective attention. *Brain Research*, *1606*, 86–94.

MacLean, M. H., Diaz, G. K., & Giesbrecht, B. (2016). Irrelevant learned reward associations disrupt voluntary spatial attention. *Attention, Perception & Psychophysics*, *78*, 2241–2252.

MacLeod, C., & Clarke, P. J. F. (2015). The attentional bias modification approach to anxiety intervention. *Clinical Psychological Science*, *3*(1), 58–78.

MacLeod, C., Grafton, B., & Notebaert, L. (2019). Anxiety-linked attentional bias: Is it reliable? *Annual Review of Clinical Psychology*, *15*(1), 529–554.

MacLeod, C., Mathews, A., & Tata, P. (1986). Attentional bias in emotional disorders. *Journal of Abnormal Psychology*, *95*(1), 15–20.

Magnée, M. J. C. M., Kahn, R. S., Cahn, W., & Kemner, C. (2011). More prolonged brain activity related to gaze cueing in schizophrenia. *Clinical Neurophysiology*, *122*, 506–511.

Mahoney, J. R., Verghese, J., Goldin, Y., Lipton, R., & Holtzer, R. (2010). Alerting, orienting, and executive attention in older adults. *Journal of the International Neuropsychological Society*, *16*(5), 877–889.

Malebranche, N. (1674/1992). The search after truth (Trans. T. H. Lennon and P. J. Obecalp). In: Nadler, S. (ed.), *Malebranche: Philosophical Selections*. Cambridge University Press.

Malinowski, P., Moore, A. W., Mead, B. R., & Gruber, T. (2017). Mindful aging: The effects of regular brief mindfulness practice on electrophysiological markers of cognitive and affective processing in older adults. *Mindfulness*, *8*(1), 78–94.

Maljkovic, V., & Nakayama, K. (1994). Priming of popout: I. Role of features. *Memory & Cognition*, *22*, 657–672.

Mangun, G. R., & Fannon, S. P. (2007). Attention: Control in the visual cortex. *Current Biology*, *17*(5), R170–R172.

Mangun, G. R., & Hillyard, S. A. (1990). Allocation of visual attention to spatial locations: Tradeoff functions for event-related brain potentials and detection performance. *Perception & Psychophysics*, *47*(6), 532–550.

Mangun, G. R., & Hillyard, S. A. (1991). Modulations of sensory-evoked brain potentials indicate changes in perceptual processing during visual-spatial priming. *Journal of Experimental Psychology: Human Perception and Performance*, *17*(4), 1057–1074.

Mangun, G. R., Buonocore, M. H., Girelli, M., & Jha, A. P. (1998). ERP and fMRI measures of visual spatial selective attention. *Human Brain Mapping*, *6*(5–6), 383–389.

Mangun, G. R., Hopfinger, J. B., & Heinze, H.-J. (1998). Integrating electrophysiology and neuroimaging in the study of human cognition. *Behavior Research Methods, Instruments, & Computers*, *30*(1), 118–130.

Mangun, G. R., Hopfinger, J. B., Kussmaul, C. L., Fletcher, E. M., & Heinze, H. (1997). Covariations in ERP and PET measures of spatial selective attention in human extrastriate visual cortex. *Human Brain Mapping*, *5*(4), 273–279.

Mansell, W., Clark, D. M., Ehlers, A., & Chen, Y. P. (1999). Social anxiety and attention away from emotional faces. *Cognition & Emotion*, *13*, 673–690.

Mansfield, P., & Grannell, P. (1975). Diffraction and microscopy in solids and liquids by NMR. *Physical Review B 12*(9), 3618–3634.

Marks, L. E., & Wheeler, M. E. (1998). Attention and the detectability of weak taste stimuli. *Chemical Senses*, *23*(1), 19–29.

Marois, R., Chun, M. M., & Gore, J. C. (2000). Neural correlates of the attentional blink. *Neuron (Cambridge, Mass.)*, *28*(1), 299–308.

Marois, R., Leung, H. C., & Gore, J. C. (2000). A stimulus-driven approach to object identity and location processing in the human brain. *Neuron*, *25*, 717–728.

Marois, R., Yi, D., & Chun, M. M. (2004). The neural fate of consciously perceived and missed events in the attentional blink. *Neuron (Cambridge, Mass.)*, *41*(3), 465–472.

Marrocco, R. T., & Davidson, M. C. (1998). *Neurochemistry of Attention*. MIT Press.

Martens, S., & Johnson, A. (2005). Timing attention: Cuing target onset interval attenuates the attentional blink. *Memory & Cognition*, *33*(2), 234–240.

Marti, S., Sigman, M., & Dehaene, S. (2012). A shared cortical bottleneck underlying attentional blink and psychological refractory period. *NeuroImage (Orlando, Fla.)*, *59*(3), 2883–2898.

Martínez, A., Anllo-Vento, L., Sereno, M. I., Frank, L. R., Buxton, R. B., Dubowitz, D. J., Wong, E. C., Hinrichs, H., Heinze, H. J., & Hillyard, S. A. (1999). Involvement of striate and extrastriate visual cortical areas in spatial attention. *Nature Neuroscience*, *2*(4), 364–369.

Martinez, A., DiRusso, F., Anllo-Vento, L., Sereno, M. I., Buxton, R. B., & Hillyard, S. A. (2001). Putting spatial attention on the map: timing and localization of stimulus selection processes in striate and extra-striate visual areas. *Vision Research*, *41*, 1437–1457.

Marzecová, A., Schettino, A., Widmann, A., SanMiguel, I., Kotz, S. A., & Schröger, E. (2018). Attentional gain is modulated by probabilistic feature expectations in a spatial cueing task: ERP evidence. *Scientific Reports*, *8*(1), 54.

Mason, D. J., Humphreys, G. W., & Kent, L. S. (2003). Exploring selective attention in ADHD: Visual search through space and time. *Journal of Child Psychology and Psychiatry*, *44*(8), 1158–1176.

Mason, D. J., Humphreys, G. W., & Kent, L. S. (2005). Insights into the control of attentional set in ADHD using the attentional blink paradigm. *Journal of Child Psychology & Psychiatry*, *46*, 1345–1353.

Mathews, A., Fox, E., Yiend, J., & Calder, A. (2003). The face of fear: Effects of eye gaze and emotion on visual attention. *Visual Cognition*, *10*(7), 823–835.

Mathewson, K. E., Fabiani, M., Gratton, G., Beck, D. M., & Lleras, A. (2010). Rescuing stimuli from invisibility: inducing a momentary release from visual masking with pre-target entrainment. *Cognition*, *115*, 186–191.

Mathis, K. I., Wynn, J. K., Breitmeyer, B., Nuechterlein, K. H., & Green, M. F. (2011). The attentional blink in schizophrenia: Isolating the perception/attention interface. *Journal of Psychiatric Research*, *45*(10), 1346–1351.

May, K. E., & Kana, R. K. (2020). Frontoparietal network in executive functioning in autism spectrum disorder. *Autism Research*, *13*, 1762–1777.

Mayall, K., & Humphreys, G. W. (2002). Presentation and task effects on migration errors in attentional dyslexia. *Neuropsychologia*, *40*(8), 1506–1515.

Mayer, A., Schwiedrzik, C. M., Wibral, M., Singer, W., & Melloni, L. (2016). Expecting to see a letter: Alpha oscillations as carriers of top-down sensory predictions. *Cerebral Cortex (New York, N.Y. 1991)*, *26*(7), 3146–3160.

Mazza, V., Turatto, M., & Caramazza, A. (2009). Attention selection, distractor suppression, and N2pc. *Cortex*, *45*(7), 879–890.

McAdams, C. J., & Reid, R. C. (2005). Attention modulates the responses of simple cells in monkey primary visual cortex. *The Journal of Neuroscience*, *25*(47), 11023–11033.

McAlonan, K., Cavanaugh, J., & Wurtz, R. H. (2006). Attentional modulation of thalamic reticular neurons. *The Journal of Neuroscience*, *26*(16), 4444–4450.

McCormick, P. A., Klein, R. M., & Johnston, S. (1998). Splitting versus sharing focal attention: Comment on Castiello and Umiltà (1992). *Journal of Experimental Psychology: Human Perception and Performance*, *24*(1), 350–357.

McDonald, J. J., Teder-Salejarvi, W. A., Heraldez, D., & Hillyard, S. A. (2001). Electrophysiological evidence for the missing link in crossmodal attention: Cognitive neuroscience. *Canadian Journal of Experimental Psychology, 55*(2), 141–149.

McDonald, J. J., Teder-Sälejärvi, W. A., & Hillyard, S. A. (2000). Involuntary orienting to sound improves visual perception. *Nature, 407*, 906–908.

McDonald, J. J., Ward, L. M., & Kiehl, K. A. (1999). An event-related brain potential study of inhibition of return. *Perception & Psychophysics, 61*, 1411–1423.

McDonough, I. M., Wood, M. M., & Miller, W. S. (2019). A review on the trajectory of attentional mechanisms in aging and the Alzheimer's disease continuum through the attention network test. *The Yale Journal of Biology & Medicine, 92*(1), 37–51.

McMains, S. A., & Somers, D. C. (2004). Multiple spotlights of attentional selection in human visual cortex. *Neuron, 42*, 677–686.

McMains, S. A., & Somers, D. C. (2005). Processing efficiency of divided spatial attention mechanisms in human visual cortex. *The Journal of Neuroscience, 25*(41), 9444–9448.

McNally, R. J. (2019). Attentional bias for threat: Crisis or opportunity? *Clinical Psychology Review, 69*, 4–13.

Medina, J., Kannan, V., Pawlak, M. A., Kleinman, J. T., Newhart, M., Davis, C., Heidler-Gary, J. E., Herskovits, E. H., & Hillis, A. E. (2009). Neural substrates of visuospatial processing in distinct reference frames: Evidence from unilateral spatial neglect. *Journal of Cognitive Neuroscience, 21* (11), 2073–2084.

Meijs, E. L., Slagter, H. A., de Lange, F. P., & van Gaal, S. (2018). Dynamic interactions between top–down expectations and conscious awareness. *The Journal of Neuroscience, 38*(9), 2318–2327.

Mesulam, M. M. (1981). A cortical network for directed attention and unilateral neglect. *Annals of Neurology, 10*(4), 309–325.

Meuter, R. F. I., & Lacherez, P. F. (2016). When and why threats go undetected: Impacts of event rate and shift length on threat detection accuracy during airport baggage screening. *Human Factors, 58*(2), 218–228.

Meyer, K. N., Du, F., Parks, E., & Hopfinger, J. B. (2018). Exogenous vs. endogenous attention: Shifting the balance of fronto-parietal activity. *Neuropsychologia, 111*, 307–316.

Meyer, K. N., Sheridan, M. A., & Hopfinger, J. B. (2020). Reward history impacts attentional orienting and inhibitory control on untrained tasks. *Attention, Perception & Psychophysics, 82*(8), 3842–3862.

Michalareas, G., Vezoli, J., van Pelt, S., Schoffelen, J.-M., Kennedy, H., & Fries, P. (2016). Alpha–beta and gamma rhythms subserve feedback and feedforward influences among human visual cortical areas. *Neuron, 89*, 384–397.

Michel, R., & Busch, N. A. (2023). No evidence for rhythmic sampling in inhibition of return. *Attention, Perception & Psychophysics, 85*, 2111–2121.

Mignard, M., & Malpeli, J. G. (1991). Paths of information flow through visual cortex. *Science (American Association for the Advancement of Science), 251*(4998), 1249–1251.

Milham, M. (2003). Practice-related effects demonstrate complementary roles of anterior cingulate and prefrontal cortices in attentional control. *NeuroImage, 18*(2), 483–493.

Miller, J. (1989). The control of attention by abrupt visual onsets and offsets. *Perception & Psychophysics, 45*(6), 567–571.

Miltner, W. H. R., Braun, C. H., & Coles, M. G. H. (1997). Event-related brain potentials following incorrect feedback in a time-estimation task: Evidence for a "generic" neural system for error detection. *Journal of Cognitive Neuroscience, 9*, 788–798.

Minagar, A., Ragheb, J., & Kelley, R. E. (2003). The Edwin Smith surgical papyrus: Description and analysis of the earliest case of aphasia. *Journal of Medical Biography, 11* (2), 114–117.

Miranda, A. T., & Palmer, E. M. (2014). Intrinsic motivation and attentional capture from gamelike features in a visual search task. *Behavior Research Methods, 46*(1), 159–172.

Mirpour, K., & Bisley, J. W. (2021). The roles of the lateral intraparietal area and frontal eye field in guiding eye movements in free viewing search behavior. *Journal of Neurophysiology, 125*(6), 2144–2157.

Mirza, M. B., Adams, R. A., Friston, K. J., & Parr, T. (2019). Introducing a Bayesian model of selective attention based on active inference. *Scientific Reports, 9*(1), 1–22.

Mishra, J., Zinni, M., Bavelier, D., & Hillyard, S.A. (2011). Neural basis of superior performance of action videogame players in an attention-demanding task. *Journal of Neuroscience, 31*, 992–998.

Mitroff, S. R., Ericson, J. M., & Sharpe, B. (2018). Predicting airport screening officers' visual search competency with a rapid assessment. *Human Factors, 60*(2), 201–211.

Mogg, K., & Bradley, B. P. (1999). Some methodological issues in assessing attentional biases for threatening faces in anxiety: A replication study using a modified version of the probe detection task. *Behaviour Research and Therapy, 37*, 595–604.

Mogg, K., & Bradley, B. P. (2018). Anxiety and threat-related attention: Cognitive-motivational framework and treatment. *Trends in Cognitive Sciences, 22* (3), 225–240.

Mogg, K., Waters, A. M., & Bradley, B. P. (2017). Attention bias modification (ABM): Review of effects of multisession ABM training on anxiety and threat-related attention in high-anxious individuals. *Clinical Psychological Science, 5*(4), 698–717.

Mollick, J. A., & Kober, H. (2020). Computational models of drug use and addiction: A review. *Journal of Abnormal Psychology, 129*, 544–555.

Monchi, O., Petrides, M., Petre, V., Worsley, K., & Dagher, A. (2001). Wisconsin card sorting revisited: Distinct neural circuits participating in different stages of the task identified by event-related functional magnetic resonance imaging. *Journal of Neuroscience, 21*(19), 7733–7741.

Moore, A., Gruber, T., Derose, J., & Malinowski, P. (2012). Regular, brief mindfulness meditation practice improves electrophysiological markers of attentional control. *Frontiers in Human Neuroscience, 6*, 18.

Moore, C., & Dunham, P. (1995). *Joint Attention: Its Origins and Role in Development*. Lawrence Erlbaum Associates.

Moore, M. J., Milosevich, E., Mattingley, J. B., & Demeyere, N. (2023). The neuroanatomy of visuospatial neglect: A systematic review and analysis of lesion-mapping methodology. *Neuropsychologia, 180*, 108470.

Moores, E., Laiti, L., & Chelazzi, L., (2003). Associative knowledge controls deployment of visual selective attention. *Nature Neuroscience, 6*(2), 182–189.

Moran, J., & Desimone, R. (1985). Selective attention gates visual processing in the extrastriate cortex. *Science, 229*, 782–784.

Moray, N. (1959). Attention in dichotic listening: Affective cues and the influence of instructions. *Quarterly Journal of Experimental Psychology, 11*, 56–60.

Morillon, B., Schroeder, C. E., & Wyart, V. (2014). Motor contributions to the temporal precision of auditory attention. *Nature Communications, 5*, 5255.

Morrisey, M. N., Hofrichter, R., & Rutherford, M. D. (2019). Human faces capture attention and attract first saccades without longer fixation. *Visual Cognition, 27*(2), 158–170.

Mort, D. J., Malhotra, P., Mannan, S. K., Rorden, C., Pambakian, A., Kennard, C., & Husain, M. (2003). The anatomy of visual neglect. *Brain, 126*, 1986–1997.

Motter, B. C. (1993). Focal attention produces spatially selective processing in visual cortical areas V1, V2 and V4 in the presence of competing stimuli. *Journal of Neurophysiology*, *70*, 909–919.

Mounts, J. R. W. (2000). Attentional capture by abrupt onsets and feature singletons produces inhibitory surrounds. *Perception & Psychophysics*, *62*(7), 1485–1493.

Mozer, M. C. (1983). Letter migration in word perception. *Journal of Experimental Psychology: Human Perception and Performance*, *9*(4), 531–546.

Mueller, A., Hong, D. S., Shepard, S., & Moore, T. (2017). Linking ADHD to the neural circuitry of attention. *Trends in Cognitive Sciences*, *21*(6), 474–488.

Muggleton, N. G., Postma, P., Moutsopoulou, K., Nimmo-Smith, I., Marcel, A., & Walsh, V. (2006). TMS over right posterior parietal cortex induces neglect in a scene-based frame of reference. *Neuropsychologia*, *44*(7), 1222–1229.

Muhl-Richardson, A., Parker, M. G., Recio, S. A., Tortosa-Molina, M., Daffron, J. L., & Davis, G. J. (2021). Improved X-ray baggage screening sensitivity with "targetless" search training. *Cognitive Research: Principles and Implications*, *6*(1), 33.

Mullane, J. C., Corkum, P. V., Klein, R. M., & McLaughlin, E. (2009). Interference control in children with and without ADHD: A systematic review of flanker and Simon task performance. *Child Neuropsychology*, *15* (4), 321–342.

Müller, H. J., & Rabbitt, P. M. (1989). Reflexive and voluntary orienting of attention: Time course of activation and resistance to interruption. *Journal of Experimental Psychology: Human Perception and Performance*, *15*, 315–330.

Müller, M. M., Andersen, S. K., & Keil, A. (2008). Time course of competition for visual processing resources between emotional pictures and foreground task. *Cerebral Cortex*, *18*(8), 1892–1899.

Müller, M. M., Gruber, T., & Keil, A. (2000). Modulation of induced gamma band activity in the human EEG by attention and visual information processing. *International Journal of Psychophysiology*, *38*(3), 283–299.

Müller, M. M., Malinowski, P., Gruber, T., & Hillyard, S. A. (2003). Sustained division of the attentional spotlight. *Nature*, *424*, 309–312.

Müller, N. G., & Kleinschmidt, A. (2003). Dynamic interaction of object- and space-based attention in retinotopic visual areas. *The Journal of Neuroscience*, *23*(30), 9812–9816.

Müller, N. G., & Kleinschmidt, A. (2007). Temporal dynamics of the attentional spotlight: Neuronal correlates of attentional capture and inhibition of return in early visual cortex. *Journal of Cognitive Neuroscience*, *19*(4), 587–593.

Muller, S., Rothermund, K., & Wentura, D. (2016). Relevance drives attention: attentional bias for gain- and loss-related stimuli is driven by delayed disengagement. *Quarterly Journal of Experimental Psychology*, *69*, 752–763.

Murphy, E. R., Norr, M., Strang, J. F., Kenworthy, L., Gaillard, W. D., & Vaidya, C. J. (2017). Neural basis of visual attentional orienting in childhood autism spectrum disorders. *Journal of Autism and Developmental Disorders*, *47*(1), 58–67.

Mutreja, R., Craig, C., & O'Boyle, M. W. (2016). Attentional network deficits in children with autism spectrum disorder. *Developmental Neurorehabilitation*, *19*, 389–397.

Näätänen, R. (1975). Selective attention and evoked potentials in humans – A critical review. *Biological Psychology*, *2*(4), 237–307.

Näätänen, R. (1982). Processing negativity: An evoked-potential reflection of selective attention. *Psychological Bulletin*, *92*(3), 605–640.

Näätänen, R., Gaillard, A. W. K., & Mäntysalo, S. (1978). Early selective-attention effect on evoked potential reinterpreted. *Acta Psychologica*, 42(4), 313–329.

Nadler, S. (ed.) (1992). *Malebranche: Philosophical Selections*. Hackett Publishing Company.

Nadra, J. G., & Mangun, G. R. (2023). Placing willed attention in context: A review of attention and free will. *Frontiers in Cognition, 2*, 1205618.

Nakayama, K., & Mackeben, M. (1989). Sustained and transient components of focal visual attention. *Vision Research, 29*(11), 1631–1647.

Nandy, A. S., Nassi, J. J., & Reynolds, J. H. (2017). Laminar organization of attentional modulation in macaque visual area V4. *Neuron, 93*(1), 235–246.

Naveh-Benjamin, M., Guez, J., & Marom, M. (2003). The effects of divided attention at encoding on item and associative memory. *Memory & Cognition, 31*(7), 1021–1035.

Naveh-Benjamin, M., Guez, J., & Sorek, S. (2007). The effects of divided attention on encoding processes in memory: Mapping the locus of interference. *Canadian Journal of Experimental Psychology, 61*, 1–12.

Nikolaou, K., Field, M., Critchley, H., & Duka, T. (2013). Acute alcohol effects on attentional bias are mediated by subcortical areas associated with arousal and salience attribution. *Neuropsychopharmacology, 38*(7), 1365–1373.

Nobre, A. C., & Van Ede, F. (2018). Anticipated moments: Temporal structure in attention. *Nature Reviews: Neuroscience, 19*(1), 34–48.

Nobre, A. C., Sebestyen, G. N., Gitelman, D. R., Mesulam, M. M., Frackowiak, R. S. J., & Frith, C. D. (1997). Functional localization of the system for visuospatial attention using positron emission tomography. *Brain, 120*(3), 515–533.

Nobre, A. C., Sebestyen, G. N., & Miniussi, C. (2000). The dynamics of shifting visuo-spatial attention revealed by event-related potentials. *Neuropsychologia, 38*, 964–974.

Noesselt, T., Hillyard, S. A., Woldorff, M. G., Schoenfeld, A., Hagner, T., Jäncke, L., Tempelmann, C., Hinrichs, H., & Heinze, H. (2002). Delayed striate cortical activation during spatial attention. *Neuron, 35*(3), 575–587.

Norman, D. A., & Shallice, T. (1986). Attention to action. In: Davidson, R. J., Schwartz, G. E., & Shapiro, D. (eds.), *Consciousness and Self-Regulation: Advances in Research and Theory* (pp. 1–18). Springer.

Norris, C. J., Creem, D., Hendler, R., & Kober, H. (2018). Brief mindfulness meditation improves attention in novices: Evidence from ERPs and moderation by neuroticism. *Frontiers in Human Neuroscience, 12*, 315.

Nummenmaa, L., Hyönä, J., & Calvo, M. G. (2006). Eye movement assessment of selective attentional capture by emotional pictures. *Emotion, 6*, 257–268.

O'Connor, D. H., Fukui, M. M., Pinsk, M. A., & Kastner, S. (2002). Attention modulates responses in the human lateral geniculate nucleus. *Nature Neuroscience, 5*(11), 1203–1209.

O'Craven, K. M., Downing, P. E., & Kanwisher, N. (1999). fMRI evidence for objects as the units of attentional selection. *Nature, 401*, 584–587.

O'Craven, K. M., Rosen, B. R., Kwong, K. K., Treisman, A., & Savoy, R. L. (1997). Voluntary attention modulates fMRI activity in human MT-MST. *Neuron, 18*, 591–598.

O'Regan, J. K. (1979). Moment to moment control of eye saccades as a function of textual parameters in reading. In: Kolers, P. A., Wrolstad, M. E., & Bouma, H. (eds.), *Processing of Visible Language* (pp. 49–60). Plenum.

Oberlin, B. G., Alford, J. L., & Marrocco, R. T. (2005). Normal attention orienting but abnormal stimulus alerting and conflict effect in combined subtype of ADHD. *Behavioural Brain Research, 165*, 1–11.

Ogden, J. A. (1985). Anterior-posterior interhemispheric differences in the loci of lesions producing visual hemineglect. *Brain and Cognition, 4*, 59–75.

Oh, S., & Kim, M. (2004). The role of spatial working memory in visual search efficiency. *Psychonomic Bulletin & Review, 11*, 275–281.

Ohman, A., Flykt, A., & Esteves, F. (2001). Emotion drives attention: Detecting the snake in the grass. *Journal of Experimental Psychology: General*, *130*, 466–478.

Oliveri, M., & Vallar, G. (2009). Parietal versus temporal lobe components in spatial cognition: Setting the mid-point of a horizontal line. *Journal of Neuropsychology*, 3(2), 201–211.

Olivers, C. N. L., Meijer, F., & Theeuwes, J. (2006). Feature-based memory-driven attentional capture: Visual working memory content affects visual attention. *Journal of Experimental Psychology: Human Perception and Performance*, *32*(5), 1243–1265.

Olson, I. R., Chun, M. M., & Allison, T. (2001). Contextual guidance of attention: Human intracranial erp evidence in an anatomically early, temporally late stage of visual processing. *Brain*, *124*, 1417–1425.

Ono, Y., & Taniguchi, Y. (2017). Attentional capture by emotional stimuli: Manipulation of emotional valence by the sample pre-rating method: Attentional capture by emotional stimuli. *Japanese Psychological Research*, *59*(1), 26–34.

Oren, N., Abecasis, D., Inbar, E., Glik, A., Steiner, I., & Shapira-Lichter, I. (2023). A new perspective on the role of the frontoparietal regions in Stroop-like conflicts. *Human Brain Mapping*, *44*(11), 4310–4320.

Orlandi, N., & Lee, G. (2019). How radical is predictive processing? In: Colombo, M., Irvine, E., & Stepleton, M. (eds.), *Andy Clark and His Critics* (pp. 206–221). Oxford University Press.

Otero-Millan, J., Macknik, S. L., Robbins, A., & Martinez-Conde, S. (2011). Stronger misdirection in curved than in straight motion. *Frontiers in Human Neuroscience*, 5, 133.

Overbeek, T., Nieuwenhuis, S., & Ridderinkhof, K. (2005). Dissociable components of error processing: On the functional significance of the Pe vis-à-vis the ERN/Ne. *Journal of Psychophysiology*, *19*, 319–329.

Oxner, M., Martinovic, J., Forschack, N., Lempe, R., Gundlach, C., & Müller, M. (2023). Global enhancement of target color – not proactive suppression – explains attentional deployment during visual search. *Journal of Experimental Psychology: General*, *152*(6), 1705–1722.

Ozaki, I., Jin, C. Y., Suzuki, Y., Baba, M., Matsunaga, M., & Hashimoto, I. (2004). Rapid change of tonotopic maps in the human auditory cortex during pitch discrimination. *Clinical Neurophysiology*, *115*(7), 1592–1604.

Padmala, S., & Pessoa, L. (2008). Affective learning enhances visual detection and responses in primary visual cortex. *Journal of Neuroscience*, *28*(24), 6202–6210.

Palmer, J., Huk, A. C., & Shadlen, M. N. (2005). The effect of stimulus strength on the speed and accuracy of a perceptual decision. *Journal of Vision*, *5*, 376–404.

Panichello, M. F., & Buschman, T. J. (2021) Shared mechanisms underlie the control of working memory and attention. *Nature*, *592*, 601–605.

Pantazis, D., & Adler, A. (2021). Meg source localization via deep learning. *Sensors (Basel, Switzerland)*, *21*(13), 4278.

Papadopoulos, N., Rinehart, N. J., Bradshaw, J. L., Taffe, J., & McGinley, J. (2015). Is there a link between motor performance variability and social-communicative impairment in children with ADHD-CT: A kinematic study using an upper limb Fitts' aiming task. *Journal of Attention Disorders*, *19*(1), 72–77.

Pardo, J. V., Pardo, P. J, Janer, K. W., & Raichle, M. E. (1990). The anterior cingulate cortex mediates processing selection in the Stroop attentional conflict paradigm. *Proceedings of the National Academy of Sciences of the United States of America*, 87 (1), 256–259.

Parkhurst, D., Law, K., & Niebur, E. (2002). Modelling the role of salience in the allocation of visual selective attention. *Vision Research*, *42*(1), 107–123.

Parks, E. L., & Hopfinger, J. B. (2008). Hold it! Memory affects attentional dwell time. *Psychonomic Bulletin & Review, 15*(6), 1128–1134.

Parks, E. L., Kim, S., & Hopfinger, J. B. (2014). The persistence of distraction: A study of attentional biases by fear, faces, and context. *Psychonomic Bulletin & Review, 21*(6), 1501–1508.

Parr, T., & Friston, K. J. (2017) Working memory, attention, and salience in active inference. *Scientific Reports, 7*, 14678.

Pashler, H. E. (1984). Processing stages in overlapping tasks: Evidence for a central bottleneck. *Journal of Experimental Psychology: Human Perception & Performance, 10*, 358–377.

Paus, T., Jech, R., Thompson, C.J., Comeau, R., Peters, T., & Evans, A.C. (1997). Transcranial magnetic stimulation during positron emission tomography: A new method for studying connectivity of the human cerebral cortex. *Journal of Neuroscience, 17*(9), 3178–3184.

Peelen, M. V., Heslenfeld, D. J., & Theeuwes, J. (2004). Endogenous and exogenous attention shifts are mediated by the same large-scale neural network. *NeuroImage, 22*, 822–830.

Pellicano, E., & Burr, D. (2012): When the world becomes "too real": A Bayesian explanation of autistic perception. *Trends in Cognitive Science, 16*, 504–510.

Peñuelas-Calvo, I., Jiang-Lin, L. K., Girela-Serrano, B., Delgado-Gomez, D., Navarro-Jimenez, R., Baca-Garcia, E., & Porras-Segovia, A. (2022). Video games for the assessment and treatment of attention-deficit/hyperactivity disorder: A systematic review. *European Child & Adolescent Psychiatry, 31*(1), 5–20.

Peters, B., Kaiser, J., Rahm, B., & Bledowski, C. (2021). Object-based attention prioritizes working memory contents at a theta rhythm. *Journal of Experimental Psychology: General, 150*(6), 1250–1256.

Petersen, S. E., Corbetta, M., Miezin, F. M., & Shulman, G. L. (1994). PET studies of parietal involvement in spatial attention: Comparison of different task types. *Canadian Journal of Experimental Psychology, 48*(2), 319–338.

Petersen, S. E., Robinson, D. L., & Morris, J. D. (1987). Contributions of the pulvinar to visual spatial attention. *Neuropsychologia, 25*(1), 97–105.

Peterson, M. S., & Kramer, A. F. (2001). Attentional guidance of the eyes by contextual information and abrupt onsets. *Perception & Psychophysics, 63*, 1239–1249.

Peterson, M. S., Kramer, A. F., & Irwin, D. E. (2004). Covert shifts of attention precede involuntary eye movements. *Perception & Psychophysics, 66*, 398–405.

Petkov, C., Kang, X., Alho, K., Bertrand, O., Yund, E., & Woods, D. (2004). Attentional modulation of human auditory cortex. *Nature Neuroscience, 7*, 658–663.

Picton, T. W., John, M. S., Dimitrijevic, A., & Purcell, D. (2003). Human auditory steady-state responses. *International Journal of Audiology, 42*, 177–219.

Pierce, K., Conant, D., Hazin, R., Stoner, R., & Desmond, J. (2011). Preference for geometric patterns early in life as a risk factor for autism. *Archives of General Psychiatry, 68*(1), 101–109.

Pietrzak, A., Marszałek, A., Kunikowska, J., Piotrowski, T., Medak, A., Pietrasz, K., Wojtowicz, J., & Cholewiński, W. (2021). Detection of clinically silent brain lesions in [18F]FDG PET/CT study in oncological patients: Analysis of over 10,000 studies. *Scientific Reports, 11*(1), 18293.

Pimenta, M. G., Brown, T., Arns, M., & Enriquez-Geppert, S. (2021). Treatment efficacy and clinical effectiveness of EEG neurofeedback as a personalized and multimodal treatment in ADHD: A critical review. *Neuropsychiatric Disease and Treatment, 17*, 637–648.

Pinker, S. (1994). *The Language Instinct: How the Mind Creates Language*. William Morrow & Company, Inc.

Plaisted, K., O'Riordan, M., & Baron-Cohen, S. (1998). Enhanced visual search for a conjunctive target in autism: A research note. *Journal of Child Psychology and Psychiatry, 39*, 777–783.

Pogosyan, A., Gaynor, L. D., Eusebio, A., & Brown, P. (2009). Boosting cortical activity at beta-band frequencies slows movement in humans. *Current Biology, 19*(19), 1637–1641.

Polich, J. (2007). Updating P300: An integrative theory of P3a and P3b. *Clinical Neurophysiology, 118*(10), 2128–2148.

Pool, E., Brosch, T., Delplanque, S., & Sander, D. (2014). Where is the chocolate? Rapid spatial orienting toward stimuli associated with primary rewards. *Cognition, 130*(3), 348–359.

Posner, M. I. (1980). Orienting of attention. *Quarterly Journal of Experimental Psychology, 32*(1), 3–25.

Posner. M. I. (2016). Orienting of attention: Then and now. *Quarterly Journal of Experimental Psychology, 69*(10), 1864–1875.

Posner, M. I., & Cohen, Y. (1984). Components of visual orienting. In: Bouma, H., & Bouwhis, D. (eds.), *Attention and Performance* (Vol. X, pp. 531–556). Erlbaum Associates.

Posner, M. I., Inhoff, A., Friedrich, F. J., & Cohen, A. (1987). Isolating attentional systems: A cognitive-anatomical analysis. *Psychobiology, 15*, 107–121.

Posner, M. I., Nissen, M. J., & Ogden, W. C. (1978). Attended and unattended processing modes: The role of set for spatial location. In: Pick, H. L., & Saltzman, N. J. (eds.), *Modes of Perceiving and Processing Information* (pp. 137–157). Lawrence Erlbaum Associates.

Posner, M. I., Rafal, R. D., Choate, L. S., & Vaughan, J. (1985). Inhibition of return: Neural basis and function. *Cognitive Neuropsychology, 2*(3), 211–228.

Potter, M. C., Chun, M. M., Banks, B. S., & Muckenhoupt, M. (1998). Two attentional deficits in serial target search: The visual attentional blink and an amodal task-switch deficit. *Journal of Experimental Psychology: Learning, Memory, & Cognition, 24*, 979–992.

Praamstra, P., Boutsen, L., & Humphreys, G. W. (2005). Frontoparietal control of spatial attention and motor intention in human EEG. *Journal of Neurophysiology, 94*, 764–774.

Praamstra, P., Kourtis, D., Kwok, H. F., & Oostenveld, R. (2006). Neurophysiology of implicit timing in serial choice reaction-time performance. *Journal of Neuroscience, 26*, 5448–5455.

Prime, D. J., & Jolicœur, P. (2009). On the relationship between occipital cortex activity and inhibition of return. *Psychophysiology, 46*(6), 1278–1287.

Prime, D. J., & Ward, L. M. (2004). Inhibition of return from stimulus to response. *Psychological Science, 15*, 272–276.

Prinzmetal, W., McCool, C., & Park, S. (2005). Attention: Reaction time and accuracy reveal different mechanisms. *Journal of Experimental Psychology: General, 134*, 73–92.

Prior, M., Sanson, A., Freethy, C., & Geffen, G. (1985). Auditory attentional abilities in hyperactive children. *Journal of Child Psychology and Psychiatry, 26*(2), 289–304.

Privitera, C. M., Carney, T., Klein, S., & Aguilar, M. (2014). Analysis of microsaccades and pupil dilation reveals a common decisional origin during visual search. *Vision Research, 95*, 43–50.

Przybylski, A. K., Weinstein, N., & Murayama, K. (2017). Internet gaming disorder: Investigating the clinical relevance of a new phenomenon. *The American Journal of Psychiatry, 174*(3), 230–236.

Ptito, A., Arnell, K., Jolicœur, P., & Macleod, J. (2008). Intramodal and crossmodal processing delays in the attentional blink paradigm revealed by event-related potentials. *Psychophysiology, 45*(5), 794–803.

Radeborg, K., Briem, V., & Hedman, L. R. (1999). The effect of concurrent task difficulty on working memory during simulated driving. *Ergonomics, 42*(5), 767–777.

Rafal, R. D., & Posner, M. I. (1987). Deficits in human visual spatial attention following thalamic lesions. *Proceedings of the National Academy of Sciences of the United States of America, 84*(20), 7349–7353.

Rafal, R. D., Posner, M. I., Friedman, J. H., Inhoff, A. W., & Bernstein, E. (1988). Orienting of visual attention in progressive supranuclear palsy. *Brain*, *111*(2), 267–280.

Ramgir, A., & Lamy, D. (2022). Does feature intertrial priming guide attention? The jury is still out. *Psychonomic Bulletin & Review*, *29*(2), 369–393.

Randall, W. M., & Smith, J. L. (2011). Conflict and inhibition in the cued-Go/NoGo task. *Clinical Neurophysiology*, *122*(12), 2400–2407.

Rao, R. P. (2005). Bayesian inference and attentional modulation in the visual cortex. *Neuroreport*, *16*(16), 1843–1848.

Rao, R. P., & Ballard, D. H. (1999). Predictive coding in the visual cortex: A functional interpretation of some extra-classical receptive-field effects. *Nature Neuroscience*, *2*, 79–87.

Rau, P. P., Zheng, J., Wang, L., Zhao, J., & Wang, D. (2020). Haptic and auditory–haptic attentional blink in spatial and object-based tasks. *Multisensory Research*, *33*(3), 295–312.

Rauss, K. S., Pourtois, G., Vuilleumier, P., & Schwartz, S. (2009). Attentional load modifies early activity in human primary visual cortex. *Human Brain Mapping*, *30*, 1723–1733.

Raymond, J., Shapiro, K., & Arnell, K. (1992). Temporary suppression of visual processing in an RSVP task: An attentional blink? *Journal of Experimental Psychology: Human Perception and Performance*, *18*, 849–860.

Rayner, K. (1998). Eye movements in reading and information processing: 20 years of research. *Psychological Bulletin*, *124*(3), 372–422.

Rayner, K., McConkie, G. W., & Ehrlich, S. (1978). Eye movements and integrating information across fixations. *Journal of Experimental Psychology: Human Perception & Performance*, *4*, 529–544.

Redden, R. S., MacInnes, W. J., & Klein, R. M. (2021). Inhibition of return: An information processing theory of its natures and significance. *Cortex*, *135*, 30–48.

Redelmeier, D. A., & Tibshirani, R. J. (1997). Association between cellular-telephone calls and motor vehicle collisions. *The New England Journal of Medicine*, *336*(7), 453–458.

Reed, J., & Johnson, P. (1994). Assessing implicit learning with indirect tests. *Journal of Experimental Psychology: Learning, Memory, & Cognition*, *20*, 585–594.

Reichle, E. D., Reineberg, A. E., & Schooler, J. W. (2010). Eye movements during mindless reading. *Psychological Science*, *21*(9), 1300–1310.

Rennie, D., Bull, R., & Diamond, A. (2004). Executive functioning in preschoolers: Reducing the inhibitory demands of the dimensional change card sort task. *Developmental Neuropsychology*, *26*, 423–443.

Rensink, R. A., O'Regan, J. K., & Clark, J. J. (1997). To see or not to see: The need for attention to perceive changes in scenes. *Psychological Science*, *8*(5), 368–373.

Reteig, L. C., Newman, L. A., Ridderinkhof, K. R., & Slagter, H. A. (2022). Effects of tDCS on the attentional blink revisited: A statistical evaluation of a replication attempt. *PLoS One*, *17*(1), e0262718.

Reynolds, J. H., Pasternak, T., & Desimone, R. (2000). Attention increases sensitivity of V4 neurons. *Neuron*, *26*(3), 703–714.

Ridderinkhof, K. R., Wylie, S. A., van den Wildenberg, W. P. M., Bashore, T. R., & van der Molen, M. W. (2021). The arrow of time: Advancing insights into action control from the arrow version of the Eriksen flanker task. *Attention, Perception & Psychophysics*, *83*(2), 700–721.

Ridding, M. C., & Rothwell, J. C. (2007). Is there a future for therapeutic use of transcranial magnetic stimulation? *Nature Reviews: Neuroscience*, *8*(1), 559–567.

Riecke, L., Peters, J. C., Valente, G., Poser, B. A., Kemper, V. G., Formisano, E., & Sorger, B. (2018). Frequency-specific attentional modulation in human primary auditory cortex and midbrain. *NeuroImage*, *174*, 274–287.

Ries, A. J., & Hopfinger, J. B. (2011). Magnocellular and parvocellular influences on reflexive attention. *Vision Research (Oxford)*, *51*(16), 1820–1828.

Rigsby, T. J., Stilwell, B. T., Ruthruff, E., & Gaspelin, N. (2023). A new technique for estimating the probability of attentional capture. *Attention, Perception & Psychophysics*, *85*(2), 543–559.

Rinne, T., Balk, M. H., Koistinen, S., Autti, T., Alho, K., & Sams, M. (2008). Auditory selective attention modulates activation of human inferior colliculus. *Journal of Neurophysiology*, *100*, 3323–3327.

Rizzolatti, G., Riggio, L., Dascola, I., & Umiltá, C., (1987). Reorienting attention across the horizontal and vertical meridians: Evidence in favour of a premotor theory of attention. *Neuropsychologia*, *25*(1A), 31–40.

Ro, T., Farnè, A., & Chang, E. (2003). Inhibition of return and the human frontal eye fields. *Experimental Brain Research*, *150*, 290–296.

Rodríguez, V., Valdés-Sosa, M., & Freiwald, W. (2002). Dividing attention between form and motion during transparent surface perception. *Cognitive Brain Research*, *13*(2), 187–193.

Roelfsema, P. R., Lamme, V. A. F., & Spekreijse, H. (1998). Object- based attention in the primary visual cortex of the macaque monkey. *Nature*, *395*, 376–381.

Rogers, R. D., Bayliss, A. P., Szepietowska, A., Dale, L., Reeder, L., Pizzamiglio, G., Czarna, K., Wakeley, J., Cowen, P. J., & Tipper, S. P. (2014). I want to help you, but I am not sure why: Gaze-cuing induces altruistic giving. *Journal of Experimental Psychology: General*, *143*(2), 763–777.

Rolls, E. T., Wirth, S., Deco, G., Huang, C., & Feng, J. (2023). The human posterior cingulate, retrosplenial, and medial parietal cortex effective connectome, and implications for memory and navigation. *Human Brain Mapping*, *44*(2), 629–655.

Ross, B., & Lopez, M. D. (2020). 40-Hz binaural beats enhance training to mitigate the attentional blink. *Scientific Reports*, *10*(1), 7002.

Ross, B., Jamali, S., Miyazaki, T., & Fujioka, T. (2013). Synchronization of beta and gamma oscillations in the somatosensory evoked neuromagnetic steady-state response. *Experimental Neurology*, *245*, 40–51.

Rossi, A. F., Bichot, N. P., Desimone, R., & Ungerleider, L. G. (2007). Top down attentional deficits in macaques with lesions of lateral prefrontal cortex. *The Journal of Neuroscience*, *27*(42), 11306–11314.

Rothkopf, C. A., Ballard, D. H., & Hayhoe, M. M. (2007). Task and context determine where you look. *Journal of Vision*, *7*(14), 1–20.

Ruchkin, D. S., Sutton, S., Mahaffey, D., & Glaser, J. (1986). Terminal CNV in the absence of motor response. *Electroencephalography and Clinical Neurophysiology*, *63*(5), 445–463.

Rueda, M. R., Rothbart, M. K., McCandliss, B. D., Saccomanno, L., & Posner, M. I. (2005). Training, maturation, and genetic influences on the development of executive attention. *Proceedings of the National Academy of Sciences of the United States of America*, *102*(41), 14931–14936.

Rutman, A. M., Clapp, W. C., Chadick, J. Z., & Gazzaley, A. (2010). Early top-down control of visual processing predicts working memory performance. *Journal of Cognitive Neuroscience*, *22*(6), 1224–1234.

Ryan, J. D., Althoff, R. R., Whitlow, S., & Cohen, N. J. (2000). Amnesia is a deficit in relational memory. *Psychological Science*, *11*, 454460.

Saalmann, Y. B., Pinsk, M. A., Wang, L., Li, X., & Kastner, S. (2012). The pulvinar regulates information transmission between cortical areas based on attention demands. *Science*, *337*, 753–756.

Sabag, M., & Geva, R. (2022). Hyper and hypo attention networks activations affect social development in children with autism spectrum disorder. *Frontiers in Human Neuroscience*, *16*, 902041.

Sàenz, M., Buracâs, G. T., & Boynton, G. M. (2003). Global feature-based attention for motion and color. *Vision Research*, *43*(6), 629–637.

Saffran, E. M., & Coslett, H. B. (1996). Attentional dyslexia in Alzheimer's disease: A case study. *Cognitive Neuropsychology*, *13*, 205–228.

Saleem, A. B., Ayaz, A., Jeffery, K. J., Harris, K. D., & Carandini, M. (2013). Integration of visual motion and locomotion in mouse visual cortex. *Nature Neuroscience*, *16*, 1864–1869.

Salvador, R., Wenger, C., & Miranda, P. C. (2015). Investigating the cortical regions involved in MEP modulation in tDCS. *Frontiers in Cellular Neuroscience*, *9*, 405–405.

Sampalo, M., Lázaro, E., & Luna, P. (2023). Action video gaming and attention in young adults: A systematic review. *Journal of Attention Disorders*, *27*(5), 530–538.

Sanabria, D., Capizzi, M., & Correa, A. (2011). Rhythms that speed you up. *Journal of Experimental Psychology: Human Perception and Performance*, *37*, 236–244.

Sanbonmatsu, D. M., Strayer, D. L., Medeiros-Ward, N., & Watson, J. M. (2013). Who multi-tasks and why? Multi-tasking ability, perceived multi-tasking ability, impulsivity, and sensation seeking. *PLoS One*, *8*(1), e54402.

Sapir, A., Soroker, N., Berger, A., & Henik, A., (1999). Inhibition of return in spatial attention: Direct evidence for collicular generation. *Nature Neuroscience*, *2*(12), 1053–1054.

Sawaki, R., & Luck, S. J. (2010). Capture versus suppression of attention by salient singletons: Electrophysiological evidence for an automatic attend-to-me signal. *Attention, Perception & Psychophysics*, *72*(6), 1455–1470.

Sawaki, R., Geng, J. J., & Luck, S. J. (2012). A common neural mechanism for preventing and terminating the allocation of attention. *Journal of Neuroscience*, *32*, 10725–10736.

Schall, J. D. (1997). Visuomotor areas of the frontal lobe. *Cerebral Cortex*, *12*, 527–638.

Scheerer, N. E., Birmingham, E., Boucher, T. Q., & Iarocci, G. (2021). Attention capture by trains and faces in children with and without autism spectrum disorder. *PLoS One*, *16*(6), e0250763.

Scheffers, M. K., Coles, M. G. H., Bernstein, P., Gehring, W. J., & Donchin, E. (1996). Event-related brain potentials and error-related processing: An analysis of incorrect responses to go and no-go stimuli. *Psychophysiology*, *33*(1), 42–53.

Schmidt, L.J., Belopolsky, A.V., & Theeuwes, J. (2015). Attentional capture by signals of threat. *Cognition & Emotion*, *29*(4), 687–694.

Schmitt, M., Postma, A., & de Haan, E. (2000). Interactions between exogenous auditory and visual spatial attention. *Quarterly Journal of Experimental Psychology*, *53A*, 105–130.

Schoenfeld, M., Hopf, J., Martinez, A., Mai, H., Sattler, C., Gasde, A., Heinze, H., & Hillyard, S. (2007). Spatio-temporal analysis of feature-based attention. *Cerebral Cortex (New York, N.Y. 1991)*, *17*(10), 2468–2477.

Schönenberg, M., Weingärtner, A., Weimer, K., & Scheeff, J. (2021). Believing is achieving – On the role of treatment expectation in neurofeedback applications. *Progress in Neuro-Psychopharmacology & Biological Psychiatry*, *105*, 110129.

Schubotz, R. I. (2007). Prediction of external events with our motor system: Towards a new framework. *Trends in Cognitive Science*, *11*, 211–218.

Schuckit, M. A. (1994). Low level of response to alcohol as a predictor of future alcoholism. *The American Journal of Psychiatry*, *151*, 184–189.

Schuckit, M. A., & Smith, T. L. (2000). The relationships of a family history of alcohol dependence, a low level of response to alcohol and six domains of life functioning to the development of alcohol use disorders. *Journal of Studies on Alcohol*, *61*, 827–835.

Schuller, A., & Rossion, B. (2001). Spatial attention triggered by eye gaze increases and speeds up early visual activity. *Neuroreport*, *12*(11), 2381–2386.

Schuller, A., & Rossion, B. (2004). Perception of static eye gaze direction facilitates subsequent early visual processing. *Clinical Neurophysiology*, *115*(5), 1161–1168.

Schwiedrzik, C. M., & Freiwald, W. A. (2017). High-level prediction signals in a low-level area of the macaque face-processing hierarchy. *Neuron (Cambridge, Mass.)*, *96*(1), 89–97.e4.

Sdoia, S., Conversi, D., Pecchinenda, A., & Ferlazzo, F. (2019). Access to consciousness of briefly presented visual events is modulated by transcranial direct current stimulation of left dorsolateral prefrontal cortex. *Scientific Reports*, *9*(1), 10950–10959.

Sengupta, B., & Stemmler, M. B. (2014). Power consumption during neuronal computation *Proceedings of the IEEE*, *102*(5), 738–750.

Serences, J. T. (2008). Value-based modulations in human visual cortex. *Neuron*, *60*, 1169–1181.

Serences, J. T., Ester, E. F., Vogel, E. K., & Awh, E. (2009). Stimulus-specific delay activity in human primary visual cortex. *Psychological Science*, *20*, 207–214.

Sereno, A. B., Briand, K. A., Amador, S. C., & Szapiel, S. V. (2006). Disruption of reflexive attention and eye movements in an individual with a collicular lesion. *Journal of Clinical and Experimental Neuropsychology*, *28*(1), 145–166.

Sereno, M. I., Dale, A. M., Reppas, J. B., Kwong, K. K., Belliveau, J. W., Brady, T. J., Rosen, B. R., & Tootell, R. B. (1995). Borders of multiple visual areas in humans revealed by functional magnetic resonance imaging. *Science*, *268*(5212), 889–893.

Servant, M., & Logan, G. D. (2019). Dynamics of attentional focusing in the Eriksen flanker task. *Attention, Perception & Psychophysics*, *81*(8), 2710–2721.

Sessa, P., Luria, R., Verleger, R., & Dell'Acqua, R. (2007). P3 latency shifts in the attentional blink: further evidence for second target processing postponement. *Brain Research*, *1137*, 131–139.

Shah-Basak, P. P., Chen, P., Caulfield, K., Medina, J., & Hamilton, R. H. (2018). The role of the right superior temporal gyrus in stimulus-centered spatial processing. *Neuropsychologia*, *113*, 6–13.

Shallice, T., & Cipolotti, L. (2018). The prefrontal cortex and neurological impairments of active thought. *Annual Review of Psychology*, *69*(1), 157–180.

Shallice, T., & Warrington, E. K. (1977). The possible role of selective attention in acquired dyslexia. *Neuropsychologia*, *15*(1), 31–41.

Shapiro, K. L., Arnell, K. M., & Raymond, J. E. (1997). The attentional blink: A view on attention and a glimpse on consciousness. *Trends in Cognitive Sciences*, *1*, 291–296.

Shapiro, K. L., Hanslmayr, S., Enns, J. T., & Lleras, A. (2017). Alpha, beta: The rhythm of the attentional blink. *Psychonomic Bulletin & Review*, *24*(6), 1862–1869.

Shen, D., Vuvan, D. T., & Alain, C. (2018). Cortical sources of the auditory attentional blink. *Journal of Neurophysiology*, *120*, 812–829.

Sherrington, C. S. (1906). *The Integrative Action of the Nervous System* (Vol. 35). Yale University Press.

Shin, E., Hopfinger, J. B., Lust, S. A., Henry, E. A., & Bartholow, B. D. (2010). Electrophysiological evidence of alcohol-related attentional bias in social drinkers low in alcohol sensitivity. *Psychology of Addictive Behaviors*, *24*, 508–515.

Shin, J. C., & Ivry, R. B. (2002). Concurrent learning of temporal and spatial sequences. *Learning & Memory*, *28*, 445–457.

Shulman, G. L., Astafiev, S. V., McAvoy, M. P., D'Avossa, G., & Corbetta, M. (2007). Right TPJ deactivation during visual search: Functional significance and support for a filter hypothesis. *Cerebral Cortex*, *17* (11), 2625–2633.

Shulman, G. L., Ollinger, J. M., Akbudak, E., Conturo, T. E., Snyder, A. Z., Petersen, S. E., & Corbetta, M. (1999). Areas involved in encoding and applying directional expectations to moving objects. *Journal of Neuroscience, 19*, 9480–9496.

Skinner, B. F. (1938). *The Behavior of Organisms: An Experimental Analysis*. B.F. Skinner Foundation.

Skinner, B. F. (1953). *Science and Human Behavior*. Macmillan.

Skinner, B. F. (1957). *Verbal Behavior*. Appleton-Century-Crofts.

Slotnick, S. D. (2012). *Controversies in Cognitive Neuroscience*. Macmillan International Higher Education.

Slotnick, S. D., Hopfinger, J. B., Klein, S. A., & Sutter, E. E. (2002). Darkness beyond the light: Attentional inhibition surrounding the classic spotlight. *Neuroreport, 13*(6), 773–778.

Slotnick, S. D., Schwarzbach, J., & Yantis, S. (2003). Attentional inhibition of visual processing in human striate and extrastriate cortex. *NeuroImage, 19*(4), 1602–1611.

Smaers, J. B., Gómez-Robles, A., Parks, A. N., & Sherwood, C. C. (2017). Exceptional evolutionary expansion of prefrontal cortex in great apes and humans. *Current Biology, 27*(10), 1549.

Smallwood, J., Brown, K. S., Tipper, C., Giesbrecht, B., Franklin, M. S., Mrazek, M. D., Carlson, J. M., & Schooler, J. W. (2011). Pupillometric evidence for the decoupling of attention from perceptual input during offline thought. *PLoS One, 6*(3), e18298.

Smith, D. T., & Archibald, N. (2019). Visual search in progressive supranuclear palsy. In: Hodgson, T. L. (ed.), *Processes of Visuo-Spatial Attention and Working Memory* (pp. 305–324). Springer.

Smith, D. T., Rorden, C., & Jackson, S. R. (2004). Exogenous orienting of attention depends upon the ability to execute eye movements. *Current Biology, 14*(9), 792–795.

Smith, H., & Milne, E. (2009). Reduced change blindness suggests enhanced attention to detail in individuals with autism. *Journal of Child Psychology and Psychiatry, 50*(3), 300–306.

Smith, R., Badcock, P., & Friston, K.J. (2021). Recent advances in the application of predictive coding and active inference models within clinical neuroscience. *Psychiatry and Clinical Neuroscience, 75*, 3–13.

Snow, J. C., Allen, H. A., Rafal, R. D., & Humphreys, G. W. (2009). Impaired attentional selection following lesions to human pulvinar: Evidence for homology between human and monkey. *Proceedings of the National Academy of Sciences of the United States of America, 106*(10), 4054–4059.

Sobel, N., Prabhakaran, V., Desmond, J. E., Glover, G. H., Goode, R. L., Sullivan, E. V., & Gabrieli, J. D. E. (1998). Sniffing and smelling: Separate subsystems in the human olfactory cortex. *Nature, 392*(6673), 282–286.

Sokolov, E. N. (1960). Neuronal models and the orienting influence. In: Brazier, M. A. (ed.), *The Central Nervous System and Behavior: III* (pp. 187–276). Macy Foundation.

Song, K., Meng, M., Lin, C., Zhou, K., & Luo, H. (2014). Behavioral oscillations in attention: Rhythmic α pulses mediated through θ band. *Journal of Neuroscience, 34* (14), 4837–4844.

Soto, D., Heinke, D., Humphreys, G. W., & Blanco, M. J. (2005). Early, involuntary top-down guidance of attention from working memory. *Journal of Experimental Psychology: Human Perception and Performance, 31*(2), 248–261.

Soto, D., Hodsoll, J., Rotshtein, P., & Humphreys, G. W. (2008). Automatic guidance of attention from working memory. *Trends in Cognitive Sciences, 12*(9), 342–348.

Soto, D., Humphreys, G. W., & Rotshtein, P. (2007). Dissociating the neural mechanisms of memory-based guidance of visual selection. *Proceedings of the National Academy of Sciences of the United States of America, 104*(43), 17186–17191.

Soto-Faraco, S., Spence, C., Fairbank, K., Kingstone, A., Hillstrom, A. P., & Shapiro, K. (2002). A crossmodal attentional blink between vision and touch. *Psychonomic Bulletin & Review, 9*(4), 731–738.

Southwell, R., Baumann, A., Gal, C., Barascud, N., Friston, K., & Chait, M.

(2017). Is predictability salient? A study of attentional capture by auditory patterns. *Philosophical Transactions of the Royal Society B: Biological Sciences, 372*(1714), 20160105.

Spelke, E., Hirst, W., & Neisser, U. (1976). Skills of divided attention. *Cognition, 4*(3), 215–230.

Spence, C. (2021). Extending the study of visual attention to a multisensory world (Charles W. Eriksen special issue). *Attention, Perception & Psychophysics, 83*(2), 763–775.

Spence, C., & Driver, J. (1997). Audiovisual links in exogenous covert spatial orienting. *Perception & Psychophysics, 59*, 1–22.

Spence, C., & Frings, C. (2020). Multisensory feature integration in (and out) of the focus of spatial attention. *Attention, Perception & Psychophysics, 82*(1), 363–376.

Spence, C., McGlone, F., Kettenmann, B., & Kobal, G. (2001). Attention to olfaction. *Experimental Brain Research, 138*(4), 432–437.

Spence, C., Nicholls, M. R., Gillespie, N., & Driver, J. (1998). Cross-modal links in exogenous covert spatial orienting between touch, audition, and vision. *Perception & Psychophysics, 60*, 544–557.

Spets, D. S., & Slotnick, S. D. (2021). Are there sex differences in brain activity during long-term memory? A systematic review and fMRI activation likelihood estimation meta-analysis. *Cognitive Neuroscience, 12*(3–4), 163–173.

Spitzer, H., Desimone, R., & Moran, J. (1988). Increased attention enhances both behavioral and neuronal performance. *Science, 240*, 338–340.

Stäblein, M., Sieprath, L., Knöchel, C., Landertinger, A., Schmied, C., Ghinea, D., Mayer, J. S., Bittner, R. A., Reif, A., & Oertel-Knöchel, V. (2016). Impaired working memory for visual motion direction in schizophrenia: Absence of recency effects and association with psychopathology. *Neuropsychology, 30*(6), 653–663.

Stammler, B., Flammer, K., Schuster, T., Lambert, M., Neumann, O., Lux, M.,

Matuz, T., & Karnath, H. (2023). Spatial neglect therapy with the augmented reality app "Negami" for active exploration training: A randomized controlled trial on 20 stroke patients with spatial neglect. *Archives of Physical Medicine and Rehabilitation, 104* (12), 1987–1994.

Stangl, M., Maoz, S. L. & Suthana, N. (2023). Mobile cognition: imaging the human brain in the 'real world'. *Nature Reviews: Neuroscience, 24*, 347–362.

Steinhauser, M., & Yeung, N. (2010). Decision processes in human performance monitoring. *Journal of Neuroscience, 30*, 15643–15653.

Sterzer, P., Adams, R. A., Fletcher, P., Frith, C., Lawrie, S. M., Muckli, L., Petrovic, P., Uhlhaas, P., Voss, M., & Corlett, P. R. (2018). The predictive coding account of psychosis. *Biological Psychiatry, 84*(9), 634–643.

Still, G. F. (1902). Some abnormal psychical conditions in children: The Goulstonian lectures. *Lancet, 1*, 1008–1012

Stitt, I., Zhou, Z. C., Radtke-Schuller, S., & Frohlich, F. (2018). Arousal dependent modulation of thalamo-cortical functional interaction. *Nature Communications, 9*, 2455.

Stone, S. P., Halligan, P. W., & Greenwood, R. J. (1993). The incidence of neglect phenomena and related disorders in patients with an acute right or left hemisphere stroke. *Age and Ageing, 22*, 46–52.

Störmer, V. S., & Alvarez, G. A. (2014). Feature-based attention elicits surround suppression in feature space. *Current Biology, 30*(19), 1985–1988.

Strayer, D. L., & Drews, F. A. (2007). Cell-phone-induced driver distraction. *Current Directions in Psychological Science, 16*(3), 128–131.

Strayer, D. L., & Johnston, W. A. (2001). Driven to distraction: Dual-task studies of simulated driving and conversing on a cellular telephone. *Psychological Science, 12*(6), 462–466.

Strayer, D. L., Drews, F. A., & Crouch, D. J. (2006). Comparing the cell-phone driver and the drunk driver. *Human Factors*, *48*, 381–391.

Stroop, J. R. (1935). Studies of interference in serial verbal reactions. *Journal of Experimental Psychology*, *18*(6), 643–662.

Stuss, D. T., Alexander, M. P., Shallice, T., Picton, T. W., Binns, M. A., Macdonald, R., Borowiec, A., & Katz, D. I. (2005). Multiple frontal systems controlling response speed. *Neuropsychologia*, *43*(3), 396–417.

Stuss, D. T., Levine, B., Alexander, M. P., Hong, J., Palumbo, C., Hamer, L., Murphy, K. J., & Izukawa, D. (2000). Wisconsin card sorting test performance in patients with focal frontal and posterior brain damage: Effects of lesion location and test structure on separable cognitive processes. *Neuropsychologia*, *38*(4), 388–402.

Stuss, D. T., Murphy, K. J., Binns, M. A., & Alexander, M. P. (2003). Staying on the job: The frontal lobes control individual performance variability. *Brain (London, England: 1878)*, *126*(11), 2363–2380.

Sumantry, D., & Stewart, K. E. (2021). Meditation, mindfulness, and attention: A meta-analysis. *Mindfulness*, *12*(6), 1332–1349.

Summerfield, C., & Koechlin, E. (2008). A neural representation of prior information during perceptual inference. *Neuron*, *59*(2), 336–347.

Sutton, S., Braren, M., Zubin, J., & John, E. R. (1965). Evoked-potential correlates of stimulus uncertainty. *Science*, *150*(3700), 1187–1188.

Swanson, J. M., Kinsbourne, M., Nigg, J., Lanphear, B., Stefanatos, G. A., Volkow, N., Taylor, E., Casey, B. J., Castellanos, F. X., & Wadhwa, P. D. (2007). Etiologic subtypes of attention-deficit/hyperactivity disorder: Brain imaging, molecular genetic and environmental factors and the dopamine hypothesis. *Neuropsychology Review*, *17*(1), 39–59.

Swanson, J. M., Posner, M., Potkin, S., Bonforte, S., Youpa, D., Fiore, C., Cantwell, D., & Crinella, F. (1991). Activating tasks for the study of visual-spatial attention in ADHD children: A cognitive anatomic approach. *Journal of Child Neurology*, *6*(1 Suppl.), S119–S127.

Tarbell, H. (1971). Magic as a science. In: Read, R. W. (ed.), *The Tarbell Course in Magic* (Vol. 1). D. Robbins. [Original work published 1927.]

Tatler, B. W., Hayhoe, M. M., Land, M. F., & Ballard, D. H. (2011). Eye guidance in natural vision: Reinterpreting salience. *Journal of Vision*, *11*(5), 1–23.

Tay, D., Harms, V., Hillyard, S. A., & McDonald, J. J. (2019). Electrophysiological correlates of visual singleton detection. *Psychophysiology*, *56*(8), e13375.

Taylor, K., Mandon, S., Freiwald, W. A., & Kreiter, A. K. (2005). Coherent oscillatory activity in monkey area V4 predicts successful allocation of attention. *Cerebral Cortex*, *15*(9), 1424–1437.

Taylor, P. C., Rushworth, M. F., & Nobre, A. C. (2008). Choosing where to attend and the medial frontal cortex: An FMRI study. *Journal of Neurophysiology*, *100*(3), 1397–1406.

Taylor, S. (1997). Isolation of specific interference processing in the Stroop task: PET activation studies. *NeuroImage*, *6*(2), 81–92.

Telford, C. W. (1931). The refractory phase of voluntary and associative responses. *Journal of Experimental Psychology*, *14*, 1–36.

Ten Oever, S., Werf, O. J., Schuhmann, T., & Sack, A. T. (2022). Absence of behavioural rhythms: Noise or unexplained neuronal mechanisms? (response to Fiebelkorn, 2021). *The European Journal of Neuroscience*, *55* (11–12), 3121–3124.

Theeuwes, J. (1991). Exogenous and endogenous control of attention: The effect of visual onsets and offsets. *Perception & Psychophysics*, *49*(1), 83–90.

Theeuwes, J. (1992). Perceptual selectivity for color and form. *Perception & Psychophysics*, *51*(6), 599–606.

Theeuwes, J. (1993). Visual selective attention: A theoretical analysis. *Acta Psychologica*, *83* (2), 93–154.

Theeuwes, J. (1994). Stimulus-driven capture and attentional set: Selective search for color and visual abrupt onsets. *Journal of Experimental Psychology: Human Perception and Performance, 20*, 799–799.

Theeuwes, J. (2004). Top-down search strategies cannot override attentional capture. *Psychonomic Bulletin and Review, 11*(1), 65–70.

Theeuwes, J., & Van der Burg, E. (2011). On the limits of top-down control of visual selection. *Attentention, Perception, & Psychophysics. 73*, 2092–2103

Theeuwes, J., & Van der Stigchel, S. (2006). Faces capture attention: Evidence from inhibition of return. *Visual Cognition, 13*(6), 657–665.

Theeuwes, J., Mathôt, S., & Kingstone, A. (2010). Object-based eye movements: The eyes prefer to stay within the same object. *Attention, Perception, & Psychophysics, 72*(3), 597–601.

Thibault, R. T., MacPherson, A., Lifshitz, M., Roth, R. R., & Raz, A. (2018). Neurofeedback with fMRI: A critical systematic review. *NeuroImage (Orlando, Fla.), 172*, 786–807.

Thompson, K. G. (2005). Neuronal basis of covert spatial attention in the frontal eye field. *Journal of Neuroscience, 25*(41), 9479–9487.

Thut, G., Nietzel, A., Brandt, S. A., & Pascual-Leone, A. (2006). α-Band electroencephalographic activity over occipital cortex indexes visuospatial attention bias and predicts visual target detection. *Journal of Neuroscience, 26*(37), 9494–9502.

Thut, G., Nietzel, A., & Pascual-Leone, A. (2005). Dorsal posterior parietal rTMS affects voluntary orienting of visuospatial attention. *Cerebral Cortex, 15*, 628–638.

Tipples, J., & Sharma, D. (2000). Orienting to exogenous cues and attentional bias to affective pictures reflect different processes. *British Journal of Psychology, 91*, 87–97.

Titchener, E. B. (1916). *A Text-Book of Psychology*. MacMillan.

Todorova, G. K., Pollick, F. E., & Muckli, L. (2021). Special treatment of prediction errors in autism spectrum disorder. *Neuropsychologia, 163*, 108070.

Tomassini, A., Spinelli, D., Jacono, M., Sandini, G., & Morrone, M. C. (2015). Rhythmic oscillations of visual contrast sensitivity synchronized with action. *The Journal of Neuroscience, 35*(18), 7019–7029.

Toplak, M. E., Dockstader, C., & Tannock, R. (2006). Temporal information processing in ADHD: Findings to date and new methods. *Journal of Neuroscience Methods, 151*(1), 15–29.

Torralba, A., Oliva, A., Castelhano, M. S., & Henderson, J. M. (2006). Contextual guidance of eye movements and attention in real-world scenes: The role of global features in object search. *Psychological Review, 113*(4), 766–786.

Trapp, S., Schütz-Bosbach, S., & Bar, M. (2018). Empathy: The role of expectations. *Emotion Review, 10*(2), 161–166.

Treisman, A. (1960). Contextual cues in selective listening. *Quarterly Journal of Experimental Psychology, 12*, 242–248.

Treisman, A. (1964). Monitoring and storage of irrelevant messages in selective attention. *Journal of Verbal Learning and Verbal Behavior, 3*, 449–459.

Treisman, A., & Gelade, G. (1980). A feature-integration theory of attention. *Cognitive Psychology, 12*, 97–136.

Treisman, A., & Schmidt, H. (1982). Illusory conjunctions in the perception of objects. *Cognitive Psychology, 14*(1), 107–141.

Treisman, A., & Souther, J. (1986). Illusory words: The roles of attention and of top-down constraints in conjoining letters to form words. *Journal of Experimental Psychology: Human Perception and Performance, 12*(1), 3–17.

Tremblay, S., Vachon, F., & Jones, D. M. (2005). Attentional and perceptual sources of the auditory attentional blink. *Perception & Psychophysics, 67*(2), 195–208.

Triviño, M., Arnedo, M., Lupiáñez, J., Chirivella, J., & Correa, Á. (2011). Rhythms can overcome temporal orienting deficit after right frontal damage. *Neuropsychologia*, *49*(14), 3917–3930.

Triviño, M., Correa, A., Arnedo, M., & Lupiáñez, J. (2010). Temporal orienting deficit after pre-frontal damage. *Brain*, *133*, 1173–1185.

Triviño, M., Correa, Á., Lupiáñez, J., Funes, M. J., Catena, A., He, X., & Humphreys, G. W. (2016). Brain networks of temporal preparation: A multiple regression analysis of neuropsychological data. *NeuroImage*, *142*, 489–497.

Turing, A. (1948). Machine intelligence. In: Copeland, B. J. (ed.), *The Essential Turing: The Ideas That Gave Birth to the Computer Age* (pp. 395–432). Oxford University Press.

Turing, A. (1950). Computing machinery and intelligence. *Mind*, *LIX*(236): 433–460.

Ullsperger, M., Harsay, H. A., Wessel, J. R., & Ridderinkhof, K. R. (2010). Conscious perception of errors and its relation to the anterior insula. *Brain Structure and Function*, *214*, 629–643.

Vachon, F., & Jolicoeur, P. (2011). Impaired semantic processing during task-set switching: Evidence from the N400 in rapid serial visual presentation. *Psychophysiology*, *48*(1), 102–111.

Valdes-Sosa, M., Bobes, M. A., Rodriguez, V., & Pinilla, T. (1998). Switching attention without shifting the spotlight: Object-based attentional modulation of brain potentials. *Journal of Cognitive Neuroscience*, *10*(1), 137–151.

Vallar, G. (1998). Spatial hemineglect in humans. *Trends in Cognitive Sciences*, *2*(3), 87–95.

Vallesi, A., Shallice, T., & Walsh, V. (2007). Role of the prefrontal cortex in the foreperiod effect: TMS evidence for dual mechanisms in temporal preparation. *Cerebral Cortex*, *17*(2), 466–474.

Van Bockstaele, B., Verschuere, B., Tibboel, H., De Houwer, J., Crombez, G., & Koster, E. H. W. (2014). A review of current evidence for the causal impact of attentional bias on fear and anxiety. *Psychological Bulletin*, *140*(3), 682–721.

van der Werf, O. J., Ten Oever, S., Schuhmann, T., & Sack, A. T. (2022). No evidence of rhythmic visuospatial attention at cued locations in a spatial cuing paradigm, regardless of their behavioural relevance. *The European Journal of Neuroscience*, *55*(11–12), 3100–3116.

van Ede, F., de Lange, F., Jensen, O., & Maris, E. (2011). Orienting attention to an upcoming tactile event involves a spatially and temporally specific modulation of sensorimotor alpha- and beta-band oscillations. *The Journal of Neuroscience*, *31*(6), 2016–2024.

Van Koningsbruggen, M. G., Gabay, S., Sapir, A., Henik, A., & Rafal, R. D. (2010). Hemispheric asymmetry in the remapping and maintenance of visual saliency maps: A TMS study. *Journal of Cognitive Neuroscience*, *22*, 1730–1738.

van Moorselaar, D., Huang, C., & Theeuwes, J. (2023). Electrophysiological indices of distractor processing in visual search are shaped by target expectations. *Journal of Cognitive Neuroscience*, *35*(6), 1032–1044.

van Veen, V., & Carter, C. S. (2002). The timing of action-monitoring processes in the anterior cingulate cortex. *Journal of Cognitive Neuroscience*, *14*(4), 593–602.

Van Velzen, J. V., & Eimer, M. (2003). Early posterior ERP components do not reflect the control of attentional shifts toward expected peripheral events. *Psychophysiology*, *40*, 827–831.

Van Voorhis, S., & Hillyard, S. A. (1977). Visual evoked potentials and selective attention to points in space. *Perception & Psychophysics*, *22*(1), 54–62.

Vangkilde, S., Petersen, A., & Bundesen, C. (2013). Temporal expectancy in the context of a theory of visual attention. *Philosophical Transactions of the Royal Society B: Biological Sciences*, *368*(1628), 20130054.

Verdon, V., Schwartz, S., Lovblad, K. O., Hauert, C. A., & Vuilleumier, P. (2010). Neuroanatomy of hemispatial neglect and its functional components: a study using voxel-based lesion-symptom mapping. *Brain*, *133* (3), 880–894.

Verhaeghen, P., Steitz, D. W., Sliwinski, M. J., & Cerella, J. (2003). Aging and dual-task performance: A meta-analysis. *Psychology and Aging*, *18*(3), 443–460.

Vidal, J. R., Chaumon, M., O'Regan, J. K., & Tallon-Baudry, C. (2006). Visual grouping and the focusing of attention induce gamma-band oscillations at different frequencies in human magnetoencephalogram signals. *Journal of Cognitive Neuroscience*, *18*, 1850–1862.

Vidyasagar, T. R. (1998). Gating of neuronal responses in macaque primary visual cortex by an attentional spotlight. *Neuroreport*, *9*, 1947–1952.

Vidyasagar, T. R., & Pammer, K. (2010). Dyslexia: A deficit in visuo-spatial attention, not in phonological processing. *Trends in Cognitive Sciences*, *14*(2), 57.

Vilotijević, A., & Mathôt, S. (2023). Emphasis on peripheral vision is accompanied by pupil dilation. *Psychonomic Bulletin & Review*, *30*(5), 1848–1856.

Vocat, R., Pourtois, G., & Vuilleumier, P. (2008). Unavoidable errors: A spatio-temporal analysis of time-course and neural sources of evoked potentials associated with error processing in a speeded task. *Neuropsychologia*, *46*(10), 2545–2555.

Vogel, E. K., & Luck, S. J. (2000). The visual N1 component as an index of a discrimination process. *Psychophysiology*, *37*, 190–123.

Vogel, E. K., & Luck, S. J. (2002). Delayed working memory consolidation during the attentional blink. *Psychonomic Bulletin & Review*, *9*(4), 739–743.

Vogel, E. K., Luck, S. J., & Shapiro, K. L. (1998). Electrophysiological evidence for a postperceptual locus of suppression during the attentional blink. *Journal of Experimental Psychology: Human Perception and Performance*, *24*(6), 1656–1674.

Vogt, B. A. (2019). The cingulate cortex in neurologic diseases: History, structure, overview. *Handbook of Clinical Neurology*, *166*, 3–21.

Vossel, S., Geng, J. J., & Fink, G. R. (2014). Dorsal and ventral attention systems: Distinct neural circuits but collaborative roles. *The Neuroscientist*, *20*(2), 150–159.

Vossel, S., Mathys, C., Daunizeau, J., Bauer, M., Driver, J., Friston, K. J., & Stephan, K. E. (2014). Spatial attention, precision, and Bayesian inference: A study of saccadic response speed. *Cerebral Cortex*, *24*(6), 1436–1450.

Vossel, S., Weidner, R., Thiel, C. M., & Fink, G. R. (2009). What is "odd" in Posner's location-cueing paradigm? Neural responses to unexpected location and feature changes compared. *Journal of Cognitive Neuroscience*, *21*(1), 30–41.

Wacongne, C., Changeux, J., & Dehaene, S. (2012). A neuronal model of predictive coding accounting for the mismatch negativity. *The Journal of Neuroscience*, *32*(11), 3665–3678.

Walter, W. G. (1936). The location of cerebral tumours by electroencephalography. *Lancet*, *2*, 305–308.

Walter, W. G., Cooper, R., Aldridge, V. J., McCallum, W. C., & Winter, A. L. (1964). Contingent Negative Variation: An electric sign of sensorimotor association and expectancy in the human brain. *Nature*, *203*(4943), 380–384.

Wang, B., & Theeuwes, J. (2020). Salience determines attentional orienting in visual selection. *Journal of Experimental Psychology: Human Perception and Performance*, *46*(10), 1051–1057.

Wang, L., Gu, Y., Zhao, G., & Chen, A. (2020). Error-related negativity and error awareness in a Go/No-go task. *Scientific Reports*, *10*(1), 4026.

Wang, L. H., Yu, H. B., & Zhou, X. L. (2013). Interaction between value and perceptual salience in value-driven attentional capture. *Journal of Vision*, *13*(3), 5.

Wang, Q., Cavanaugh, P., & Green, M. (1994). Familiarity and pop-out in visual search. *Perception and Psychophysics*, *56*, 495500.

Ward, L. M. (1994). Supramodal and modality-specific mechanisms for stimulus-driven shifts of auditory and visual attention. *Canadian Journal of Experimental Psychology*, *48*, 242–259.

Warnking, J., Dojat, M., Guérin-Dugué, A., Delon-Martin, C., Olympieff, S., Richard, N., Chéhikian, A., & Segebarth, C. (2002). fMRI retinotopic mapping – Step by step. *NeuroImage*, *17*(4), 1665-1683.

Warren, J., & Griffiths, T. (2003). Distinct mechanisms for processing spatial sequences and pitch sequences in the human auditory brain. *Journal of Neuroscience*, *23*, 5799–5804.

Warren, R. M., & Warren, R. P. (1968). *Helmholtz on Perception: Its Physiology and Development*. John Wiley.

Wascher, E., & Tipper, S. P. (2004). Revealing effects of noninformative spatial cues: An EEG study of inhibition of return. *Psychophysiology*, *41*, 716–728.

Waters, A. M., Bradley, B. P., & Mogg, K. (2014). Biased attention to threat in paediatric anxiety disorders (generalized anxiety disorder, social phobia, specific phobia, separation anxiety disorder) as a function of "distress" versus "fear"' diagnostic categorization. *Psychological Medicine*, *44*(3), 607–616.

Watson, J. M., & Strayer, D. L. (2010). Supertaskers: Profiles in extraordinary multitasking ability. *Psychonomic Bulletin & Review*, *17*(4), 479–485.

Watson, P., Pearson, D., Theeuwes, J., Most, S. B., & Le Pelley, M. E. (2020). Delayed disengagement of attention from distractors signalling reward. *Cognition*, *195*, 1–13.

Weaver, M. D., van Zoest, W., & Hickey, C. (2017). A temporal dependency account of attentional inhibition in oculomotor control. *NeuroImage*, *147*, 880–894.

Weissman, D. H., Woldorff, M. G., & Mangun, G. R. (2002). A role for top-down attentional orienting during interference between global and local aspects of hierarchical stimuli. *NeuroImage*, *17*, 1266–1276.

Welford, A. T. (1952). The "psychological refractory period" and the timing of high-speed performance – A review and a theory. *British Journal of Psychology*, *43*, 2–19.

Werling, D. M., & Geschwind, D. H. (2013). Sex differences in autism spectrum disorders. *Current Opinion in Neurology*, *26*(2), 146–153.

Wernicke, C. (1874). *Der aphaische symptomencomplex; eine psychologische studie auf anatomischer basis*. Springer-Verlag.

Westerhausen, R., Kompus, K., & Hugdahl, K. (2011). Impaired cognitive inhibition in schizophrenia: A meta-analysis of the Stroop interference effect. *Schizophrenia Research*, *133*(1), 172–181.

Wetter, O. E. (2013). Imaging in airport security: Past, present, future, and the link to forensic and clinical radiology. *Journal of Forensic Radiology and Imaging*, *1*(4), 152–160.

Whitmarsh, S., Udden, J., Barendregt, H., & Petersson, K. M. (2013). Mindfulness reduces habitual responding based on implicit knowledge: Evidence from artificial grammar learning. *Consciousness & Cognition*, *22*(3), 833–845.

Wiart, L., Côme, A. B. S., Debelleix, X., Petit, H., Joseph, P. A., Mazaux, J. M., & Barat, M. (1997). Unilateral neglect syndrome rehabilitation by trunk rotation and scanning training. *Archives of Physical Medicine and Rehabilitation*, *78*(4), 424–429.

Wiener, M., Turkeltaub, P., & Coslett, H. B. (2010). The image of time: A voxel-wise meta-analysis. *NeuroImage*, *49*, 1728–1740.

Wiese, W., & Metzinger, T. (2017). Vanilla PP for philosophers: A primer on predictive processing. In: Metzinger, T., & Wiese, W. (eds.), *Philosophy and Predictive Processing* (pp. 1–18). MIND Group.

Wijers, A. A., Mulder, G., Okita, T., Mulder, L. J. M., & Scheffers, M. K. (1989). Attention to color: An analysis of selection, controlled search, and motor activation, using event-related potentials. *Psychophysiology*, *26*(1), 89–109.

Willcutt, E. G. (2012). The prevalence of DSM-IV attention-deficit/hyperactivity disorder: A meta-analytic review. *Neurotherapeutics, 9*(3), 490–499.

Witkiewitz, K., Bowen, S., Douglas, H., & Hsu, S. H. (2013). Mindfulness-based relapse prevention for substance craving. *Addictive Behaviors, 38*(2), 1563–1571.

Woldorff, M. G. (1993). Distortion of ERP averages due to overlap from temporally adjacent ERPs: Analysis and correction. *Psychophysiology, 30*(1), 98–119.

Woldorff, M. G., & Hillyard, S. A. (1991). Modulation of early auditory processing during selective listening to rapidly presented tones. *Electroencephalography and Clinical Neurophysiology, 79*(3), 170–191.

Woldorff, M. G., Fox, P. T., Matzke, M., Lancaster, J. L., Veeraswamy, S., Zamarripa, F., Seabolt, M., Glass, T., Gao, J. H., Martin, C. C., & Jerabek, P. (1997). Retinotopic organization of early visual spatial attention effects as revealed by PET and ERPs. *Human Brain Mapping, 5*(4), 280–286.

Woldorff, M. G., Gallen, C. C., Hampson, S. A., Hillyard, S. A., Pantev, C., Sobel, D., & Bloom, F. E. (1993). Modulation of early sensory processing in human auditory cortex during auditory selective attention. *Proceedings of the National Academy of Sciences of the United States of America, 90*(18), 8722–8726.

Wolfe, J. M. (2020). Forty years after feature integration theory: An introduction to the special issue in honor of the contributions of Anne Treisman. *Attention, Perception & Psychophysics, 82*(1), 1–6.

Wood, N., & Cowan, N. (1995). The cocktail party phenomenon revisited: How frequent are attention shifts to one's name in an irrelevant auditory channel? *Journal of Experimental Psychology: Learning, Memory, and Cognition, 21*(1), 255–260.

Woodman, G. F., & Luck, S. J. (2004). Visual search is slowed when visuospatial working memory is occupied. *Psychonomic Bulletin & Review, 11*(2), 269–274.

Woodman, G. F., & Luck, S. J. (2007). Do the contents of visual working memory automatically influence attentional selection during visual search? *Journal of Experimental Psychology: Human Perception & Performance, 33*, 363–377.

Woodman, G. F., Arita, J. T., & Luck, S. J. (2009). A cuing study of the N2pc component: An index of attentional deployment to objects rather than spatial locations. *Brain Research, 1297*, 101–111.

Worden, M., & Schneider, W. (1996). Visuospatial attentional selection examined with functional magnetic resonance imaging. *Society for Neuroscience Abstracts, 22*, 1856.

Worden, M. S., Foxe, J. J., Wang, N., & Simpson, G. V. (2000). Anticipatory biasing of visuospatial attention indexed by retinotopically specific alpha-band electroencephalography increases over occipital cortex. *Journal of Neuroscience, 20*, 1–6.

World Health Organization. (2022). *ICD-11: International Classification of Diseases* (11th revision). https://icd.who.int/

Wright, R. D., Richard, C. M., & McDonald, J. J. (1995). Neutral location cues and cost/benefit analysis of visual attention shifts. *Canadian Journal of Experimental Psychology, 49*(4), 540–548.

Wrobel, A., Ghazaryan, A., Bekisz, M., Bogdan, W., & Kaminski, J. (2007). Two streams of attention-dependent beta activity in the striate recipient zone of cat's lateral posterior- pulvinar complex. *Journal of Neuroscience, 27*, 2230–2240.

Wu, E. X. W., Liaw, G. J., Goh, R. Z., Chia, T. T. Y., Chee, A. M. J., Obana, T., Rosenberg, M. D., Yeo, B. T. T., & Asplund, C. L. (2020). Overlapping attentional networks yield divergent behavioral predictions across tasks: Neuromarkers for diffuse and focused attention? *NeuroImage, 209*, 116535.

Wu, S., Cheng, C. K., Feng, J., D'Angelo, L., Alain, C., & Spence, I. (2012). Playing a first-person shooter video game induces neuroplastic change. *Journal of Cognitive Neuroscience*, *24*, 1286–1293.

Wu, S.-C., Remington, R. W., & Folk, C. L. (2014). Onsets do not override top-down goals, but they are responded to more quickly. *Attention, Perception, & Psychophysics*, *76*(3), 649–654.

Yadav, S. K., Bhat, A. A., Hashem, S., Nisar, S., Kamal, M., Syed, N., Temanni, M., Gupta, R. K., Kamran, S., Azeem, M. W., Srivastava, A. K., Bagga, P., Chawla, S., Reddy, R., Frenneaux, M. P., Fakhro, K., & Haris, M. (2021). Genetic variations influence brain changes in patients with attention-deficit hyperactivity disorder. *Translational Psychiatry*, *11*(1), 349.

Yamaguchi, S., Tsuchiya, H., & Kobayashi, S. (1994). Electroencephalographic activity associated with shifts of visuospatial attention. *Brain*, *117*(3), 553–562.

Yang, P. F., Phipps, M. A., Newton, A. T., Chaplin, V., Gore, J. C., Caskey, C. F., & Chen, L. M. (2018). Neuromodulation of sensory networks in monkey brain by focused ultrasound with MRI guidance and detection. *Scientific Reports*, *8*(7993), 1–9.

Yantis, S. (2000). Goal-directed and stimulus-driven determinants of attentional control. In: Monsell, S., & Driver, J. (eds.), *Attention and Performance XVIII* (pp. 73–103). MIT Press.

Yantis, S., & Jonides, J. (1984). Abrupt visual onsets and selective attention: Evidence from visual search. *Journal of Experimental Psychology: Human Perception and Performance*, *10*(5), 601–621.

Yantis, S., & Jonides, J. (1990). Abrupt visual onsets and selective attention: Voluntary versus automatic allocation. *Journal of Experimental Psychology: Human Perception and Performance*, *16*(1), 121–134.

Yantis, S., & Jonides, J. (1996). Attentional capture by abrupt onsets: New perceptual objects or visual masking? *Journal of Experimental Psychology: Human Perception and Performance*, *22*(6), 1505–1513.

Yarbus, A. L. (1967). *Eye Movements and Vision*. Plenum.

Yiend, J., & Mathews, A. (2001). Anxiety and attention to threatening pictures. *Quarterly Journal of Experimental Psychology*, *54A*, 665–681.

Yiend, J., Mathews, A., Burns, T., Dutton, K., Fernández-Martín, A., Georgiou, G. A., Luckie, M., Rose, A., Russo, R., & Fox, E. (2015). Mechanisms of selective attention in generalized anxiety disorder. *Clinical Psychological Science*, *3*(5), 758–771.

Young, A., & Wimmer, R. D. (2017). Implications for the thalamic reticular nucleus in impaired attention and sleep in schizophrenia. *Schizophrenia Research*, *180*, 44–47.

Young-Bernier, M., Tanguay, A. N., Tremblay, F., & Davidson, P. S. R. (2015). Age differences in reaction times and a neurophysiological marker of cholinergic activity. *Canadian Journal on Aging*, *34*(4), 471–480.

Yovel, G., & Kanwisher, N. (2005). The neural basis of the behavioral face-inversion effect. *Current Biology*, *15*(24), 2256–2262.

Yu, C., Li, Y., Stitt, I. M., Zhou, Z. C., Sellers, K. K., & Frohlich, F. (2018). Theta oscillations organize spiking activity in higher-order visual thalamus during sustained attention. *eNeuro*, *5*(1), ENEURO.0384-17.2018.

Zanto, T. P., Rubens, M. T., Thangavel, A., & Gazzaley, A. (2011). Causal role of the prefrontal cortex in top-down modulation of visual processing and working memory. *Nature Neuroscience*, *14*(5), 656–661.

Zastrow, M. (2017). News feature: Is video game addiction really an addiction? *Proceedings of the National Academy of Sciences of the United States of America*, *114*(17), 4268–4272.

Zauner, A., Fellinger, R., Gross, J., Hanslmayr, S., Shapiro, K., Gruber, W., Müller, S., & Klimesch, W. (2012). Alpha entrainment is responsible for the attentional blink phenomenon. *NeuroImage*, *63*(2), 674–686.

Zhang, D., Zhang, R., Zhou, L., Zhou, K., & Chang, C. (2023). The brain network underlying attentional blink predicts symptoms of attention deficit hyperactivity disorder in children. *Cerebral Cortex (New York, N.Y. 1991)*, *33*(6), 2761–2773.

Zhang, W., & Luck, S. J. (2009). Feature-based attention modulates feedforward visual processing. *Nature Neuroscience*, *12*, 24–25.

Zhang, Y., Chen, Y., Bressler, S. L., & Ding, M. (2008). Response preparation and inhibition: The role of the cortical sensorimotor beta rhythm. *Neuroscience*, *156*(1), 238–246.

Zhang, Y., Du, G., Yang, Y., Qin, W., Li, X., & Zhang, Q. (2015). Higher integrity of the motor and visual pathways in long-term video game players. *Frontiers in Human Neuroscience*, *9*, 98.

Zhao, J., Al-Aidroos, N., & Turk-Browne, N. B. (2013). Attention is spontaneously biased toward regularities. *Psychological Science*, *24*(5), 667–677.

Zhou, H., Schafer, R. J., & Desimone, R. (2016). Pulvinar-cortex interactions in vision and attention. *Neuron*, *89*, 209–220.

Zhou, Y., Curtis, C. E., Sreenivasan, K. K., & Fougnie, D. (2022). Common neural mechanisms control attention and working memory. *Journal of Neuroscience*, *42*(37), 7110–7120.

Zikopoulos, B., & Barbas, H. (2006). Prefrontal projections to the thalamic reticular nucleus form a unique circuit for attentional mechanisms. *Journal of Neuroscience*, *26*(28), 7348–7361.

Zilverstand, A., Sorger, B., Slaats-Willemse, D., Kan, C. C., Goebel, R., & Buitelaar, J. K. (2017). fMRI neurofeedback training for increasing anterior cingulate cortex activation in adult attention deficit hyperactivity disorder. An exploratory randomized, single-blinded study. *PLoS One*, *12*(1), e0170795.

Zivan, M., Morag, I., Yarmolovsky, J., & Geva, R. (2021). Hyper-reactivity to salience limits social interaction among infants born pre-term and infant siblings of children with ASD. *Frontiers in Psychiatry*, *12*, 650.

Zsadanyi, S. E., Kurth, F., & Luders, E. (2021). The effects of mindfulness and meditation on the cingulate cortex in the healthy human brain: A review. *Mindfulness*, *12*(10), 2371–2387.

Index

For EU product safety concerns, contact us at Calle de José Abascal, 56–1°,
28003 Madrid, Spain or eugpsr@cambridge.org.

www.ingramcontent.com/pod-product-compliance
Ingram Content Group UK Ltd.
Pitfield, Milton Keynes, MK11 3LW, UK
UKHW050903071225
465726UK00007B/283